Joshua L. Posner
Agronomy Department
Univ. of WI-Madison

3/18/92

MODELING PLANT AND SOIL SYSTEMS

AGRONOMY

A Series of Monographs

The American Society of Agronomy and Academic Press published the first six books in this series. The General Editor of Monographs 1 to 6 was A. G. Norman. They are available through Academic Press, Inc., 111 Fifth Avenue, New York, NY 10003.

1. C. EDMUND MARSHALL: The Colloid Chemical of the Silicate Minerals. 1949
2. BYRON T. SHAW, *Editor*: Soil Physical Conditions and Plant Growth. 1952
3. K. D. JACOB, *Editor*: Fertilizer Technology and Resources in the United States. 1953
4. W. H. PIERRE and A. G. NORMAN, *Editors*: Soil and Fertilizer Phosphate in Crop Nutrition. 1953
5. GEORGE F. SPRAGUE, *Editor*: Corn and Corn Improvement. 1955
6. J. LEVITT: The Hardiness of Plants. 1956

The Monographs published since 1957 are available from the American Society of Agronomy, 677 S. Segoe Road, Madison, WI 53711.

7. JAMES N. LUTHIN, *Editor*: Drainage of Agricultural Lands. 1957
8. FRANKLIN A. COFFMAN, *Editor*: Oats and Oat Improvement. 1961
9. A. KLUTE, *Editor*: Methods of Soil Analysis. 1986
   Part 1—Physical and Mineralogical Methods. Second Edition.
   A. L. PAGE, R. H. MILLER, and D. R. KEENEY, *Editor*: Methods of Soil Analysis. 1982
   Part 2—Chemical and Microbiological Properties. Second Edition.
10. W. V. BARTHOLOMEW and F. E. CLARK, *Editors*: Soil Nitrogen. 1965
    (Out of print; replaced by no. 22)
11. R. M. HAGAN, H. R. HAISE, and T. W. EDMINSTER, *Editors*: Irrigation of Agricultural Lands. 1967
12. FRED ADAMS, *Editor*: Soil Acidity and Liming. Second Edition. 1984
13. E. G. HEYNE, *Editor*: Wheat and Wheat Improvement. Second Edition. 1987
14. A. A. HANSON and F. V. JUSKA, *Editors*: Turfgrass Science. 1969
15. CLARENCE H. HANSON, *Editor*: Alfalfa Science and Technology. 1972
16. J. R. WILCOX, *Editor*: Soybeans: Improvement, Production, and Uses. Second Edition. 1987
17. JAN VAN SCHILFGAARDE, *Editor*: Drainage for Agriculture. 1974
18. G. F. SPRAGUE and J. W. DUDLEY, *Editors*: Corn and Corn Improvement, Third Edition. 1988
19. JACK F. CARTER, *Editor*: Sunflower Science and Technology. 1978
20. ROBERT C. BUCKNER and L. P. BUSH, *Editors*: Tall Fescue. 1979
21. M. T. BEATTY, G. W. PETERSEN, and L. D. SWINDALE, *Editors*: Planning the Uses and Management of Land. 1979
22. F. J. STEVENSON, *Editor*: Nitrogen in Agricultural Soils. 1982
23. H. E. DREGNE and W. O. WILLIS, *Editors*: Dryland Agriculture. 1983
24. R. J. KOHEL and C. F. LEWIS, *Editors*: Cotton. 1984
25. N. L. TAYLOR, *Editor*: Clover Science and Technology. 1985
26. D. C. RASMUSSON, *Editor*: Barley. 1985
27. M. A. TABATABAI, *Editor*: Sulfur in Agriculture. 1986
28. R. A. OLSON and K. J. FREY, *Editors*: Nutritional Quality of Cereal Grains: Genetic and Agronomic Improvement. 1987
29. A. A. HANSON, D. K. BARNES, and R. R. HILL, JR., *Editors*: Alfalfa and Alfalfa Improvement. 1988
30. B. A. STEWART and D. R. NIELSEN, *Editors*: Irrigation of Agricultural Crops. 1990
31. JOHN HANKS and J. T. RITCHIE, *Editors*: Modeling Plant and Soil Systems, 1991

MODELING PLANT AND SOIL SYSTEMS

John Hanks and J. T. Ritchie, co-editors

Managing Editor: S. H. Mickelson

Editor-in-Chief ASA Publications: G. A. Peterson

Editor-in-Chief CSSA Publications: C. W. Stuber

Editor-in-Chief SSSA Publications: R. J. Luxmoore

Number 31 in the series
AGRONOMY

American Society of Agronomy, Inc.
Crop Science Society of America, Inc.
Soil Science Society of America, Inc.
Publishers
Madison, Wisconsin USA
1991

Copyright © 1991 by the American Society of Agronomy, Inc.
Crop Science Society of America, Inc.
Soil Science Society of America, Inc.

ALL RIGHTS RESERVED UNDER THE U.S. COPYRIGHT LAW
OF 1978 (P. L. 94-553)

Any and all uses beyond the limitations of the "fair use" provision of the law require written permission from the publishers and/or author(s); not applicable to contributions prepared by officers or employees of the U.S. Government as part of their official duties.

American Society of Agronomy, Inc.
Crop Science Society of America, Inc.
Soil Science Society of America, Inc.
677 South Segoe Road, Madison, WI 53711 USA

Library of Congress Cataloging-in-Publication Data

Modeling plant and soil systems / John Hanks and J.T. Ritchie, co-editors.
    p.   cm. — (Agronomy ; no. 31)
  Includes bibliographical references and index.
  ISBN 0-89118-106-7
  1. Crops and soils—Computer simulation. 2. Crops—Computer simulation. 3. Soils—Computer simulation. 4. Cropping systems—Computer simulation. I. Hanks, R. J. (Ronald John), 1927- . II. Ritchie, J. T. (Joe T.), 1937- . III. Series.
S5967.M62   1991
631.5 '01 '13—dc20   91-19991
                                                            CIP

Printed in the United States of America

CONTENTS

PREFACE xi
CONTRIBUTORS xiii
CONVERSION FACTORS FOR SI AND NON-SI UNITS xv

1 Introduction 1
R. J. HANKS AND J. T. RITCHIE

2 Temperature and Crop Development 5
J. T. RITCHIE AND D. S. NeSMITH

    I. Terminology 6
   II. Sources of Error in Thermal Time Calculations 7
  III. Determining Temperature Response Functions 9
   IV. Idealized Response Function 11
    V. Timing of Developmental Stages 13
   VI. Models for Daily Cycle of Temperature 22
  VII. Photoperiod Influence on Crop Development 23
VIII. Vernalization Influence on Winter Cereal Development 24
   IX. Maturity Type Influence on Crop Development 25
    X. Influence of Temperature During Floral Induction on
        Leaf Number 25
   XI. Conclusions 26
References 27

3 Wheat Phasic Development 31
J. T. RITCHIE

    I. Simulating Phasic Development 32
   II. Cold Hardening and Winter Kill 40
  III. Other Process Level Wheat Models 42
   IV. Model Validation 43
    V. FORTRAN Program for the Wheat Development Model 43
   VI. Appendix 44
References 54

4 Maize Phasic Development 55
J. R. KINIRY

    I. Model Descriptions 56
   II. Model Validation 63

	III. Research Needs	68
	IV. Summary	69
	References	69

5 Soybean Development — 71

JAMES W. JONES, K. J. BOOTE, S. S. JAGTAP, AND J. W. MISHOE

I. Review of Previous Models	72
II. The Phenology Model in SOYGRO	74
III. Estimation of Coefficients for the Model	79
IV. Testing the Model	83
V. Summary	89
References	89

6 Simulation of Root Growth — 91

C. ALLAN JONES, W. L. BLAND, J. T. RITCHIE, AND J. R. WILLIAMS

I. Soil Factors Affecting Rooting	92
II. Root Distribution	98
III. Model Inputs	103
IV. Model Behavior	104
V. Conclusion	110
VI. Appendix	111
References	120

7 Predicting Canopy Light-Use Efficiency from Leaf Characteristics — 125

J. M. NORMAN AND T. J. ARKEBAUER

I. Canopy Light-Use Efficiency Definitions	126
II. Model Descriptions	127
III. Model Comparisons with Measurements	135
IV. Review of Light-Use Efficiency Measurements	136
V. Interpretations from the Cupid Model	137
VI. Summary	141
References	141

8 Soil Spatial Variability — 145

ESHEL BRESLER

I. Modeling Infiltration and Redistribution	145
II. Modeling Solute Transport	151
III. Critical Assumptions	161
IV. Research Needs	162
V. Summary	163
VI. Listings	164
References	179

9 Infiltration and Redistribution — 181

R. J. HANKS

  I. Infiltration Equations 181
 II. General Equation for Unsaturated Flow 182
III. Numerical Model 182
 IV. Model Verifications 186
  V. Examples 186
 VI. Input and Output Data 190
VII. Critical Assumptions 190
VIII. Other Models 191
 IX. Summary 191
  X. Research Needs 192
 XI. Appendix 192
References 203

10 Drainage — 205

R. WAYNE SKAGGS

  I. DRAINMOD 206
 II. Field Evaluation of the Model 224
III. Water Management System Objectives 225
 IV. Crop Yield 228
  V. Application of the Yield Version of DRAINMOD—
     An Example 232
 VI. Research Needs 239
VII. Summary 240
References 240

11 Soil Evaporation and Transpiration — 245

R. J. HANKS

  I. Modeling Evaporation 246
 II. Modeling Transpiration 248
III. Model Verification 252
 IV. Critical Assumptions 255
  V. Computer Requirements 255
 VI. Summary 256
VII. Appendix 256
References 271

12 Simulation of Water Uptake by Plant Roots — 273

G. S. CAMPBELL

  I. Water Uptake from a Uniformly Rooted Soil Layer 273
 II. Water Uptake from Soil with Spatially Varying
     Root Density 275
III. Numerical Algorithms for Root Water Uptake 277
 IV. Model Input-Output and Validation 280
  V. Critical Assumptions 282

13 Nitrogen Dynamics in Soil-Plant Systems 287
D. C. GODWIN AND C. ALLAN JONES

VI. Research Needs	283
VII. Summary	283
VIII. Acknowledgment	284
References	284
I. The CERES-N Model	289
II. Model Inputs	304
III. Model Evaluation	304
IV. Model Limitations and Uses	304
V. Acknowledgments	308
VI. Appendix 1	308
VII. Appendix 2	316
References	319

14 Modeling Phosphorus Dynamics in the Soil-Plant System 323
C. ALLAN JONES, ANDREW N. SHARPLEY, AND J. R. WILLIAMS

I. Previous Models	323
II. Model Structure	326
III. Model Inputs	327
IV. Model Performance	328
V. Conclusions	334
VI. Appendix	335
References	337

15 Nitrogen Solute Transport 341
K. K. TANJI AND M. NOUR EL DIN

I. Model Description	341
II. Model Validation	347
III. Summary	350
IV. Appendix 1	351
V. Appendix 2	361
References	363

16 Solute Transport and Reactions in Salt-Affected Soils 365
CHARLES W. ROBBINS

I. Modeling Solute Transport	365
II. Modeling Lime and Gypsum Solubility Reactions	367
III. Modeling Cation Exchange	368
IV. Critical Assumptions	370
V. Model Validation	371
VI. Summary	372
VII. Additional Research Needs	376

VIII. Appendix 1	377
IX. Appendix 2	391
X. Appendix 3	392
References	394

17 Soil Heat Flow   397

ROBERT HORTON AND SANG-OK CHUNG

I. Heat Flow Model	398
II. Energy Partitioning at the Soil Surface	399
III. Finite Difference Equations	401
IV. Computer Program Description	401
V. Test of Model	404
VI. Comparison between Measured and Predicted Temperatures	407
VII. Conclusion	414
VIII. Acknowledgment	415
IX. Appendix	415
References	437

18 Runoff and Water Erosion   439

J. R. WILLIAMS

I. Runoff Model	440
II. Water Erosion Model	445
III. Model Tests	449
IV. Summary and Conclusions	449
V. Appendix	450
References	454

19 Modified EPIC Wind Erosion Model   457

EDWARD L. SKIDMORE AND J. R. WILLIAMS

I. Model Description	458
II. Vegetative Factor	462
III. Summary	466
References	467

20 Promises and Problems of Extending Models to the Grass Roots Level   471

V. PHILIP RASMUSSEN

I. Technology Transfer	472
II. Bridging the Model/User Gap—Closing the Loop	473
III. Natural Language Systems	477
IV. Appendix 1	480
V. Appendix 2	484
VI. Appendix 3	487
References	488

21 Irrigation Scheduling   491
ROBERT W. HILL

I. Recent Experience in Irrigation Scheduling Activities 492
II. Soil Water Budget Models 493
III. Operation of Irrigation Scheduling Programs 496
IV. Comparison with Field Data 505
V. Other Similar Models 505
VI. Critical Assumptions 507
VII. Summary 508
VIII. Acknowledgments 508
References 509

22 Biophysical Simulation of Wheat and Soybean to Assess the Impact of Timeliness on Double-Cropping Economics   511
LUCAS D. PARSCH, MARK J. COCHRAN, KALVEN L. TRICE, AND H. DON SCOTT

I. Simulation Models 512
II. Experimental Design 518
III. Results 520
IV. Conclusions 533
References 533

Subject Index 535

PREFACE

This monograph is the first effort by the American Society of Agronomy, Crop Science Society of America, and the Soil Science Society of America to publish a book on the use of computer simulation to evaluate agronomic systems. It is a testament to the desire of agronomists to enter a new era of agricultural research and development; a time in which simulation partially substitutes for experiments to determine recommendations for various agrotechnology packages. Improved accuracy in simulating crop and soil systems has convinced a growing number of researchers and technologists of the importance of being able to predict outcomes needed in agricultural decision making. Combining the soil, plant, and climate system into quantitative terms that lead to accurate predictions of outcomes is needed as agronomists enter the information age.

The main focus of agronomic research and development efforts in the past few decades has been related to food and feed production. The quality of the environment has now become an imperative issue that crop and soil scientists must address, in addition to maintaining adequate food production. The research needed to find management strategies that optimize high production and minimize environmental degradation will need to include simulation of all or part of the soil-plant atmosphere system. Experimental research alone would require more than an order of magnitude of effort and resources as compared to present levels of production and agricultural activity to adequately reach these goals. An era of low funding for agricultural research and extension makes the use of simulation models even more of a necessity as a major assistance tool for helping with decision making in sustainable agricultural systems. Progress made by using models is much faster and less expensive than through experimental research alone. Key simulated results, however, must continue to be evaluated with field experiments.

The chapters in this monograph have been written to provide examples of models of component processes necessary to simulate the dynamics of crop and soil systems. The information is intended to be useful to researchers, extension and other professional technologists, and teachers. For researchers, the models of processes will define details regarding information necessary from experimental trials to adequately simulate results of those experiments. Because every step in a simulation must be clearly defined, the lack of accuracy in simulating various processes also helps to identify gaps where additional research is needed. For professional agronomists who recommend technological packages to farmers and other users of crop and soil information, models could provide reasonable and quick answers to the "what if" questions for specific soil, crop, and management combinations. For teachers, the use of computer simulation models can provide stimulating opportunities for students to learn about details of dynamic processes and how the processes can be linked together to produce a desired system. Models of crop and soil systems also provide a means of doing simulated experiments in the classroom setting that would be impossible to do in the field within the time constraint of a crop year.

Chapters in this monograph have generally been organized to cover the plant processes first, then the soil processes, and finally to give some examples of how simulation models can be applied to problem-solving situations. Users of the information should realize there are different levels of detail in simulation models and different methods to approach the modeling of a process. Most of the authors have identified models other than their own from the literature that simulate the same process. Our examples of models of processes in the book are not necessarily at the same level of detail. Thus, the use of each modelled process to form a system might lead to an unbalanced analysis of the system. The user should seek models of processes treated at approximately the same level of detail for building a complete system model. We have intentionally provided some variation in level of detail in modelled processes to provide a range of examples of approaches to modeling.

J. T. Ritchie, *co-editor*
Dep. of Soil Science
Michigan State University
East Lansing, Michigan

R. J. Hanks, *co-editor*
Soils and Biometeorology Dep.
Utah State University
Logan, Utah

CONTRIBUTORS

T. J. Arkebauer	Assistant Professor, Department of Agronomy, University of Nebraska, Lincoln, NE 68583-0817
W. L. Bland	Assistant Professor, Blackland Research Center, Texas Agricultural Experiment Station, 808 East Blackland Road, Temple, TX 76502
K. J. Boote	Professor of Agronomy, Agronomy Department, University of Florida, Gainesville, FL 32611
Eshel Bresler	Professor and Director, Institute of Soils and Water, ARO, Volcani Center, P.O. Box 6, Bet Dagan 50-250, ISRAEL
G. S. Campbell	Professor of Soils, Department of Agronomy and Soils, Washington State University, Pullman, WA 99164
Sang-Ok Chung	Post-Doctoral Researcher, Agronomy Department, Iowa State University, Ames, IA 50011. Currently Assistant Professor, Department of Agricultural Engineering, Kyungpook National University, SOUTH KOREA
Mark J. Cochran	Associate Professor of Agricultural Economics, Department of Agricultural Economics, University of Arkansas, Fayetteville, AR 72701
M. Nour El Din	Postgraduate Researcher, Department of Land, Air, and Water Resources, University of California, Davis, CA 95616
D. C. Godwin	International Fertilizer Department Center, Muscle Shoals, AL 35660. Currently Alton Park, MS-2, Dubbo, NSW 2830, AUSTRALIA
R. J. Hanks	Professor of Soil Physics, Plant, Soil, and Biometeorology Department, Utah State University, Logan, UT 84322-4840
Robert W. Hill	Professor, Agricultural and Irrigation Engineering, Utah State University, Logan, UT 84322-4105
Robert Horton	Associate Professor of Soil Physics, Agronomy Department, Iowa State University, Ames, IA 50011
S. S. Jagtap	Agroclimatologist, International Institute of Tropical Agriculture, Oyo Road, PMB 5230, Ibadan, NIGERIA
C. Allan Jones	Resident Director of Research and Professor, Blackland Research Center, Texas Agricultural Experiment Station, 808 East Blackland Road, Temple, TX 76502
James W. Jones	Professor of Agricultural Engineering, Agricultural Engineering Department, University of Florida, Gainesville, FL 32611
J. R. Kiniry	Research Agronomist, USDA-ARS, SPA, 808 East Blackland Road, Temple, TX 76502
J. W. Mishoe	Professor of Agricultural Engineering, Agricultural Engineering Department, University of Florida, Gainesville, FL 32611
D. S. NeSmith	Graduate Research Assistant, Department of Crop and Soil Sciences, A572 Plant and Soil Sciences Building, Michigan State University, East Lansing, MI 48824-1325

CONTRIBUTORS

J. M. Norman — Professor of Soil Science, Department of Soil Science, University of Wisconsin, 1525 Observatory Drive, Madison, WI 53706

Lucas D. Parsch — Associate Professor, Department of Agricultural Economics, 225 Agriculture Building, University of Arkansas, Fayetteville, AR 72701

V. Philip Rasmussen — Associate Professor and Assistant Department Head, Plant, Soil and Biometeorology Department, Utah State University, Logan, UT 84322-4820

J. T. Ritchie — Professor, Homer Nowlin Chair, Department of Crop and Soil Sciences, A570 Plant and Soil Sciences Building, Michigan State University, East Lansing, MI 48824-1325

Charles W. Robbins — USDA-ARS, 3793 North, 3600 East, Kimberly, ID 83341

H. Don Scott — Professor of Agronomy, Department of Agronomy, 115 Plant Science, University of Arkansas, Fayetteville, AR 72701

Andrew N. Sharpley — Soil Scientist, USDA-ARS, Water Quality and Watershed Research Laboratory, P.O. Box 1430, Durant, OK 74702

R. Wayne Skaggs — William Neal Reynolds Professor, Department of Biological and Agricultural Engineering, North Carolina State University, Box 7625, Raleigh, NC 27695-7625

Edward L. Skidmore — Soil Scientist, USDA-ARS, Wind Erosion Research, Department of Agronomy, Kansas State University, Manhattan, KS 66506

K. K. Tanji — Professor of Water Science, Department of Land, Air, and Water Resources, University of California, Davis, CA 95616

Kalven L. Trice — Graduate Research Assistant, Department of Agricultural Economics, University of Arkansas, Fayetteville, AR 72701. Currently Agricultural Economist, USDA-SCS, 10417 Campus Way South, Upper Marlboro, MD 20772

J. R. Williams — Hydraulic Engineer, USDA-ARS, Grassland, Soil and Water Research Laboratory, 808 East Blackland Road, Temple, TX 76502

Conversion Factors for SI and non-SI Units

Conversion Factors for SI and non-SI Units

To convert Column 1 into Column 2, multiply by	Column 1 SI Unit	Column 2 non-SI Unit	To convert Column 2 into Column 1, multiply by
Length			
0.621	kilometer, km (10^3 m)	mile, mi	1.609
1.094	meter, m	yard, yd	0.914
3.28	meter, m	foot, ft	0.304
1.0	micrometer, μm (10^{-6} m)	micron, μ	1.0
3.94×10^{-2}	millimeter, mm (10^{-3} m)	inch, in	25.4
10	nanometer, nm (10^{-9} m)	Angstrom, Å	0.1
Area			
2.47	hectare, ha	acre	0.405
247	square kilometer, km^2 (10^3 m)^2	acre	4.05×10^{-3}
0.386	square kilometer, km^2 (10^3 m)^2	square mile, mi^2	2.590
2.47×10^{-4}	square meter, m^2	acre	4.05×10^3
10.76	square meter, m^2	square foot, ft^2	9.29×10^{-2}
1.55×10^{-3}	square millimeter, mm^2 (10^{-3} m)^2	square inch, in^2	645
Volume			
9.73×10^{-3}	cubic meter, m^3	acre-inch	102.8
35.3	cubic meter, m^3	cubic foot, ft^3	2.83×10^{-2}
6.10×10^4	cubic meter, m^3	cubic inch, in^3	1.64×10^{-5}
2.84×10^{-2}	liter, L (10^{-3} m^3)	bushel, bu	35.24
1.057	liter, L (10^{-3} m^3)	quart (liquid), qt	0.946
3.53×10^{-2}	liter, L (10^{-3} m^3)	cubic foot, ft^3	28.3
0.265	liter, L (10^{-3} m^3)	gallon	3.78
33.78	liter, L (10^{-3} m^3)	ounce (fluid), oz	2.96×10^{-2}
2.11	liter, L (10^{-3} m^3)	pint (fluid), pt	0.473

CONVERSION FACTORS FOR SI AND NON-SI UNITS

Mass

To convert Column 1 into Column 2, multiply by	Column 1 SI Unit	Column 2 non-SI Unit	To convert Column 2 into Column 1, multiply by
2.20×10^{-3}	gram, g (10^{-3} kg)	pound, lb	454
3.52×10^{-2}	gram, g (10^{-3} kg)	ounce (avdp), oz	28.4
2.205	kilogram, kg	pound, lb	0.454
0.01	kilogram, kg	quintal (metric), q	100
1.10×10^{-3}	kilogram, kg	ton (2000 lb), ton	907
1.102	megagram, Mg (tonne)	ton (U.S.), ton	0.907
1.102	tonne, t	ton (U.S.), ton	0.907

Yield and Rate

0.893	kilogram per hectare, kg ha^{-1}	pound per acre, lb acre^{-1}	1.12
7.77×10^{-2}	kilogram per cubic meter, kg m^{-3}	pound per bushel, bu^{-1}	12.87
1.49×10^{-2}	kilogram per hectare, kg ha^{-1}	bushel per acre, 60 lb	67.19
1.59×10^{-2}	kilogram per hectare, kg ha^{-1}	bushel per acre, 56 lb	62.71
1.86×10^{-2}	kilogram per hectare, kg ha^{-1}	bushel per acre, 48 lb	53.75
0.107	liter per hectare, L ha^{-1}	gallon per acre	9.35
893	tonnes per hectare, t ha^{-1}	pound per acre, lb acre^{-1}	1.12×10^{-3}
893	megagram per hectare, Mg ha^{-1}	pound per acre, lb acre^{-1}	1.12×10^{-3}
0.446	megagram per hectare, Mg ha^{-1}	ton (2000 lb) per acre, ton acre^{-1}	2.24
2.24	meter per second, m s^{-1}	mile per hour	0.447

Specific Surface

10	square meter per kilogram, m^2 kg^{-1}	square centimeter per gram, cm^2 g^{-1}	0.1
1000	square meter per kilogram, m^2 kg^{-1}	square millimeter per gram, mm^2 g^{-1}	0.001

Pressure

9.90	megapascal, MPa (10^6 Pa)	atmosphere	0.101
10	megapascal, MPa (10^6 Pa)	bar	0.1
1.00	megagram per cubic meter, Mg m^{-3}	gram per cubic centimeter, g cm^{-3}	1.00
2.09×10^{-2}	pascal, Pa	pound per square foot, lb ft^{-2}	47.9
1.45×10^{-4}	pascal, Pa	pound per square inch, lb in^{-2}	6.90×10^3

(continued on next page)

Conversion Factors for SI and non-SI Units

To convert Column 1 into Column 2, multiply by	Column 1 SI Unit	Column 2 non-SI Unit	To convert Column 2 into Column 1, multiply by
Temperature			
1.00 (K − 273)	Kelvin, K	Celsius, °C	1.00 (°C + 273)
(9/5 °C) + 32	Celsius, °C	Fahrenheit, °F	5/9 (°F − 32)
Energy, Work, Quantity of Heat			
9.52×10^{-4}	joule, J	British thermal unit, Btu	1.05×10^3
0.239	joule, J	calorie, cal	4.19
10^7	joule, J	erg	10^{-7}
0.735	joule, J	foot-pound	1.36
2.387×10^{-5}	joule per square meter, $J\ m^{-2}$	calorie per square centimeter (langley)	4.19×10^4
10^5	newton, N	dyne	10^{-5}
1.43×10^{-3}	watt per square meter, $W\ m^{-2}$	calorie per square centimeter minute (irradiance), $cal\ cm^{-2}\ min^{-1}$	698
Transpiration and Photosynthesis			
3.60×10^{-2}	milligram per square meter second, $mg\ m^{-2}\ s^{-1}$	gram per square decimeter hour, $g\ dm^{-2}\ h^{-1}$	27.8
5.56×10^{-3}	milligram (H_2O) per square meter second, $mg\ m^{-2}\ s^{-1}$	micromole (H_2O) per square centimeter second, $\mu mol\ cm^{-2}\ s^{-1}$	180
10^{-4}	milligram per square meter second, $mg\ m^{-2}\ s^{-1}$	milligram per square centimeter second, $mg\ cm^{-2}\ s^{-1}$	10^4
35.97	milligram per square meter second, $mg\ m^{-2}\ s^{-1}$	milligram per square decimeter hour, $mg\ dm^{-2}\ h^{-1}$	2.78×10^{-2}
Plane Angle			
57.3	radian, rad	degrees (angle), °	1.75×10^{-2}

CONVERSION FACTORS FOR SI AND NON-SI UNITS

Electrical Conductivity, Electricity, and Magnetism

To convert Column 1 into Column 2, multiply by	Column 1 SI Unit	Column 2 non-SI Unit	To convert Column 2 into Column 1, multiply by
10	siemen per meter, S m^{-1}	millimho per centimeter, mmho cm^{-1}	0.1
10^4	tesla, T	gauss, G	10^{-4}

Water Measurement

9.73 × 10^{-3}	cubic meter, m^3	acre-inches, acre-in	102.8
9.81 × 10^{-3}	cubic meter per hour, m^3 h^{-1}	cubic feet per second, ft^3 s^{-1}	101.9
4.40	cubic meter per hour, m^3 h^{-1}	U.S. gallons per minute, gal min^{-1}	0.227
8.11	hectare-meters, ha-m	acre-feet, acre-ft	0.123
97.28	hectare-meters, ha-m	acre-inches, acre-in	1.03 × 10^{-2}
8.1 × 10^{-2}	hectare-centimeters, ha-cm	acre-feet, acre-ft	12.33

Concentrations

1	centimole per kilogram, cmol kg^{-1} (ion exchange capacity)	milliequivalents per 100 grams, meq 100 g^{-1}	1
0.1	gram per kilogram, g kg^{-1}	percent, %	10
1	milligram per kilogram, mg kg^{-1}	parts per million, ppm	1

Radioactivity

2.7 × 10^{-11}	becquerel, Bq	curie, Ci	3.7 × 10^{10}
2.7 × 10^{-2}	becquerel per kilogram, Bq kg^{-1}	picocurie per gram, pCi g^{-1}	37
100	gray, Gy (absorbed dose)	rad, rd	0.01
100	sievert, Sv (equivalent dose)	rem (roentgen equivalent man)	0.01

Plant Nutrient Conversion

	Elemental	Oxide	
2.29	P	P$_2$O$_5$	0.437
1.20	K	K$_2$O	0.830
1.39	Ca	CaO	0.715
1.66	Mg	MgO	0.602

1 Introduction

R. J. HANKS
Utah State University
Logan, Utah

J. T. RITCHIE
Michigan State University
East Lansing, Michigan

A model has been defined as a small imitation of the real thing or as a system of postulates, data and inferences presented as a mathematical description of an entity or state of affairs. Models can also be used as a technique to organize what is known about a subject into a system showing the effect of the interrelation of many factors on some desired result. Thus, in this book, we are attempting to present useful representations of plant and soil systems.

Plant and soil systems are very complicated with numerous factors influencing any desired end result. Recent advances in computer technology have created the possibility of considering the influence of several variables combined in various interactions. Consequently, numerous plant and soil system models of varying complexity have been proposed. The purpose of this book is to consider some of those models that appear to have practical utility to users other than modelers.

Plant and soil system models have advanced rapidly in the recent past because of the tremendous power they give the user. For example, we no longer need to answer the farmer's inquiry, "How much will one less irrigation influence the yield of my corn?" with "It depends..." We can now sit down with the farmer, consider many of the factors involved, and answer the question with a reasonable estimate. We can also try several other possibilities that are of interest. This has the advantage of calling attention to the factors involved as well as indicating the importance of various components. Using models to test the sensitivity of the end results to any one variable is also valuable. A plant and soil system model is essentially a collection of information, much of which has been known for years, that has been organized into a coherent arrangement. When needed information is somewhat lacking, approximations or empirical equations are used to estimate the real system. Almost all present practical models use empirical approximations because of the lack of better information and the need for model simplicity. Models should be used with caution, taking into account the

---

Copyright © 1991 ASA-CSSA-SSSA, 677 S. Segoe Rd., Madison, WI 53711, USA. *Modeling Plant and Soil Systems*—Agronomy Monograph no. 31.

assumptions or limitations of each particular model. The model user should have enough experience to tell when model simulations are nonsense and when they are reasonable. Thus models are not substitutes for experience, but rather a tool to use with experience and investigation.

For a model to be of practical value the data requirements must be reasonable. Ideally, known crop, soil, and climatic characteristics should be all the data needed. However, the use of models has shown us that, in some instances, additional data not previously available are needed. This illustrates another valuable attribute of models, to point out needed or more accurate data requirements.

Model construction is a good educational experience for the aspiring model builder. Students now being trained should be able to construct their own models and be familiar with models that are available in their chosen field. Models will undoubtedly be one of the standard tools of the future scientist. They allow scientists to see how their field of specialty fits into a larger system and should lead them to areas of future important research. Models will also be of assistance if scientists are required to work with people from other disciplines, because they will be able to simulate the influence of their field or discipline into a larger context.

ATTRIBUTES OF MODELS TO BE COVERED

The emphasis in this book will be on a few of the models available that have wide applicability to soil and crop scientists. It has been necessary to severely limit the number of models discussed to a few crops and a few situations. Thus, the reader should be aware that there are many models, such as those discussed by Whisler et al. (1986), that would meet the criteria set forth herein. The models included in this book are relatively simple and orientated toward problem solving, as contrasted with detailed research models, which are oriented toward understanding the many processes involved. The models covered here should have the following requirements:

1. Commonly available data should be required.
    a. Climate—daily temperature, precipitation, wind, humidity, pan evaporation, radiation.
    b. Soil—texture, depth, water availability.
    c. Crop—type, cultivars, planting date, growth stages.
    d. Management—fertilizer application, tillage, irrigation.
2. The input of management factors should be emphasized and easily accomplished.
3. The model must maintain a balance of all parts such as water balance, salt balance, carbon balance, nitrogen balance, etc. . .
4. Plasticity of plants (genetic difference and adaptability).
5. Computer requirements should be reasonable. The model should preferably be capable of running on microcomputers or minicomputers. The FORTRAN or BASIC computer codes should be available as part of each model. Estimates of computer time should be made.

6. General variables should be required. Models should have the minimum of site-specific factors. The model should be capable of giving reasonable answers at a different location from where it was developed using local data.
7. The model should be stable and not include hunting routines that could hang up the computer. If hunting routines are used they should be automatically terminated after several cycles.
8. The model must have been tested under field conditions on a data set other than that on which it was calibrated.
9. Main assumptions and simplifications should be clearly indicated. The principle assumptions should be clearly outlined so the user is well aware of the range of applicability of the model.
10. The potential and actual data output that comes from the model should be given.

REFERENCE

Whisler, F.D., B. Acock, D.N. Baker, R.E. Fry, H.F. Hodges, J.R. Lambert, H.E. Lemmon, J.M. McKinion, and V.R. Reedy. 1986. Crop simulation models in agronomic systems. Adv. Agron. 40:141-208.

2 Temperature and Crop Development

J. T. RITCHIE AND D. S. NeSMITH

Michigan State University
East Lansing, Michigan

The state of a plant is determined by both growth and developmental processes. In modeling crop systems, separating the two processes is important because they are affected by different environmental variables. Development refers to the timing of critical events in the life cycle of a plant. Growth refers to the increase in weight, volume, length, or area of some part or all of the plant. In more mechanistic crop modeling, considerably more research and emphasis has been placed on plant growth through the photosynthesis process than on development.

The potential biomass yield of a crop can be thought of as the product of the rate of mass accumulation multiplied by the duration of growth. The rate of biomass accumulation is principally influenced by the amount of light intercepted by plants over a fairly wide optimum temperature range. However, the duration of growth for a particular cultivar is usually almost directly proportional to temperature, over a wide range of temperatures. Highest potential yields of a particular annual crop are obtained in regions where the season duration is maximized because of relatively low temperatures. In regions such as the tropics, where the temperature is usually relatively high, potential yield levels can reach those of cooler temperature regions only by combining yields from two or more crops in sequence so that the duration of the total growth periods is about the same in both regions. This suggests that the potential rate of biomass growth is relatively constant over space and time when the temperatures are within the range for growth, whereas the duration of growth is more variable in space and time. Thus, the modeling of crop duration is critical in order to predict crop potential productivity.

Predicting crop growth duration is necessary to find genotypes with a desired growth period that enables farmers to optimize yields. These yields are produced within the constraints of a soil water supply or a favorable thermal environment. Also, the ability to predict the stage of crop development is important for such management decisions as timing of pesticide application, scheduling the orderly harvest of crops, or synchronizing the flowering of cross-pollination crops for hybrid seed production.

Reaumur first suggested in 1735 that the duration of particular stages of growth was directly related to temperature and that this duration for a particular species could be predicted using the sum of mean daily air tem-

---

Copyright © 1991 ASA-CSSA-SSSA, 677 S. Segoe Rd., Madison, WI 53711, USA. *Modeling Plant and Soil Systems*—Agronomy Monograph no. 31.

peratures (Wang, 1960). This procedure for normalizing time with temperature to predict plant development rates has been widely used in the 20th century. Investigators who have studied the use of the system on the same cultivar in different environments have found several inaccuracies in the system. Most attention given to improving the system has focused on determining the low temperature at which development is zero (the base temperature) and the high temperature at which development ceases to increase or begins to decline. This paper is intended to provide insight into the value of using temperature for predicting plant development and to discuss some of the sources of uncertainties when using a temperature summation system.

I. TERMINOLOGY

Several synonymous terms have been used to describe the process of summation of temperatures to predict plant growth duration (Nuttonson, 1955). These include the terms degree-days (°Cd), day-degrees, heat units, heat sums, thermal units, and growing degree-days. Other terms that, in addition to a temperature summation, also account for photoperiod or give different weighting to night and day temperatures have been used for specific crops. These terms include Soybean Development Units (Brown & Chapman, 1961), Ontario Corn Heat Units (Brown, 1975), Biometeorological Time Scale (Robertson, 1968), and Photothermal Units (Nuttonson, 1948). A term that we believe is most appropriate in describing plant development is thermal time, as suggested by Gallagher (1979). Using the term thermal as an adjective for time appropriately refers to a state of matter dependent on temperature. The noun heat implies that energy is added, causing substances to rise in temperature. Heat does not seem as appropriate as thermal in describing plant development. Because a plant's time scale is closely coupled with its thermal environment, it is appropriate to think of thermal time as a plant's view of time. Likewise, if the photoperiod is used to modify thermal time, the resulting term is photothermal time.

Thermal time has the convenient units °Cd. The most simple useful definition of thermal time (t_d) is

$$t_d = \sum_{i=1}^{n} (\bar{T}_a - T_b)$$

where $\bar{T}_a$ is daily mean air temperature, T_b is the base temperature at which development stops, and n is the number of days of temperature observations used in the summation. The calculation of $\bar{T}_a$ is usually performed by averaging the daily maximum and minimum temperatures. This calculation of thermal time is appropriate for predicting plant development if several conditions are met:

1. The temperature response of the development rate is linear over the range of temperatures experienced.

2. The daily temperature does not fall below T_b for a significant part of the day.
3. The daily temperature does not exceed an upper threshold temperature for a significant part of the day.
4. The growing region of the plant has the same mean temperature as $\bar{T}_a$.

II. SOURCES OF ERROR IN THERMAL TIME CALCULATIONS

Most available daily maximum and minimum temperature records have been observed from liquid-in-glass thermometers exposed in standard white, louvered instrument shelters (Epperson & Dale, 1983). The observations are made once daily using separate thermometers for measuring maximum or minimum temperatures. A possible bias in these temperature records can be caused by the time of observation (Schaal & Dale, 1977). Minimum thermometer readings for the next observational day cannot be higher than the temperature at the time of observation. Thus, a set minimum temperature at early morning observation times can carry over to the next morning, causing a downward bias in the minimum and average temperature. Similarly it is possible to have an upward bias in maximum temperature readings when the once per day readings are taken in the afternoon. This bias can be the most significant in summer months when daylight savings time is used, forcing readings taken at the end of a normal working day to be quite close to the time when the maximum temperature is expected. Biases of this type were found to average about 0.5 to 1.0 °C/d for several months of the year at West Lafayette and Whiteston, IN (Epperson & Dale, 1983). Thus, the thermal time biases for a crop season could be in the range of 100 °Cd for this error. This type of measurement error is becoming less of a problem because manual weather stations are being replaced with automatic electronic data acquisition systems.

An error can even be introduced into thermal time calculations when the maximum and minimum temperatures are accurate because the daily mean temperature will likely differ from the average of the maximum and minimum temperatures. The major discrepancy will occur when a cold or warm air mass moves into the area near the beginning or ending of the daily temperature record. A comparison of the sum of the differences between the average temperatures calculated using the daily maximum and minimum values and the average temperature taken by averaging values recorded every minute are shown in Fig. 2-1 for 3 yr of the maize (*Zea mays* L.) growing season in East Lansing, MI. The recordings were taken with an automatic weather station, recording on a midnight to midnight basis. The results demonstrate that there were between-year and within-year differences in the bias. Cumulative biases ranged from about 25 °C for 100 d in 1987 to almost no bias in 1985. Even in the worst case, this error in mean temperature calculation would cause only a small bias in plant development prediction, because a typical thermal time calculation in 100 d can be about 2000 °Cd from

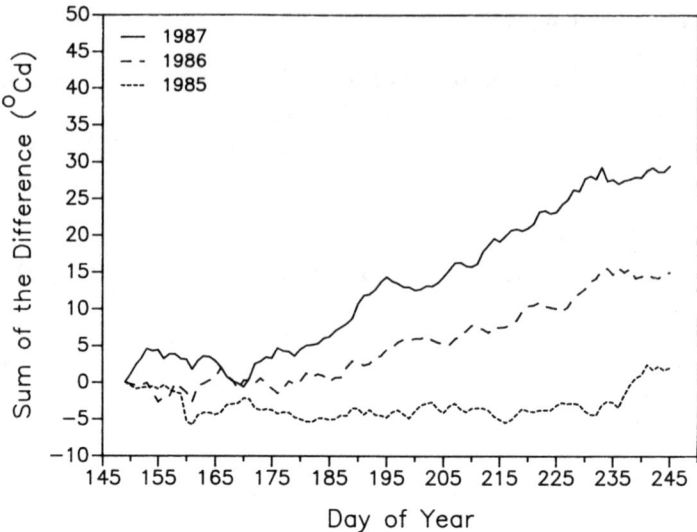

Fig. 2-1. Sum of the difference of thermal time calculated from daily maximum and minimum temperatures and daily average temperature calculated from 1 min readings. Data were taken during 1985, 1986, and 1987, near East Lansing, MI.

a crop such as maize in the Midwest. Thus, this error is relatively small when compared with other sources of uncertainty in thermal calculations.

Another source of temperature error can be encountered because most thermometers are not located in or near the field where plant development is to be evaluated. This error could be quite large, especially in regions where elevation is different between the sensor and the field.

Even when temperature sensors are in the field where the crop is growing, the air temperature at shelter height may be different from the temperature sensed by the plant. The specific location on a plant where temperature influences development is at the developing points, the zones where plant cell division and expansion are occurring. Watts (1972) measured rates of extension of maize leaves under varying root, shoot, and apical meristem temperature regimes and found that the base of the apical meristem was where temperature most strongly affected the plant's extension growth. Earlier work of Beauchamp and Lathwell (1966) demonstrated similar results. Law and Cooper (1976) and Cooper and Law (1978) found that maize plants were more sensitive to soil temperature than to air temperature until the 12th leaf tip was visible. After this, the apical meristem emerged from below the soil and became sensitive to air temperature. Swan et al. (1987) found that development rates of maize were best described using soil temperature until the 6th leaf was full size, after which air temperature was used.

For most monocot crop plants in the seedling stage, the apical meristem growth zone is often about 1 cm below the soil surface. Thus, during times when the plant's meristem is at or near the soil surface, the near surface soil temperature is more appropriate for predicting plant development than air temperature.

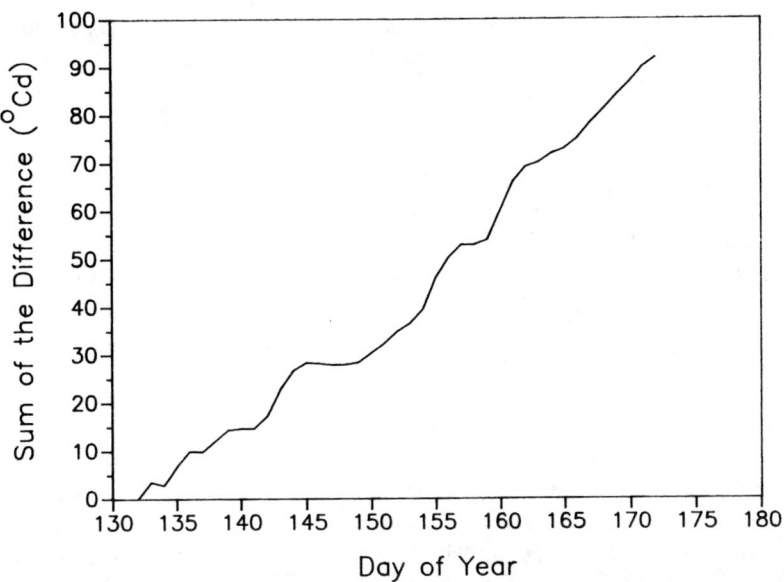

Fig. 2-2. Sum of the difference of thermal time calculated from average air temperature taken at a height of 1.5 m and average soil temperature taken at a depth of 2.5 cm. Data were taken during 1987 near Kalamazoo, MI.

Figure 2-2 shows the sum of the difference between thermal time calculated using soil temperature at 2.5 cm depth as compared with air temperature for Kalamazoo, MI during 1987. The maize planting date and initialization of thermal time was 12 May (Day of Year 132). On most days, the practically bare soil was warmer than the air temperature with an average daily difference of about 2.2 °C. The error associated with using air temperature for thermal time calculations was about 85 °C at the time when the tip of the 12th leaf had appeared. This error would cause a bias of 4 to 6 d in the calculation of tasseling time if the average air temperatures were in the range of 20 to 28 °C, assuming a base temperature of 8 °C. Soil and air temperature differences can be greater than in this example, depending on factors such as location, time of year, and tillage and residue management practices (Wierenga et al., 1982; NeSmith et al., 1987; Swan et al., 1987). Thus, for the greatest accuracy, the soil temperature at the level of the apical meristem should be used for the prediction of plant development until the internodes elongate enough for the meristem to be some distance above the soil surface.

III. DETERMINING TEMPERATURE RESPONSE FUNCTIONS

As mentioned previously, the emphasis of this chapter is the duration of events, not the rates of growth. The lack of distinction between these two important physiological processes is likely the principal reason why there are differences in opinion regarding the nature and value of temperature response functions. For example, Gilmore and Rogers (1958) used a curve of maize seedling growth rates at various temperatures reported by Lehenbauer (1914) to establish the rationale for using temperature accumulation with a linear scale up to 32 °C, after which there was no additional accumulation when the maximum temperature exceeded 32 °C. Others have argued for a nonlinear temperature development curve on the basis of the Lehenbauer (1914) data (Coelho & Dale, 1980).

The contrast between growth and development can be demonstrated from a leaf growth experiment reported by Grobellaar (1963). Maize plants were grown in a similar thermal environment until about the time the fourth leaf tip appeared. Then plants were placed into thermal environments ranging from 5 to 40 °C. Leaf linear extension growth was measured daily for the fourth, fifth, and sixth leaf. Figure 2-3 shows the results for the fifth leaf. We obtained the average extension rates and final leaf length from the solid lines shown for each temperature. The leaf length was extrapolated to zero to determine the duration of growth. The final leaf length had to be approximated for the 15, 20, and 40 °C treatments because there was no clear ending growth point when the measurements were discontinued. The 5 °C treatment had no reported growth.

The leaf length of the 15 °C treatment was assumed to be the same as the length of the 20 °C treatment, which was found for the fourth leaf (not shown). Figure 2-4a presents the derived leaf development rate as the inverse of the duration of the leaves' extension growth. The single leaf develop-

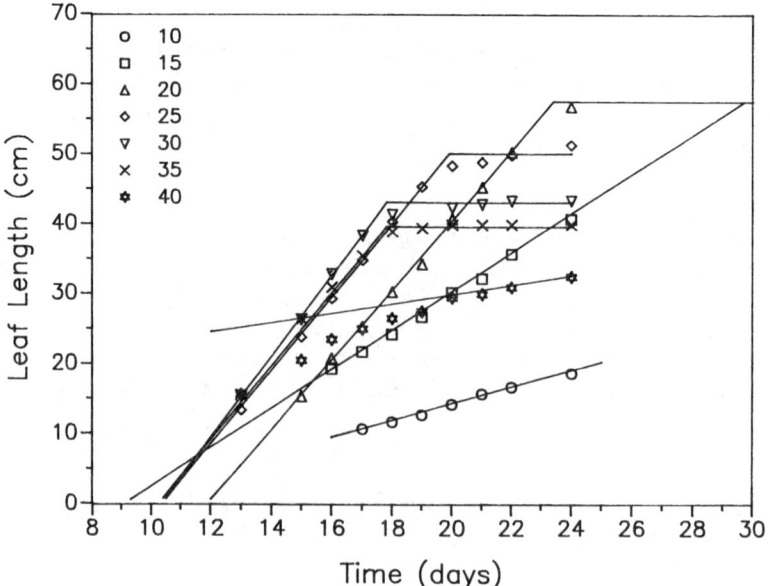

Fig. 2-3. Extension of the fifth leaf of maize grown at various temperatures from Grobellaar (1963). The solid lines were fitted by the authors.

ment curve fits, quite well, a typical linear development system with T_b at 8 °C, T_p at 33 °C, and T_x at 41 °C. The average extension growth rate at the various temperatures is given in Fig. 2-4b. This curve, typical of other growth curves, such as the gross photosynthesis rate, has a fairly well-defined optimum across the temperature range of about 22 to 32 °C. The final leaf size is the product of the duration of growth and the mean growth rate. The final length of the sixth leaf as influenced by temperature is shown in Fig. 2-4c. The fact that the leaves vary in size by a factor of almost two clearly demonstrates that temperature affects leaf development rate with a different response function than it affects growth rate. Leaf size data for maize presented by Hesketh and Warrington (1989) with those in Fig. 2-4c indicate, that a reduction in size begins at about 13 °C. If the growth and development responses were similar, the final leaf length would be the same for all temperature treatments. Sofield et al. (1974) demonstrated a similar principle for grain-mass accumulation rate. Grain growth rate at various temperatures was far less variable than the duration of grain filling.

The principal focus of this demonstration of the difference between growth and development is that when crop response to temperature is observed, there is little point in describing the growth rate without defining the durations. As with individual leaf sizes, the product of the rate and duration usually leads to larger leaf size, grain size, or grain number when the temperatures at which the crop is growing are considerably below T_p. Respiration rate is often given as the reason for low yields at high temperatures without any consideration being given to how high temperatures reduce growth duration.

When plants' development rates are linear over a wide range of existing temperatures for a region, the use of thermal time to describe the duration of growth events is certainly reasonable, so long as the limitations to the approach are understood.

IV. IDEALIZED RESPONSE FUNCTION

For thermal time to be appropriate as a predictor of plant development rates, there should be a linear relationship between the development rate and plant temperature over a well-defined range of temperatures, as demonstrated in the leaf development data of Fig. 2-4a. The low temperature at which the development rate equals zero is usually termed the base temperature (T_b). Because development is an irreversible process, development rate is not negative below the T_b, but remains at zero. Plant death resulting from cold temperatures will usually not occur for several degrees below T_b if the exposure is brief. The high temperature at which the linear relationship no longer holds can be appropriately termed the peak development temperature (T_p). For temperatures above T_p, the development rates usually decrease rapidly in another somewhat linear relationship.

Fewer studies have addressed plant development rates at temperatures above T_p, partly because there is probably little interest in production of

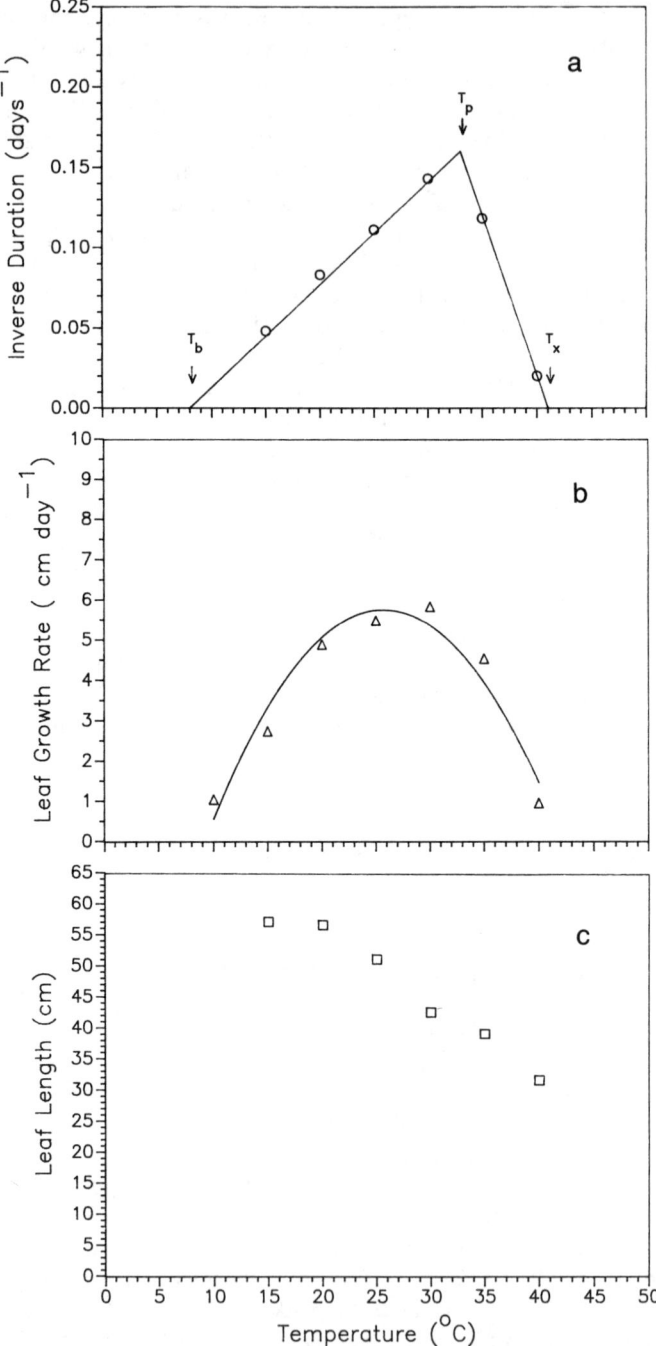

Fig. 2-4. Leaf development rate (a), average leaf extension rate (b), and final leaf size (c) of the fifth leaf of maize at various temperatures. Data were calculated with graphical data from Grobellaar (1963). Arrows indicate the temperatures taken to be T_b, T_p, and T_x, which are 8, 33, and 41 °C, respectively.

crops when maximum temperatures exceed T_p for much of the day. To our knowledge, crop plants do not survive the high temperatures that would be required to cause development rate to go to zero at unusually high temperatures. Thus, the T_x threshold can only be obtained through extrapolation of development rates measured at temperatures above T_p, but lower than T_x. The development rate between T_b and T_p is measured in units of the inverse of the time (usually days) used to measure duration of developmental events. The inverse of the slope of the development rate-temperature relationship, thus, is recorded in time-temperature units, which we refer to as the thermal time required for the developmental event to occur. For temperatures out of the T_b to T_p range, other equations are needed to calculate thermal time.

The prediction of development using thermal time has usually focused around the timing of important events in the phasic development of determinant crops, such as time to anthesis and physiological maturity. The timing of events such as germination, seedling emergence, leaf primordia appearance, leaf tip or ligule appearance, and leaf growth duration are also developmental events for which thermal time can be an appropriate concept.

V. TIMING OF DEVELOPMENTAL STAGES

A. Anthesis and Maturity

Quantitative relationships used frequently to calculate phasic developmental events in maize have often referred to Gilmore and Rogers (1958) and Arnold (1959). These authors specifically researched, under field conditions the determination of T_b or T_p. In both instances, planting dates were used to obtain variations in crop temperature regimes and development times. The study of Gilmore and Rogers (1958) was performed in Texas by planting on five dates between 17 Feb. and 8 May 1956. Arnold (1959) performed eight and nine plantings in Illinois in 1954 and 1955, respectively, between early May and early July. Arnold (1959) specifically focused on determination of T_b using several statistical procedures. The development process considered was the time from planting to harvest. Distinctly different values for T_b resulted, about 6 °C in 1954 and 4.3 °C in 1955, regardless of the statistical procedure used to estimate T_b. The Arnold data are shown graphically in Fig. 2-5. The data demonstrate the problems associated with an accurate determination of T_b in field studies because there is a lack of information involving the temperature response function near T_b. Although the data show a clear temperature-dependent response, extrapolation of the T_b value in this way can lead to a high degree of uncertainty. The value of T_b obtained by the best fit of a linear equation to the data of Arnold (1959) extrapolates to about 4.5 °C for the 2 yr combined data. However, using a T_b value of 8 °C, which we believe to be more appropriate on the basis of other developmental processes measured over most of the range of temperatures between T_b and T_p, also fit the Arnold data quite well as shown by the solid line in Fig. 2-5.

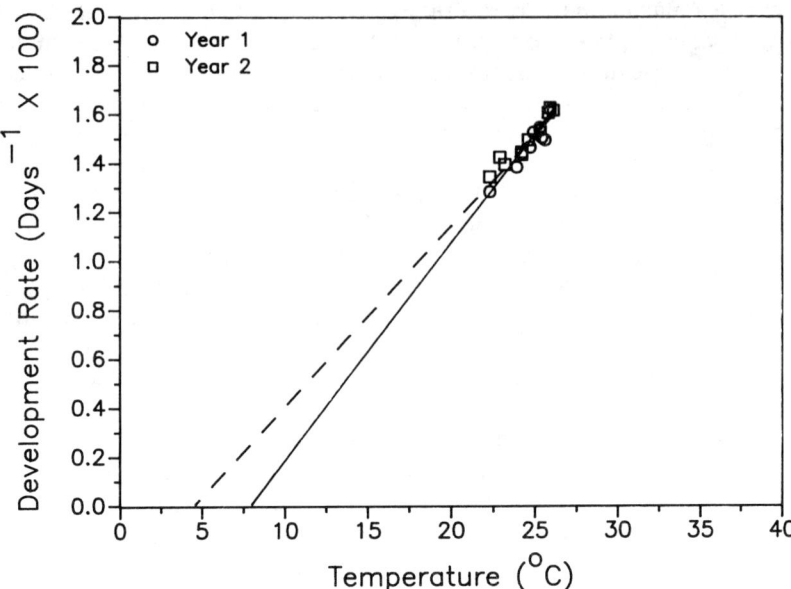

Fig. 2–5. Maize development rate, calculated as days from sowing to harvest, as a function of average seasonal temperature. Data from Arnold (1959). The dashed line represents least squares and the solid line is forced through 8 °C.

Gilmore and Rogers (1958) measured silking in maize. From their tabular information it was possible to calculate the average temperatures measured in their studies. The near linearity of the development rate as influenced by average temperature is shown graphically in Fig. 2–6. The line fit through the data was extrapolated to 8 °C to demonstrate the reasonableness of that value for T_b.

Gilmore and Rogers (1958) used the value of 10 °C for T_b without attempting to use other values as was done by Arnold (1959). They evaluated the use of 30 and 32.2 °C as possible T_p values and found that 32.2 °C minimized errors in the prediction of silking dates. They believed this maximum development value to be appropriate because it corresponded to the maximum rate of seedling elongation in the Lehenbauer (1914) study. They did not evaluate T_p as a single peak development rate, but rather as the beginning of an optimum, considering any temperature above a threshold high temperature to have the same development response as that at the threshold temperature. This evaluation of the upper temperature threshold is similar to the way T_b is calculated by taking any temperature below T_b as zero development rate. The consideration of T_p as an optimum does not agree with the Lehenbauer seedling expansion rate data, which rapidly decreased at temperatures above 32 °C. The 32.2 °C upper threshold temperature found by Gilmore and Rogers (1958) was likely the result of their particular data set containing maximum temperatures on both sides of a more appropriate T_p value higher than 32.2 °C. We assume for argument this

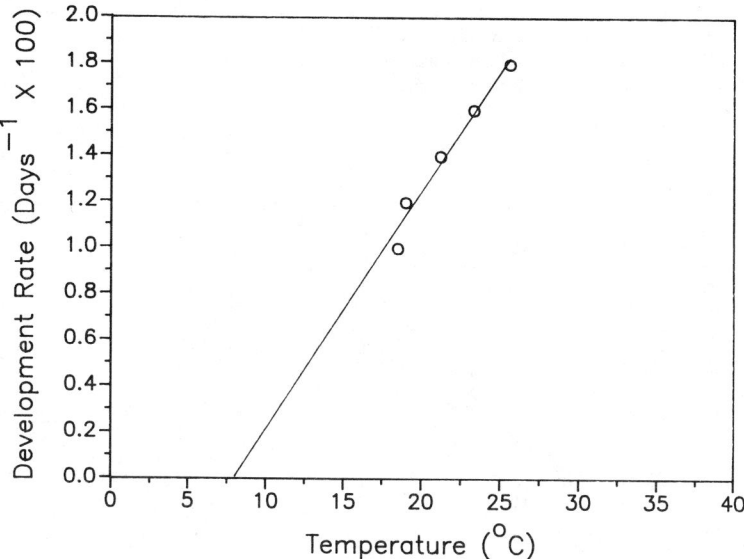

Fig. 2-6. Maize development rate, calculated as days from sowing to silking, as a function of average seasonal temperature. Data from Gilmore and Rogers (1958).

higher value to be 24 °C, so that errors of underprediction cancelled errors of overprediction. If this is true, then the use of temperatures above 32.2 °C as the temperature of maximum development is not a good generalization and is likely to create a thermal time bias downward if there is a predominance of maximum daily temperatures between 33 to 35 °C and biased upward for maximum temperatures above 38 °C. The assumption of a T_p value of 34 °C for maize with a rapid developmental rate decrease for temperatures above T_p will be discussed later.

B. Germination and Emergence

Germination has been used often for evaluating developmental temperature response functions. Although seeds do not germinate uniformly at constant temperature, Garcia-Huidobro et al. (1982), Ellis et al. (1986), and Covell et al. (1986) found that a linear relation resulted when the inverse of the duration to a specified fraction of the seeds germinated was related to temperature of the seeds during germination. Figure 2-7 presents the results of data reported by Covell et al. (1986) on soybean germination for several percentages of germination. The results imply a typical development relationship with relatively well defined constant T_b and T_p values. The extrapolated T_x values, however, were not constant for all germination percentages. Because the seed does not germinate uniformly, the use of thermal time to predict germination must be done as a probability, as shown in Fig. 2-8 for the soybean data of Covell et al. (1986). For low temperatures near T_b and high temperatures above T_p, there is often a fraction of the seed that does not

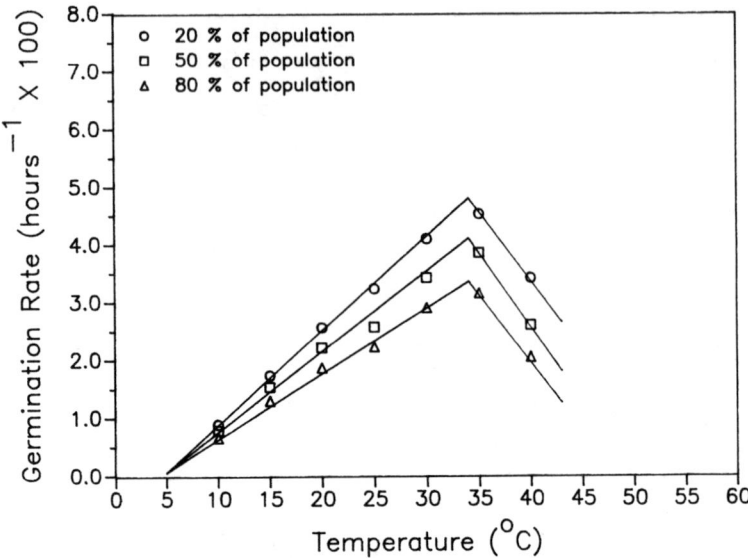

Fig. 2-7. Soybean germination rate (inverse of duration) for 20, 50, and 80% of the population at various temperatures. Data from Covell et al. (1986).

germinate when compared with the middle range of temperatures (Garcia-Huidobro et al., 1982). Thus, cumulative germination curves similar to the one in Fig. 2-8 will not reach 100% germination for all temperatures in the range of T_b to T_x.

Determination of the rate at which seedlings emerge through the ground surface can also be used to evaluate development rate. This phenomenon is much like germination with one exception; the elongation rate of the shoot from the seed level to the surface is principally a growth process and would not be expected to have a temperature response function similar to the developmental process of germination. Maize shoot and radicle elongation rates reported by Blacklow (1972) show evidence of a nonlinear relationship with temperature and an optimum rate between 28 and 34 °C. Depth of planting also influences the time to emergence and creates a level of uncertainty in predicting development rates. Thus, seedling emergence is not recommended as the most accurate means of evaluating temperature response functions for plant development.

C. Grain Filling

The duration of grain filling is also a developmental event that can be used for establishing the temperature development curves. Grain filling data of Spiertz (1977) for wheat grown under controlled temperature conditions indicates that the inverse of the duration of grain filling is a linear function of temperature. The value of T_b, however, appears to be about 3 °C, which is somewhat higher than zero, which is thought to be most appropriate for vegetative development in wheat. This higher T_b temperature for grain fill-

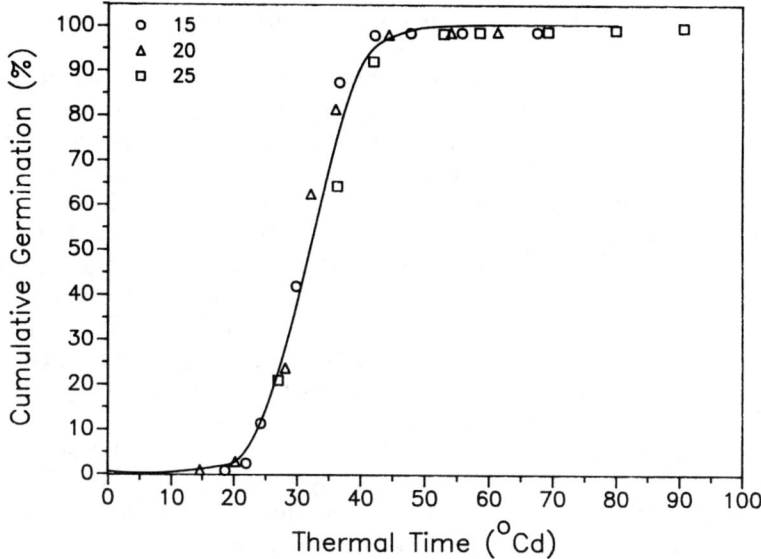

Fig. 2-8. Cumulative soybean germination at 15, 20, and 25 as a function of thermal time. Data from Covell et al. (1986).

ing in other crop species may also hold. The grain filling data of Sofield et al. (1974) also indicate a temperature for T_b for wheat. Johnson and Tanner (1972) reported that variations exist in grain filling duration between several genotypes of maize. They reported a 21- to 32-d variation in grain filling for field temperatures that were almost the same for the period between silking and maturity at Guelph, Ontario, Canada.

Grain filling duration is more difficult to quantify than visual development events such as leaf appearance or time to flowering. Grain filling for many crops has a lag period after anthesis before active filling occurs. After this, the filling rate is almost constant if average temperatures are relatively constant until the grain is practically filled, unless there is a shortage of assimilate or stored carbohydrate available for grain filling. Determining the beginning and ending of grain filling requires destructive sampling, thus creating a higher level of variability in the data than for non destructive type observations. Determining the timing of grain filling from visual observations such as flowering or silking to physiological maturity often leads to a great deal of uncertainty in assessing grain filling duration. McGarrahan and Dale (1984) reported that, in maize, government officials who estimate grain yields have used various criteria for maturity determination, such as the color change of the husk from green to white, kernel hardness, ease of shelling, percent grain moisture, and black layer formation. Although black layer formation has been used in recent years, it is not a highly accurate indication of end of grain filling, although it provides qualitative evidence of differences between genotypes that would be needed in modeling (Daynard, 1972).

D. Leaf Initiation and Appearance

Leaf appearance as a measure of the plant development rate has a distinct advantage for determining the temperature response function for plant development. Measurements are simple and nondestructive, and the event of leaf appearance occurs many times during a season. Leaf appearance is also quite logical as a developmental measurement because many crop plants produce leaves or nodes on a main stem at a predictable rate, after which they flower and then begin seed growth. Thus if the rate of leaf appearance is predictable along with the total number of leaves that will appear, then plant ontogeny during vegetative growth can be reliably predicted when the plant temperature is known. The total number of leaves or nodes on a main stem can be influenced by environmental factors other than temperatures, such as photoperiod (Hesketh et al., 1969) or vernalization (Chujo, 1966), and there are large differences in total leaf numbers within many species because of different maturity genotypes (Hesketh et al., 1969; Quinby, 1972).

Leaf appearance has been described in three ways for monocotyledon crop species; primordia, tip, and collar appearance. For dicotyledon crop species, leaf unfolding or leaf at full size can be used. Leaf primordia appearance is more difficult to determine because plants have to be dissected and experience is needed in making the primordia count with the assistance of a low-power microscope. Timing of a full-size leaf is also difficult to determine without making quantitative assessment of leaf dimensions. For monocots, however, the leaf collar appearance occurs approximately when the leaf is full size.

Leaf primordia of grass-type crops initiate at a constant rate when grown in a constant temperature (Warrington & Kanemasu, 1983a; Zur et al., 1989; Stern & Kirby, 1979). When the rate constant is plotted against temperature of the plant, the relationship has a typical development relationship. However, the variance in the measurement creates considerable uncertainty, thus making it difficult to establish critical temperature thresholds as closely as can be done with germination and visible leaf appearance rates. The large variation in such measurements arises from the difficulty of obtaining enough samples because the counting of primordia is time consuming and the measurement destroys the plant. When measurements are destructive, the variation among plants, caused partly by the expected germination and emergence variation, is considerably larger than when nondestructive measurements, such as leaf appearance, can be made on the same plant. When the temperature response function is determined from primordia observations in the field, the variance is extremely large (Baker & Gallagher, 1983b), making a reasonable and reproducible determination of T_b, and T_p practically impossible.

The rate of appearance of full-size leaves has been used primarily in monocot cereal crops to determine temperature response functions. The appearance of the ligule occurs at a rather distinct time, thus making such observations attractive for determining temperature response functions. The principal problem in using ligule appearance is that for plants with relatively large leaves, such as maize, sorghum (*Sorghum vulgare* L.), and pearl millet

(*Pennisetum glaucum* L.), the appearance of the last few leaf ligules occurs at accelerated rates due to more rapid expansion of internodes. Also the last few leaves decrease in length relative to the leaves in the middle section of the plant (Warrington & Kanemasu, 1983b; Muldoon et al., 1984). When temperature response functions are determined from leaf ligule appearance, care must be taken to choose the range of leaves used for establishing the appearance rate information (Hesketh & Warrington, 1989). Figure 2-9 demonstrates from data of Zur et al. (1989) how the appearance of the last few leaf ligules increases rapidly relative to the others, which appear at relatively constant rates.

Leaf-tip appearance is linear throughout practically all the range of leaf numbers. The appearance of the tip is also a fairly simple and objective observation. The slope of the leaf-tip appearance rate for the data of Tollenaar et al. (1979) is approximately 0.0265 leaves/°Cd between 8 and 34°C (Fig. 2-10). Leaf appearance rates from this study were observed for most of the possible range of expected temperatures and clearly indicate the linear response and well-defined T_b. The highest temperature in the study was 35°C, thus making it difficult to draw conclusions about a clear T_p or T_x value. Several studies on maize leaf appearance reported by Hesketh and Warrington (1989) indicate a rather consistent T_b value near 8°C and a range in the slope of both field data and growth chamber data of only 0.022 to 0.026 leaves/°Cd.

The inverse of the slope of the temperature-leaf appearance rate curve provides a convenient method to describe plant vegetative development. This

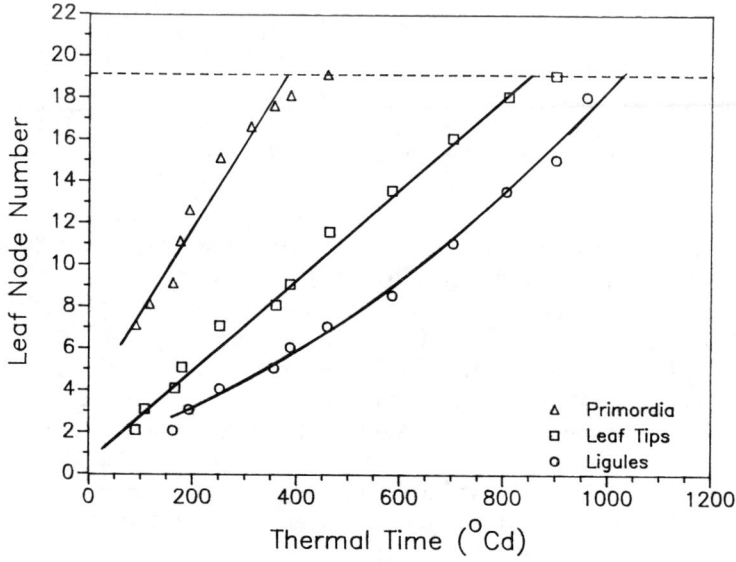

Fig. 2-9. Leaf primordia, leaf tip, and ligule numbers of maize as a function of thermal time. Data from Zur et al. (1989).

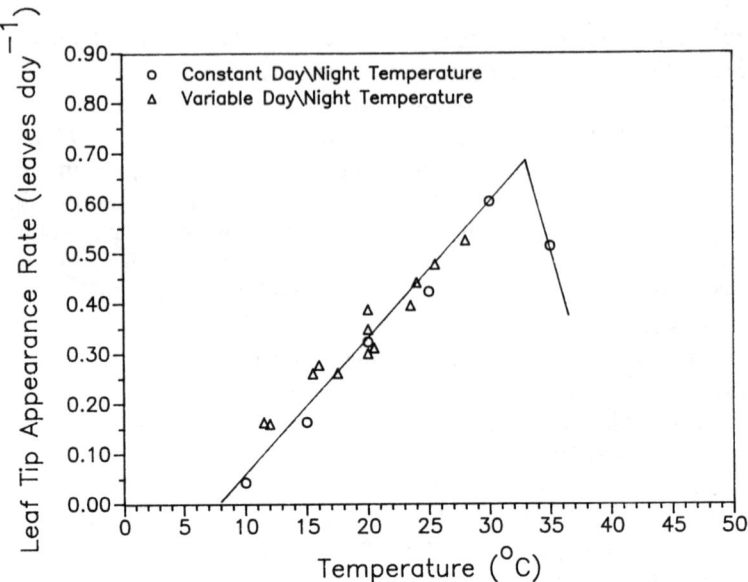

Fig. 2-10. Maize leaf-tip appearance rate at constant and variable day and night temperatures. Data from Tollenaar et al. (1979).

is done by describing the thermal time required for each leaf to appear, which is often termed a phyllochron. For maize, the rather consistently measured phyllochron is in the narrow range of 38 to 45 °Cd/leaf-tip appearance using $T_b = 8$ °C.

For dicot crop plants, the leaf unfolding rate has also provided temperature response functions. The leaf unfolding rate on the main stem of soybean was reported for a wide range of temperatures by Hesketh et al. (1973). A linear developmental response function was obtained with $T_b = 6$ °C. Sinclair (1984) reported general agreement with this finding for soybean [*Glycine max* (L.) Merr.], however the value of T_b was not as well defined as in the Hesketh et al. (1973) study.

Leaf appearance rates of winter cereals have proven to be a more complex response function than those reported warm-season type crops like maize and soybean. Most studies have found that T_b is about zero to 2 °C for wheat and barley, but the slope of the leaf appearance-temperature relationship determined under field conditions has been reported to vary by as much as a factor of two (Baker et al., 1980; Kirby et al., 1985; Kirby & Perry, 1987). For those studies, dates of sowing experiments provided the desired growing season temperature variations. Measured phyllochron values for wheat ranged from 75 to 140 °Cd, assuming T_b was zero. The fast development (low phyllochron) found in spring sowings and the slow development found in autumn sowings led the investigators to conclude that the phyllochron value was correlated with the rate at which daylength was changing when the plants emerged. The correlation coefficient between the inverse of the phyllochron and the mean rate of changes of daylength was rather low for Baker et al. (1980) in the United Kingdom and for Kirby and Perry (1987) in western Australia. The regression equations obtained in the two studies were approximately the same, with phyllochron values of about 100 °Cd at the intercept and approximately similar slopes. Data reported for Kansas by Baker et al. (1986) showing that wheat phyllochron emerged a few days after the autumn equinox agreed qualitatively with the Australia and U.K. data, having a value of about 106 °Cd. However, for wheat emerging at Arizona in late December when the rate of change in daylength was practically zero, the phyllochron for leaves 1 through 8 were about 80 °Cd (Baker et al., 1986), or 20 less than the United Kingdom and Australia results indicated. Wheat phyllochron data reported by Bauer et al. (1984) for plants emerging about the middle of May in North Dakota, averaged about 75 °Cd, whereas the United Kingdom and Australia phyllochron previously mentioned were about 90 °Cd at similarly changing daylength conditions at plant emergence. This evidence suggests that the use of rate of change of daylength at emergence does not consistently provide a means to predict the phyllochron.

Growth chamber research on the phyllochron in wheat and barley by Cao and Moss (1989a) indicated that the temperature at emergence had a distinct influence on the phyllochron, with wheat phyllochron values of about 60 °Cd at 7.5 °C and 120 °Cd at 24 °C. Barley showed similar patterns of response with somewhat lower phyllochron for all conditions. Cao and Moss (1989b) also found that the photoperiod itself caused the phyllochron to vary somewhat, with low values for short days and high values for long days. Their work indicates that this photoperiod influence could change the phyllochron after the third leaf had emerged, indicating that the environmental conditions after emergence were also important in determining the phyllochron.

The Cao and Moss data (1989a, b) were taken principally during the expansion of this second leaf. Evidence cited by Rickman et al. (1985) suggests that the development of the first two or three leaves may be influenced by the seed size because they depend on stored material for growth. Thus the absolute values of the phyllochron measured by Cao and Moss may be somewhat biased. In their photoperiod experiment, they also departed from usual growth chamber procedures by providing full light intensity for the entire photoperiod. The usual procedure is to provide all treatments with the same amount and duration of full light and then extend the daylength with low level light to the desired daylength. This treatment could cause a bias in the phyllochron results of Cao and Moss when compared with other work.

The leaf appearance rate research of Cao and Moss (1989a) covered a range of temperatures from 7.5 to 25 °C. Results averaged for four wheat varieties are shown in Fig. 2-11. The results indicate a lack of linearity for the temperature-development relationship. Growth chamber data of Chujo (1966) for wheat have similar nonlinearity, although the leaf appearance rates are considerably lower than the Cao and Moss (1989a) rates at comparable temperatures. If winter cereal plants have this nonlinear response in the field, then the thermal time concept will not be applicable as it has proven to be

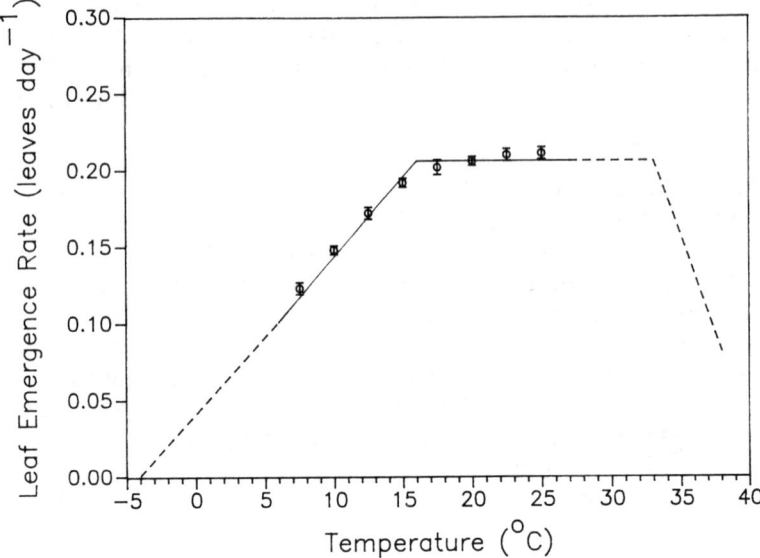

Fig. 2-11. Average leaf emergence rate of four wheat cultivars at various temperatures. Data from Cao and Moss (1989a). The dashed line represents an estimate for data outside of the range of the experiment.

for warm-season crops like maize and soybean. An alternative to the nonlinearity could be a linear response in the range of -4 to $16\,°C$, after which there is no T_p, rather an optimum range between $16\,°C$ to at least $25\,°C$. Results of J.T. Ritchie (1984, unpublished data) indicate an optimum wheat leaf appearance rate in the range 24 to about $31\,°C$, with rapid reduction at $36\,°C$. Thus, the development response function for winter cereals might be qualitatively similar to the dashed line shown in Fig. 2-11, with considerable variance possible in the slope between T_b and $16\,°C$ related to other environmental factors. The uncertainty of the temperature development relationship clearly indicates that more research needs to be done in establishing a temperature response function for winter cereal development.

VI. MODELS FOR DAILY CYCLE OF TEMPERATURE

In cases in which the range of daily temperatures fall completely or partially outside the range of T_b to T_p in temperature-dependent crop development relationships, some additional procedures are needed for calculating thermal time for that day, since the average of the maximum and minimum is not satisfactory. Several models have been used to estimate the diurnal variation in the temperatures that lie between the values of the measured maximum and minimum temperatures. Although several relatively simple procedures have been provided in the literature as reviewed by Arnold (1960), the use of a sine-exponential type model is suggested as the most appropriate for use in crop models. Two separate research reports concluded that this type of model provided the least error (Parton & Logan, 1981; Wann et al., 1985). The sine-exponential model describes the diurnal temperature variation by a truncated sine function during the day and an exponential function at night. It requires three fitted parameters that need to be fit from recorded diurnal temperature data. Procedures are needed to initiate the time of sunrise and sunset for the sine-exponential model, but those are available from information on day of year and latitude of the site. Wann et al. (1985) reported that the three fitted parameters varied somewhat among locations in North Carolina. The accuracy of the model was influenced little by use of average parameters. Thus the use of the averaged parameters they found will probably suffice in most instances for calculating thermal time for one day.

When the day is separated into 1- to 3-h segments, the degree-hours can be calculated for the 24-h period and then divided by 24 to get the appropriate thermal time for one day.

VII. PHOTOPERIOD INFLUENCE ON CROP DEVELOPMENT

Photoperiod response of several crops is discussed in subsequent chapters in detail. However, the influence of photoperiod on the prediction of crop development and temperature relationships needs to be addressed here. As discussed previously, leaf-tip appearance rate has a linear relationship with temperature between T_b and T_p. Therefore, rate of leaf-tip appearance can be satisfactorily predicted from temperature alone. However, problems arise when trying to predict total leaf number and timing of other developmental events such as floral induction from temperature alone (Baker et al., 1980; Baker & Gallagher, 1983a; Warrington & Kanemasu, 1983b).

Many plant species respond to photoperiod or, more specifically, to night length. In general, short-day species have a longer developmental duration when exposed to long photoperiods. Research has shown that photoperiod significantly affects leaf number in crops similar to maize and that this influences the duration of developmental phases (Hesketh et al., 1969; Stevenson & Goodman, 1972; Warrington & Kanemasu, 1983c). For example, even though leaves are initiated at a constant rate with temperature, if the total number of leaves to be produced is not known, thermal time to floral induction cannot be predicted accurately. Hence, events such as tassel emergence are a function of leaf number as well as leaf development rate. Figure 2–12 demonstrates this for a photoperiod-sensitive tropical cultivar, X304C, grown at three locations with large contrasts in daylength (J.T. Ritchie, U. Singh, and J. Kiniry, 1985, unpublished data).

At the Hawaii site, the photoperiod (daylength and civil twilight) at tassel initiation was about 13.5 h, at Texas about 14.2 h and at Michigan about 16.3 h. The leaf number varied from 18 at Hawaii to 28 in Michigan. However, the rates at which the leaf tips appeared when expressed on a thermal time basis, all fit approximately the same curve, with the short day environment plants terminating leaf appearance at an earlier thermal time. The slope

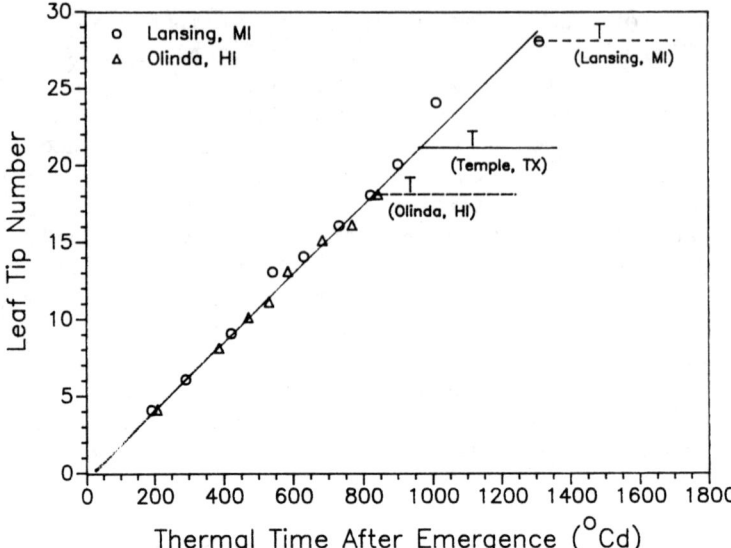

Fig. 2-12. Maize leaf tip number as a function of thermal time after emergence at Olinda, HI; Temple, TX; and East Lansing, MI (U. Singh et al., 1985, unpublished data). Final leaf number is indicated by horizontal lines and time of tasseling is indicated by T.

of the linear relationship shown in Fig. 2-12 is 0.0226 leaf tips/°Cd, which fits well within the range of values found from growth chamber data. The time of tasseling after the final leaf tip appearance for all three locations was approximately 130°Cd. These data clearly demonstrate that for photoperiod-sensitive genotypes, thermal time to tasseling is not a constant. Thermal time would only be constant for plants grown in similar photoperiod environments during the time when plants respond to photoperiod during floral induction (Kiniry et al., 1983b). This type of photoperiod control on leaf number should be obtained for all photoperiod-sensitive plants.

Winter cereal plants respond to photoperiod in the opposite direction as most of the warm-season crops. They begin having photoperiod sensitivity to floral induction at emergence (Ellis et al., 1986). Leaf number is influenced both by photoperiod and vernalization. Winter cereals are long-day plants, meaning that they develop more rapidly toward reproduction, making fewer leaves, during long days. Halloran (1975) reported a variation between 7 and 14 leaves on vernalized wheat plants (var. Robin) grown in photoperiods of 18 and 8 h, respectively.

VIII. VERNALIZATION INFLUENCE ON WINTER CEREAL DEVELOPMENT

As with photoperiod, vernalization also has an influence on leaf numbers. Vernalization is a response to relatively cold temperatures in some species that must occur before reproductive growth will begin. For wheat, temperatures above zero to about 8 °C seem to be the most effective for vernalization (Ahrens & Loomis, 1963; Chujo, 1966). Vernalization also can occur with lower effectiveness in temperatures up to about 15 °C (Chujo, 1966). Winter wheat plants of variety Norin 27 grown for 60 d at 1 or 4°C and then transferred to a warmer glasshouse environment, developed only seven leaves, whereas plants grown at 18 °C for 60 d developed 16 leaves (Chujo, 1966). In the same experiment, different durations of vernalizing temperatures between 20 and 60 d caused variation in leaf numbers between 7 and 16. Spring-type winter cereals have little sensitivity to vernalization, which is the principal difference between them and the winter types. There are also intermediate types that have varying degrees of sensitivity to vernalization. These are often referred to as semi-winter types (Kirby and Perry, 1987).

IX. MATURITY TYPE INFLUENCE ON CROP DEVELOPMENT

A final factor influencing crop development and temperature response is crop maturity type. Warm-season cultivars similar to maize within a species have a wide variability in the number of days to maturity, even when grown under the same photoperiod (Kiniry et al., 1983a). While phyllochron are quite predictable using temperature information, there has to be allowance for maturity differences for models in order to accurately determine when leaf initiation stops and leaf number is determined.

Leaf primordia are produced during a juvenile phase that can vary in thermal time by more than a factor of 2 between genotypes in maize (Kiniry et al., 1983a). Figure 2-13 demonstrates how maturity genotypes can influence leaf tip numbers for three maize genotypes grown at East Lansing, MI (un published data). Each genotype had the same sowing date. The early maturity variety LG11 produced 15 leaves, the intermediate type B73XMO17 produced 20 leaves, and X304C produced 28 leaves. Their leaf tips appeared at almost the same rate; only the number differed. Leaf number differences between several maturity genotypes of sorghum varied between 16 to 29 (Quinby, 1972). Rice genotype variation in maturity type is also common. (Varga and Chang, 1985).

X. INFLUENCE OF TEMPERATURE DURING FLORAL INDUCTION ON LEAF NUMBER

A number of research reports quoted in Warrington and Kanemasu (1983c) indicated that the temperature during leaf initiation also influences the leaf number in maize of the same genotype. Practically all studies on this topic have indicated an increase in leaf numbers with increasing temperatures in the range 15 to 32 °C. The average leaf number increase between those temperature ranges averaged about 1 leaf/4 °C temperature increase. This phenomenon may have as much influence on leaf number variation for the same genotypes for field grown maize as photoperiod because of the year-to-year temperature variation at the time of tassel induction.

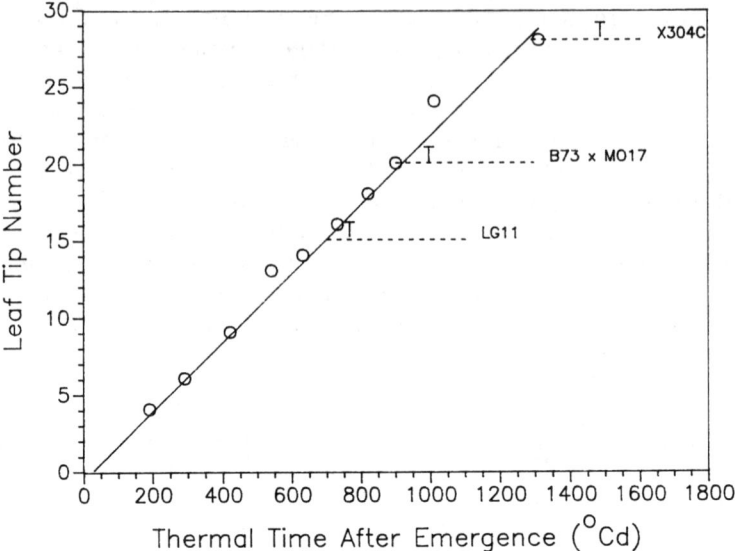

Fig. 2-13. Leaf tip number as a function of thermal time after emergence for maize cultivars LG11, B 73 × MO17, and X304C grown near East Lansing, MI (J.T. Ritchie, 1985, unpublished data). Final leaf number is indicated by horizontal lines and time to tasseling is indicated by T.

The difference in leaf number of a genotype with temperature during induction could be because the leaf primordia initiation rate is controlled by temperature (Warrington & Kanemasu, 1983b), but the rate of induction is a real-time process because it is influenced by daylength. If this is true, plants growing at higher temperatures during floral induction would produce more primordia during a fixed initiation time and, thus, produce more leaves.

Temperatures below about 14 °C have also caused an increase in leaf numbers in maize (Warrington & Kanemasu, 1983c; Stevenson & Goodman, 1972). This indicates that, at these low temperatures, the rate of induction may be quite strongly reduced. More research is needed to clarify this influence of temperature that causes different leaf numbers of similar genotypes, since the leaf number quite strongly determines the duration of the vegetative stage.

XI. CONCLUSIONS

Understanding temperature response of crop development is important in building crop models. The thermal time concept is useful, but can be misused as well. Considerable evidence has shown that leaf appearance rate is one of the most consistent developmental processes that can be used to determine the temperature response function. Thus, this is the most appropriate application for thermal time. Efforts need to be made to determine correct response curves for several crop species over the full range of temperatures expected to influence development. Also, the conventional base temperature of 10°C used for many warm-season crops when calculating °Cd and crop response curves is probably somewhat too high for crops like maize. Compounding factors such as photoperiod, vernalization and crop maturity type, and temperature during floral induction need to be considered when predicting crop development using temperature. When properly used within the appropriate constraints, thermal time can be a powerful, yet simple tool for modeling crop development.

REFERENCES

Ahrens, J.F., and W.E. Loomis. 1963. Floral induction and development in winter wheat. Crop Sci. 3:463-466.

Arnold, C.Y. 1959. The determination and significance of the base temperature in a linear heat unit system. Proc. Am. Soc. Hortic. Sci. 74:430-455.

Arnold, C.Y. 1960. Maximum-minimum temperatures as a basis for computing heat units. Proc. Am. Soc. Hortic. Sci. 76:682-692.

Baker, C.K., and J.N. Gallagher. 1983a. The development of winter wheat in the field. 1. Relation between apical development and plant morphology within and between seasons. J. Agric. Sci. (Cambridge) 101:327-336.

Baker, C.K., and J.N. Gallagher. 1983b. The development of winter wheat in the field. 2. The control of primordium initiation rate by temperature and photoperiod. J. Agric. Sci. (Cambridge) 101:337-344.

Baker, C.K., J.N. Gallagher, and J.L. Monteith. 1980. Daylength change and leaf appearance in winter wheat. Plant Cell Environ. 3:285-287.

Baker, J.T., P.J. Pinter, Jr., R.J. Reginato, and E.T. Kanemasu. 1986. Effects of temperature on leaf appearance in spring and winter wheat cultivars. Agron. J. 78:605-613.

Bauer, A., C. Fanning, J.W. Enz, and C.V. Eberlein. 1984. Use of growing-degree days to determine spring wheat growth stages. North Dakota State Univ. Agric. Ext. Bull. EB-37.

Beauchamp, E.G., and D.J. Lathwell. 1966. Root-zone temperature effects on the early development of maize. Plant Soil (26(2):224-234.

Blacklow, W.M. 1972. Influence of temperature on germination and elongation of the radicle and shoot of corn (*Zea mays* L.). Crop Sci. 12:647-650.

Brown, D.M. 1975. Heat units for corn in southern Ontario. Ontario Agric. College, Ministry of Agric. and Food Factsheet 75-077.

Brown, D.M., and L.J. Chapman. 1961. Soybean ecology. III. Soybean development units for zones and varieties in the Great Lakes Region. Agron. J. 53:306-308.

Cao, W., and D.N. Moss. 1989a. Temperature effect on leaf emergence and phyllochron in wheat and barley. Crop Sci. 29:1018-1021.

Cao, W., and D.N. Moss. 1989b. Daylength effect on leaf emergence and phyllochron in wheat and barley. Crop Sci. 29:1021-1025.

Chujo, H. 1966. Difference in vernalization effect in wheat under various temperatures. Proc. Crop Sci. Soc. Jpn. 35:177-186.

Coelho, D.T., and R.F. Dale. 1980. An energy-crop growth variable and temperature function for predicting corn growth and development: Plant to silking. Agron. J. 72:503-510.

Cooper, P.J.M., and L.R. Law. 1978. Enhanced soil temperature during very early growth and its association with maize development and yield in the Highlands of Kenya. J. Agric. Sci. (Cambridge) 89:569-577.

Covell, S., R.H. Ellis, E.H. Roberts, and R.J. Summerfield. 1986. The influence of temperature on seed germination rate in grain legumes. J. Exp. Bot. 37(178):705-715.

Daynard, T.B. 1972. Relationships among black layer formation, grain moisture percentage, and heat unit accumulation in corn. Agron. J. 64:716-719.

Ellis, R.H., S. Covell, E.H. Roberts, and R.J. Summerfield. 1986. The influence of temperature on seed germination rate in grain legumes. J. Exp. Bot. 37(183):1503-1515.

Epperson, D.L., and R.F. Dale. 1983. Comparison of climatologies for existing heterogeneous temperature records with those for adjusted records for West Lafayette and Whitestown, Indiana. Proc. Indiana Acad. 92:395-404.

Gallagher, J.N. 1979. Field studies of cereal leaf growth. J. Exp. Bot. 30(117):625-636.

Garcia-Huidobro, J., J.L. Monteith, and G.R. Squire. 1982. Time, temperature, and germination of pearl millet (*Pennisetum typhoides* S. & H.). J. Exp. Bot. 33(133):288-296.

Gilmore, E.C., Jr., and J.S. Rogers. 1958. Heat units as a method of measuring maturity in corn. Agron. J. 50:611-615.

Grobellaar, W.P. 1963. Response of young maize plants to root temperatures. Meded. Landbouwhogesch. Wageningen. 63:1-71.

Halloran, G.M. 1975. Genotype differences in photoperiod sensitivity. Ann. Bot. 39:845-851.

Hesketh, J.D., S.S. Chase, and D.K. Nanda. 1969. Environmental and genetic modification of leaf number in maize, sorghum, and Hungarian millet. Crop Sci. 9:460-463.

Hesketh, J.D., D.L. Myhre, and C.R. Wiley. 1973. Temperature control of time intervals between vegetative and reproductive events in soybeans. Crop Sci. 13:250-254.

Hesketh, J.D., and I.J. Warrington. 1989. Corn growth response to temperature: Rate and duration of leaf emergence. Agron. J. 81:696-701.

Johnson, D.R., and J.W. Tanner. 1972. Calculation of the rate and duration of grain filling in corn (*Zea mays* L.). Crop Sci. 12:485-486.

Kiniry, J.R., J.T. Ritchie, and R.L. Musser. 1983a. Dynamic nature of the photoperiod response in maize. Agron. J. 75:700-703.

Kiniry, J.R., J.T. Ritchie, R.L. Musser, E.P. Flint, and W.C. Iwig. 1983b. The photoperiod sensitive interval in maize. Agron. J. 75:687-690.

Kirby, E.J.M., M. Appleyard, and G. Fellowes. 1985. Effect of sowing date and variety on main shoot leaf emergence and number of leaves of barley and wheat. Agronomie (Paris) 5(2):117-126.

Kirby, E.J.M., M. Appleyard, and G. Fellowes. 1985. Leaf emergence rates of wheat in a Mediterranean environment. Aust. J. Agric. Res. 38:455-464.

Law, R., and P.J.M. Cooper. 1976. The effect and importance of soil temperature in determining the early growth rate and final grain yields of maize in Western Kenya. East Afr. Agric. For. J. 41(3):189-200.

Lehenbauer, P.A. 1914. Growth of maize seedlings in relation to temperature. Physiol. Res. 1:247-288.

McGarrahan, J.P., and R.F. Dale. 1984. A trend toward a longer grain-filling period for corn: A case study in Indiana. Agron. J. 76:518-522.

Muldoon, J.F., T.B. Daynard, B. Van Diunen, and M. Tollenaar. 1984. Comparisons among rates of appearance of leaf tips, collars, and leaf area in maize (*Zea mays* L.) Maydica 29:109-120.

NeSmith, D.S., W.L. Hargrove, D.E. Radcliffe, E.W. Tollner, and H. Arioglu. 1987. Tillage and residue management effects on properties of an Ultisol and double-cropped soybean production. Agron. J. 79:570-576.

Nuttonson, M.Y. 1948. Some preliminary observations of phenological data as a tool in the study of photoperiodic and thermal requirements of various plant materials. p. 129-143. *In* Vernalization and photoperiodism. Chronica Botanica, Waltham, MA.

Nuttonson, M.Y. 1955. Wheat-climate relationships and the use of phenology in ascertaining the thermal and photo-thermal requirements of wheat. Am. Inst. Crop Ecol., Washington, DC.

Parton, W.J., and J.A. Logan. 1981. A model for diurnal variation in soil and air temperature. Agric. Meteorol. 23:205-216.

Quinby, J.R. 1972. Influence of maturity genes on plant growth in sorghum. Crop Sci. 12:490-492.

Rickman, R.W., B.L. Klepper, and C.M. Peterson. 1985. Wheat seedling growth and development response to incident photosynthetically active radiation. Agron. J. 77:283-287.

Robertson, G.W. 1968. A biometeorological time scale for a cereal crop involving day and night temperatures and photoperiod. Int. J. Biometeorol. 12(3):191-223.

Schaal, L.A., and R.F. Dale. 1977. Time of observation temperature bias and "climate change." J. Appl. Meteorol. 16:215-222.

Sinclair, T.R. 1984. Leaf area development in field-grown soybeans. Agron. J. 76:141-146.

Sofield, I., L.T. Evans, and I.F. Wardlaw. 1974. The effects of temperature and light on grain filling in wheat. R. S. N.Z. Bull. 12:909-915.

Spiertz, J.H.J. 1977. The influence of temperature and light intensity on grain growth in relation to the carbohydrate and nitrogen economy of the wheat plant. Neth. J. Agric. Sci. 25:182-197.

Stern, W.R., and E.J.M. Kriby. 1979. Primordium initiation at the shoot apex in four contrasting varieties of spring wheat in response to sowing date. J. Agric. Sci. (Cambridge) 93:203-215.

Stevenson, J.C., and M.M. Goodman. 1972. Ecology of exotic races of maize. I. Leaf number and tillering of 16 races under four temperatures and two photoperiods. Crop Sci. 12:864-868.

Swan, J.B., E.C. Schneider, J.F. Moncrief, W.H. Paulson, and A.E. Peterson. 1987. Estimating corn growth, yield, and grain moisture from air growing degree days and residue cover. Agron. J. 79:53-60.

Tollenaar, M., T.B. Daynard, and R.B. Hunter. 1979. Effect of temperature on rate of leaf appearance and flowering date in maize. Crop Sci. 19:363-366.

Varga, B.S., and T.T. Chang. 1985. The flowering response of the rice plant to photoperiod. 4th ed. Int. Rice Res. Inst., Los Banos, Laguna, Philippines.

Wang, J.Y. 1960. A critique of the heat unit approach to plant response studies. Ecology 41:785-790.

Wann, M., D. Yen, and H.J. Gold. 1985. Evaluation and calibration of three models for daily cycle of air temperature.

Warrington, I.J., and E.T. Kanemasu. 1983a. Corn growth response to temperature and photoperiod. I. Seedling emergence, tassel initiation, and anthesis. Agron. J. 75:749-754.

Warrington, I.J., and E.T. Kanemasu. 1983b. Corn growth response to temperature and photoperiod. II. Leaf-initiation and leaf appearance rates. Agron. J. 75:755-761.

Warrington, I.J., and E.T. Kanemasu. 1983c. Corn growth response to temperature and photoperiod. III. Leaf number. Agron. J. 75:762-766.

Watts, W.R. 1972. Leaf extension in *Zea mays*. II. Leaf extension in response to independent variation of the temperature of the apical meristem, of the air around the leaves, and of the root-zone. J. Exp. Bot. 23(76):713-721.

Wierenga, P.J., D.R. Nielsen, R. Horton, and B. Kies. 1982. Tillage effects on soil temperature and thermal conductivity. p. 69-90 *In* Unger and Van Doren (ed.) Predicting tillage effects on soil physical properties and processes. ASA Spec. Publ. 44. ASA, CSSA, SSSA, Madison, WI.

Zur, B., J. Reid, and J. Hesketh. 1989. The dynamics of a maize canopy development. 1. Leaf ontogeny. Field Crop Res. (in press).

3 Wheat Phasic Development

J. T. RITCHIE

Michigan State University
East Lansing, Michigan

Simulation of crop yield focuses around three important areas: growth rate, growth duration, and the extent to which stresses influence these two processes. Growth rate simulation requires further partitioning of the assimilates into the plant organs that are growing during the specific plant growth phase. Soil water and nutrient deficiencies or extremes in temperature can trigger stress. Growth duration is important in the determination of potential crop yields. In general, the longer the growth duration for the crop, the higher the yield potential. In wheat (*Triticum aestivum* L.), this is especially true from the time that the stem and inflorescence start to grow until the end of grain filling.

The duration of different growth phases is referred to as phasic development. Phasic development is affected primarily by genetic and environmental factors. The genetic diversity of wheat sensitivity to photoperiod, vernalization, and cold temperatures has allowed plant breeders to select wheat cultivars that can produce grain in environments as far north as Alaska to as far south as southern Argentina, and on most arable land in between where the supply of water is adequate. Because wheat has been a relatively low-value crop during most of this century, it is often grown under rainfed conditions where precipitation is marginal for part or all of the season. In such instances, cultivars have to be developed that complete their life cycle soon enough to avoid complete crop failure. Because of this genetic diversity and the diversity among the regions where wheat is grown, it is essential to include quantitative aspects of phasic development in simulation models to make them useful for many applications.

To avoid site-specific crop models, the growth processes need to be evaluated separately from the development processes. Phasic development is used to determine the duration of the major stages of plant growth. Morphological development is used to determine the appearance and number of leaves on the main stem, the number of tillers, and the number of grains on a plant. Plant morphological development is somewhat independent of phasic development, although it is closely coupled with phasic development and plant growth. Potential expansion growth of such plant parts as leaves and stems need to be calculated separately from mass growth (photosynthesis). Expansion growth is considered as a sink that is driven primarily by temperature

Copyright © 1991 ASA-CSSA-SSSA, 677 S. Segoe Rd., Madison, WI 53711, USA. *Modeling Plant and Soil Systems*—Agronomy Monograph no. 31.

of the expanding tissue. Mass growth is the source necessary to fill and maintain the expanding tissue, and also to provide assimilate to the root system for expansion and maintenance. Potential mass growth is influenced primarily by plant radiation interception.

By separately evaluating these four aspects of plant development and growth, the logic for partitioning assimilates into different plant parts can be accommodated in simulation models. Two of the major principles follow.

1. During vegetative growth, shoots have a higher priority than roots for assimilates as long as the supply of water and nutrients from the soil is adequate. When water or nutrients are limited during vegetative growth, roots have a higher priority for assimilates than shoots.
2. During the grain filling period, the grains are the dominant sink for assimilates. Material for filling the grains can be derived from photosynthesis and stored assimilates. Water and nutrient deficiencies have little effect on the ability of material to be transported to the grain.

I. SIMULATING PHASIC DEVELOPMENT

The model described in this chapter uses weather and genetic information to calculate the dates of various phases of wheat development. Although growth and partitioning are not included in this paper, the growth of various plant parts can be logically calculated when phasic development is accurately simulated. The growth stages of wheat are organized around the plant's life cycle, when changes occur in the partitioning of assimilate among the various plant organs. For example, prior to terminal spikelet formation, practically all assimilate is partitioned between the leaves and the roots. After terminal spikelet formation, stems become a major sink for assimilates. Later, the ear becomes the major growing organ.

For this model, the growth stages of wheat are numbered from 1 to 9 (Table 3-1). Stages 1 to 5 are the active, aboveground growing stages; Stages 6 to 9 describe other events in a cycle of crop management. The stages are divided this way to provide a means of logically partitioning biomass during the season when various plant parts are growing. The control of the duration is calculated, taking into account weather and genetic information provided by the user.

The model is a daily incrementing type and requires daily information on variable maximum and minimum temperature and precipitation from a weather file. The precipitation data are optional, but can be used in a simplified snow-depth submodel. Seeding depth, sowing date, and latitude are also required inputs. A constant to express the duration between leaf appearance intervals is also necessary. This constant and others related to genotype characteristics will be discussed in a later section.

Table 3-1. Growth stages of wheat needed for simulating plant development and management.

Stage	Event	Growing plant parts
7	Fallow or presowing	--
8	Sowing to germination	--
9	Germination to emergence	Roots, coleoptile
1	Emergence to terminal spikelet initiation	Roots, leaves
2	Terminal spikelet to end of leaf growth and beginning of ear growth	Roots, leaves, stems
3	End of leaf growth and beginning of ear growth to end of preanthesis ear growth	Roots, leaves, ear
4	End of preanthesis ear growth to beginning of grain filling	Roots, stems
5	Grain filling	Roots, stems, grain
6	End of grain filling to harvest	--

A. Temperature and Plant Development

Plant development is primarily controlled by the thermal environment of the growing part of the plant. The development rate is assumed to be directly proportional to temperature in the range from the base temperature (0 °C) to a maximum temperature of 26 °C. This linear relationship between temperature and development permits the use of the concept of thermal time as discussed in Chapter 2. Thermal time is the accumulation of daily temperature above 0 °C, a concept proven to work well in predicting the ontogeny of field grown winter cereals (Gallagher, 1979). The exception to the 0 °C base temperature occurs during the emergence stage, when the base temperature is assumed to be 2 °C.

When the daily minimum temperature is above 0 °C and the maximum is below 26 °C, thermal time for a day is assumed to be the mean of the maximum and minimum values. For temperatures above 0 °C, the mean daily air temperature is considered to be equal to the mean temperature of the plant crown, where cell division and expansion occur prior to stem elongation. While this assumption is subject to error for diurnal temperature variations, these errors usually cancel one another when longer period averages are used. Special procedures are used to calculate plant crown temperature when the air temperature drops below 0 °C. Details for the procedure are provided in a later section.

B. Stages Prior to Seedling Emergence

Fallow or presowing—Stage 7. If a whole crop model requires information on soil-water or nutrient balance during fallow periods, it can be derived during this stage.

Seed germination—Stage 8. Although called germination, the process that is simulated is technically water imbibition that starts the germination process. This process is assumed to occur in one day, unless the soil water content in the top layer is below a threshold value or if the temperature is below 0 °C. In some instances wheat is sown in dry soil and will not germinate until the soil becomes wet from a rain or irrigation. Information about soil water would be derived from a submodel of a complete growth model, details of which are not provided for this chapter.

Germination to emergence—Stage 9. In this stage, the simulation estimates the time to emergence of the seedling. Two factors that affect seedling emergence, temperature and the depth of sowing, are accounted for. The soil water condition is assumed to be sufficient for emergence if it was sufficient for germination. Temperature effects on emergence are expressed as thermal time. The depth of sowing influences plant emergence because it increases the time necessary for the coleoptile to expand to the soil surface. The duration of Stage 9 (P9) is expressed by

$$P9 = 40 + 10.2 \times SDEPTH, \qquad [1]$$

where P9 is the thermal time for Stage 9 and SDEPTH is the depth of sowing (cm), an input in the model.

C. Stages after Seedling Emergence

Emergence to terminal spikelet—Stage 1. The thermal time for this growth stage is highly dependent on the genotype and environment. Vernalization, photoperiod, and genetic characteristics cause the total thermal time from emergence to terminal spikelet to vary considerably. Two factors that determine the length of this growth phase are leaf appearance rate, based on the thermal time per leaf (the phyllochron), and initiation of flowering, which is determined by vernalization and photoperiod.

1. Vernalization

Winter wheat varieties usually require exposure to relatively low temperatures before spikelet formation can begin. This low temperature requirement for flowering, called vernalization, begins at germination.

Vernalization is assumed to occur at temperatures between 0 and 18 °C (Ahrens & Loomis, 1963; Trione & Metzger, 1970). The optimum temperature for vernalization is assumed to be in the range of 0 to 7 °C, with temperatures between 7 and 18 °C having a decreasing influence on the process. Minimum and maximum daily temperatures are used to calculate a daily vernalization effectiveness factor with a value between 0 and 1 (Fig. 3-1). The daily relative vernalization effectiveness factor (RVE) is then totaled to determine what is termed vernalization days. Even though there is genetic variability in sensitivity to vernalization between cultivars, 50 vernalization days are assumed to be sufficient to completely vernalize all cultivars (Table 3-2, Fig. 3-2). This variability is considered by the use of a genetic specific coefficient (P1V) to calculate the influence of vernalization on Stage 1 growth. The relative development rates as influenced by vernalization for the varieties shown in Table 3-2 were determined from comparisons made of plant development in growth cabinet experiments (Ritchie, 1980, unpublished data).

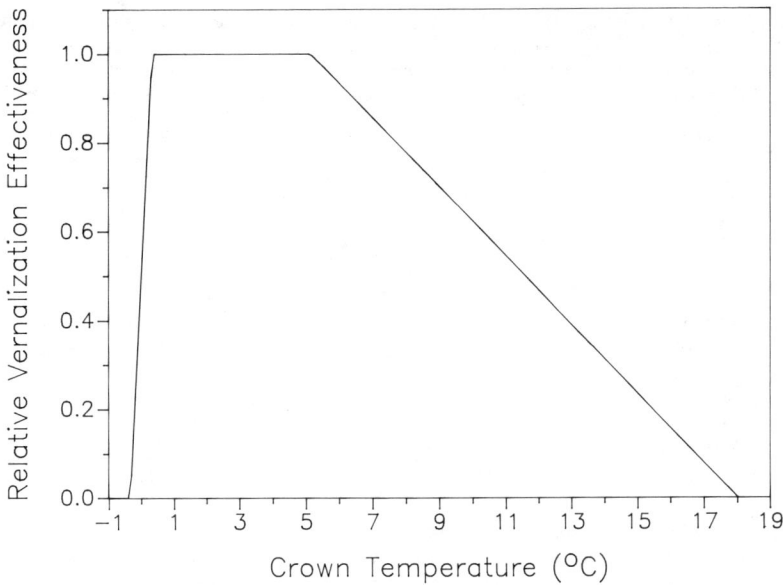

Fig. 3-1. The assumed relationship between the plant crown temperature and the relative degree of vernalization that wheat plants receive each day.

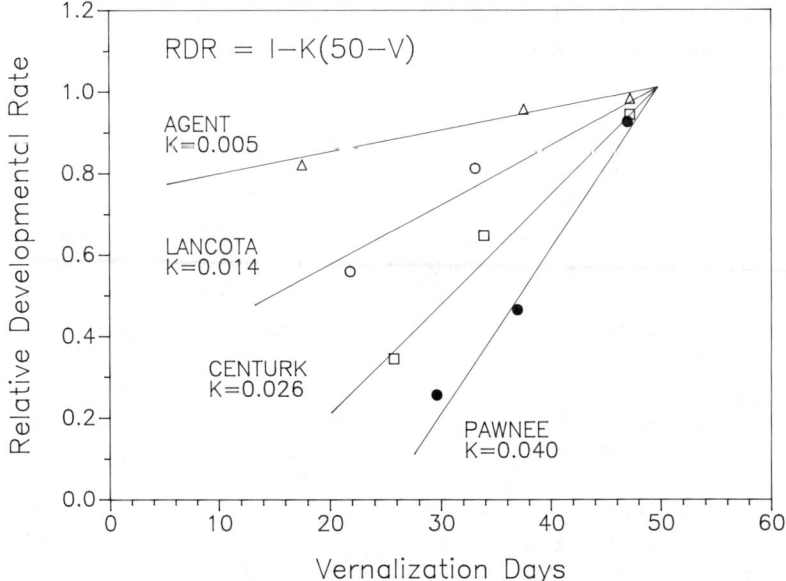

Fig. 3-2. The relationships used to predict the influence of vernalization days for specific genotypes on the relative development rate of wheat during Stage 1 growth.

The experiment conducted to determine the vernalization data in Table 3-2 was done on seedling plants that were allowed to emerge at 15 °C. The plants were then placed in a growth chamber set at 2 °C for the duration of

Table 3-2. The genetic vernalization coefficients for 12 winter wheat varieties. Values are for the vernalization constant (K) in the equation RDR = $1 - K(50 - V)$. The scaled values are the ones used as model inputs to provide a more useful scale of 0 to 8. The transformation equation used is P1V = $K \times 183 - 0.55$.

Variety	$K \times 10^{-2}$	Scaled value P1V
Agent	0.5	0.86
Lancota	1.4	2.51
Centurk	2.6	4.71
Sage	2.7	4.89
Scout 66	2.9	5.26
Sturdy	3.0	5.44
Nugaines	3.0	5.44
Triumph	3.1	5.62
Bezastaya	3.1	5.62
Coker 68-15	3.1	5.62
Arthur 71	3.2	5.81
Pawnee	4.0	7.27

the vernalization period which ranged from 0 to 50 d. After the vernalization period, they were grown for a week at 15 °C to prevent devernalization and then grown until the ear emerged at 20 °C. The photoperiod, 18 h, was considered long enough to avoid delay in development. The number of days for the plant to develop to ear emergence was determined for several vernalization day treatments. The time from terminal spikelet to ear emergence was found by dissection to be 18 d at 20 °C. Thus, 18 d was subtracted from the days to ear emergence to determine the date of terminal spikelet. The ratio of days to terminal spikelet for the 50 vernalization days treatment to the days to terminal spikelet for the treatments < 50 d provided the relative development rate. Thus, a relative development of 0.5 would indicate that it takes twice as long for a plant to develop to the terminal spikelet stage with less than full vernalization than it did for one with 50 vernalization days. Results of the study are presented for four varieties in Fig. 3-3 to demonstrate the validity of the linear equation shown to express differences between genotypes. Summarized results for winter wheat genotypes are given in Table 3-2.

Spring wheat varieties, which have a low sensitivity to vernalization, are incorporated in the model in the same way as winter wheat varieties by expressing the differences in vernalization through the input coefficient P1V. Spring wheats should have P1V values <0.5.

Devernalization can occur when young seedlings are exposed to high temperature. In the model, if the number of vernalization days is <10 and the maximum temperature >30 °C, then the number of vernalization days decreases by 0.5 d per degree above 30 °C. If the number of vernalization days is >10, no devernalization is assumed to be able to occur.

2. Photoperiod

A short photoperiod can delay Stage 1 plant development. The delay depends on the photoperiod sensitivity of the genotype, which is expressed

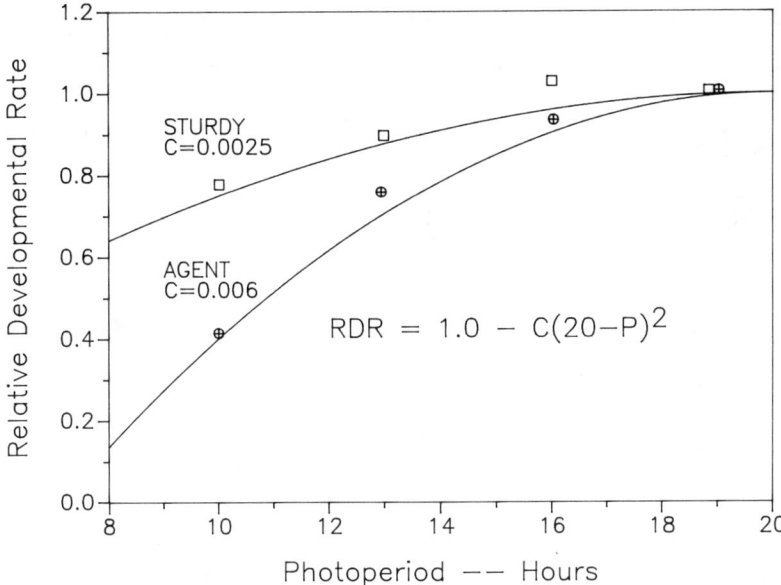

Fig. 3-3. The relationship used to predict the influence of photoperiod for specific genotypes or the relative development rate of wheat during Stage 1 growth.

in a genetic-specific characteristic (P1D) (Table 3-3, Fig. 3-3). Calculated photoperiods include civil twilight. Latitude, a required input, and the time of year are used to calculate daylength. The relative development rates as influenced by photoperiod for the varieties shown in Table 3-3 were determined from experimental comparisons of a similar type to those for determination of the vernalization genetic coefficients. For the study, seedling plants were vernalized for 50 d at 2 °C. After the vernalization they were placed in growth chambers under various photoperiod treatments; first at 15 °C for 1 wk and then at 20 °C for the remainder of the period until ear emergence occurred. The photoperiods were obtained by providing a 9 h period with full light intensity for all treatments and then extending the photoperiod with appropriate low light levels for the remainder of the time. The low light exposures were distributed evenly on either side of the full light intensity period to obtain the desired photoperiod. Thus a 16-h photoperiod treatment included 8 h of darkness, then 3.5 h of low light, then 9 h of full light, then 3.5 h of low light to complete a 24-h cycle. The relative development rate was calculated from the time to terminal spikelet for the various treatments using a procedure similar to that used for vernalization treatment evaluation. The relative development rate for the 19-h treatment was assumed to be adequate for the maximum development rate and all other treatments were referenced to it. The relative development rate for photoperiod was found to be nonlinear as indicated by the results shown in Fig. 3-4. Summarized results for 12 winter wheat genotypes are given in Table 3-3.

Table 3-3. The genetic photoperiod coefficients for 12 winter wheat varieties. Values are for the photoperiod constant (C) in the equation RDR = $1 - C(20 - P)^2$. The scaled values are the ones used as model inputs to provide a more useful scale of 0 to 3. The scaling equation used is P1D = $C \times 500$.

Variety	$C \times 10^{-3}$	Scaled value P1D
Sturdy	2.1	1.05
Coker 68-15	2.6	1.30
Bezastaya	3.2	1.60
Arthur 71	3.4	1.70
Centurk	3.7	1.85
Triump	3.9	1.95
Lancota	4.0	2.00
Nugaines	4.4	2.20
Scout 66	4.9	2.45
Pawnee	5.2	2.60
Sage	5.5	2.75
Agent	6.0	3.00

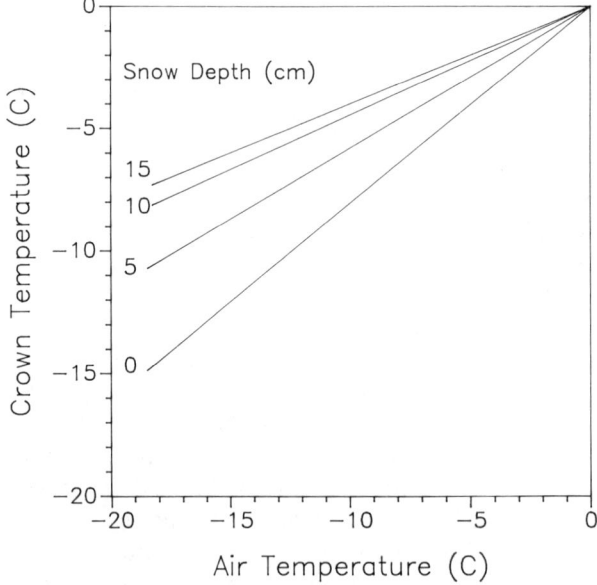

Fig. 3-4. The influence of air temperature and snow depth on the plant crown temperature as approximated from the data of Aase and Siddoway (1979).

3. Phyllochron

In determining the vegetative development of wheat, a definition of leaf appearance rate is necessary. A phyllochron is defined as the interval of time between leaf tip appearance PHINT (degree-days). Tests of models on a global scale have shown that some apparent environmental stimulus, in addition to temperature, causes the interval between leaf appearance to vary. In England, winter wheat sown in the autumn has a considerably longer phyllochron (100 degree-days) than spring-sown winter wheat (75 degree-days), as reported by Baker et al. (1980). This phenomenon was explained on the basis of the rate of change in photoperiod. In North Dakota, wheat sown in late spring has relatively short phyllochrons, usually ranging around 75 degree-days (Bauer et al., 1984). Causes of this phenomenon, however, are not clearly understood and my efforts at reproducing the effect in controlled climate chambers have mostly failed. This technical problem creates a degree of uncertainty in the phasic and morphological development part of the model, making an accurate prediction of crop duration difficult unless the phyllochron value is known. If the sowing date for a crop occurs about the same time each year, the phyllochron should be constant for a location from year to year.

The phyllochron can be measured by determining the date of appearance of leaf tips on the main stem for several plants and graphing the cumulative number of leaf tips on the Y axis against the cumulative thermal time on the X axis. The inverse of the slope of this line is the appropriate phyllochron. When the phyllochron is not known, a good estimate is 95 degree-days. This value should be approximately correct except for spring-sown wheat in latitudes greater than 30°N and 30°S, in which case a phyllochron of 75 degree-days should be appropriate.

4. Combined Influence of Photoperiod, Vernalization, and Phyllochron

Vernalization days, photoperiod, and the phyllochron are used to modify the thermal time required for Stage 1 development. Vernalization and photoperiod factors, which are calculated from the relative development rate information for vernalization days (VF) and photoperiod (DF), are calculated using the P1V and P1D genetic coefficients. The minimum value of VF and DF is then multiplied by the thermal time to reduce the rate of thermal time accumulation. When this reduced thermal time accumulation (TDU) reaches a threshold value (TTS), Stage 1 development is assumed to end.

$$\text{TDU} = \sum_{0}^{\text{TTS}} \text{DTT} \times [\min(\text{VF}, \text{DV})] \qquad [2]$$

where TDU = thermal development units, DTT = daily thermal time, and TTS = 400 × PHINT/95. The thermal time of 400 degree-days is assumed to be the time of terminal spikelet when plants are grown in long days, are fully vernalized, and have PHINT values of 95 degree-days.

Terminal spikelet initiation to the end of leaf growth—Stage 2. This stage is considered to be strictly under temperature control and takes three phyllochrons from terminal spikelet to the appearance of the final leaf. Details of this evaluation are based on the work of Kirby and Appleyard (1984). Thus, if the phyllochron is 95 degree-days, the duration of Stage 2 is 285 degree-days.

Preanthesis ear growth—Stage 3. The ear develops very rapidly in this stage and is a major sink for assimilates. This is probably the most important stage determining grain numbers per plant expected to develop into full size kernels. The duration of Stage 3 is the equivalent of two phyllochrons, even though no new leaves are developed.

Preanthesis ear growth to the beginning of grain filling—Stage 4. During this phase flowering takes place. Several measurements have indicated that it takes approximately 200 degree-days during this stage to go from the maximum ear size and volume to the time when linear grain mass accumulation begins. There are no apparent aboveground sinks for assimilates during this development stage, although the peduncle may continue to expand somewhat. This stage is also considered to have a major impact on the number of grains per plant because the total biomass production during this stage depends on the duration of the phase. Much of the assimilate produced during this stage likely is stored in the stem and other organs for later translocation to the kernels.

Grain filling—Stage 5. The size of the grain is determined during this stage. The thermal time for Stage 5 varies among genotypes and is determined by the input genetic-specific constant P5. Although the thermal time is not constant for all genotypes, all values for it are near 500 degree-days. This stage begins after flowering, 2 to 10 d after anthesis, with a rapid, usually linear, increase in kernel weight. To transform the thermal time for maturity (TTM) into a scaled value (P5) of 0 to approximately 8, the equation that follows is used.

$$P5 = (0.05 \times TTM) - 21.5. \qquad [3]$$

Physiological maturity to harvest—Stage 6. Although this stage is not considered within the context of this paper, it is available for users who may wish to consider possible yield reductions related to an inability to harvest the crop. A value of 250 degree-days can be used to approximate the thermal time from physiological maturity to harvest.

II. COLD HARDENING AND WINTER KILL

A. Estimating Crown Temperatures During Subfreezing Conditions

Although plant development is considered to be inactive for temperatures $<0\,°C$, simulation of the death of the crop at subfreezing temperatures is necessary in the crop models. When mean air temperatures are $<0\,°C$ for long periods (i.e. frozen soils) the plant crown temperature is usually higher than the air temperature because of heat that is retained in the soil. Snow depth has an important influence on plant hardening and survival when air temperatures reach the -10 to $-30\,°C$ range. An empirical submodel to predict the influence of snow depth and air temperature on plant crown temperature was developed from data of Aase and Siddoway (1979). A graphical representation of the relationships used to predict crown temperature is provided in Fig. 3-4. In the Aase and Siddoway (1979) study, the soil temperature was measured at 3 cm (the assumed crown temperature) and the air temperature was measured at 150 cm. The equation used to predict crown temperature (T_{cr}) when air temperature (T_a) $<0\,°C$ is

$$T_{cr} = 2.0 + T_a [0.4 + 0.0018 (D_s - 15)^2];$$

$$T_a < 0.0,\ 0.0 \leq D_s \leq 15. \quad [4]$$

where D_s is snow depth (cm). If $D_s > 15$, a D_s value of 15 is used. For calculating the mean daily estimated T_{cr}, the minimum T_a is used to calculate the minimum T_{cr} and the maximum T_a is used to calculate the maximum T_{cr}. The maximum and minimum calculated T_{cr} values are then averaged to obtain the estimated daily T_{cr} value.

B. Estimating Snow Depth

The development model version provided in Appendix 3-1 contains a relatively simple model for calculating D_s from temperature and precipitation input data in the weather file. Precipitation is assumed to be in the form of snow if the maximum air temperature is $\leq 1\,°C$. The increment of daily snow accumulation is assumed to be 1 cm/1 mm precipitation. Snow melt is assumed to occur when the daily value of maximum air temperatures is $>1\,°C$. The snow melt increment is assumed to be 1 cm/degree above $0\,°C$ plus 0.4 cm/1 mm precipitation. Thus, on a day with a maximum temperature of $10\,°C$ and no rain, the quantity of melted snow would be 10 cm. Similarly with a $10\,°C$ maximum temperature and 20 mm of rainfall, the snow melt would be 18 cm.

The empiricism used for calculating D_s are highly oversimplified, but provide reasonable approximations for regions where snow is likely to occur and influence plant response. The principal problem with predicting D_s is how wind causes a great deal of blowing snow and thus, there is seldom a uniform layer of snow in a field. Measurements of D_s suffer from the same problem of spatial variance within a field.

C. Estimating Plant Survival

The extent of cold-temperature damage to a wheat plant is related to the hardening that occurs prior to a low temperature event. A submodel was developed for use with plant development to predict the hardening of plants under low temperatures, and the threshold low temperature that partially or totally causes plant death. Most of the principles and information for the hardening and dehardening concepts used in the submodel were obtained from Gusta and Fowler (1976).

Hardening is assumed to occur in two phases. In the first phase, hardening occurs when the soil crown temperature is in the range between -1 and $8\,°C$. Ten days in this temperature range completes the first phase of hardening. In the second phase, hardening occurs when the temperature is <0 °C. Twelve days of the second phase condition is assumed to result in a fully hardened plant.

Dehardening is assumed to occur when the maximum crown temperature is >10 °C. Only the maximum temperature is used to calculate dehardening. The daily increment of dehardening is used to lower the hardening index. During the second phase of hardening, the daily dehardening increment is assumed to be 0.04 times the number of degrees the maximum temperature rises above 10 °C. Thus if the maximum temperature is 22.5 °C, second-phase dehardening will occur in 2 d to bring down the hardening index of a fully hardened plant from 2 to 1. During the first phase of hardening, the dehardening increment is half that of the second-phase hardening.

The threshold temperature at which plants or tillers of plants begin to die is a function of a hardening index. The daily increment of the hardening index in the first phase is 0.1 for each day the plant is in the 1 to 8 °C range. If the temperature during the day is outside that range for part of the day, a proportionally smaller hardening increment is calculated. Second-phase hardening is not begun until the first phase is completed. The second-phase hardening index increment is 0.083 per day for each day the temperature is <0 °C. There is also a partial increment possible for days in which the mean temperature is <0 °C, but the maximum temperature is above that level.

The threshold killing temperature is -6, -12, and -18 °C for hardening index values of 0, 1, and 2, respectively. The equation for the threshold killing temperature (T_k, °C) related to the hardening index (HI) is

$$T_k = -6. (1. + \text{HI})$$

When $T_{cr} \leq T_k$ all or a fraction of the plant population is assumed to be dead. If $T_{cr} < 7$ °C colder than T_k, at least 95% of the entire plant population is assumed to be dead. Although, in this model, the value of T_k is a fixed function of HI, in reality it will be a function of both the genotype and the time in the spring after full hardening has occurred and dehardening has started (Gusta & Fowler, 1976). Once dehardening has started, the value of T_k for fully rehardened plants becomes gradually lower as cycles of dehardening increase in the spring. The values of T_k in the analysis given above is for winter-hardy varieties grown in quite cold regions. Varieties with less winterhardiness would have lower T_k values. It may be possible to use laboratory and greenhouse tests, using the temperature at which 50% of the plants are killed, to quantitatively evaluate a T_k relationship to hardening for any genotype, using coefficients similar to those used to describe genotypic variation in plant ontogeny.

III. OTHER PROCESS LEVEL WHEAT MODELS

The phasic development model described in this chapter is adapted from the CERES-Wheat model described by Ritchie and Otter (1985). That model also contains details on morphogenesis, biomass accumulation and partitioning, soil water balance, and soil and plant N dynamics.

A wheat model developed in England by Weir et al. (1984) calculates photosynthesis, phenology, respiration, and dry matter partitioning. It assumes adequate N and water supplies. Van Keulen and Seligman (1987) fully documented a model for spring wheat. It calculates the water use, N balance, and growth of the crop. Two wheat models developed along a similar level of detail to CERES-Wheat are reported by Maas and Arkin (1980) and Stapper (1984). Called TAMW and SIMTAG, respectively, the models calculate phenology, growth, yield, and the water balance of various wheat genotypes. They are similar in detail to CERES-Wheat and have undergone impressive testing from several experimental trials from various parts of the world. Baker et al. (1985) describe a winter wheat model that is built on principles similar to the cotton model GOSSYM. It is a more mechanistic model that calculates photosynthesis, respiration, transpiration, growth, and morphogenesis. It incorporates the root growth from another mechanistic model named RHIZOS and is intended to predict winter wheat growth and yield.

IV. MODEL VALIDATION

The phasic development of CERES-Wheat was tested by Otter-Nacke et al. (1987) using 113 independent data sets from a diversity of locations throughout the world. These data sets contained information on sowing, anthesis, and maturity dates as well as the needed weather data. The regression of the measured and predicted yields had slope coefficients of 1.002 for date of anthesis and 0.974 for date of maturity, with intercept values of 2.46 and 1.56 d for the two stages, respectively. A root mean-square error was calculated for differences between measured and predicted dates of anthesis and maturity, resulting in values of 8.1 and 12.2 d, respectively.

The CERES-Wheat model as presented herein was tested for accuracy of predicting dates of various phenological events for winter wheat in the Kansas Crop Reporting District for several years by French and Hodges (1985). The performance was compared with the TAMW model (Maas & Arkin, 1980) and a regression type model of Feyerherm et al. (1977). The CERES model reportedly provided consistently more accurate estimates of various phenological events than the other models for all stages of development. The root mean-square error for the prediction of heading date was 3.3 d with a standard deviation of 3.2 d.

V. FORTRAN PROGRAM FOR THE WHEAT DEVELOPMENT MODEL

A program within FORTRAN source code for a version of the model described in this chapter is available in the appendix. The model is intended for use on a personal computer. It has interactive capabilities for input information, except for the weather data. The daily weather data are read from a file named by the user. The model output contains the predicted dates of each stage of development.

VI. APPENDIX

```
C     CERES Wheat Model -- Stand alone phasic development.
C                       With interactive user interface.

      PROGRAM MAIN

          REAL              LAT
          INTEGER           DOY,DOYX,OLDMSOW,OLDDSOW,RUNNO
          INTEGER           MSOW,DSOW,OPENSTA
          INTEGER           IDIM(12)
          CHARACTER*1       ANS
          CHARACTER*3       MONTH
          CHARACTER*12      FILE1,OLDFILE1
          CHARACTER*7       OUT1
          CHARACTER*30      TITLE
          LOGICAL           FEXIST,EOWF

          COMMON /PARAM/ ISOW,SDEPTH,LAT,PHINT
          COMMON /DATEC/ MO,ND,IYR,DOY,DOYX,MONTH
          COMMON /COLDC/ SNOW,TEMPCR,TDU,VF,CUMVD,HI
          COMMON /GENET/ P1V,P1D,P2,P3,P4,P5
          COMMON /PHENL/ P9,CUMDTT,TBASE,SUMDTT,DF,S1,C1,ISTAGE,DTT
          COMMON /CLIMT/ TEMPMN,TEMPMX,RAIN,SOLRAD
          COMMON /FILES1/ FILE1,NOUT1

          DATA IDIM/31, 28, 31, 30, 31, 30, 31, 31, 30, 31, 30, 31/

          NOUT1 = 41
C
          INQUIRE(FILE='BATCH.$$$',EXIST=FEXIST)
          IF (FEXIST) THEN
             OPEN(5,FILE='BATCH.$$$',STATUS='OLD')
          ENDIF
C
          WRITE(*,6100)
          READ(5,'(A1)') A

          RUNNO = 0
          OLDSDEPTH = 3.0
          OLDPHINT = 95.0
          OLDP1V = 5.0
          OLDP1D = 3.0
          OLDP5 = 5.0

          WRITE(*,'(///,3X,A,$)') 'Type the output file name : '
          READ(*,'(12A)') OUT1
          OPEN(NOUT1,FILE=OUT1,STATUS='UNKNOWN')
          WRITE(*,'(3X,A,A,$)') 'Type experiment title ',
         &                     '(at most 30 characters) : '
          READ(*,'(30A)') TITLE
```

```
      100 RUNNO = RUNNO + 1
          EOWF = .FALSE.
      C *** Read in weather file name

          WRITE(*,'(/3X,A,$)') 'Type weather file name '
          IF (RUNNO .NE. 1) THEN
             WRITE(*,'(A,A,A,$)') '(default = ',OLDFILE1,') : '
          ELSE
             WRITE(*,'(A,$)') ': '
          ENDIF
          READ(5,'(12A)') FILE1
          IF (FILE1 .EQ. ' ') THEN
             FILE1 = OLDFILE1
          ELSE
             OLDFILE1 = FILE1
          ENDIF

      C *** Check whether the weather file exists and read the first date weather
          OPEN (11,FILE = FILE1,STATUS = 'OLD',IOSTAT = OPENSTA)
          IF (OPENSTA .NE. 0) THEN
             WRITE(*,5300) FILE1
             STOP 'Weather file not found, terminated in main program.'
          ENDIF
          READ (11,2400,ERR=5200,END=5200) LAT
          READ (11,2500,ERR=5200,END=5200) IYR,INITDA
          BACKSPACE(11)
          IF (MOD(IYR,4) .EQ. 0) THEN
             IDIM(2) = 29
          ENDIF

      C *** Read in the sowing date
      200 WRITE(*,'(3X,A,$)') 'Sowing month (1 ~ 12) '
          IF (RUNNO .NE. 1) THEN
             WRITE(*,'(A,I2,A,$)') '(default = ',OLDMSOW,') : '
          ELSE
             WRITE(*,'(A,$)') ': '
          ENDIF
          READ (5,'(I3)') MSOW
          IF (MSOW .GT. 12 .OR. MSOW .LT. 0) THEN
             WRITE (*,2000) 'Data is out of range, input again.'
             GOTO 200
          ENDIF
          IF (MSOW .EQ. 0) THEN
             MSOW = OLDMSOW
          ELSE
             OLDMSOW = MSOW
          ENDIF

      250 WRITE(*,'(3X,A,I2,A,$)') 'Sowing date (1 ~ ',IDIM(MSOW),') '
          IF (RUNNO .NE. 1) THEN
             WRITE(*,'(A,I3,A,$)') '(default = ',OLDDSOW,') : '
          ELSE
             WRITE(*,'(A,$)') ': '
          ENDIF
          READ (5,'(I2)') DSOW
          IF (DSOW .GT. IDIM(MSOW) .OR. DSOW .LT. 0) THEN
             WRITE (*,2000) 'Data is out of range, input again.'
             GOTO 250
          ENDIF
          IF (DSOW .EQ. 0) THEN
             DSOW = OLDDSOW
          ELSE
```

```
            OLDDSOW = DSOW
         ENDIF

C *** Transalate sowing date into day-of-year format
         ISOW = 0
         I = 1
C *** While loop
 260     IF (MSOW .EQ. I) GOTO 280
         ISOW = ISOW + IDIM(I)
         I = I + 1
         GOTO 260
 280     ISOW = ISOW + DSOW

C *** Check whether the sowing date is before the beginning of weather
         IF (ISOW .LT. INITDA) THEN
            WRITE(*,5400) ISOW,INITDA
            GOTO 200
         ENDIF

C *** Read in experiment management data
         WRITE(*,'(3X,A,F5.1,A,$)') 'Sowing depth (default =',
     &                    OLDSDEPTH,' cm) : '
         READ(5,'(F5.0)') SDEPTH
         IF (SDEPTH .EQ. 0) THEN
            SDEPTH = OLDSDEPTH
         ELSE
            OLDSDEPTH = SDEPTH
         ENDIF

         WRITE(*,'(3X,A,F5.1,A,$)') 'Phyllochron (default =',
     &                    OLDPHINT,' degree days) : '
         READ(5,'(F5.0)') PHINT
         IF (PHINT .EQ. 0) THEN
            PHINT = OLDPHINT
         ELSE
            OLDPHINT = PHINT
         ENDIF

C *** Read in development genetic coefficients
         WRITE(*,'(/,3X,A)') 'Input Genetic Coefficients : '
         WRITE(*,'(5X,A,A,F4.1,A,$)') 'P1V (vernalization,1 for spring ',
     &            'type,5 for winter type,default=',OLDP1V,') : '
         READ(5,'(F5.0)') P1V
         IF (P1V .EQ. 0) THEN
            P1V = OLDP1V
         ELSE
            OLDP1V = P1V
         ENDIF

         WRITE(*,'(5X,A,A,F5.1,A,$)') 'P1D (1 ~ 6, low ~ high sensitive ',
     &            'to daylength, default =',OLDP1D,') : '
         READ(5,'(F5.0)') P1D
         IF (P1D .EQ. 0) THEN
            P1D = OLDP1D
         ELSE
            OLDP1D = P1D
         ENDIF

         WRITE(*,'(5X,A,A,F5.1,A,$)')'P5 (grain filling degree days,',
     &            ' default =',OLDP5,') :'
         READ(5,'(F5.0)') P5
```

WHEAT PHASIC DEVELOPMENT

```
      IF (P5 .EQ. 0) THEN
        P5 = OLDP5
      ELSE
        OLDP5 = P5
      ENDIF

C *** Repeat until find the weather data of the sowing day
  300 READ (11,7100,ERR=5000,IOSTAT=IOS)
     &       IYR,DOY,SOLRAD,TEMPMX,TEMPMN,RAIN

      IF (SOLRAD .LE. 0.0) THEN
        WRITE (*,7400) FILE1
        STOP 'Weather data missing, terminate in main program'
      ENDIF
      IF (DOY .NE. ISOW) GOTO 300

C *** Initialization of variables
      S1 = SIN(LAT * 0.01745)
      C1 = COS(LAT * 0.01745)
      ISTAGE = 7
      TBASE = 2.
      HI = 0.
      SNOW = 0.
      DOYX = 367
      CUMDTT = 0.
      SUMDTT = 0.
      DTT = 0.

      IF (DOY .EQ. ISOW) BACKSPACE 11

C *** Print out input data
      WRITE (NOUT1,2100) TITLE,RUNNO
      WRITE (NOUT1,2200) LAT,SDEPTH,PHINT
      WRITE (NOUT1,2300) P1V,P1D,P5
      WRITE (*,2100) TITLE,RUNNO
      WRITE (*,2200) LAT,SDEPTH,PHINT
      WRITE (*,2300) P1V,P1D,P5

C *** Set genetic coefficients to appropriate units
      P1V = P1V * 0.0054545 + 0.0003
      P1D = P1D * 0.002
      P5 = 430. + P5 * 20.
      IRET = 0
C *** Repeat until the end of Stage 6
  400 READ (11,7100,ERR=5000,IOSTAT=IOS)
     &       IYR,DOY,SOLRAD,TEMPMX,TEMPMN,RAIN
C *** If run out of weather data then open the weather file of next year
C *** (Assume the 11th and 12th char of weather file name is the year no.)
  500 IF (IOS .LT. 0) THEN
        CLOSE(11)
        READ(FILE1(11:12),'(I2.2)') KYR
        KYR = KYR + 1
        IF (KYR .GT. 99) THEN
          KYR = 10
        ENDIF
        WRITE(FILE1(11:12),'(I2.2)') KYR
        INQUIRE(FILE=FILE1,EXIST=FEXIST)
        IF (FEXIST) THEN
          OPEN(11,FILE=FILE1,STATUS='UNKNOWN')
          READ(11,7200,ERR=5000) LAT
          READ (11,7100,END=500,ERR=5000,IOSTAT=ISO)
```

```
     &          IYR,DOY,SOLRAD,TEMPMX,TEMPMN,RAIN
          ELSE
             WRITE (NOUT1,7300)
             WRITE (*,7300)
             WRITE(*,'(6X,A,A12,A)')
     &          'Assumed next weather file ',FILE1,' does not exist'
             EOWF = .TRUE.
          ENDIF
        ENDIF

        IF (.NOT. EOWF) THEN
           IF (SOLRAD .LE. 0.0) THEN
              WRITE (*,7400) FILE1
              STOP 'Weather data missing, terminate in main program'
           ENDIF

C ***      Cold weather handling routine --- from watbal subroutine
           IF (TEMPMX .LE. 1) THEN
              SNOW = SNOW + RAIN
           ELSE IF (SNOW .NE. 0) THEN
              SNOMLT = TEMPMX + RAIN * 0.4
              IF (SNOMLT .GT. SNOW) SNOMLT = SNOW
              SNOW = SNOW - SNOMLT
           ENDIF

           IF (DOY .EQ. ISOW .OR. ISTAGE .NE. 7) THEN
              CALL PHENOL(IRET)
           ENDIF
           IF (IRET .NE. 1) THEN
              CUMDTT = CUMDTT + DTT
           ENDIF
        ENDIF
        IF (.NOT. EOWF .AND. IRET .NE. 1) GOTO 400

        CLOSE(11)
        WRITE(*,'(/7X,A,$)') 'Run again (Y/N) ? (default = Yes) '
        READ(5,'(A)') ANS
        IF (ANS .NE. 'N' .AND. ANS .NE. 'n') GOTO 100

        ENDFILE(NOUT1)
        CLOSE(NOUT1)

C ***   Skip the error routine when there is no error.
        GOTO 5500

C ***   Error handling routine
 5000   WRITE(*,5100) FILE1,DOY
 5100   FORMAT(10X,'IO Error in weather file : ',A12,'.'./
     &         10X,'Error may occur in or just after day ',I3,'.')
        STOP 'Weather file IO error, terminate in main program.'
 5200   WRITE (*,1000) FILE1
 1000   FORMAT(/5X,'Error! Format data mismatch in first two lines of ',
     &         'weather file:',/7X,A12,'.')
        STOP 'Error in weather file, terminated in main program.'
 5300   FORMAT(/5X,'Error! Weather file: ',A12,' not found.')
 5400   FORMAT(/5X,'Error! Planting day ',I3,', is before',
     1         /7X,'the first available weather day ',I3,'.',
     2         /7X,'Input sowing day again!',/)

 5500   CONTINUE
```

```
      2000 FORMAT(8X,A40,/)
      2100 FORMAT(/10X,A30,10X,'RUN ',I3)
      2200 FORMAT(1X,'Latitude=',F5.1,',   Sowing depth=',F5.1,
     &           ' cm ,',' Phyllochron = ',F6.2)
      2300 FORMAT(1X,'Genetic specific constants :',3X,'P1V = ',F3.1,2X,
     1           'P1D = ',F3.1,2X,'P5 = ',F3.1)
      2400 FORMAT(5X,F6.2)
      2500 FORMAT(5X,I2,1X,I3,26X)
      2600 FORMAT(8X,I5,9X,F5.0)
      2800 FORMAT(1X,F5.0,'-',F5.0,F9.3,1X,4(1X,F6.3),1X,F6.3,2F7.1)
      6100 FORMAT(//////////,
     1     9x,'Welcome to the C E R E S  W H E A T Model Version 2.10',/,
     2     9X,'This modified version only calculates the PHENOLOGY part',//,
     8     /////////,15x,' Please press <Enter> to start ',$)
      7100 FORMAT(5X,I2,1X,I3,F6.2,2(1X,F5.1),1X,F5.1,1X,F6.2)
      7200 FORMAT(5X,F6.2)
      7300 FORMAT(6X,'END OF WEATHER DATA')
      7400 FORMAT(2x,' Please correct your weather file ',A12,
     &           /,2x,' Missing solar radiation data.')
           END
C
C     ***** SUBROUTINE TO CALCULATE PHENOLOGICAL STAGE ******
C
      SUBROUTINE PHENOL(IRET)
C
      INTEGER          DOY,DOYX
      CHARACTER*3      MONTH
      CHARACTER*12     FILE1
      COMMON /PARAM/ ISOW,SDEPTH,LAT,PHINT
      COMMON /DATEC/ MO,ND,IYR,DOY,DOYX,MONTH
      COMMON /COLDC/ SNOW,TEMPCR,TDU,VF,CUMVD,HI
      COMMON /GENET/ P1V,P1D,P2,P3,P4,P5
      COMMON /PHENL/ P9,CUMDTT,TBASE,SUMDTT,DF,S1,C1,ISTAGE,DTT
      COMMON /CLIMT/ TEMPMN,TEMPMX,RAIN,SOLRAD
      COMMON /FILES1/ FILE1,NOUT1

      IRET=0
      TEMPCN=TEMPMN
      TEMPCX=TEMPMX
      XS=SNOW
      IF (XS.GT.15.) XS=15.
      IF (TEMPMN.LT.0.) TEMPCN=2.+TEMPMN*(0.4+0.0018*(XS-15.)**2)
      IF (TEMPMX.LT.0.) TEMPCX=2.+TEMPMX*(0.4+0.0018*(XS-15.)**2)
      TEMPCR=(TEMPCX+TEMPCN)/2.
C     ************ CALCULATES THERMAL TIME ******************
      DTT=TEMPCR-TBASE
      TDIF=TEMPCX-TEMPCN
      IF (TDIF.EW.0) TDIF=1.0
      IF (TEMPCX.GE.TBASE) GOTO 100
      DTT=0.
      GOTO 400
  100 IF (TEMPCN.GT.TBASE) GOTO 200
      TCOR=(TEMPCX-TBASE)/TDIF
      DTT=(TEMPCX-TBASE)/2.*TCOR
  200 IF (TEMPCX.LE.26.) GOTO 400
      TCOR=(TEMPCX-26.)/TDIF
      DTT=13.*(1.+TCOR)+TEMPCN/2.*(1.-TCOR)
      IF(TEMPCN.LE.26.) GOTO 300
      DTT=26.
  300 IF ( TEMPCX.LT.34.) GOTO 400
```

```
      TCOR=(TEMPCX-34.)/TDIF
      DTT=(60.-TEMPCX)*TCOR+26.*(1.-TCOR)
      IF (TEMPCN.GE.26.) GOTO 400
      TCOR=(26.-TEMPCN)/TDIF
      DTT=DTT*(1.-TCOR)+(TEMPCN+26.)/2.*TCOR
400   SUMDTT=SUMDTT+DTT
500   GO TO (1500,1800,2000,2200,2600,3500,600,1100,1300), ISTAGE
C     ****************DETERMINE SOWING DATE******************
600   CALL CALDAT
      WRITE (NOUT1,3900)
      WRITE(*,3900)
      WRITE (NOUT1,3800) ND,MONTH,IYR,DOY,CUMDTT
      WRITE(*,3800) ND,MONTH,IYR,DOY,CUMDTT
      ISTAGE=8
      RETURN
C     ***********DETERMINE GERMINATION DATE******************
1100  CALL CALDAT
      WRITE (NOUT1,1200) ND,MONTH,IYR,DOY,CUMDTT
      WRITE (*,1200) ND,MONTH,IYR,DOY,CUMDTT
1200  FORMAT(4X,I2,1X,A3,1X,I2,5X,'(',I3,')',6X,F6.0,10X,'GERMINATION')
      ISTAGE=9
      P9=40.+10.2*SDEPTH
      CUMDTT=0.
      SUMDTT=0.
      VF=0.
      CUMVD=0.
      TBASE=2.
      RETURN
C     ***********DETERMINE SEEDLING EMERGENCE DATE********
1300  CALL COLD
      IF (SUMDTT.LT.P9) RETURN
      CALL CALDAT
      WRITE (*,1400)ND,MONTH,IYR,DOY,CUMDTT
      WRITE (NOUT1,1400)ND,MONTH,IYR,DOY,CUMDTT
1400  FORMAT(4X,I2,1X,A3,1X,I2,5X,'(',I3,')',6X,F6.0,10X,'EMERGENCE')
      ISTAGE=1
      SUMDTT=SUMDTT-P9
      DTT=SUMDTT
      TDU=0.0
      DF=0.01
      TBASE=0.
      RETURN
C     *********DETERMINE DURATION OF VEGETATIVE PHASE*****
1500  CALL COLD
      IF (VF.LT.0.3) GO TO 1600
      DEC=0.4093*SIN(0.0172*(DOY-82.2))
      DLV=((-S1*SIN(DEC)-0.1047)/(C1*COS(DEC)))
      IF(DLV.LT.-0.87)DLV=-0.87
      HRLT=7.639*ACOS(DLV)
      DF=1.-P1D*(20.-HRLT)**2
1600  TDU=TDU+DTT*AMIN1(VF,DF)
      IF (TDU.LE.400.*(PHINT/95.)) RETURN
      CALL CALDAT
      WRITE (NOUT1,1700) ND,MONTH,IYR,DOY,CUMDTT,CUMVD
      WRITE (*,1700) ND,MONTH,IYR,DOY,CUMDTT,CUMVD
1700  FORMAT(4X,I2,1X,A3,1X,I2,5X,'(',I3,')',6X,F6.0,10X,
     &       'T SPKLT VER DAYS=',F3.0)
      ISTAGE=2
      SUMDTT=0.
      P2=PHINT*3.
      RETURN
```

WHEAT PHASIC DEVELOPMENT

```
C    **********END VEGETATIVE AND BEGIN EAR GROWTH*******
1800 IF (SUMDTT.LT.P2) RETURN
     CALL CALDAT
     WRITE (NOUT1,1900) ND,MONTH,IYR,DOY,CUMDTT
     WRITE (*,1900) ND,MONTH,IYR,DOY,CUMDTT
1900 FORMAT(4X,I2,1X,A3,1X,I2,5X,'(',I3,')',6X,F6.0,10X,
    &    'END VER BEGIN EAR GROWTH')
     ISTAGE=3
     P3=PHINT*2.
     SUMDTT=SUMDTT-P2
     RETURN
C    *****DETERMINE END OF PRE-ANTHESIS EAR GROWTH*******
2000 IF (SUMDTT.LT.P3) RETURN
     CALL CALDAT
     WRITE (NOUT1,2100)ND,MONTH,IYR,DOY,CUMDTT
     WRITE (*,2100)ND,MONTH,IYR,DOY,CUMDTT
2100 FORMAT(4X,I2,1X,A3,1X,I2,5X,'(',I3,')',6X,F6.0,10X,
    &    'END EAR GROWTH')
     ISTAGE=4
     P4=200.
     SUMDTT=SUMDTT-P3
     RETURN
C    ******DETERMINE BEGINNING OF GRAIN FILL****************
2200 IF(SUMDTT.GT.80)GOTO 2300
2300 IF (SUMDTT.LT.P4) RETURN
     CALL CALDAT
     WRITE (NOUT1,2500)ND,MONTH,IYR,DOY,CUMDTT
     WRITE (*,2500)ND,MONTH,IYR,DOY,CUMDTT
2500 FORMAT(4X,I2,1X,A3,1X,I2,5X,'(',I3,')',6X,F6.0,10X,
    &    'BEGIN GRAIN FILL')
     ISTAGE=5
     TBASE=1.
     SUMDTT=SUMDTT-P4
     RETURN
C    ************DETERMINE PHYSIOLOGICAL MATURITY*********
2600 IF (SUMDTT.LT.P5) RETURN
     CALL CALDAT
     WRITE (NOUT1,2900)ND,MONTH,IYR,DOY,CUMDTT
     WRITE (*,2900)ND,MONTH,IYR,DOY,CUMDTT
2900 FORMAT(4X,I2,1X,A3,1X,I2,5X,'(',I3,')',6X,F6.0,10X,
    &    'MATURITY')
     IRET=1
     RETURN
C    ************* STAGE 6 in wheat is useless ******
3500 RETURN
C
3800 FORMAT(4X,I2,1X,A3,1X,I2,5X,'(',I3,')',6X,F6.0,10X,'SOWING')
3900 FORMAT (/,7X,
    1 'DATE   (Day of year) Cum. DTT      PHENOLOGICAL STAGE')
     END
C
C*** SUBROUTINE TO CALCULATE DAMAGE TO CROP DUE TO COLD WEATHER ***
C
     SUBROUTINE COLD
C
C    Called by: PHENOL
C
     INTEGER            DOY,DOYX
     CHARACTER*12       FILE1

     COMMON /DATEC/ MO,ND,IYR,DOY,DOYX,MONTH
```

```
      COMMON /COLDC/ SNOW,TEMPCR,TDU,VF,CUMVD,HI
      COMMON /GENET/ P1V,P1D,P2,P3,P4,P5
      COMMON /PHENL/ P9,CUMDTT,TBASE,SUMDTT,DF,S1,C1,ISTAGE,DTT
      COMMON /CLIMT/ TEMPMN,TEMPMX,RAIN,SOLRAD
      COMMON/FILES1/ FILE1,NOUT1
C
      IF (VF.NE.1.) THEN
        VD = 0.
      ENDIF
      IF (VF .NE. 1 .AND. TEMPMX .GT. 0.) THEN
        IF (TEMPMN .LE. 15.) THEN
          VD1 = 1.4-0.0778*TEMPCR
          VD2 = 0.5+13.44/(TEMPMX-TEMPMN+3.)**2*TEMPCR
          VD = AMIN1(1.,VD1,VD2)
          IF (VD .LT. 0.) VD = 0.
          CUMVD = CUMVD + VD
        ENDIF
        IF (CUMVD .LT. 10 .AND. TEMPMX .GT. 30.) THEN
          CUMVD = CUMVD - .5 * (TEMPMX - 30.)
        ENDIF
        IF (CUMVD .LT. 0.) CUMVD = 0.
        IF (ISTAGE .NE. 9) THEN
          VF = 1. - P1V * (50. - CUMVD)
          IF (VF .LE. 0.) THEN
            VF=0.
          ELSE IF (VF .GT. 1.) THEN
            VF=1.0
          ENDIF
        ENDIF
      ENDIF

      IF (ISTAGE .LE. 7. .AND.
     &   (TEMPMN .LE. TBASE-3. .OR. HI .NE. 0.)) THEN

        HTI = 1.0
        IF (HI .GE. HTI) THEN
          IF (TEMPCR .LE. TBASE + 0.) THEN
            HI = HI + 0.083
            IF (HI .GT. HTI*2.) HI = HTI * 2.
          ENDIF
          IF (TEMPMX .GE. TBASE + 10.) THEN
            HI = HI + 0.2 - 0.02 * TEMPMX
            IF (HI .GT. HTI) HI = HI + 0.2 - 0.02 * TEMPMX
            IF (HI .LT. 0.) HI = 0.
          ENDIF
        ELSE IF (TEMPCR .GE. TBASE - 1.) THEN
          IF (TEMPCR .LE. TBASE + 8.) THEN
            HI = HI + 0.1 - (TEMPCR - (TBASE + 3.5))**2/506.
            IF (HI .GE. HTI .AND. TEMPCR .LE. TBASE + 0.) THEN
              HI = HI + 0.083
              IF (HI .GT. HTI*2.) HI = HTI * 2.
            ENDIF
          ENDIF
          IF (TEMPMX .GE. TBASE + 10.) THEN
            HI = HI + 0.2 - 0.02 * TEMPMX
            IF (HI .GT. HTI) HI = HI + 0.2 - 0.02 * TEMPMX
            IF (HI .LT. 0.) HI = 0.
          ENDIF
        ENDIF
        IF (TEMPMN .LE. -6.) THEN
C          ******* CALCULATES PLANT DEATH ********
          TEMKIL = TBASE - 6. - 6. * HI
```

```
              IF (TEMKIL .GE. TEMPCR) THEN
C        ***** Assume the plant population = 250 ******
              IF ((0.95-0.02*(TEMPCR-TEMKIL)**2) .GE. 0.02) THEN
                 WRITE(NOUT1,200)DOY,TEMKIL,TEMPCR,HI
                 WRITE(*,200)DOY,TEMKIL,TEMPCR,HI
              ELSE
                 WRITE(7,300)DOY,TEMKIL,TEMPCR,HI
                 WRITE(*,300)DOY,TEMKIL,TEMPCR,HI
                 ISTAGE=5
              ENDIF
            ENDIF
          ENDIF
        ENDIF
      RETURN
 200  FORMAT(' CROP WAS DAMAGED BY COLD TEMPERATURE ',
     1       'ON DAY',I5,5X,
     2       'TEMKIL=',F5.1,5X,'TEMPCR=',F5.1,5X,'HI=',F5.2,5X)
 300  FORMAT(' AT LEAST 95% OF CROP KILLED BY COLD TEMP ',
     1       'ON DAY',I5,5X,
     2       'TEMKIL=',F5.1,5X,'TEMPCR=',F5.1,5X,'HI=',F5.2)
      END
C
C***  SUBROUTINE TO CONVERT DAY OF YEAR TO CALENDAR *** C*** DATE
      ***
C
      SUBROUTINE CALDAT
C
C    Called by:  PHENOL
C
      CHARACTER*3   MON(12)
      INTEGER       IDIM(12)
      INTEGER       DOY,DOYX
      CHARACTER*3   MONTH

      COMMON /DATEC/ MO,ND,IYR,DOY,DOYX,MONTH

      DATA MON/'Jan','Feb','Mar','Apr','May','Jun','Jul'
     1 ,'Aug','Sep','Oct','Nov','Dec'/
      DATA IDIM/31, 28, 31, 30, 31, 30, 31, 31, 30, 31, 30, 31/
C
      IF (DOY .LT. DOYX) THEN
         IF (MOD(IYR,4) .EQ. 0) THEN
            IDIM(2) = 29
         ELSE
            IDIM(2) = 28
         ENDIF
      ENDIF
      MO = 0
      ND = 0
C
C    Repeat until month is found
C
 100  MO = MO+1
      ND = ND+IDIM(MO)
      IF (ND .LT. DOY) GOTO 100
      ND = DOY-ND+IDIM(MO)
      DOYX = DOY
      MONTH = MON(MO)
      RETURN
      END
```

REFERENCES

Aase, J.K., and F.H. Siddoway. 1979. Crown-depth soil temperatures and winter protection for winter wheat survival. Soil Sci. Soc. Am. J. 43:1229–1233.

Ahrens, J.F., and W.E. Loomis. 1963. Floral induction and development in winter wheat. Crop Sci. 3:463–466.

Baker, C.K., J.N. Gallagher, and J.L. Monteith. 1980. Daylength change and leaf appearance in winter wheat. Plant Cell Environ. 3:285–287.

Baker, D.N., F.D. Whisler, W.J. Parton, E.L. Klepper, C.V. Cole, W.O. Willis, D.E. Smika, A.L. Black, and A. Bauer. 1985. The development of winter wheat: A physical physiological process model. p. 176–187. *In* ARS Wheat Yield Project. ARS 38. Natl. Technical Info. Serv., Springfield, VA.

Bauer, A., C. Fanning, J.W. Enz, and C.V. Eberlein. 1984. Use of growing-degree days to determine spring wheat growth stages. North Dakota State Univ. Agric. Ext. Bull. EB-37.

Feyerherm. A.M., E.T. Kanemasu, and G.M. Paulsen. 1977. Response of winter wheat grain yields to meteorological variation. Final Contract Rep. NASA-14282, Natl. Aeronautics and Space Admin., Houston, TX.

French, V., and T. Hodges. 1985. Comparison of crop phenology models. Agron. J. 77:170–171.

Gallagher, J.N. 1979. Field studies of cereal leaf growth. I. Initiation and expansion in relation to temperature and ontogeny. J. Exp. Bot. 30(117):625–636.

Gusta, L.V., and D.B. Fowler. 1976. Effects of temperature on dehardening and rehardening of winter cereals. Can. J. Plant Sci. 56:673–678.

Kirby, E.J.M., and M. Appleyard. 1984. Cereal development guide. 2nd ed. Arable Unit, Natl. Agric. Ctr., Warwickshire, England.

Maas, S.J., and G.F. Arkin. 1980. TAMW: A wheat growth and development simulation model. Program and Model Documentation 30-3.

Otter-Nacke, S., D.C. Godwin, and J.T. Ritchie. 1987. Testing and validating the CERES-Wheat model in diverse environments. AgRISTARS Publ. No. YM-15-00407. NTIS, Springfield, VA.

Ritchie, J.T., and S. Otter. 1985. Description and performance of CERES-Wheat: A user-oriented wheat yield model. p. 159–175. *In* ARS Wheat Yield Project. ARS-38. Natl. Tech. Info. Serv., Springfield, VA.

Stapper, M. 1984. SIMTAG: A simulation model of wheat genotypes. Univ. of New England, Dep. of Agronomy and Soil Sci., Armidale, New South Wales, Australia.

Trione, E.J., and R.J. Metzger. 1970. Wheat and barley vernalization in a precise temperature gradient. Crop Sci. 10:390–392.

Van Keulen, H., and W.G. Seligman. 1987. Simulation of water use, nitrogen nutrition and growth of a spring wheat crop. PUDOC, Wageningen, The Netherlands.

Weir, A.H., P.L. Bragg, J.R. Porter, and J.H. Rayner. 1984. A winter wheat crop simulation model without water or nutrient limitations. J. Agric. Sci. 102:371–382.

4 Maize Phasic Development

J. R. KINIRY
USDA-ARS
Temple, Texas

Grain yields of maize (*Zea mays* L.) vary greatly across locations and among years within locations. Mathematical modeling of maize yield has been popular because of the desire to predict grain yields across these variable environments. Maize production models include the Runge-Benci Model (Runge & Benci, 1975), the Splinter Model (Splinter, 1974), SIMAIZ (Duncan, 1975), the Bio-photo-thermal Model (Coligado and Brown, 1975), the Energy-Crop Growth Model (Coelho & Dale, 1980), CORNF (Stapper & Arkin, 1980), and CERES-Maize (Jones & Kiniry, 1986).

Most maize production models are designed to predict grain yield response to environment, but they differ in the types and complexity of biological processes involved. These differences are especially evident in techniques of predicting phenology. The Runge-Benci Model makes no attempt at predicting development. The Splinter Model and SIMAIZ both predict growth stages by accumulating degree-days. When the number of degree-days reaches a specified value, the plant is assumed to reached the next growth stage. No attempt is made to quantify photoperiod sensitivity. The Energy-Crop Growth Model also predicts phenology based solely on temperature. The temperature function is a series of four lines fit to growth rate data. The Bio-photo-thermal Model combines genetic, photoperiodic, and thermal factors to predict number of days to tassel initiation. The CORNF and CERES-Maize Models both use photoperiod and temperature to predict development. In both, photoperiods > 12.5 h delay tassel initiation in sensitive genotypes and increase the final number of leaves. However, CERES-Maize provides a more detailed system of predicting stages and number of leaves, and its components can be more easily tested and validated. Recently, a phenology model was described in Japan (Torigoe, 1986; Torigoe et al., 1986) that includes leaf initiation, leaf-collar appearance, and developmental stages similar to CERES-Maize. In this model, development rates are temperature dependent and photoperiod sensitivity is ignored.

The objective of this chapter is to describe a model for predicting maize phenology based on photoperiod and temperature. Tassel and leaf primordia initiation are simulated to predict total number of leaves (TLNO). Rate of leaf-tip appearance is simulated to predict when the last leaf will emerge. Silking is predicted to occur soon thereafter. Three phases between silking

---

Copyright © 1991 ASA-CSSA-SSSA, 677 S. Segoe Rd., Madison, WI 53711, USA. *Modeling Plant and Soil Systems*—Agronomy Monograph no. 31.

and physiological maturity are simulated. Two processes involved in phenology, which are not included, are the dependence of seed germination on soil water and the dependence of physiological maturity on assimilate supply. By omitting these, neither soil water balance nor assimilate allocation are required.

I. MODEL DESCRIPTIONS

A. Model Inputs and Operation Details

The model described in this chapter is designed to run with a minimal amount of weather data and input variables. The model is written in FORTRAN. The only required climatic data are daily minimum and maximum air temperatures (TEMPMN, TEMPMX). The latitude (LAT) and the day of the year of sowing (ISOW) are also required. There are three parameters that describe a cultivar: (i) the daily thermal time from seedling emergence to the end of the juvenile phase (P1); (ii) photoperiod sensitivity measured in days of tassel initiation delay per hour of photoperiod increase (P2); and (iii) the daily thermal time from silking to physiological maturity (P5). These can be independently derived as described in the text or can be estimated from values for similar cultivars for the location. The average computer time required to execute the model for one cultivar and one season on a large mainframe computer is 3 s. Simulations involving several years of weather data can be run with the same cultivar and planting date.

Phenological phases described in the model represent plant growth intervals delineated by distinct physiological events. The system of numbering these phases is circular as described in Table 4-1. An identifying integer (ISTAGE) is given for all phases, including the phase between physiological maturity and sowing (ISTAGE = 7) when the soil is fallow.

The model operates with a daily incrementing loop, which is executed until the end of the weather data is reached (Fig. 4-1). First the daily temperatures are read. Next, variables pertinent to the present phase are calculated. Finally, the decision is made as to whether the next phenological phase has been reached. This design is efficient and makes the model easy to understand and test.

Table 4-1. Description of the phenological phases used in the model.

Phase no.	Phase description
7	Prior to sowing (fallow)
8	Sowing to germination
9	Germination to seedling emergence
1	Seedling emergence to end of juvenile phase
2	End of juvenile phase to tassel initiation (photoperiod-sensitive phase)
3	Tassel initiation to silking
4	Silking to beginning of effective filling period of grain (lag phase)
5	Effective filling period of grain
6	End of effective filling period to physiological maturity (black layer)

MODELING PLANT AND SOIL SYSTEMS

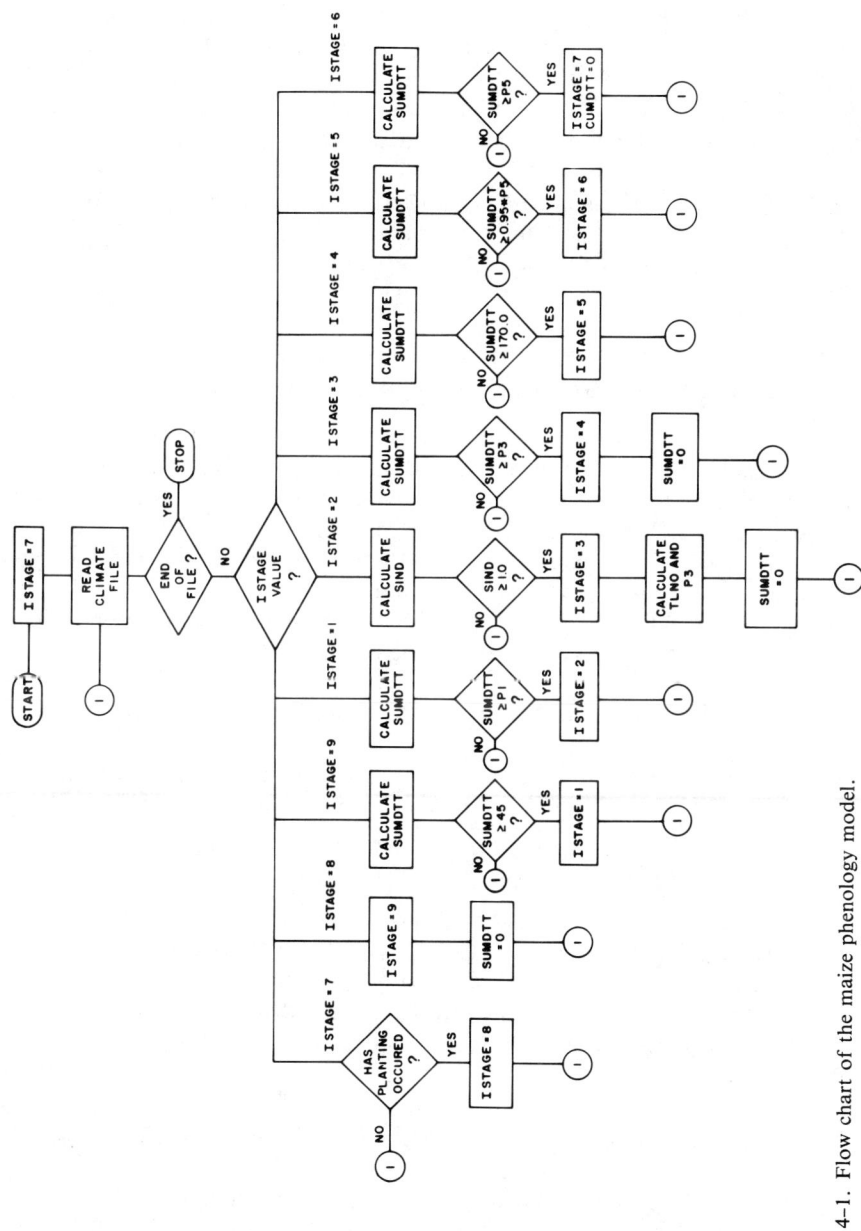

Fig. 4-1. Flow chart of the maize phenology model.

B. Model Assumptions Concerning Maize Phenology

The critical assumptions, as described in detail in the following sections, relate to the development response to temperature for the apical meristem and leaves, and the rate of photoperiodic induction. The model assumes that the rate of development in various stages increases linearly above the base temperature up to 34 °C, and then decreases linearly to zero as temperatures increase from 34 to 44 °C. Similarly, rates of leaf initiation and leaf-tip appearance are assumed to change linearly in these two ranges of temperature. Photoperiodic induction is assumed to decrease with increasing photoperiod for photoperiods greater than 12.5 h. The number of days of tassel initiation delay for each hour increase in photoperiod is assumed to be a constant for any given photoperiod-sensitive cultivar.

C. Maize Development

1. Thermal Response of Maize Development Rates

A growing degree-day or daily thermal time (DTT) system is used to simulate all processes except photoperiodic induction. Daily thermal time, calculated in the present model, is similar to the heat stress equation (Gilmore & Rogers, 1958), except that a base temperature of 8 °C is used for most processes, a high temperature cutoff is activated at 34 °C, and values are decreased linearly from their maximum at 34 °C to zero at 44 °C. Using the symbolism DTT_{TBASE} as the DTT for base temperature (TBASE), DTT is calculated from the mean daily temperature (TEMPM) as

$$DTT_{TBASE} = TEMPM - TBASE, \quad TEMPM > TBASE \quad [1]$$

The DTT_{TBASE} is set to zero if daily maximum temperature (TEMPMX) is less than TBASE. This approach is altered if one of two conditions exist:

1. TEMPMX is greater than and daily minimum temperature (TEMPMN) is less than TBASE, or
2. TEMPMX exceeds 34 °C.

In such cases, eight values between TEMPMX and TEMPMN are interpolated with a zero-to-one factor (TMFAC) calculated with a polynomial fit to a sine wave curve. These values are substituted for TEMPM in Eq. [1] for temperatures between TBASE and 34 °C and in Eq. [2] for temperatures between 34 and 44 °C.

$$DTT_{TBASE} = [(44 - TEMPM)/10](34 - TBASE) \quad [2]$$

The mean of these interpolated values is the value for DTT.

a. Base Temperatures of Development. The base temperature of 8 °C is used for all phenological phases except seedling emergence. This value came from a linear fit to rates of leaf tip appearance measured in controlled-temperature growth chambers (Tollenaar et al., 1979; Kiniry & Ritchie, 1983, unpublished data; Fig. 4-2). A linear fit to predictions of leaf initiation rate (Warrington & Kanemasu, 1983) in the midrange of temperature also had a TBASE close to 8 °C (Fig. 4-3). The decrease in DTT above 34 °C, down to zero at 44 °C, was derived from the leaf-tip appearance rate data in Fig. 4-2.

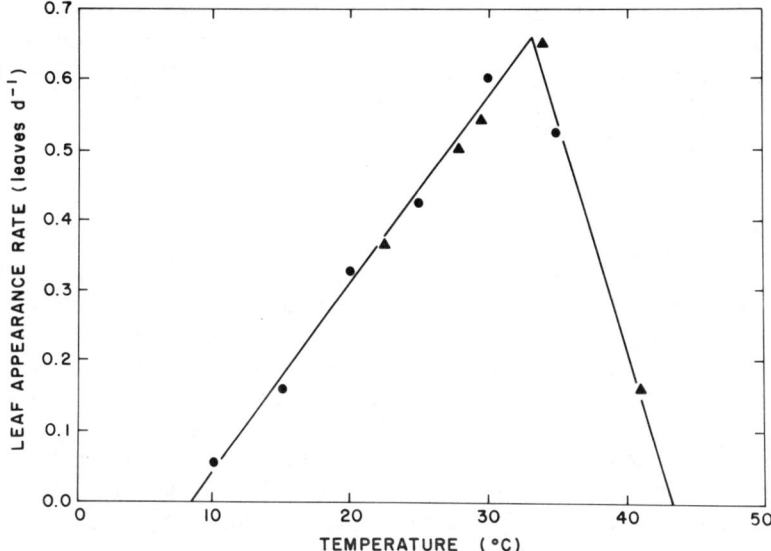

Fig. 4-2. Leaf-tip appearance rate of maize as a function of temperature. (● Tollenaar et al., 1979; ▲ Kiniry & Ritchie, 1983, unpublished data.)

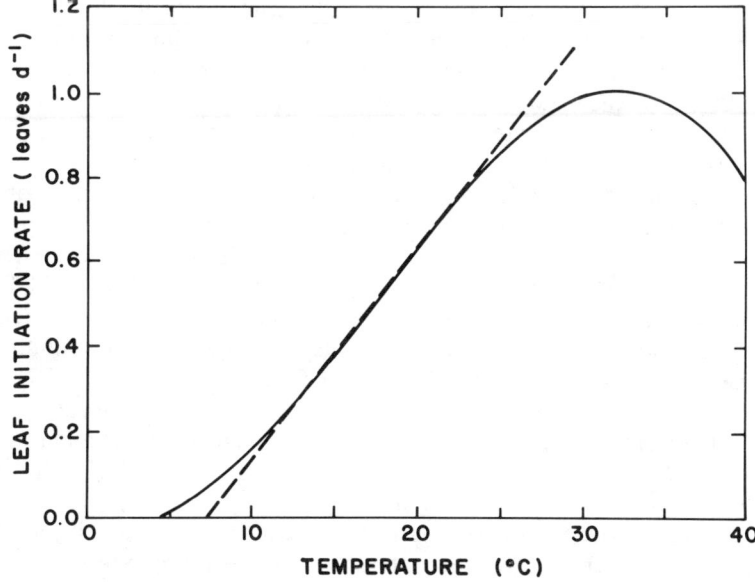

Fig. 4-3. Function for leaf primordium initiation rate of maize as a function of temperature (Warrington & Kanemasu, 1983).

The DTT from germination to seedling emergence has a TBASE value of 10 °C. Coleoptile elongation rate has been shown to linearly increase from a value near zero at 10 °C to a maximum value at 30 °C (Blacklow, 1972).

2. Planting to Seedling Emergence Interval

In the model, planting to seedling emergence requires one day for germination and then 45 DTT_{10} for coleoptile elongation. This assumes adequate soil moisture for germination at the time of planting. Results from unpublished field experiments conducted by the author have shown that there was an average of 45 DTT_{10} in the interval from one day after planting to seedling emergence when planting depth was 5 cm.

3. Seedling Emergence to Silking Interval

The period from seedling emergence to silking involves three separate but related systems. The total number of leaves (TLNO) is determined from the number of leaf primordia initiated between seedling emergence and tassel initiation. Date of tassel initiation is determined using both DTT_8 and photoperiod. Silking, or end of leaf growth, is determined from TLNO and the rate of leaf-tip appearance. These three systems, leaf initiation, leaf-tip appearance, and the induction of tassel initiation, were developed independently and can be tested separately. They not only provide a framework for this model, but also one for further research in maize phenology.

The same DTT system described above with a base of 8 °C is used to predict both leaf primordia initiation and leaf-tip appearance. The difference between the two rates is in the DTT_8 required per leaf tip or primordium. A field experiment at Temple, TX, (Kiniry and Ritchie, 1981 unpublished data) which used destructive sampling to count leaf primordia, showed that 21 DTT_8 were required for each leaf primordium to initiate and 38.9 DTT_8 for each leaf tip to appear. These values allow for prediction of leaf development, but date of tassel initiation is also required to determine TLNO.

Tassel initiation is the stage when leaf primordia initiation ends and branches of the tassel begin to develop. Work in the field at Temple, TX has shown that there are six primordia present at seedling emergence. Dividing the DTT_8 total from seedling emergence to tassel initiation by 21 DTT_8 allows prediction of the number of new primordia that were initiated.

Prediction of tassel initiation is critical to the system. The method used is based on work by Rood and Major (1980), and Kiniry et al. (1983a, b). While the plant is in the juvenile phase immediately following seedling emergence, the plant's development rate is dependent on temperature and independent of photoperiod. After a genotype-specific sum of DTT_8 (P1) has been reached, the plant's apical development is assumed to be independent of temperature and in the photoperiod-sensitive phase. In the model, all genotypes initiate their tassel 4 d after the start of this phase in photoperiods $\leq$ 12.5 h. For photoperiods > 12.5 h, the rate of induction (RATEIN) is:

MODELING PLANT AND SOIL SYSTEMS

$$\text{RATEIN} = 1/(4 + P2 \times (\text{HRLT} - 12.5)] \qquad [3]$$

where HRLT is the number of daylight hours including civil twilight and P2 is a genotype-specific parameter for sensitivity. In the model, plants reach the tassel initiation stage when the total of daily values of RATEIN is $\geq$ 1.0. Values for P1 and P2 have been determined for a wide range of genotypes (Table 4-2). Values for other cultivars can be calculated from controlled-environment experiments. The P1 value is the DTT_8 total from seedling emergence to 4 d prior to tassel initiation in photoperiods of $\leq$ 12.5 h. The P2 value is the number of days delay in tassel initiation for each hour increase in photoperiod > 12.5 h.

This system of predicting date of tassel initiation provides a method whereby high temperatures on different dates prior to tassel initiation can have variable effects on the final number of leaves (TLNO). Tollenaar and Hunter (1983) found that high temperatures immediately prior to tassel initiation increased TLNO. In the present model, duration of the inductive phase is strictly photoperiod dependent. However, initiation of leaf primordia is still temperature dependent. High temperatures during this phase will increase the rate of leaf initiation, but will not increase the rate of development of the apex. This results in a greater TLNO and delayed silking. Likewise, in the juvenile phase, both leaf initiation and phase duration are temperature dependent. High temperatures during this phase will increase both development rates and cause no change in TLNO.

In this model, silking and the end of leaf growth are assumed to occur on the same day. To determine the DTT_8 from seedling emergence to the end of growth of the last leaf, two factors must be taken into account. The second leaf tip emerges from the leaf whorl about 20 DTT_8 after seedling emergence. There are about 76 DTT_8 from last leaf-tip appearance to the ligule appearance of that leaf. Thus, the total DTT_8 from seedling emergence to emergence of the last leaf ligule is

$$(\text{TLNO} - 2) \times 38.9 + 96.0. \qquad [4]$$

Subtracting the DTT_8 at tassel initiation yields P3; the DTT_8 total from tassel initiation to the end of leaf growth. Analysis of the data of Kiniry (1979) shows that, across three hybrids, four planting dates, two locations, and two years, the mean DTT_8 from tassel emergence to silking was 62 DTT_8. Assuming the last leaf tip emerges at tassel emergence, there are 14 DTT_8, or about 1 d between silking and collaring of the last leaf.

4. Silking to Maturity Interval

Grain development is the major phenomenon in the period from silking to physiological maturity. The three phases defined in the present system are those described by Johnson and Tanner (1972). These phases include a lag phase, a period of nearly linear grain filling called the effective fill period, and the period from the end of the effective fill period to physiological maturity (black layer).

Table 4-2. Genotype-specific values for the daily thermal time with an 8°C base temperature (DTT_8) from seedling emergence to the end of the juvenile phase (P1), the days delay in tassel initiation for each hour increase in photoperiod (P2), and the DTT_8 from silking to physiological maturity (P5).

Cultivar	P1	P2	P5
	DTT_8	d delay h^{-1}	DTT_8
Southern Canada			
CORNL281	110	0.30	--
CP170	120	0.00	680†
F7 × F2	125	0.00	732†
LG11	125	0.00	737†
PIO 3995	130	0.30	--
Northern USA			
INRA 260	135	0.00	739†
EDO	135	0.30	--
A654 × F2	135	0.00	751†
DEKALB XL71	140	0.30	--
F478 × W705A	140	0.00	670†
PIO 3901	144	0.30	--
Northern Nebraska, N. Iowa, N. Illinois, and N. Indiana			
PIO 3720	180	0.80	685
A632 × W117	187	0.00	730†
PIO 3382	200	0.70	--
PIO 3780	200	0.76	685
C281	202	0.30	685
Southern Nebraska, S. Iowa, S. Illinois, and S. Indiana			
PIO 511A	220	0.30	685
PIO 3183	260	0.50	750
W69A × F546	240	0.30	--
A632 × VA26	240	0.30	--
W64A × W117	245	0.00	--
NEB 611	260	0.30	720
B14 × OH43	265	0.80	665
B8 × 153R	218	0.30	760
Central Missouri and Kansas to North Carolina and Southward			
PIO 3147	255	0.76	685
WF9 × B37	260	0.80	710
PV82S	280	0.50	750
PV76S	260	0.50	750
B56 × C131A	318	0.50	700
B73 × Mo17	220	0.52	880
NC + 59	280	0.30	750
McCurdy 67-14	265	0.30	825
Tropical Cultivars			
H610	340	0.52	840
PIO X304C	390	0.52	940

† Values (Derieux & Bonhomme, 1982) assume maturity occurs at 30% grain moisture.

Kiniry and Keener (1982) found that the summed DTT_{10} were less variable than days for the interval between silking and physiological maturity for various planting dates. The present model assumes a genotype-specific number of DTT_8 from silking to maturity. Future experimental work may indicate a different base temperature.

Work by Cross (1975) shows that the DTT for the lag phase did not differ significantly among genotypes. In the present model, this was determined to be 170 DTT_8 (Kiniry, 1985). The effective fill period begins at the end of the lag phase and ends when 95% of the total DTT_8 from silking to physiological maturity have been accumulated. The final developmental phase, from the end of the effective fill period to physiological maturity, requires the remaining 5%. A statement has been added to the model to prevent delayed maturity if cool temperatures prevent DTT accumulation during this final, short stage. If the DTT is ≤ 2.0 on a day during this last stage, maturity is assumed to occur.

Values for P5, the required summation of DTT_8 from silking to maturity, ranged from 665 for 'B14 × OH43' to 940 for 'PIO X304C' (Table 4-2). These values were derived from field data that included both silking and maturity measurements.

II. MODEL VALIDATION

A. Planting to Silking Interval

Predictions of interval lengths were tested for planting to silking and silking to physiological maturity. Testing for the first interval was done on four hybrids, Pioneer 3780, B73 × Mo17, H610, and McCurdy 67-14, each with several measured dates of silking. Another test was performed using 10 entries grown at eight locations in Europe (Derieux & Bonhomme, 1982).

Locations for Pioneer 3780 ranged from as far north as University Park, PA to as far south as Temple, TX (Table 4-3). The mean error in prediction of silking was 0.1 d early, with a standard deviation (SD) of 7.3 d. The greatest error occurred for data from Greeley, CO, with date of silking underestimated by 17 d. This could have been due to dry soil conditions delaying germination or because soil temperatures were cooler than air temperatures early in the season, causing delayed emergence. Deleting this one observation, the mean error was 2.0 d early with an SD of 3.8 d.

The hybrid with the most extensive test data was B73 × Mo 17 (Table 4-4). It was grown in nine plantings at Columbia, MO in 1978 and 1979 (Kiniry, 1979; Griffin, 1980), and was included in a multilocation study (Stapper & Arkin, 1980). There were plantings in 1982, 1983, and 1984 in Temple, TX (Kiniry, 1985) and a multilocation study in Europe in 1977 and 1978 (Derieux & Bonhomme, 1982).

Errors were similar in magnitude to those for Pioneer 3780. When large errors occurred, they were underpredictions of the days to silking. This may have been due to errors in dates of germination or seedling emergence. The mean error was −2.7 d and the SD was 5.6 d. Deleting the three data sets with large negative errors, the mean error was −1.2 d and the SD was 4.2 d.

The hybrid H610 was planted on several dates and locations in Hawaii (Table 4-5). It failed to have the large negative errors associated with some of the data sets of the previous two hybrids. This may have been due to the lack of cool soil temperatures. The mean error was 2.2 d and the SD was 4.3 d.

Table 4-3. Predicted and measured dates of silking of maize hybrid Pioneer 3780.

Location	Year	Silking predicted	Date measured	Difference
		day of year		d
Pennsylvania State Univ.†	1979	217	214	3
		222	222	0
		240	232	8
		250	243	7
Greeley, CO‡	1976	197	214	-17
Tyron, NE§	1978	211	211	0
Temple, TX¶	1982	149	151	-2
	1983	160	158	2
	1984	158	160	-2
$\bar{x}$				-0.1
SD				7.3

† Yao (1980).
‡ Cuany et al. (1977).
§ Clawson (1980).
¶ Kiniry (1985).

Table 4-4. Predicted and measured dates of silking of maize hybrid B73 × Mo17.

Location	Year	Silking predicted	Date measured	Difference
		day of year		d
Columbia, MO	1978	194	203	-9
Location 1†	1978	202	207	-5
	1979	197	195	2
	1979	210	211	-1
Location 2†	1978	224	221	3
	1978	228	226	2
	1979	198	205	-7
	1979	220	223	-3
Location 3‡	1979	210	207	3
Swift Current, Saskatchewan§	1979	233	246	-13
Bloomington, IL§	1979	205	209	-4
Temple, TX§	1979	187	198	-11
Temple, TX¶	1982	150	157	-7
	1983	161	164	-3
	1984	155	157	-2
Europe#	1977-1978			
Mons, France		235	248	-13
Fuchs, France		215	220	-5
Rome, Italy		204	202	2
Martonvasar, Hungary		223	220	3
Debrecen, Hungary		220	219	1
Zajecar, Yugoslavia		215	209	6
Radzikow, Poland		238	236	2
$\bar{x}$				-2.7
SD				5.6

† Kiniry and Keener (1982).
‡ Griffin (1980).
§ Stapper and Arkin (1980). Cooperators included H.R. Davidson, Canadian Dep. Agric., Swift Current, Saskatchewan and J.L. Malcolm, Funk Seeds Int., Bloomington, IL.
¶ Kiniry (1985).
# Derieux and Bonhomme (1982).

Table 4-5. Predicted and measured dates of silking of maize hybrid H610 in Hawaii. †

Location	Year	Silking predicted	Date measured	Difference
		——— Day of year ———		d
Waipio, HI	1983	43	43	0
Iole, HI	1982	218	208	10
Iole, HI	1979	134	130	4
Iole, HI	1978	114	110	4
Iole, HI	1978	239	235	4
Kukaiau, HI	1978	125	123	2
Kukaiau, HI	1979	206	209	−3
Kukaiau, HI	1978	91	87	4
Kukaiau, HI	1979	113	118	−5
Kukaiau, HI	1979	97	94	3
Halawa, HI	1978	265	259	6
Kukaiau, HI	1978	240	243	−3
$\bar{x}$				2.2
SD				4.3

† U. Sing and G. Uehara (1983, unpublished data).

Table 4-6. Predicted and measured dates of silking of maize hybrid McCurdy 67-14.

Location	Year	Silking predicted	Date measured	Difference
		——— Day of year ———		d
Columbia, MO†				
Location 1	1978	200	207	−7
	1978	206	210	−4
	1979	202	199	3
	1979	213	214	−1
Location 2	1978	227	225	2
	1978	229	231	−2
	1979	206	208	−2
	1979	225	225	0
Temple, TX‡	1981	179	174	5
	1982	140	145	−5
$\bar{x}$				−1.1
SD				3.7

† Kiniry and Keener (1982).
‡ C.A. Jones, 1982, unpublished data.

The final hybrid tested was McCurdy 67-14 (Table 4-6). There were eight plantings in Columbia, MO and two in Temple, TX. The mean error of these predictions was −1.1 d and the SD was 3.7 d.

Considering all four hybrids, the average mean error was −0.4 d and the average SD was 5.2 d. Large errors were usually underpredictions of the days to silking.

Another analysis was done with data from Europe consisting of 10 entries grown at eight locations (Derieux & Bonhomme, 1982; Table 4-7). Genetic parameters for all entries were derived using data from Fuchs, France. Data from this location were not included in the testing. Mean errors ranged

Table 4-7. Measured subtracted from predicted silking dates for 10 entries grown at seven locations in Europe in 1977 and 1978.[†]

Location	Year	Inra 260	F7 × F2	CP170	LG11	A654 × F2	F478 × W705A	A632 × W117	F16 × F19	W69A × F546	A632 × Va 26
Aubia, France	1977	4	--	6	4	6	3	0	4	--	--
Mons, France	1978	-1	-3	-6	-5	-2	-3	-1	-2	-3	-4
Rome, Italy	1978	2	2	3	2	4	3	4	5	7	6
Martonvasar, Hungary	1978	6	5	6	5	--	5	0	--	--	13
Debrecen, Hungary	1978	7	6	6	5	--	8	6	--	--	9
Zajecar, Yugoslavia	1978	6	6	4	6	7	7	6	9	15	15
Radzikow, Poland	1978	-4	0	1	-3	-5	-9	4	-3	6	6
x̄		2.9	2.7	2.9	2.0	2.0	2.0	2.7	2.6	6.25	9.2
SD		4.1	3.7	4.3	4.3	5.2	6.0	3.0	5.0	7.4	7.5

[†] All data from Derieux and Bonhomme (1982).

from 2.0 d for cultivars LGH, A654 × F2, and F478 × W705A to 9.2 d for A632 × Va26. Standard deviations ranged from 3.0 d for A632 × W1117 to 7.5 d for A632 × Va26. Pooling all entries in all locations, the mean error was 3.2 d and the SD was 4.9 d. This SD was comparable to the mean SD for the previous tests. It appears that errors in this model's predictions can be expected to have a SD of 4 to 5 d.

B. Silking to Maturity Interval

Tests on the growth phase from silking to physiological maturity were conducted with B73 × Mo17 and McCurdy 67-14. Both hybrids were grown with multiple plantings in Missouri and Texas. The required DTT_8 to complete this growth interval was determined using a subset of the data. All the data were included in the test results. Errors for B73 × Mo17 (Table 4-8) had a SD of 8.0 d. This was comparable to errors for the planting to silking interval of the same hybrid. The two greatest errors overpredicted the interval duration. By deleting the two plantings with large errors, the mean error became -0.1 d and the SD was 3.3 d. These two plantings did not appear to be stressed; kernel dry weights were 0.32 and 0.34 g kernel^{-1} and grain yields were 8220 and 8370 kg ha^{-1}, respectively.

McCurdy 67-14 (Table 4-9) had a SD of 8.3 d, which was greater than the SD associated with the errors in silking date prediction. For this hybrid, the three largest error values underpredicted the duration of the growth phase. This hybrid, when grown in Missouri, experienced low temperatures when approaching physiological maturity. The present system of using DTT_8 may be too simple to predict physiological maturity as temperatures approach freezing. It is possible that photosynthetic rate or carbohydrate availability may influence the time of maturity under cool conditions.

Table 4-8. Predicted and measured days from silking to physiological maturity for hybrid B73 × Mo17.

Location	Year	Interval		Difference
		Predicted	Measured	
		d		
Columbia, MO				
Location 1†	1978	52	52	0
	1978	53	53	0
	1979	53	55	−2
	1979	59	66	−7
Location 2†	1978	64	47	17
	1978	69	51	18
	1979	57	56	1
	1979	76	74	2
Temple, TX‡	1979	46	45	1
Temple, TX§	1984	47	43	4
x̄				3.4
SD				8.0

† Kiniry and Keener (1982).
‡ Stapper and Arkin (1980).
§ Kiniry (1985).

Table 4-9. Predicted and measured days from silking to physiological maturity for hybrid McCurdy 67-14.

Location	Year	Interval Predicted	Measured	Difference
		d		
Columbia, MO				
Location 1†	1978	50	53	−3
	1978	50	52	−2
	1979	50	66	−16
	1979	56	71	−15
Location 2†	1978	61	52	9
	1978	68	68	0
	1979	71	72	−1
	1979	56	72	−16
Temple, TX‡	1981	41	41	0
	1982	44	50	−6
x̄				−5.0
SD				8.3

† Kiniry and Keener (1982).
‡ C.A. Jones (1982, unpublished data).

III. RESEARCH NEEDS

A possible source of uncertainty in the model is inconsistency in the DTT_8 per leaf primordium or leaf tip. Warrington and Kanemasu (1983) reported the leaf initiation rate in a 12 h photoperiod at 28 °C was 95% as great as in a 14 or 16 h photoperiod at the same temperature. In contrast, Gmelig-Meyling (1973) found that the rate of leaf appearance was not photoperiod dependent for plants grown in photoperiods ranging from 9 to 17 h. Also, in contrast with our findings in the field, Tollenaar et al. (1984) found that leaf appearance rates of different hybrids varied by as much as 14%. Another deviation occurred in a field experiment planted in October in Temple, TX, when leaf primordia required 27 instead of 21 DTT_8.

Rate of daylength change has been proposed as one factor that alters the rate of leaf appearance (Baker et al., 1980). These two rates were shown to be correlated in wheat (*Triticum aestivum* L.) grown in the field. However, in growth chambers with controlled temperatures, leaf appearance rate was not affected by rate of daylength change for the first 2 wk after seedling emergence (Kiniry & Ritchie, 1983, unpublished data). In this experiment, two wheat and two maize cultivars were grown in a constant-temperature environment with either + 5, − 5, or zero minutes of change in photoperiod per day.

The silking to maturity interval also needs further research. An obstacle to defining the DTT system for this interval lies in the problems associated with growing large maize plants in growth cabinets. Maize grain development is seldom normal in such controlled environments. Likewise, problems with light quality and temperature control make work in greenhouses equally difficult. At the present time, the author believes a DTT_8 with the required sum calculated from field data for each cultivar is sufficient.

IV. SUMMARY

The model presented here was shown to function in a wide range of environments and with a wide range of genotypes. Similar to the traditional GDD_{10} sums that are widely used, it requires only daily maximum and minimum temperatures. The model's advantage is its ability to account for photoperiod sensitivity and the effects of high temperatures on final leaf number. In addition, three components of vegetative development, leaf initiation, leaf-tip appearance, and tassel initiation, can be independently tested.

REFERENCES

Baker, C.K., J.N. Gallagher, and J.L. Monteith. 1980. Daylength change and leaf appearance in winter wheat. Plant Cell Environ. 3:285-287.
Blacklow, W.M. 1972. Influence of temperature on germination and elongation of the radicle and shoot of corn (*Zea mays* L.). Crop Sci. 12:647-650.
Clawson, K.L. 1980. Irrigation scheduling of corn using infrared thermometry. Univ. Nebraska Agric. Meteorol. Progress Rep. 80-4.
Coelho, D.T., and R.F. Dale. 1980. An energy-crop growth variable and temperature function for predicting corn growth and development: Planting to silking. Agron. J. 72:503-510.
Coligado, M.C., and D.M. Brown. 1975. A bio-photo-thermal model to predict tassel initiation time in corn (*Zea mays* L.). Agric. Meteorol. 15:11-31.
Cross, H.Z. 1975. Diallel analysis of duration and rate of grain filling of seven inbred lines of corn. Crop Sci. 15:532-535.
Cuany, R.L., J.F. Swink, J.C. Keenan, E.W. McCord, and V.A. Wegrzyn. 1977. Performance test of corn hybrids in Colorado in 1976. Colorado State Univ. Exp. Stn. General Ser. 964.
Derieux, M., and R. Bonhomme. 1982. Heat unit requirements for maize hybrids in Europe, results of the European FAO subnetwork. Maydica 27:59-77.
Duncan, W.G. 1975. SIMAIZ: A model simulating growth and yield in corn p. 32-48. *In* D.N. Baker et al. (ed.) The application of systems methods to crop production. Proc. Symp., Mississippi State, MS. 7-8 June 1973. Mississippi Agric. For. Exp. Stn., Mississippi State, MS.
Gilmore, E.C., Jr., and J.S. Rogers. 1958. Heat units as a method of measuring maturity in corn. Agron. J. 50:611-615.
Gmelig-Meyling, H.D. 1973. Effect of light intensity, temperature and daylength on the rate of leaf appearance of maize. Neth. J. Agric. Sci. 21:68-76.
Griffin, J.L. 1980. Quantification of the effects of water stress on corn growth and yield. M.S. thesis. Univ. of Missouri, Columbia.
Johnson, D.R., and J.W. Tanner. 1972. Calculation of the rate and duration of grain filling in corn (*Zea mays* L.). Crop Sci. 12:485-486.
Jones, C.A., and J.R. Kiniry. 1986. CERES-Maize: A simulation model of maize growth and development. Texas A&M Univ. Press, College Station, TX.
Kiniry, J.R. 1979. Application of an enzyme kinetic equation to maize (*Zea mays* L.) development rates. M.S. thesis. Univ. of Missouri, Columbia.
Kiniry, J.R. 1985. Responses of maize kernel number to shading stress: timing of sensitivity in the reproductive stage and characteristics of genotypes differing in this sensitivity. Ph.D. diss. Texas A&M Univ., College Station (Diss. Abstr. 85-14258).
Kiniry, J.R., and M.E. Keener. 1982. An enzyme kinetic equation to estimate maize development rates. Agron. J. 74:115-119.
Kiniry, J.R., J.T. Ritchie, and R.L. Musser. 1983a. Dynamic nature of the photoperiod response in maize. Agron. J. 75:700-703.
Kiniry, J.R., J.T. Ritchie, R.L. Musser, E.P. Flint, and W.C. Iwig. 1983b. The photoperiod sensitive interval in maize. Agron. J. 75:687-690.
Rood, S.B., and D.J. Major. 1980. Responses of early corn inbreds to photoperiod. Crop Sci. 20:679-682.

Runge, E.C.A., and J.F. Benci. 1975. Modeling corn production-estimating production under variable soil and climatic conditions. p. 195-214. *In* Proc. 30th Ann. Corn and Sorghum Res. Conf., Chicago, IL. 9-11 Dec. 1975. Am. Seed Trade Assoc., Washington, DC.

Splinter, W.E. 1974. Modeling of plant growth for yield prediction. Agric. Meteorol. 14:243-253.

Stapper, M., and G.F. Arkin. 1980. CORNF: A dynamic growth and development model for maize (*Zea mays* L.). Texas Agric. Exp. Stn. Res. Ctr. Program and Model Documentation 80-2.

Tollenaar, M., T.B. Daynard, and R.B. Hunter. 1979. Effect of temperature on rate of leaf appearance and flowering date in maize. Crop Sci. 19:363-366.

Tollennaar, M., and R.B. Hunter. 1983. A photoperiod and temperature sensitive period for leaf number of maize. Crop Sci. 23:457-460.

Tollenaar, M., J.F. Muldoon, and T.B. Daynard. 1984. Differences in rates of leaf appearance among maize hybrids and phases of development. Can. J. Plant Sci. 64:759-763.

Torigoe, Y. 1986. A conceptual model of developmental phases of maize on the basis of the relationship between differentiation and growth of vegetative and reproductive organs. Jpn. J. Crop Sci. 55:465-473.

Torigoe, Y., H. Watanabe, and H. Kurihara. 1986. Varietal differences in morphological and physiological characters of the developmental phases of maize. Jpn. J. Crop Sci. 55:474-482.

Warrington, I.J., and E.T. Kanemasu. 1983. Corn growth response to temperature and photoperiod. II. Leaf initiation and leaf appearance rates. Agron. J. 75:755-761.

Yao, N.R. 1980. Vegetative and reproductive development of corn (*Zea mays* L.) at four spring planting dates. M.S. thesis. Pennsylvania State Univ., University Park.

5 Soybean Development

JAMES W. JONES, K. J. BOOTE, S. S. JAGTAP, AND J. W. MISHOE
University of Florida
Gainesville, Florida

Simulation models have been developed for studying the effects of environment and management on soybean [*Glycine max* (L.) Merr.] crop growth and yield (Meyer et al., 1979; Wilkerson et al., 1983; Acock et al., 1983). One of the major challenges faced by researchers who are developing these models is predicting the timing of vegetative and reproductive events during the plant's life cycle. These events, called stages, include emergence, first flower appearance, and physiological maturity. In this chapter, the first occurrence of a reproductive event will be referred to as a stage whereas the time intervals between stages will be called phases. Soybean yield is highly sensitive to the timing of stages and to the durations of these phases. Furthermore, there are considerable differences in soybean development depending on cultivar and environment. A cultivar that has first flower appearance 40 d from planting in one location may never flower at another location even though dry matter growth may follow similar patterns at the two locations. These complex cultivar and environment interactions must be predicted if crop models are to be useful across a range of locations.

Two environmental variables, photoperiod and temperature, strongly affect soybean development. Soybean is a quantitative, short-day plant. Most cultivars flower sooner under long nights than under short nights (Borthwick & Parker, 1938), although development toward flowering can be decreased by very low light levels (Summerfield & Roberts, 1987). Most research on photoperiod effects has focused on flowering. Later-maturing cultivars are more sensitive to photoperiod than early maturing ones, with some early maturing cultivars showing insensitivity to photoperiod (Criswell & Hume, 1972).

Soybean response to photoperiod, however, is not restricted to flower development. Fisher (1963) showed that at least three short days were required after flowering for fruit to set. The cultivar Biloxi flowered in 14- and 15-h photoperiods, but never developed fruit. Johnson et al. (1960) found significant effects on the durations of phases from flowering, pod set, and end of flowering to maturity when photoperiod was increased at the beginning of each of those phases. Rates of seed filling may also be affected by photoperiod (Thomas & Raper, 1976; Cure et al., 1982).

Copyright © 1991 ASA-CSSA-SSSA, 677 S. Segoe Rd., Madison, WI 53711, USA. *Modeling Plant and Soil Systems*—Agronomy Monograph no. 31.

Temperature affects both vegetative and reproductive development in soybean. Parker and Borthwick (1943) found that floral induction was optimal for foliage temperatures between 21 and 27 °C at night. Above 27 °C, fewer floral primordia were formed. Similar responses to temperature were found for the durations of first flower to first pod phase and the first to last flower phase. These results are similar to those later reported by Hesketh et al. (1973) for flowering and maturity. Their work showed that reproductive development was insensitive to temperatures between 22 and 28 °C, but that it slowed for temperatures outside this range. Temperature can also affect fruit set. Thomas and Raper (1981) found that temperatures below 22 °C caused a decreasing carpel initiation rate with no carpel initiation occurring at temperatures below 14 °C. Hesketh et al. (1973) found that no pods were formed at 14 or 18 °C in their study.

Vegetative development is highly sensitive to temperature (Hesketh et al. 1973; Thomas & Raper, 1977). Hesketh et al. (1973) showed a linear relationship between node formation rate and temperature between about 8 and 30 °C. They also showed that the effect of temperature on reproductive development was in fact different from its effect on vegetative development. Vegetative and reproductive stages are not independent, however. Many soybean cultivars are determinate, with the final mainstem node terminating in a flower. Other cultivars are indeterminate, but Sinclair (1984) found a strong correlation between the end of node formation and the timing of pod formation at upper nodes. Thus, the final number of main stem nodes is affected by photoperiod and temperature interactions.

Several soybean phenology models have been developed to predict the occurrence of stages of growth for combinations of photoperiod and temperature. One of these models is a component of the soybean growth and yield model SOYGRO V5.4 (Jones et al., 1987), an updated version of the model was published by Wilkerson et al. (1983). In this chapter, four other soybean phenology models will be reviewed, followed by a more thorough presentation of the development model used in SOYGRO.

I. REVIEW OF PREVIOUS MODELS

Major et al. (1975) assumed the following equation to describe the rate of reproductive development (R) for each day as a function of daylength (L) and average daily temperature (T):

$$R = F_1(L) \times F_2(T) \qquad [1]$$

where F_1 and F_2 are functions of daylength and temperature, respectively. The occurrence of a reproductive stage was predicted by integrating R daily to obtain the cumulative development between stages of growth, M,

$$M = \int_{S_1}^{S_2} R \, dt \qquad [2]$$

where S_1 is the starting or reference stage, S_2 is the stage to be predicted, dt is the differential time used to integrate the rates, and $M = 0.0$ at S_1. Stage S_2 was reached when $M = 1.0$.

Major et al. (1975) used field experiments with various cultivars planted on different dates and at different locations to provide a range of photoperiod and temperature environments. They then assumed quadratic forms of F_1 and F_2 as follows:

$$F_1(L) = a(L - a_0) + a_2(L - a_0)^2 \quad [3]$$

$$F_2(T) = b(T - b_0) + b_2(T - b_0)^2 \quad [4]$$

An iterative regression analysis method was used to estimate the six parameters in Eq. [3] and [4] for each variety and each of five growth phases: planting to emergence, emergence to flowering, flowering to beginning pod fill, flowering to termination of flowering, and flowering to physiological maturity. Their work confirmed the need to use photoperiod and temperature for all growth phases except emergence and the need for a critical daylength (a_o) in the model. Their work did not include later-maturing cultivars.

Jones and Liang (1978) used a different approach to model flowering in soybean by considering three phases of development: planting to primary leaf, primary leaf to flower initiation, and flower initiation to first flower appearance. They used a variable X_i to represent the balance between a promotion and inhibition system, where X_i increases in the dark and decreases in daylight:

$$X_i = (X_{i-1} + N_i)K \quad [5]$$

where N_i is night length and K is a temperature-dependent variable ≤ 1.0. When X_i exceeded R, another temperature-dependent variable, flowering was initiated. Jones and Liang (1978) found that the rate of change of night length affected their parameter estimates, so they multiplied Eq. [5] by ($1 - A + \Delta N$) and used a statistical parameter estimation procedure to fit A, K, and R as quadratic functions of temperature, where A is a coefficient that reduces development rate under increasing night lengths. The duration of time from flower initiation to flowering was also fit using linear functions of temperature and photoperiod. One good feature of their model concerned treatment of flower initiation and appearance as two separate events. This model did not consider stages beyond flowering.

Hadley et al. (1984) developed a model for predicting flowering dates in soybean. They grew nine cultivars in growth cabinets in five photoperiods ranging from 8 h 20 min. to 15 h using three night temperatures (14, 19, and 24 °C) combined with one daytime temperature of 30 °C. One photoperiod-insensitive cultivar, Fiskeby 5, was included with eight sensitive cultivars. The inverse of the duration of planting to first flower (representing the average rate of development, $R = 1/f$) was computed for each treatment combination. Hadley et al. (1984), like Major et al. (1975), used a critical photoperiod. When photoperiods were longer than the critical value, they found that the rate of progress toward flowering was linear with photoperiod. Their equation was:

$$1/f = a' + b'(L) + c'(T) \qquad [6]$$

where a', b', and c' are statistically fit coefficients.

They also found that a base temperature of 7.8 °C was similar for all varieties, but that the critical night length varied with cultivar and with temperature.

Hodges and French (1985) presented a soybean phenology model (SOYPHEN) that computed daily development rate as a product of three functions, similar to the way the Major et al. (1975) model did with two functions. Hodges and French (1985) included a water stress function in addition to photoperiod and temperature. Rather than use quadratic regression, they assumed piecewise linear functions and used a statistical procedure to fit the breakpoints of the functions. They defined seven phases encompassing planting to maturity and included juvenile phase in their model to account for a period of time after emergence when the plants do not respond to photoperiod. Data from the Major et al. (1975) and McBlain et al. (1987) studies were used to estimate coefficients for each phase. Their functions for temperature and photoperiod were sufficiently general to include a base temperature, a linear increase in development rate with temperature, which is equivalent to degree days, and a critical photoperiod and linear change as a result of increasing daylength. In addition, the functions were able to include nonlinear changes such as decreased development rate at high temperatures. Their statistical procedure, which allowed estimation of coefficients for each stage, provided flexibility, but did not ensure physiological reasonableness of coefficients.

None of these soybean phenology models have been integrated into crop growth and yield models. Two of the models do not predict the occurrence of stages beyond flowering and, thus, would not be sufficient in a whole-season model. The other two models (Major et al., 1975; Hodges & French, 1985) could perhaps be used in whole-season crop growth models. The phenology model to be described below was developed specifically for integration into the SOYGRO growth and yield model. It has several features that are similar to the Hodges and French (1985) and Major et al. (1975) models. Differences exist, however, particularly in the methods used to estimate the coefficients. These differences will be discussed below.

II. THE PHENOLOGY MODEL IN SOYGRO

SOYGRO (Wilkerson et al., 1983; Jones et al., 1987) is a physiological crop model that simulates growth and yield of soybean under various soil, weather, and management conditions. The model includes a component that predicts development of the crop, which is then the basis for changes in partitioning of dry matter and, ultimately, yield. The overall goal of the development model in SOYGRO is to predict the durations of important growth phases of soybean cultivars growing in various climates. The dry matter growth and partitioning in SOYGRO is controlled to a large extent by the plant's physiological time scale. The development model, called GPHEN, has been adapted to run independently or as an integral part of SOYGRO. The model relies heavily on concepts of thermal and photothermal time, with coefficients that express each cultivar's sensitivity to photoperiod and temperature. These concepts will be discussed and the procedures used for estimating the coefficients will be described.

A. Photothermal Time

The basic form of the phenology model used in SOYGRO is a multiplicative relationship to compute development rate in each reproductive growth phase as a function of temperature and night length:

$$R(t) = F_r(T) \times F(N) \quad [7]$$

where $R(t)$ is development rate on day t, $F_r(T)$ is the temperature function for phases sensitive to both temperature and photoperiod, and $F(N)$ is the night length function, with both T and N as functions of time at a specific location [$T = T(t)$ and $N = N(t)$]. Progress toward Stage S_2 from Stage S_1 is computed by integrating $R(t)$ starting from S_1 or

$$t_p = \int_{S_1}^{t} R(t)\, dt \quad [8]$$

where t_p is the photothermal time from S_1, which depends on both temperature and photoperiod.

The function $F_r(T)$ is a piecewise linear function of nighttime temperature (Fig. 5-1). This function was obtained directly from the data present-

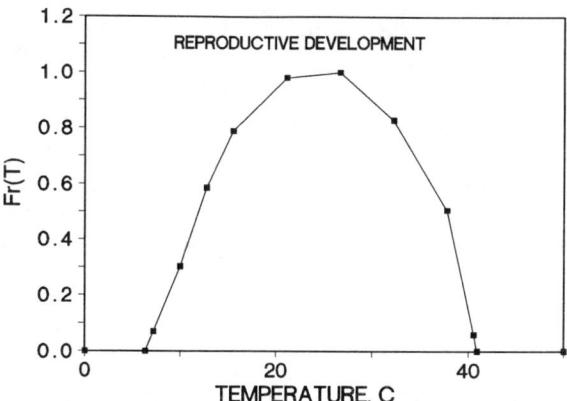

Fig. 5-1. Relationship between reproductive development rate of soybean and temperature [$F_r(T)$] based on the work of Parker and Borthwick (1943).

ed by Parker and Borthwick (1943, Table 2) on the relative effect of foliage temperature at night on initiation of floral primordia when nighttime lasts 16-h. Normalized over all points by dividing by the peak value, the function ranges between 0 and 1. When the value of $F_r(T) = 0.5$ on a given day, development rate is one-half the maximum rate at the given photoperiod. This can be interpreted as being equivalent to having a plant that develops only 0.5 d if temperature is optimal. Another way of expressing this development is that $R(t)$ is the number of photothermal days occurring during a real day. At the temperature and length of nighttime when $F_r(T)$ and $F(N)$ are both 1.0, development progresses at its fastest rate and 1.0 photothermal day occurs during the actual day. When t_p reaches a threshold of f_{min}, then Stage S_2 occurs. Using notation similar to that of Hadley et al. (1984), $F_r(T)$ can be computed from constant-temperature, development-time studies as:

$$F_r(T) = f_{min}/f \qquad [9]$$

where f_{min} is the minimum duration of a growth phase over all temperatures at optimal photoperiod, and f is the duration of the phase at temperature T.

We assumed that $F_r(T)$ was the same for all growth phases sensitive to photoperiod. Hadley et al. (1984) showed that the threshold temperature of 7.8 °C was the same for all cultivars, so the same function was also used for all cultivars.

The night length function was developed by assuming a linear change in the duration of photoperiod-sensitive phases with night length (Fig. 5-2). This curve is similar to those of Cregan and Hartwig (1984), except that Fig. 2 is linear below a critical night length, whereas their data were nonlinear for some cultivars. Hodges and French (1985) and Hadley et al. (1984) assumed that $1/f$ was linear, whereas $1/f$ in our formulation is nonlinear. Early attempts to estimate parameters from field data in our study showed that results were better by making f a linear function of night length rather than $1/f$.

Figure 5-2 shows f_N, the normalized development duration obtained by dividing each f value by f_{min}, the minimum duration of a phase. Thus, f_N has a minimal value of 1.0. A f_N value of 2.0 means that, at the night length in question, the duration of a phase is twice as long as the minimum duration of the phase at optimal temperature and night length. Since f_N is a linear function of night length below a critical night length, $N\phi$, $1/f_N$ is nonlinear and ranges between $1/TH$ and 1, where TH is the maximum value of f_N. Many cultivars will develop toward flowering at some minimum rate even during long days (Roberts & Summerfield, 1987). The function $F(N) = 1/f_N$ and can be interpreted similarly to $F_r(T)$. At a constant temperature, development proceeds at its maximum rate when $F(N) = 1.0$. Roberts and Summerfield (1987) suggest that $F(N)$ would be linear to some night length, then equal to a minimum rate below this night length. Our experience also suggests that this form of $F(N)$ may prove useful if sufficient data were available for parameter estimates.

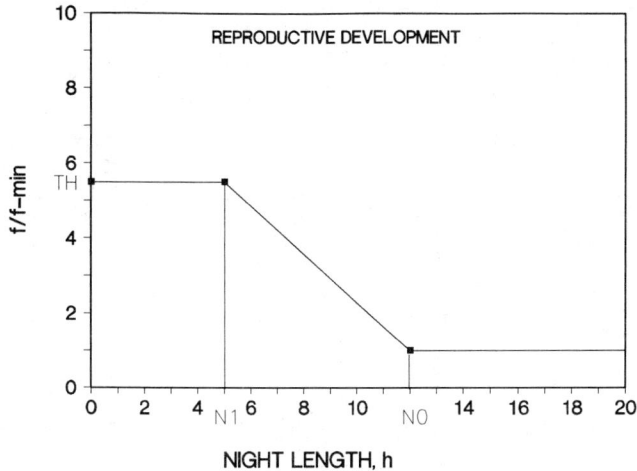

Fig. 5-2. Relationship between the relative duration of a reproductive phase and night length used in SOYGRO V5.4 ($f_N = f/f_{min}$). The night length function $F(N) = 1/f_N$ as shown.

By multiplying $F_r(T)$ and $F(N)$, the maximum value is still 1.0, which means that at optimal temperature and night length, a photothermal day $[R(t)] = 1.0$ real day. Thus, the threshold for development for any phase can be found experimentally by holding T and N at optimal values and measuring real days. In principle, $F_r(T)$ can be obtained by holding N at its optimal value and varying T. Conversely, $F(N)$ can be found by holding T at its optimal level and varying N. Since $F_r(T)$ was assumed to be the same for all cultivars, the problem in our work became one of estimating the values of $N\phi$, $N1$, and TH in Fig. 5-2 for describing f_N. Based on earlier work, these values were expected to vary considerably among cultivars.

There is some controversy as to whether day and night or just night temperatures affect development, although data of Parker and Borthwick (1939; 1943) suggest a more dominant effect of night temperature. In our model, photothermal time accumulates hourly during nighttime hours only. Numerically, $R(t)/h_n$ is computed for each hour in the night and summed until the threshold f_{min} is reached for the stage, and then S_2 occurs (h_n is the number of hours during the night). Hourly temperatures are computed from daily maximum and minimum temperatures and photoperiod using a sine wave during daylight hours, starting 2 h after sunrise, and a linear decrease at night to the minimum temperature of the next morning. The photoperiod is computed using the model presented by Goudriaan and van Laar (1978).

B. Thermal Time

For stages insensitive to photoperiod, a separate physiological or thermal time scale was developed based simply on temperature. Hesketh et al. (1973) showed a linear increase in development rate $(1/f)$ for leaf appearance for temperatures between 8 and 30 °C. A third function, $F_v(T)$, was

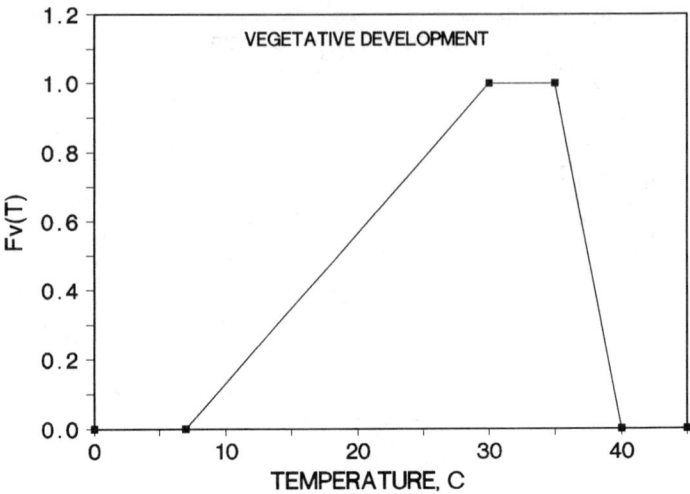

Fig. 5-3. Relationship between development rate and temperature for soybean phases insensitive to photoperiod, [$F_v(T)$], based on work of Hesketh et al. (1973).

based on their data, normalized to range between 0.0 and 1.0 and extended to higher temperatures (Fig. 5-3). Thus, $F_v(T)$ is the number of thermal days per actual day, and the threshold for a stage insensitive to photoperiod is the minimum number of days needed to develop at the optimal temperature. Thermal time is computed by

$$t_T = \int_{S_1}^{t} F_v(T)dt \qquad [10]$$

and stage S_2 occurs when t_T reaches the threshold thermal time for the stage, f_{min}. In the model, thermal time accumulates hourly, day and night. Numerically, $F_v(T)/24$ is summed each hour until the threshold f_{min} for the stage is reached, then S_2 occurs.

C. Description of Development Phases

Table 5-1 defines the growth phases used in the SOYGRO V5.4 model. These phases are based in part on the soybean stages of growth published by Fehr and Caviness (1977). Other stages were defined because of the need to predict dry matter growth and partitioning in the crop.

The model starts at planting and predicts the duration of Phase 1 (emergence) using thermal time. Then, V1 (first true leaf) is predicted by accumulating thermal time during Phase 2 (emergence to V1) until a threshold is reached. Subsequently, the appearances of additional leaves, V2, V3, etc., are predicted using thermal time until the NDLEAF stage (time of last leaf expansion) occurs. Plants may have a juvenile phase that extends beyond V1. By definition, the juvenile phase ends when the plants first become sensitive to

Table 5-1. Description of growth phases used in the development model (GPHEN) of SOYGRO V5.4, based in part on the stage descriptions of Fehr and Caviness (1977).

Growth phase	Description	Environmental sensitivity†
0	Before planting (fallow)	--
1	Planting to emergence (V0)‡	T
2	V0 to first unifoliolate fully expanded (V1)	T
3	Juvenile phase (V1 to end of juvenile phase)	T
4	Floral induction phase (end of juvenile phase to flower induction, R0)‡	T, N
5	R0 to flower apperarance (R1)	T, N
6	R1 to first pod > 2 cm in length (NPOD0)	T, N
7	R1 to first pod > 2 cm in the upper 4 nodes (R4)	T, N
8	R1 to last leaf expansion (NDLEAF)	T, N
9	R1 to last pod formation (NDSET)	T, N
10	R1 to physiological maturity (R7)	T, N
11	R7 to harvest maturity (R8)	T

† T indicates that the phase duration is sensitive to temperature, whereas N indicates sensitivity to night length.
‡ The V and R prefixes indicate vegetative and reproductive stages as defined by Fehr and Caviness (1977).

photoperiod. Wilkerson et al. (1988) found that most of the cultivars they tested were sensitive to photoperiod by the time V1 was reached. Only two of the cultivars that were texted exhibited juvenile phases.

The induction phase duration is sensitive to both photoperiod and temperature. The duration of the next phase (R0 to R1) is partially a function of the thermal time required to express the node of the first flower, but also partially dependent on photoperiod as shown by Fisher (1963). Based on the work of Johnson et al. (1960) and on our own attempts to characterize the duration of various phases, all phases from R1 to R7 are sensitive to both photoperiod and temperature. The final phase, R7 to R8, is basically a maturation period during which time leaves abscise and seeds begin to dry. Thermal time is used to describe the final phase.

The timing of the different growth stages is important in establishing the changes in dry matter partitioning in the plant. Beginning at a stage referred to as NPOD0, the crop begins to add pods and seeds, which have priority for assimilates. Dry matter partitioning to vegetative growth decreases and approaches zero when a full pod load is set. Changes in temperature and photoperiod would change the duration of real time between stages, and stretch or shrink the growth and partitioning pattern over longer or shorter real-time periods.

III. ESTIMATION OF COEFFICIENTS FOR THE MODEL

A. Thermal Thresholds

Four of the eleven phases used to describe soybean development (Table 5-1) are dependent only on temperature. The function in Fig. 5-3, based on the work of Hesketh et al. (1973), was used to describe sensitivity to temperature in each of these four phases. Emergence is assumed to occur five thermal days after planting and V1 at nine thermal days after planting for all cultivars. There is no water stress effect in the stand-alone GPHEN model on any development phases. The time between R7 and R8 (Phase 11) is assumed to be 8.5 thermal days for all cultivars. Analysis of data from Quincy, FL showed that thermal time was better than photothermal time in predicting this phase duration, although an actual-time average of 12.0 d predicted this phase duration just as well as thermal time. Although some differences occurred among cultivars, we assumed this R7 to R8 threshold to be constant to minimize the number of cultivar-specific coefficients and because the SOYGRO model is relatively insensitive to this phase. The other phase sensitive to thermal time is the juvenile phase. Wilkerson et al. (1988) found that most cultivars in their experiment were sensitive to photoperiod by the V1 stage, but Jupiter had a 5-d juvenile phase (beyond V1). The breeding line F85-1226J supplied by Kuell Hinson of the Agronomy Dep. at the University of Florida had a juvenile phase that extended to V5 in that study. In the GPHEN model, the thermal time required for the juvenile phase is a required cultivar-dependent coefficient. When there is no previous evidence of a juvenile phase in a cultivar, the duration of Phase 3 is best set equal to zero.

B. Photothermal Coefficients

The photoperiod switching experiments reported by Wilkerson et al. (1988) were used as the basis for setting the threshold for Phase 5, R0 to R1, to 12.86 thermal days. Their work showed that all cultivars tested had a similar minimum time; (14 d or 12.86 thermal days as defined by Fig. 5-3). In addition, their work showed that at least five short days were required. Therefore, R1 is predicted in the model by accumulating both thermal and photothermal time from R0 for all cultivars, and requiring that the thermal time threshold of 12.86 thermal days and the photothermal time threshold of 5.0 be met before R1 occurs. The most difficult parameters to estimate for a given cultivar were the critical night length (NO) and the increase in duration of phases depending on night length, or the TH value shown in Fig. 5-2. The slope of the curve in Fig. 5-2 could have been used, but we arbitrarily defined TH as the relative effect of photoperiod on phase duration at a 5-h night length ($N1 = 5$). Data taken by D.C. Herzog (1980, unpublished data) in Quincy, FL (30.6 °N Lat) planting date experiments were used to estimate NO and TH for this model. An iterative optimization model was developed to estimate NO and TH by minimizing the error sum of squares between predicted and observed dates of stages for soybean cultivars planted on 9 to 11 dates in each of 3 yr at Quincy, FL. Figure 5-4 shows the number of days between R1 and R7 for three cultivars and nine planting dates in 1979, ranging from Maturity Group III to VII. The parameters NO, TH, and f_{min} were estimated based on the duration of the R1 to R7 phase. The lines in Fig. 5-4 are predictions for 1979 after using all 3 yr of data (1978-1980) to estimate the parameters.

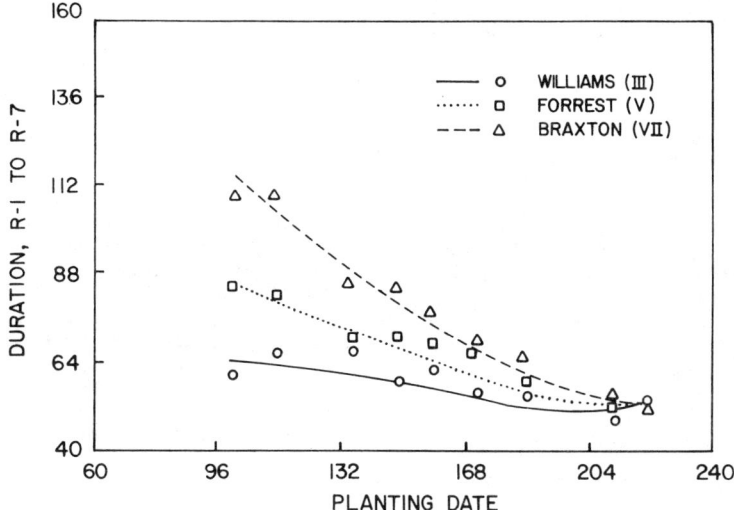

Fig. 5-4. Simulated flowering dates compared with observed values for three cultivars for 1 yr (1979) out of the 3 yr of data from Quincy, FL, used to estimate the coefficients.

To determine the thresholds for other photoperiod-sensitive phases, such as R1 to R4, it was assumed that the parameters *NO* and *TH* were the same as for R1 to R7. Then photothermal thersholds (f_{min}) were computed from the field data. Comparisons of standard deviations (SD) and coefficients of variability (CV) for phase durations in actual, thermal, and photothermal time verified that all phases after the end of juvenile phase through R7 were best described by photothermal time. For example, CV values for the R1 to R4 phase were 41.9, 42.3, and 12.3% for actual, thermal, and photothermal time, respectively, for Bragg.

Table 5-2 summarizes the thermal and photothermal thresholds for each growth phase that are common to all cultivars for the GPHEN model. Table 5-3 shows cultivar dependent thresholds and the *NO* and *TH* coefficients that were estimated from the Quincy FL data for 16 cultivars.

The major differences in parameters between cultivars in Table 5-3 were in the *TH* values, which express the sensitivity to photoperiod below critical night lengths. When considering the parameters for all cultivars, standard deviations in the induction, R1 to R4, R1 to R5, and R1 to R7 phase thresholds were 2.17, 1.10, 1.39, and 2.16 photothermal days, respectively. Thus, average threshold values were used in SOYGRO V5.4 (Jones et al., 1987) to extend the cultivar list by computing general cultivar coefficients for Maturity Groups (MG) 00, 0, I, II, and IV. However, there was a high correlation between *TH* and maturity group ($r = 0.83$), and the values of *NO* and *TH* were not averaged across maturity groups. In contrast, *NO* did not vary as much for the higher maturity groups (VX), but apparently decreased with lower maturity groups (Cregan & Hartwig, 1984). We used the data of Cregan and Hartwig (1984) to estimate *TH* and *NO* for Evans, a MG-0 cultivar, which resulted in values of 2.0 and 9.5 h, respectively. The

Table 5-2. Thresholds for soybean growth phases that did not vary among cultivars.

	Growth phases										
	1	2	3	4	5	6	7	8	9	10	11
Thermal time (d)	5.0	9.0	†‡		12.86						8.50
Photothermal time (d)				†	5.0	†§	†	†¶	34.0	†	

† Cultivar dependent.
‡ The juvenile phase for many cultivars is zero days. If there is no information for a cultivar, it is assumed that the duration of this Phase 3 is zero thermal days.
§ Threshold for Phase 7 multiplied by 0.58, based on data from Gainesville, FL.
¶ For determinate cultivars (MG-V through MG-X), this threshold is the same as that for Phase 7 (R1 to R4). For indeterminate cultivars, it is set equal to the duration of R1 to R5, a phase not identified in the model as such, but computed from field data.

Table 5-3. Genetic coefficients for various cultivars based on data from Quincy, FL, 1978 to 1980.

			Growth phases				
Cultivars	N0	TH	Juvenile (3)	Induction (4)	R1-R4 (7)	R1-R5	R1-R7 (10)
	h	f/f_{min}	Thermal time (d)	Photothermal time (d)			
MG-3							
Williams	10.43	6.00	0.0	4.54	11.44	14.71	45.45
MG-5							
Davis	11.82	7.12	0.0	11.61	11.73	14.55	45.98
Essex	11.88	4.62	0.0	7.99	10.41	13.63	43.25
Forrest	12.09	4.62	0.0	8.81	12.46	15.17	42.91
Hill	11.58	5.75	0.0	11.24	11.31	14.07	39.23
MG-6							
Tracy	11.86	8.06	0.0	6.36	11.05	13.37	43.23
MG-7							
Bragg	11.80	13.50	0.0	5.01	9.74	12.23	44.78
Braxton	12.13	11.62	0.0	5.24	9.70	11.74	40.65
Lee-68	12.10	7.20	0.0	6.71	10.21	12.52	41.84
Lee-74	12.09	7.60	0.0	6.89	10.17	23.53	41.65
Govan	12.20	10.40	0.0	6.01	9.40	11.37	39.14
Ransom	12.12	6.75	0.0	7.61	12.07	13.95	46.18
MG-8							
Cobb	11.96	17.62	0.0	4.68	9.57	11.73	43.93
Hardee	11.90	22.85	0.0	4.65	8.84	11.19	42.98
MG-9							
Jupiter	12.00	18.62	5.0	7.77	11.33	15.22	45.84
MG-10							
Vicoja	11.81	31.00	0.0	4.65	8.84	11.19	42.98
$\bar{x}$	11.86	11.45	--	6.85	10.52	13.07	43.13
SD	0.40	7.28	--	2.17	1.10	1.39	2.16

thresholds for Evans were first estimated as the average thresholds in Table 5-3 for each phase. These were later modified to be 4.0, 7.0, 10.67, 17.0, and 44.0 for Phases 4, 6, 7, 8, and 10, respectively, to fit data supplied by Don Reicosky (1985, personal communication) on Evans in Morris, MN. We later confirmed the validity of the Evans parameters by comparing simulated with observed development for Evans in Gainesville, FL. The other general MG coefficients for MG-00, I, II, and IV have not yet been tested.

IV. TESTING THE MODEL

Tests were conducted to evaluate the accuracy of the model and to determine limitations in predicting soybean development at various locations. Development stage data and weather data were available from the Major et al. (1975) phenology modeling work. Their data were obtained for independent testing of our model for which the coefficients had been developed using data from Quincy, FL. Comparisons were possible for two cultivars that were common to these two data sets.

Figure 5-5 shows predicted and observed days to flower for one location, Columbia, MO, out of the Major et al. (1975) data. There was a change in days to flower from about 55 to 35 d as planting date increased from Day of Year 112 to over Day of Year 200 for Williams (Fig. 5-5a). The model also predicted about a 20-d change in days to flower, but with a bias of about 5 d. For the cultivar Hill (Fig. 5b), the days to flower ranged from about 84 to 50 as planting date increased. The development model predicted the longer duration of days to flower for Hill when compared with Williams, although there was a slight trend for the model to underpredict by 3 to 5 d for early planting dates and overpredict time to flower by 2 to 3 d for late planting days.

Table 5-4 summarizes the differences between predicted and observed durations of days to flower and days from flower to physiological maturity for the two cultivars tested using the Major et al. (1975) data. For each site, averages and standard deviations were computed for the differences to determine the bias in the model and its variability. The bias in predicting the days to flower ranged from 1.44 to -4.50 d for Hill and 2.43 to 9.33 d for Williams. The model predicted the duration of flowering to maturity to be 0.43 to 5.25 d shorter than observed for Hill, on average, for the three sites. Predicted durations of flowering to physiological maturity averaged from 0.75 d longer to 8.38 d shorter than observed for Williams. However, there were problems in predicting maturity for some late planted crops. For example, at one location (Mt. Vernon, MO), Hill was planted on Day or Year 298 (25 October). The phenology model did not predict maturity before freeze occurred. In another case, Williams planted on July 13 in Mt. Vernon reached maturity on Day of Year 277 (4 October), whereas the model predicted maturity 18 d later. The slowing effect of cold temperatures on development seems to be too severe in the model. Otherwise, model predictions were mostly within

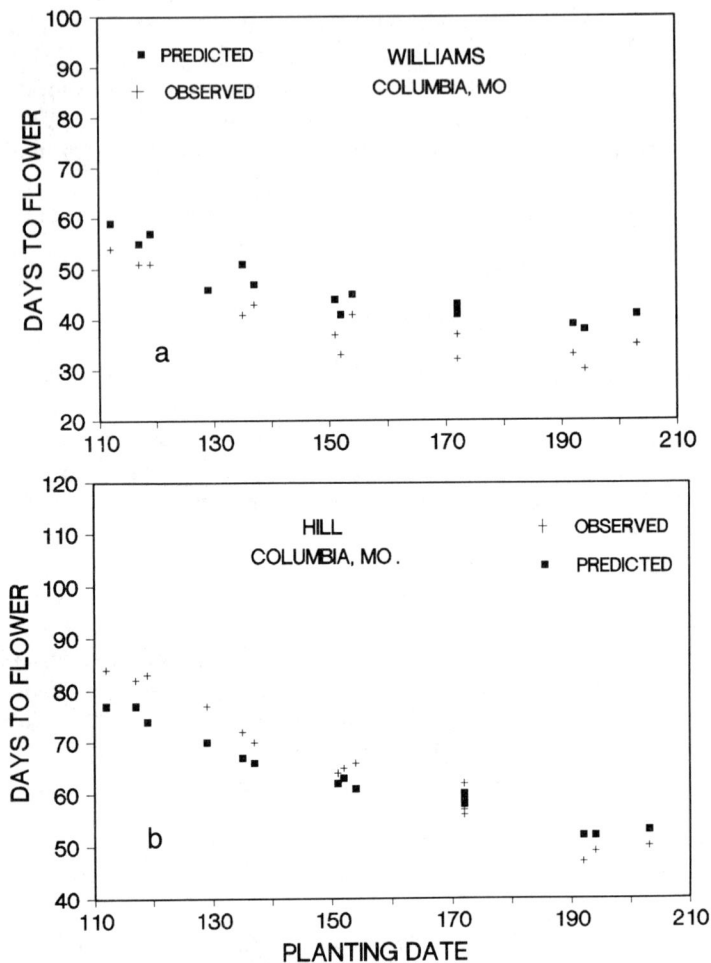

Fig. 5–5. Comparisons of predicted (■) and observed (+) days to flower for soybean planted on different days from Exp. 1 of the Major et al. (1975) study in Colombia, MO: (a) Williams cultivar (MG-III) and (b) Hill cultivar (MG-V).

5 to 10 d of observed flowering and maturity dates for both cultivars over three seasons, three locations, and various planting dates.

Tests were also conducted on Bragg (MG-VII) using data taken by G.G. Wilkerson (1986, unpublished data) and R.W. McClendon (1986, unpublished data), who are working with SOYGRO. Flowering and physiological maturity dates were simulated for 14 seasons to compare with observed dates for Bragg. Data for nine seasons were available from Gainesville, FL (29.6°N Lat), one from Quincy, FL (30.6°N Lat), two from Watkinsville, GA (34.0°N Lat), and two from Clayton NC (35.9°N Lat). Actual weather data and planting dates were input into the model, and predicted minus observed flowering and physiological maturity dates were computed for each case (Fig. 5–6). On average, the model predicted flowering 2.36 d earlier than observed

Table 5-4. Comparison between observed and predicted flowering and maturity dates for three cultivars. Predictions were based on coefficients computed from the data from Quincy, FL.

Location (latitude)	Phase					
	Planting to flowering (R1)			Flowering (R1) to physiological maturing (R7)		
	Average difference	SD	n	Average difference	SD	n
			d			
	Bragg (VII)					
†	−2.36	4.20	14	1.00	6.59	14
	Hill (V)					
Spickard, MO‡ (40.23°N)	−4.50	4.52	14	−5.25	7.14	9
Columbia, MO‡ (38.95°N)	−2.20	4.25	15	−3.38	4.94	13
Mt. Vernon, MO‡ (37.10°N)	1.44	3.80	9	−0.43	3.50	7
Columbia, MO§ (38.95°N)	−0.17	4.64	35			
	Williams (III)					
Spickard, MO‡ (40.23°N)	2.43	3.98	14	−8.38	2.74	9
Columbia, MO‡ (38.95°N)	5.87	2.36	15	0.13	5.77	15
Mt. Vernon, MO‡	9.33	2.05	9	0.75	1.56	8
Columbia, MO§ (38.95°N)	6.00	3.72	35			

† Gainesville, FL ($n = 10$), Watkinsville, GA ($n = 2$), and Raleigh, NC ($n = 2$).
‡ Data from Major (1975), Exp. 1.
§ Data from Major (1975), Exp. 2.

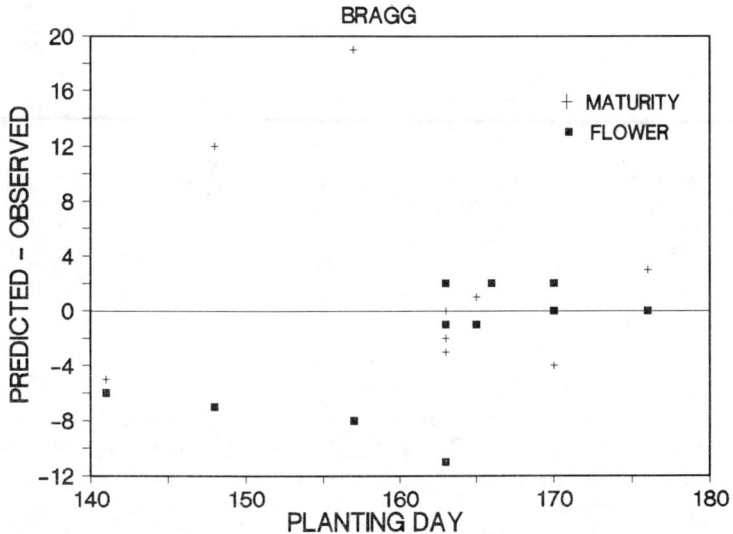

Fig. 5-6. Comparison of observed subtracted from predicted dates of flowering (■) and physiological maturity (+) for Bragg soybean (MG-VII) planted on various dates in four locations (Gainesville, FL, 29.6°N Lat; Quincyk FL, 30.6°N Lat, Watkinsville, GA, 34.0°N Lat, and Clayton, NC, 35.9°N Lat).

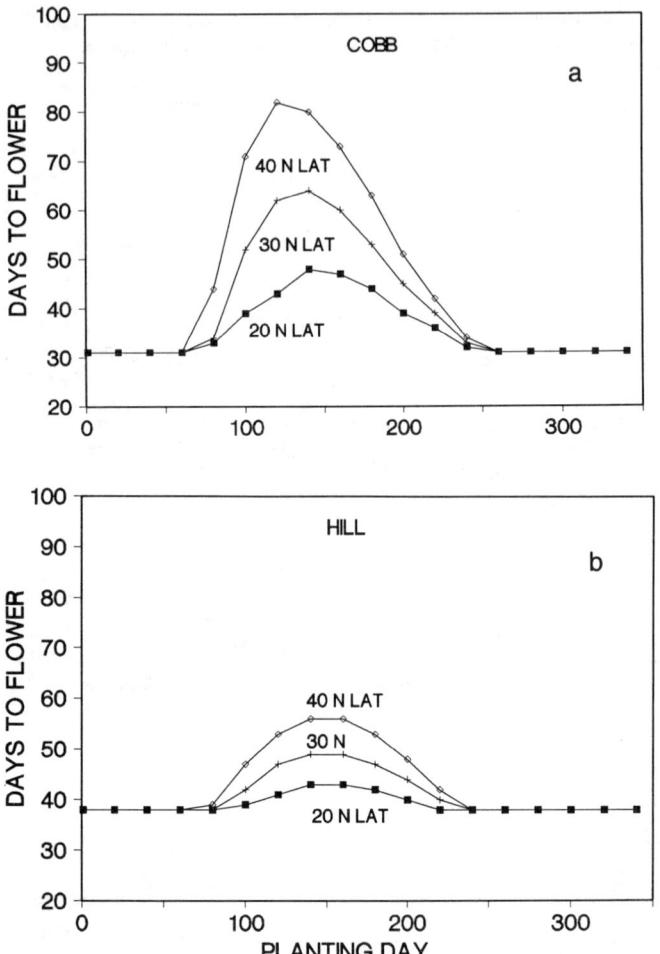

Fig. 5-7. Simulated days to flower for cultivars grown at three latitudes at a constant temperature of 27 °C: (a) Cobb (MG-VIII) and (b) Hill (MG-V).

and maturity 1.0 d earlier. The variability in time to maturity was greater than that for flowering. In one case in North Carolina, the predicted maturity was 19 d late, which again shows that there is some problem in the temperature response. Mostly, errors were < 10 d.

Sensitivity tests were conducted with the model to study the effects of latitude and cultivar interactions, and to qualitatively evaluate the results. Three cultivars from MG VIII (Cobb), V (Hill) and 0 (Evans) were used to demonstrate the effect of planting date on soybean development at various latitudes. A constant temperature of 27 °C was used for all planting dates, and latitudes of 20, 30, and 40 °N were used. Figure 5-7 shows days to flower for Cobb (a) and Hill (b) for each latitude vs. planting day. Cobb required

SOYBEAN DEVELOPMENT

Table 5-5. Locations and year of weather data used in simulating soybean development at various latitudes.

Location	Latitude	Year of weather data used
Cia-el Tigre, Venezuela	8.5°N	1986
Gainesville, FL	29.6°N	1981
Quincy, FL	30.6°N	1979
Starkville, MS	33.5°N	1985
Watkinsville, GA	34.0°N	1986
Clayton, NC	35.9°N	1985
Bedford, IN	38.5°N	1985
Urbana, IL	40.1°N	1985
Castana, IA	42.2°N	1979
Morris, MN	45.6°N	1985

31 d to flower and 77 d to mature under short days. Increases in days to flower occurred, with a peak of 85 d at 40°N lat when planted on Day of Year 120 (30 April). The increase in time to flower was skewed slightly to the left and is similar to the data on cultivar Biloxi (also MG-VIII) shown by Garner and Allard (1930). Days to maturity reached 210 for Cobb and were skewed more than flowering. Similar patterns occurred for Hill (Fig. 5-7b), although the magnitudes were reduced. Evans showed a constant 33 d to flower regardless of planting date at latitudes below 40°N and a maturity time of 77 d (data not shown). These simulations were performed at the optimal temperature (27°C) so that times to flower and maturity were minimal when days were sufficiently short or nights were above the critical night lengths.

The same three cultivars were used to study the effects of latitude on development using actual weather data. Ten locations ranging from 8.5 to 45.6°N lat (Table 5-5) were used with a planting date of June 10. Figure 5-8 shows the days to flower and to physiological maturity for Cobb (a), Hill (b), and Evans (c). Days to flower in Cobb were increased from about 35 to 90 d under the combinations of temperature and photoperiod from 8.5 to 45.6°N lat. Physiological maturity in Cobb was not reached at locations with latitudes higher than 35.9°N. Hill behaved similarly, with days to flower increasing from about 40 to 90 d and maturity not occuring for locations higher in latitude than 38.5°N. Evans, which is much less sensitive to photoperiod than the other cultivars, showed an increase in days to flower from 31 to 47 d as latitude increased (Fig. 5-8c). Days to physiological maturity increased from 77 d at the lowest latitude to 121 d at the highest latitude where freezing temperatures occurred prior to maturity. Most of these increases in days to flower and maturity by Evans were due to temperatures for latitudes below 40°N. In the latitude study described above, with 27°C constant temperatures, days to flower and maturity for Evans were constant regardless of planting date for latitudes of 40°N and lower.

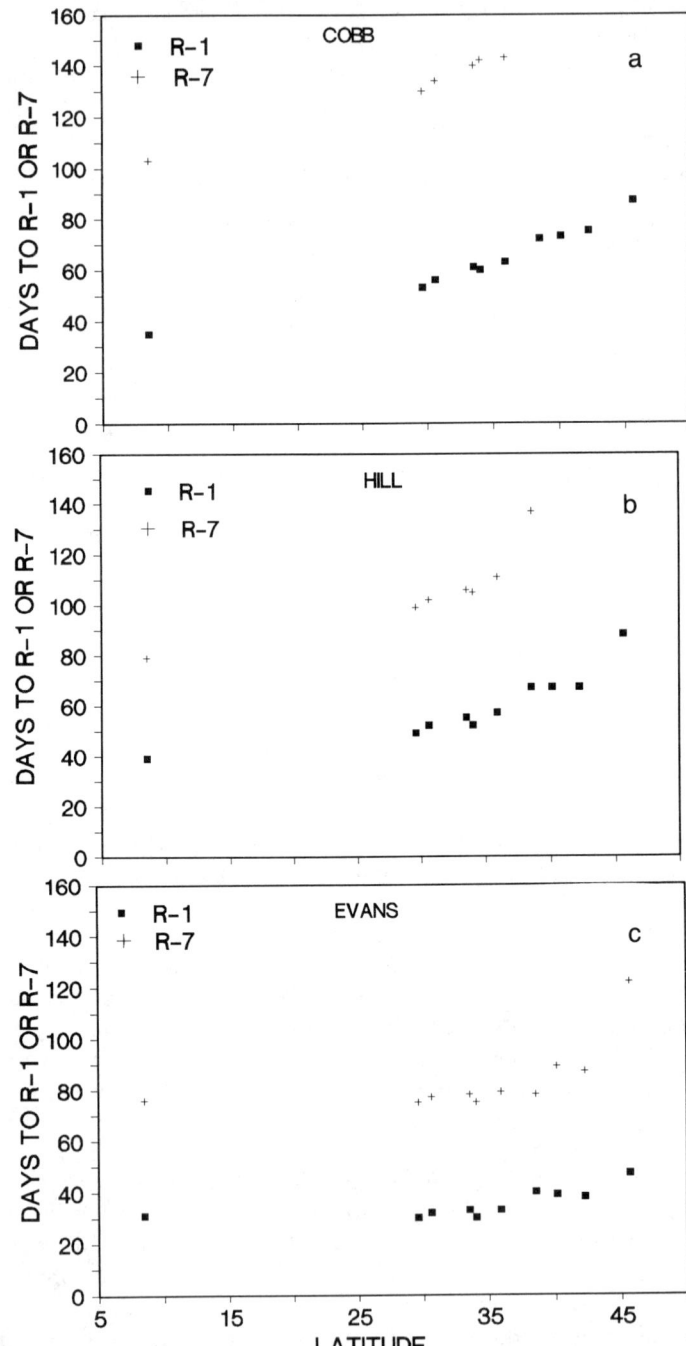

Fig. 5-8. Simulation of soybean development to flowering (■) (R1) and to maturity (+) (R7) using a 10 June planting date and actual weather data from ten locations varying in latitude from 8.5 to 45.6 °N: (a) Cobb (MG-VIII), (b) Hill (MG-V) and (c) Evans (MG-0).

V. SUMMARY

Crop simulation models are being developed for predicting growth and yield of soybean under various climate, soil, and management regimes. Areas of adaptation of soybean cultivars are largely determined by photoperiod and temperature, which control the development of the plant from sowing to maturity. Therefore, a necessary component of the crop simulation models predicts the development of the crop and the timing of important growth stages. In this chapter, a soybean development model was described. This model accumulates thermal and photothermal time based on temperature and night length functions. Stages occur when thresholds of thermal or photothermal time are reached. Differences in cultivar responses to temperature and photoperiod are predicted by determining cultivar-dependent night length coefficients and thresholds. It was assumed that all cultivars respond the same to temperature and that reproductive development responds to a different temperature function than vegetative development. Data from previous studies were used to establish both temperature functions and the shape of the night length function. Data from 3 yr of planting date studies in Quincy, FL, were used to compute the night length coefficients and thresholds for 16 cultivars using an interactive statistical procedure. The major difference among cultivars was in the night length coefficients. The model was tested for two cultivars using data from 3 yr from Missouri reported by Major et al. (1975). It was also tested for one cultivar using data from 14 crops planted in Florida, Georgia, and North Carolina. The model predicted days to flower, with average errors ranging from 1 to 9 d for the varieties and locations tested. When soybean was planted late and maturity occurred during the cool weather of late fall, the model tended to predict maturity late by as much as 19 d or at the first freeze. Improvements in the model could be made by additional research on the effects of temperature and photoperiod on development after flowering. The period of dominant seed fill is suspected to be less sensitive to photoperiod and temperature than earlier reproductive processes.

Coefficients for additional cultivars are also needed and these need to be tested at various locations. This model is a part of the soybean growth and yield model SOYGRO, but also operates alone. Previous modeling efforts, as well as the model examined here, have confirmed that predicting the development of soybean under combinations of temperature and photoperiod is possible. Refinements in these models are likely to occur and should result in even better accuracy. Improvements in the model may be obtained in several ways: (i) by new research on temperature and photoperiod effects on development after flowering, and (ii) by expanding the optimization approach to include a larger data base (more locations for each cultivar).

REFERENCES

Acock, B., V.R. Reddy, F.D. Whisler, D.N. Baker, J.M. McKinion, H.F. Hodges, and K.J. Boote. 1983. The soybean crop simulator, GLYCIM: Model documentation. U.S. Dep. of Energy Response of vegetation to carbon dioxide. Series 2, U.S. Dep. of Energy, Washington, DC.

Borthwick, H.A., and M.W. Parker. 1938. Photoperiodic perception in Biloxi soybeans. Bot. Gaz. 100:374–387.

Cregan, P.B., and E.E. Hartwig. 1984. Characterization of flowering response to photoperiod in diverse soybean genotypes. Crop Sci. 24:659–662.

Criswell, J.G., and D.J. Hume. 1972. Variation in sensitivity to photoperiod among early maturing soybean strains. Crop Sci. 12:657–660.

Cure, J.D., R.P. Patterson, C.D. Raper, Jr., and W.A. Jackson. 1982. Assimilate distribution in soybeans as affected by photoperiod during seed development. Crop Sci. 22:1245–1250.

Fehr, W.R., and C.H. Caviness. 1977. Stages of soybean development. Iowa Coop. Ext. Serv. Spec. Rep. 80.

Fisher, J.E. 1963. The effects of short days on fruitset as distinct from flower formation in soybeans. J. Exp. Bot. 41:871–873.

Garner, W.W., and H.A. Allard. 1930. Photoperiodic response of soybeans in relation to temperature and other environmental factors. J. Agric. Res. (Washington, DC) 41:719–735.

Goudriaan, J., and H.H. van Laar. 1978. Calculation of daily totals of the gross assimilation of leaf canopies. Neth. J. Agric. Sci. 26:373–382.

Hadley, P., E.H. Roberts, R.J. Summerfield, and F.R. Minchin. 1984. Effects of temperature and photoperiod on flowering in soybean: A quantitative model. Ann Bot. 53:669–681.

Hesketh, J.D., D.L. Myhre, and C.R. Willey. 1973. Temperature control of time intervals between vegetative and reproductive events in soybeans. Crop Sci. 13:250–254.

Hodges, T., and V. French. 1985. SOYPHEN: Soybean growth stages modeled from temperature, day length and water availability. Agron. J. 77:500–505.

Johnson, H.W., H.A. Borthwick, and R.C. Leffel. 1960. Effects of photoperiod and time of planting on rates of development of the soybean in various stages of the life cycle. Bot. Gaz. 122:77–95.

Jones, J.W., K.J. Boote, S.S. Jagtap, G.G. Wilkerson, G. Hoogenboom, and J.W. Mishoe. 1987. SOYGRO V5.4 Technical Documentation. Agric. Eng. Dep. Res. Rep., Univ. of Florida, Gainesville, FL.

Jones, P.G., and D.R. Liang. 1978. Simulation of the phenology of soybeans. Agric. Systems 3:295–311.

Major, D.J., D.R. Johnson, J.W. Tanner, and I.C. Anderson. 1975. Effects of daylength and temperature on soybean development. Crop Sci. 15:174–179.

McBlain, B.A., J.D. Hesketh, and R.L. Bernard. 1987. Genetic effects on reproductive phenology in soybean isolines differing in maturity genes. Can. J. Plant Sci. 67:105–116.

Meyer, G.E., R.B. Curry, J.G. Streeter, and H.J. Mederski. 1979. SOYMOD/OARDC: A dynamic simulator of soybean growth, development, and seed yield: I. Theory, structure and validation. Ohio Agric. Res. Devel. Ctr. Res. Bull. 1113.

Parker, M.W., and H.A. Borthwick. 1939. Effect of variation in temperature during photoperiodic induction upon initiation of flower primordia in Biloxi soybean. Bot. Gaz. 101:145–167.

Parker, M.W., and H.A. Borthwick. 1943. Influence of temperature on photoperiodic reactions in leaf blades of Biloxi soybeans. Bot. Gaz. 104:612–619.

Roberts, E.H., and R.J. Summerfield. 1987. Measurement and prediction of flowering in annual crops. p. 17–50. In J.G. Atherton (ed.) Manipulation of flowering. Butterworths, London.

Sinclair, T.R. 1984. Cessation of leaf emergence in indeterminate soybeans. Crop Sci. 24:483–486.

Summerfield, R.J., and E.H. Roberts. 1987. Effects of illuminance on flowering in long- and short-day grain legumes: A reappraisal and unifying model. p. 203–223. In J.G. Atherton (ed.) Manipulation of flowering. Butterworths, London.

Thomas, J.F., and C.D. Raper, Jr. 1976. Photoperiodic control of seed filling for soybeans. Crop Sci. 16:667–672.

Thomas, J.F., and C.D. Raper, Jr. 1977. Morphological response of soybeans as governed by photoperiod, temperature, and age at treatment. Bot. Gaz. 138:321–328.

Thomas, J.F., and C.D. Raper, Jr. 1981. Day and night temperature influence on carpel initiation and growth in soybean. Bot. Gaz. 142:183–187.

Wilkerson, G.G., J.W. Jones, K.J. Boote, and G.S. Buol. 1989. Photoperiodically sensitive interval in time to flower of soybean. Crop Sci. 29:721–726.

Wilkerson, G.G., J.W. Jones, K.J. Boote, K.T. Ingram, and J.W. Mishoe. 1983. Modeling soybean growth for crop management. Trans. ASAE 26:63–73.

6 Simulation of Root Growth

C. ALLAN JONES
Texas Agricultural Experiment Station
Temple, Texas

W. L. BLAND
Texas Agricultural Experiment Station
Temple, Texas

J. T. RITCHIE
Michigan State University
East Lansing, Michigan

J. R. WILLIAMS
USDA-ARS
Temple, Texas

Computer simulation of agronomic processes has become an important component of agricultural research and technology transfer. Plant root growth is one such process for which a number of models, ranging in purpose and complexity, have been developed.

Several categories of root growth models are currently available. The simplest are empirical descriptions of root distribution (Borg & Grimes, 1986; Gerwitz & Page, 1974) and predetermined patterns of root system growth (Goodwin et al., 1982). Crop models may simulate downward growth of the root system at a predetermined rate (Hansen, 1975) or include the effects of ecological factors such as soil temperature (Brouwer & deWit, 1968; Porter et al., 1986; Stone et al., 1983) and soil water potential (Hillel & Talpaz, 1976; Hoogenboom & Huck, 1986; Huck & Hillel, 1983). In contrast, the hierarchical structure of root growth in ideal physical environments has been addressed (see review in Rose, 1983). Finally, Dexter and colleagues (see citations in Dexter, 1986) have modeled the behavior of roots with respect to soil aggregates and pores in zones affected by tillage. Despite these advances in simulation of root growth, we know of no model that considers the major soil properties and crop characteristics affecting root growth. The simulation model described here is such an attempt.

This chapter provides a set of algorithms, in an operational FORTRAN program (see Appendix), that can be incorporated into existing crop simulation models to increase their sensitivity to several soil, crop, and environ-

---

Copyright © 1991 ASA-CSSA-SSSA, 677 S. Segoe Rd., Madison, WI 53711, USA. *Modeling Plant and Soil Systems*—Agronomy Monograph no. 31.

mental factors. The program is compatible with crop growth models that simulate daily root growth of an annual crop growing in a layered soil, and it uses readily available profile characteristics. The present model is, at best, applicable to mineral soils of temperate regions. Many soil situations are not addressed, such as variable-charge ion exchange capacity and wetness-dependent bulk density. The effects of soil fertility on root growth are not simulated (Barber, 1984; Troughton, 1980). Many aspects of the influence of soil characteristics on the regulation of plant root growth are not well documented in the literature, but first approximations are offered here. We hope they stimulate further experimentation and model improvement.

I. SOIL FACTORS AFFECTING ROOTING

Root growth and distribution can be limited by several properties of the soil environment. Some of these, such as soil temperature, strength, and aeration, are dynamic, changing significantly from day to day. Others, such as the presence of cemented or toxic horizons, are relatively static and may not change significantly during the growing season. In this model, effects of both dynamic and static parameters on root growth are expressed as stress factors, which range in value from 0.0 (no growth) to 1.0 (no stress). In cases in which more than one property can affect a plant process, the property with the most unfavorable stress factor is considered limiting.

A. Static Factors

1. Aluminum Toxicity

Aluminum (Al) toxicity results in swollen, stubby roots and can limit root proliferation in some acid soil horizons. Many studies have shown that the percentage of the effective cation exchange capacity occupied by Al is a good index of Al toxicity (Abruna et al., 1982; Brenes & Pearson, 1973; Farina et al., 1980; Gonzalez-Erico et al., 1979; Pavan et al., 1982). Even when the plow layer has been limed, Al toxicity in subsoil horizons can limit root depth and render crops susceptible to drought stress (Adams & Moore, 1983; Bouldin, 1979; Brenes & Pearson, 1973; Pavan et al., 1982). Because crops and cultivars vary in their response to Al toxicity (Fageria, 1982; Foy et al., 1972, 1974, b; McLean & Gilbert, 1927; Mugwira et al., 1980; Reid et al., 1969), root growth models should consider both Al saturation (ALS) and crop sensitivity to Al toxicity.

Aluminum saturation is calculated as KCl-extractable Al (EAL) divided by effective cation exchange capacity (CEC). The CEC is calculated as the sum of NH_4OAC-extracted bases (SMB) plus EAL (Soil Survey Staff, 1972). Estimates of EAL and SMB for each soil layer must be provided. The stress factor for Al saturation (SAL) is calculated as

$$SAL = \frac{ALX - ALS}{ALX - ALA}, \quad ALA \geq ALS \geq ALX \qquad [1]$$

$$SAL = 0, \quad ALS > ALX \qquad [2]$$

$$SAL = 1, \quad ALS < ALA \qquad [3]$$

where ALA and ALX are the threshold values of Al saturation at which root growth is first affected and is completely inhibited, respectively. Table 6-1 provides representative values of ALA and ALX for maize (*Zea mays* L.), sorghum [*Sorghum bicolor* (L.) Moench], and wheat (*Triticum aestivum* L.).

2. Calcium Deficiency

Calcium deficiency results in decreased root growth (Ritchey et al., 1982) and, in contrast to Al toxicity, straight, small-diameter roots with brown tips (Howard & Adams, 1965). Subsoils of highly weathered acid soils are often low in exchangeable Ca, resulting in reduced root depth and root proliferation in these soil layers because plants cannot translocate Ca downward in the phloem. Cotton root growth is reduced when Ca saturation (CAS), calculated as exchangeable Ca/CEC, is below about 15% (Adams and Moore, 1983; Howard and Adams, 1965). The relative sensitivity of root growth in a layer to Ca (SCA) is then calculated as

$$SCA = \frac{CAS - CAX}{CAA - CAX}, \quad CAA \geq CAS \geq CAX \qquad [4]$$

$$SCA = 0, \quad CAS < CAX \qquad [5]$$

$$SCA = 1, \quad CAS > CAA \qquad [6]$$

where CAA and CAX are user-specified threshold values of Ca saturation at which root growth begins to be affected and is completely inhibited, respectively. Although little is known about genotypic variation in these parameters, the existence of calcicoles and calciphobes suggests interspecific variation in sensitivity to Ca nutrition.

Table 6-1. Representative values of Al saturation above which root growth is reduced (ALA) and completely inhibited (ALX) for maize, sorghum, and wheat.

Crop	ALA	ALX	Source
	%		
Maize	55	95	Brenes and Pearson, 1973
	43	90	Gonzalez-Erico et al., 1979
Sorghum	40	90	Brenes and Pearson, 1973
Wheat			C.D. Foy, 1984, pers. comm.
Sensitive	15	40	
Tolerant	50	100	

3. Coarse Fragments

Coarse fragments (2–250 mm diam.) reduce volumetric water holding capacity and nutrient availability in approximate proportion to their volumetric fraction in the layer. The effects of coarse subsoil horizons on rooting have been quantitatively evaluated in only a few studies (Babalola & Lal, 1977; Vine et al., 1981). Results suggested that root proliferation in layers with varying fractions of coarse fragments is approximately proportional to the volume fraction of soil material <2 mm. Therefore, the relative effect of coarse fragments in a soil layer (SCF) on root growth is calculated as

$$SCF = 1.0 - ROK \qquad [7]$$

where ROK is the volume fraction of particles >2 mm diam.

4. Qualitative Constraints

Quantitative information on bulk density and coarse fragment content is sometimes unavailable, yet qualitative descriptions of a horizon may indicate that roots will have great difficulty penetrating it. For example, we can reasonably assume that horizons described as lithic, paralithic, petrocalcic, petroferric, petrogypsic, duripan, or fragipan will greatly inhibit rooting (Grossman & Berdanier, 1982). Even in the absence of such horizon designations a layer described as structureless, massive, with extremely firm consistence will likely slow root penetration and proliferation.

If such horizons occur, the user can reduce root growth in the layer by setting the soil horizon code (SCD) to a value near zero, rather than its normal value of 1.

B. Dynamic Factors

Soil water content and temperature can change dramatically during the course of the growing season, so their effects on root system growth should be estimated frequently. For the program described here, soil layer water contents and temperatures are input daily and used to estimate soil strength, aeration, and temperature effects on root growth. In a whole-plant model, soil temperature and wetness would likely be calculated daily for use in the algorithms discussed here.

1. Strength

Numerous studies have shown that root growth is approximately linearly related to soil strength (Cockroft et al., 1969; Taylor & Gardner, 1963; Gerard et al., 1982; Taylor et al., 1966; Vepraskas & Miner, 1986). Three important determinants of soil strength are bulk density, texture, and water content (Eavis, 1972; Gerard et al., 1982; Monteith & Banath, 1965; Taylor et al., 1966). The model calculates a strength stress factor from these three parameters.

Jones (1983) has shown that at near-optimal water contents, soil texture can be used to estimate the bulk densities at which root growth is either unaffected or is severely affected. For ten studies of root growth in compacted soils at near-optimum soil water contents, sand content (SAN) of the soil, measured as percent by weight, was a good predictor of the moist bulk density at which rooting was severely impaired (BDX). Roots were not usually observed in soil layers with moist bulk densities exceeding BDX, calculated as

$$\text{BDX} = 1.6 + 0.004 \times \text{SAN} \qquad [8]$$

Several studies have shown that, at near-optimum water contents, soil bulk densities must be increased by at least 0.4 Mg/m^3 in order to reduce root growth from its optimum to near zero (Asady et al., 1985; Monteith & Banath, 1965; Tackett & Pearson, 1964; Trouse, 1965; Voorhees et al., 1975; Zimmerman & Kardos, 1961). Based on this observation and data from Fig. 2 of Jones (1983), the bulk density at which there is no inhibition due to this parameter in moist soil (BDO) is calculated as

$$\text{BDO} = 1.1 + 0.005 \times \text{SAN} \qquad [9]$$

In the model, a factor for the interacting effects of texture and bulk density of moist soil (SBD) is calculated as a scaled, linear function (Gerard et al., 1982; Taylor et al., 1966; Taylor & Gardner, 1963) of the bulk density of the layer (BD), BDO, and BDX as

$$\text{SBD} = \frac{\text{BDX} - \text{BD}}{\text{BDX} - \text{BDO}}, \qquad \text{BDO} \leq \text{BD} \leq \text{BDX} \qquad [10]$$

$$\text{SBD} = 0, \qquad \text{BD} > \text{BDX} \qquad [11]$$

$$\text{SBD} = 1, \qquad \text{BD} < \text{BDO} \qquad [12]$$

Drying of the soil below the moist condition for which SBD was determined will increase the soil strength (Bar-Yosef & Lambert, 1981; Cassel, 1982; Cruz et al., 1986; Taylor & Gardner, 1963) and thereby decrease rooting. We assume that, at a given BD, maximum root growth occurs at the drained upper limit (UL) of plant-extractable water and no growth occurs at the lower limit (LL). In the early stages of drying, there is little growth inhibition, but the effect becomes more important as soil water decreases; a sine function is proposed as a first approximation of the relationship. The following equations are used to calculate a fractional-layer water content (LWF) from the UL, LL, and the soil water content of the layer (LW). The stress factor for soil strength (SST) is calculated from relative root growth at near-optimal water contents (SBD) and LWF, as

$$\text{LWF} = \frac{\text{LW} - \text{LL}}{\text{UL} - \text{LL}}, \qquad \text{UL} \geq \text{LW} \geq \text{LL} \qquad [13]$$

$$\text{LWF} = 0, \quad \text{LW} < \text{LL} \qquad [14]$$

$$\text{LWF} = 1, \quad \text{LW} > \text{UL} \qquad [15]$$

$$\text{SST} = \text{SBD} \times \sin(1.57 \times \text{LWF}) \qquad [16]$$

where (1.57 × LWF) is a radian measure.

We note that this treatment of the effects of soil strength on root penetration does not address the role of soil structure. In structured soils with highly cohesive peds, structure and macroporosity may greatly influence root penetration (Khalifa & Buol, 1969; Wang et al., 1986). The model does, however, recognize that structure can have an impact on progression of the root front (see Eq. [24]).

2. Aeration

Numerous studies have shown that root growth is sensitive to soil aeration. Soil characteristics such as porosity, water content, biotic activity, temperature, surface water movement, and continuity of air-filled pores affect the development of suboptimal oxygen concentrations (Drew, 1983; Grable, 1966). Additionally, crops differ in their ability to tolerate soil conditions unfavorable to adequate aeration in the rhizosphere. For example, rice (*Oryza sativa* L.) and many marsh plants develop continuous air spaces (aerenchyma) in the root cortex, which facilitate axial movement of O_2 from the atmosphere to the living root tissue and the rhizosphere. The extent of aerenchyma formation, its effects on axial O_2 diffusion, the diffusion of O_2 out of the root, and the differences in metabolic adaptation to poor aeration all affect genotypic response to poorly aerated soils (Jackson & Drew, 1984). A comprehensive model of the effects of soil aeration on root growth would be very complex; the equations proposed here consider only the effects of soil water content, bulk density, texture, and genotype.

The effect of genotype on reduction of root growth due to flooding is introduced with a coefficient describing the relative root growth of the plant under completely saturated soil conditions (SFT). It is analogous to other stress factors in that if SFT = 1.0, there is no effect of flooding on root growth. There is no growth in saturated soil when SFT = 0. As a first approximation, SFT can be set to zero for flood-sensitive crops such as maize, cotton (*Gossypium hirsutum* L.), soybean [*Glycine max* (L.) Merr.], wheat, and sunflower (*Helianthus annus* L.), and to 1.0 for rice and other paddy crops. Sensitivity of a genotype to flooding may depend on the stage of growth (Cannell & Jackson, 1981), but this is not recognized in our treatment.

The fraction of total soil volume occupied by air has been used as an index of soil aeration, primarily because it is easily calculated from soil bulk density and water content, and because of its relationships to soil redox potential, O_2 concentration, and O_2 diffusivity. It is often well correlated with root growth rates in a particular soil at a particular bulk density (Grable & Siemer, 1968; Voorhees et al., 1975). However, this relationship varies significantly with bulk density and among soils.

Linn and Doran (1984) found that the fraction of water-filled pores (WFP, defined as water-filled porosity/total porosity) was well correlated with rates of several aerobic microbiological processes for soils differing significantly in either bulk density or water content. These processes were inhibited when WFP exceeded about 0.6. A similar critical water-filled porosity (CWP) can be obtained for relative root growth from studies on a variety of crops and soils (Fig. 6-1). Root growth is apparently restricted by poor aeration at slightly higher WFP in clayey soils than in sandy soils, as indicated by Fig. 6-1, perhaps because a large fraction of the water in fine-textured soils does not directly participate in plant growth. As a result, we calculate a layer aeration factor (SAI) based on WFP, CWP, and SFT as

$$CWP = 0.4 + 0.004 \times CLA \qquad [17]$$

$$SAI = SFT + (1 - WFP) \frac{(1 - SFT)}{(1 - CWP)} \qquad WFP \geq CWP \qquad [18]$$

$$SAI = 1, \qquad WFP < CWP \qquad [19]$$

3. Temperature

Low soil temperature may limit root growth, especially at locations where subsoil layers warm slowly in the spring (Taylor, 1983). The program simulates this effect by calculating a stress factor for temperature (STP) for each soil layer. The calculation is derived from the temperature of the layer (LT) and two genotype-dependent parameters; the minimum temperature at which growth occurs (TBS) and the optimum temperature for root growth (TOP) as

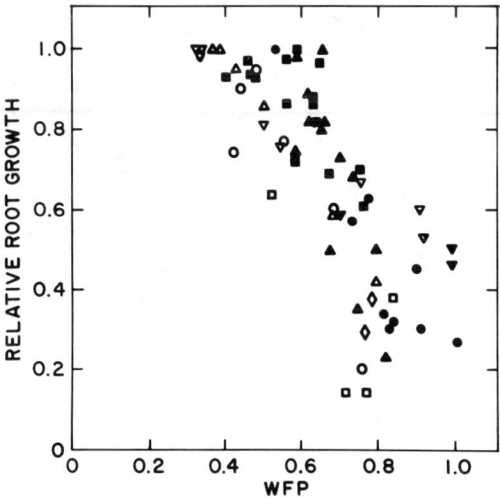

Fig. 6-1. Effect of water-filled pore space (WFP) on relative root growth in several studies (● Voorhees et al., 1975, pea [*Pisum sativa* L.], sandy soil; ○ pea, clayey soil; ◇ Bar-Yosef and Lambert, 1981, maize, sandy soil; □ cotton, sandy soil; △ Cornish et al., 1984, rye [*Lolium perenne* L.], sandy soil; ■ Grable and Siemer, 1968, corn, clayey soil; ▲ Pearson et al., 1970, corn, clayey soil; ▽ Eavis, 1972, pea, sandy soil).

$$\text{STP} = \sin 1.57 \times \frac{\text{LT} - \text{TBS}}{\text{TOP} - \text{TBS}}, \quad \text{LT} \geq \text{TBS} \quad [20]$$

$$\text{STP} = 0, \quad \text{LT} < \text{TBS} \quad [21]$$

Choice of the sine function was based on data from Stone and Taylor (1983) and the discussion in Voorhees et al. (1981).

II. ROOT DISTRIBUTION

The model simulates four processes that determine root distribution in the soil: (i) the increasing depth of the rooting front; (ii) the length/weight ratio of new roots; (iii) proliferation within soil layers; and (iv) senescence. Growth stage (GS) is used as a time base of the model which varies from 0.0 at planting to 1.0 at maturity. In this model, GS is supplied daily, but in a whole-plant model it would probably be calculated as a function of accumulated thermal time or some other expression of ontogenetic stage.

A. Depth of Rooting Front

The rooting front descends from the depth of planting (PD) at germination to an ultimate depth that is dependent on genotype and the rooting environment. Genotype-specific model inputs include the maximum rooting depth (RDX) and the growth stage at which the root system ceases to increase in depth in deep soils without rooting constraints (GSR). Borg and Grimes (1986) provided ranges of maximum rooting depths for 55 crops. Cereals usually reach their maximum rooting depth during grain filling (GSR = 0.6 to 0.9), although the root systems of crops such as sorghum may continue to descend until harvest (GSR = 1.0). We emphasize that RDX may be greater for long-season than for short-season cultivars because long-season genotypes require more thermal time prior to grain filling.

The potential daily increase in root depth (DDI) is calculated as

$$\text{DDI} = \text{RDX} \times \frac{\text{GS} - \text{GSY}}{\text{GSR}}, \quad \text{GS} < \text{GSR} \quad [22]$$

where GSY is the previous day's GS value. Some users may wish to modify Eq. [22] to make DDI a function of daily thermal time (Jones & Kiniry, 1986). This would eliminate the need to specify RDX and allow photoperiod-sensitive crops to develop deeper root systems when daylength increases the duration of the vegetative growth.

The actual daily increase in root depth (DRD) depends on physical and chemical constraints in the soil layer in which the root front is growing. Stress factors are weighted according to how they are expected to impact vertical penetration. The factors are calculated using two active stress factors (ASF1, ASF2) as

$$ASF1 = \min(STP, SCA, SAL, SCD) \qquad [23]$$

$$ASF2 = \min(SST, SAI, SCF)^{0.5} \qquad [24]$$

where the function min(arguments) returns the smallest value in the argument list. The stresses SST, SAI, and SCF are given reduced influence because of the possible compensating effects of vertical cracks and pores, which may facilitate downward growth of the root front (Taylor, 1983). The value of DRD is calculated as

$$DRD = DDI \times \min(ASF1, ASF2) \qquad [25]$$

and root depth (RD) is incremented by DRD. If the sum of RD + DRD reaches beyond the bottom of the soil layer containing the rooting front, DRD is calculated again using the unused portion of DDI and ASF1 and ASF2 from the next layer, until DDI is exhausted. This treatment of the rooting front rate of downward movement does not address possible plant-related phenomena. When soil is dry near the surface, some crops may accelerate downward root growth to collect more water (Taylor & Klepper, 1974). Conversely, the rate of cotton downward root growth may slow, perhaps as a survival mechanism (Ritchie and Burnett, 1971).

B. Proliferation

Root proliferation in a soil layer is assumed to be governed by the crop's genetically determined tendency to distribute roots in the soil, the length/weight ratio of new roots, the root length presently in the layer, the amount of dry matter translocated from the shoot, and soil physical or chemical constraints in the layer.

Even in the absence of soil physical and chemical constraints, the root system is often concentrated near the surface. This is due to the longer time available to roots to grow and branch near the soil surface, the development of nodal roots in monocots, and the often more favorable chemical and physical environment near the surface. However, plants also display inherent root growth habits (Kramer, 1983). Several classical papers have described genotypic differences in root system morphology (Evans, 1936; Venkatraman & Thomas, 1922; Weaver, 1958), and the subject was recently reviewed by O'Toole and Bland (1987). In this model, the tendency of a genotype to distribute roots is expressed by a layer-weighting factor (WFL) calculated for each rooted layer as

$$WFL = (1.0 - ZA/3.0)^{WCG} \qquad [26]$$

where ZA is the depth to the middle of the layer and WCG is a genotype-specific coefficient. If the crop tends to distribute root growth equally among all soil layers, WCG = 0. We find that WCG values between 1.0 and 3.0 provide realistic estimates of the normally exponential decrease of rooting with depth (Gerwitz & Page, 1974). We have used values of 1.0 for crops such as sunflower with relatively deep, uniformly distributed root systems, 2.0 for maize and soybean, and 3.0 for sorghum, with roots often concentrated near the soil surface. The value of 3.0 in Eq. [26] is also a shape factor that can be varied to alter the predicted rooting habit. This value must be greater than RDX.

The length to weight ratio (LWA) is a function of crop species, growth stage, soil depth, and soil physical and chemical properties (Anderson, 1987; Barber, 1971). For dicots the length/weight ratio in a layer decreases over time, due to secondary thickening as well as the death of fine branches. Even grasses that have fibrous roots with no secondary thickening produce roots of varying thicknesses and branching, depending on the nodes from which they arise, their age, and other factors (Evans, 1935). On average, the lowest length/weight ratios are found near the soil surface where large-diameter roots that have lost their finer branches are often found (Derera et al., 1969; Follett et al., 1974; Retta et al., 1984). Representative values of root length/weight ratios are given in Table 6-2. The extreme variation in this ratio within and among studies suggests that more careful field measurements are needed.

Table 6-2. Representative values of root length/weight ratios for maize, sorghum, soybean, rice, and wheat.

Crop	Cultural condition	Length/weight	Source
		m/g	
Maize	Field grown, various ages	3-173	Allmaras et al. (1975)
		28-83	Anderson (1987)
	Field grown, silking	1-29	Follett et al. (1974)
	Field grown	18-53	Barber (1971)
	Solution culture, 20-22 d	118-218	Nielsen and Barber (1978)
	16 d	141-290	
Soybean	Field grown, various ages	6-80	Allmaras et al. (1975)
	Solution culture, 18 d	164	Barber (1979)
	Sand culture, 18 d	119	
Sorghum	Field grown, irrigated, 50% bloom	34-116	Retta et al. (1984)
	Field grown, unirrigated, 50% bloom	61-150	
	Sand culture, 12-24 d	128-230	
Rice	Paddy, 4 wk		J.C. O'Toole (1986, unpublished data)
	Total roots	204-334	
	Roots below 0.30 m	355-651	International Rice Research Institute (1978)
	Paddy, heading		
	Total roots	163-440	
	Roots below 0.30 m	285-973	
Wheat	Irrigated, anthesis		Derera et al. (1969)
	0-0.15 m	180-514	
	0.15-0.30 m	187-615	
	0.30-0.61 m	236-609	
	0.61-0.92 m	300-914	
	0.92-1.22 m	275-836	
	Field grown, various ages		Gregory et al. (1978)
	0-0.5 m	240	

The model simulates these processes by allowing the user to specify two genotype-specific variables, the length/weight ratio of seedling roots (LWS) and that of the mature root system (LWM) growing in the topsoil under ideal conditions. Because they represent root growth under ideal conditions, we have chosen values similar to those measured in solution culture (Nielsen & Barber, 1978). The normal length/weight ratio (LWN) of roots growing in a layer is calculated from growth stage and the average depth of the layer (ZA) relative to root system depth (RD) (Derera et al., 1969; Nielsen & Barber, 1978) as

$$LWN = LWS - GS \times (LWS - LWM) \times (1.0 - ZA/RD) \quad [27]$$

Thus, the normal length/weight ratio approaches that of the seedling root when the growth stage is small or when the depth to the middle of the layer approaches the root depth; the normal length/weight ratio goes to the ratio of the root system near the soil surface near maturity when the growth stage = 1.0.

High soil strength and coarse fragment content physically constrain the root system and cause the plant to produce poorly branched, thickened roots. High Al saturation has a similar effect, as mentioned earlier. The model accounts for these effects when it calculates the actual length/weight ratio (LWA) of roots growing in a layer as

$$LWA = \frac{LWN}{1.0 + 3.0(1.0 - ASF3)} \quad [28]$$

where ASF3 is the minimum of SST, SAL, and SCF for the layer. Note that the denominator can reduce the actual length/weight ratio to as little as 0.225 of normal. This is equivalent to doubling root diameter due to soil physical or chemical constraints. Greater effects can be observed in individual roots, and this may be a conservative estimate of the plasticity of root diameter.

The daily potential for root weight increase in a layer (GPL) is a function of the present root length density (RLV), modified by the minimum of all the stress factors in the layer (ASF), the layer-weighting factor, the actual length/weight ratio, and the thickness of the soil layer (DZ) as

$$GPL = \frac{\{[5.0 \times RLV/(0.025 + RLV)] \times ASF \times WFL \times DZ\}}{LWA} \quad [29]$$

In this expression, potential rate of root length increase is predicted by a term in the form of the Michaelis-Menton equation, so the rate of root proliferation in a layer is a function of the present root length density (RLV) up to a maximum rate, here 5.0 km root/m^3 soil. The factors 5.0 and 0.025 could be adjusted to provide more accurate simulations, although reasonable results have been obtained with a range of values.

The values of the daily potential for root weight increase in a layer (GPL) are summed over the rooted profile (GPS) and root growth in the layer (GAL) is determined by partitioning the dry matter allocation to the root system for the day (DMD) as

$$GAL = DMD \times (GPL/GPS) \times LWA, \qquad GPS > DMD \qquad [30]$$

$$GAL = GPL \times LWA, \qquad GPS \leq DMD \qquad [31]$$

When soil stresses are severe throughout much of the profile, actual root system growth (the sum of GAL/LWA for all layers) may be less than the dry matter partitioned to the root system (DMD). This apparent loss of carbon could be partially ascribed to the extra energy costs of growing and maintaining roots in stress environments (Smucker, 1984). In a whole-plant model, this excess allocation could be used as feedback to the shoot to reduce dry matter partitioned to the root system in the future.

C. Senescence

Death of some roots during the life of a crop is a frequently observed phenomenon. We have included this loss of carbon in the model because of its implications for living root length. Except for growth, other photoassimilate costs of roots (Lambers, 1987) have not been explicitly addressed, since the emphasis of this model is on predicting the root length in given layers of soil. As a first approximation for the normal rate of root system death, we have used 1.0% of the dry weight in a layer per day. In addition, the model allows low soil moisture (Taylor & Klepper, 1974) or poor aeration (Jackson & Drew, 1984) to increase the rate of root death by a factor up to 2.0. The combined effects of the layer soil-water factor (Eq. [13-15]), the layer aeration factor (Eq. [18-19]) and root weight in the layer (RWL) on root death (DWL) are calculated as

$$DWL = 0.01 \times RWL \times [1 + \max(1 - LWF, 1 - SAI)] \qquad [32]$$

where max (arguments) is the maximum value in the argument list.

The user also specified the growth stage at which normal root senescence due to whole-plant senescence begins in determinate crops (GSD). When the growth stage is greater than GSD, an additional increment of root death is added to the calculation as

$$DWL = DWL + RWL + \left(\frac{GS - GSD}{1.0 - GSD}\right)^{3.0} \qquad [33]$$

This function causes rapid root system senescence as the crop reaches physiological maturity (Mengel & Barber, 1974). Experiments are needed to differentiate this proposed acceleration of senescence from a drastic decline in dry matter partitioned to the root system.

The total root length (RLL) and weight (RWL) in each layer are then incremented by the day's growth and decremented by senescence. Because the actual length/weight ratio varies daily, the length/weight ratio of all roots in the layer (LWR) is then updated as RLL/RWL.

III. MODEL INPUTS

Three types of model inputs are required: (i) plant characteristics; (ii) the soil's chemical and physical characteristics by horizon; and (iii) daily inputs describing crop development, dry matter partitioned to the root system, soil water, and soil temperature. Planting depth (PD) is also required.

A. Crop Characteristics

The user specifies the following plant characteristics of the root system, most of which are genetically regulated to some extent: ALA, ALX, CAA, CAX, GSD, GSR, LWM, LWS, RDX, SFT, TBS, TOP, and WCG.

B. Soil Characteristics

The model requires two types of soil data: (i) stable soil characteristics, which are assumed to remain constant during the cropping season; and (ii) dynamic properties (temperature and water content by layer), which can change during the season. Stable soil characteristics are listed below and must be provided for each soil layer (I).

Z(I)   Depth to the bottom of the soil layer (m)
BD(I) Bulk density (Mg/m^3)
SAN(I) Sand content (percent by weight of fine earth material)
SIL(I) Silt content (percent by weight of fine earth material)
ROK(I) Coarse fragment content (volume fraction of total soil volume of particles between 2 and 250 mm diam.)
SMB(I) Sum of bases (cmol/kg)
EAL(I) Extractable Al (cmol/kg)
CA(I) Exchangeable Ca (cmol/kg)
UL(I) Drained upper limit of soil-water holding capacity (m/m)
LL(I) Lower limit of plant-extractable water (m/m)
SCD(I) Code to indicate root-limiting horizon. Can be set to zero if the horizon is classified as duripan, fragipan, lithic, paralithic, petrocalcic, petroferric, petrogypsic, or skeletal. Can be set to low value if structure and consistency (e.g., massive and extremely firm) suggest minimal root penetration. Otherwise, it is set to 1.

C. Daily Inputs

Brouwer (1962, 1963) proposed a theory to explain the functional equilibrium between aboveground and below-ground organs of plants. It is used to explain why removal of parts of the shoot, low light intensity, and nitrogen and phosphorus deficiencies reduce root growth more than shoot growth. Although recent work casts doubt on the causal relationships suggested by Brouwer (Lambers, 1983; Simpson et al., 1982), there is little doubt that a functional equilibrium exists and regulates the relative magnitudes of shoot and root growth.

Several root growth models simulate the functional equilibrium (Hoogenboom & Huck, 1986; Huck & Hillel, 1983; Reynolds & Thornley, 1982; Thornley, 1972a, b). Most crop growth simulation models recognize the phenomenon and attempt to simulate its effects on dry matter partitioning. We have assumed that dry matter partitioning is simulated by the shoot growth component of any crop growth model that may use the algorithms described here. Therefore, this root growth model requires an input of daily dry matter allocation to the root system (DMD). It then partitions that dry matter or, if soil conditions are unfavorable, a part of that dry matter throughout the soil profile.

The model requires the following daily inputs:

JDATE Day of the year (1-365).
GS Crop growth stage as a fraction of the total phenological (thermal) time from germination to physiological maturity (0-1).
DMD Dry matter allocation to the root system on the day (kg/ha).
LT(I) Mean temperature at the center of soil layer I (°C).
LW(I) Volumetric water content of layer I (m^3/m^3).

IV. MODEL BEHAVIOR

This section describes simulation of the root behavior of a standard maize crop on several soils with a variety of chemical and physical characteristics. It is not possible to evaluate the root submodel rigorously or independent of the total model of which it would be a component. However, we can demonstrate model response to a few important soil properties and management practices known to affect root growth. For all simulations, the daily estimates of crop growth stage (Fig. 6-2) and potential dry matter allocation to the root system (Fig. 6-3) were identical. The CERES-Maize model (Jones & Kiniry, 1986) was used to simulate soil layer temperatures and water contents for each soil using Temple, TX weather data from 1982. Soil input data for each soil were obtained from the USDA Soil Conservation Service National Soil Survey Pedon Data Base (Nat. Soil Survey Lab., Lincoln, NE) and are given in Table 6-3. The genotype-specific inputs used for the simulations are given in Table 6-4.

A. Root System Depth

Root system depth is simulated for Siliwa soil (fine-loamy, siliceous, thermic Ultic Haplustalfs) at its measured bulk densities and with bulk den-

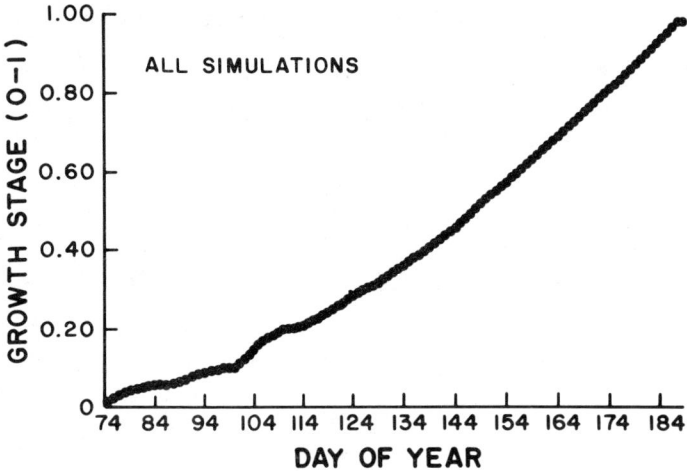

Fig. 6-2. Growth stage over time for all simulations described.

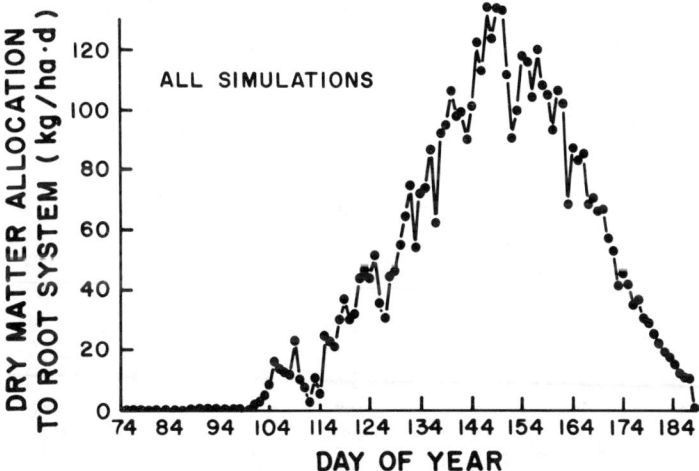

Fig. 6-3. Daily dry matter allocation to the root system for all simulations described.

sities reduced to that at the surface (Fig. 6-4). Simulated root system depth increases slowly at first due to low soil temperatures, increases more rapidly as the profile warms, then stops when growth stage reaches GSR. The resulting increase of root depth with time is similar to that described by Borg and Grimes (1986) for many crops. When the surface layer bulk density of the Siliwa soil profile is used in place of the greater actual values, simulated root depth increases more rapidly and reaches a greater maximum depth.

B. Root Length Density and Length/Weight Ratio

Maize root length densities (RLV) normally increase with time up to a maximum before physiological maturity of the crop. They may decline as

Table 6-3. Model inputs for Siliwa, Houston Black, Norfolk, and Chenango soils.

Layer	Depth to bottom	Sand	Silt	Bulk density	Coarse fragments	Sum of bases	Extractable Al	Exchangeable Ca	Drained upper limit	Lower limit
	m	% fine earth		mg/m^3	% volume	cmol/kg			m/m	
					Siliwa					
1	0.13	65.4	16.1	1.50	0.0	10.1	0.0	8.1	0.260	0.140
2	0.38	51.1	15.2	1.58	0.0	15.1	0.0	13.0	0.328	0.231
3	0.63	54.4	18.4	1.61	0.0	11.7	0.0	13.0	0.300	0.198
4	0.89	54.4	18.4	1.61	0.0	11.7	0.0	13.0	0.300	0.198
5	1.22	60.4	17.9	1.70	0.0	9.7	0.0	7.7	0.279	0.181
6	1.52	66.6	17.9	1.74	0.0	6.8	0.0	5.1	0.252	0.151
7	1.83	70.0	17.0	1.72	0.0	6.8	0.0	4.4	0.237	0.131
8	2.13	76.4	14.3	1.68	0.0	5.5	0.0	3.8	0.199	0.069
					Houston Black					
1	0.18	7.3	35.7	1.10	0.0	48.5	0.0	45.6	0.447	0.327
2	0.48	5.4	39.3	1.20	0.0	48.5	0.0	45.6	0.433	0.302
3	0.71	4.9	37.1	1.25	0.0	48.7	0.0	45.6	0.444	0.323
4	0.91	3.8	36.8	1.30	0.0	48.7	0.0	46.0	0.447	0.333
5	1.12	6.0	35.1	1.26	0.04	49.2	0.0	45.9	0.443	0.325
6	1.35	6.4	38.2	1.30	0.01	44.1	0.0	40.9	0.423	0.304
7	1.60	5.7	40.2	1.36	0.01	42.9	0.0	39.9	0.419	0.305
8	1.88	6.6	41.9	1.32	0.01	40.7	0.0	39.9	0.404	0.281
9	2.13	7.4	45.9	1.30	0.01	33.0	0.0	39.9	0.477	0.244
10	2.87	11.6	50.4	1.53	0.0	20.2	0.0	18.7	0.354	0.238
					Norfolk					
1	0.23	86.2	11.0	1.20	0.0	0.65	1.2	0.9	0.169	0.027
2	0.30	80.7	12.3	1.75	0.0	0.4	0.4	0.2	0.177	0.050
3	0.43	80.7	12.3	1.56	0.0	0.4	0.4	0.2	0.177	0.050
4	0.67	70.3	10.7	1.63	0.0	1.3	0.6	0.8	0.260	0.154
5	0.91	70.3	10.7	1.63	0.0	1.3	0.6	0.8	0.260	0.154
6	1.27	58.5	9.4	1.62	0.0	1.4	0.9	1.0	0.318	0.225
7	1.52	60.0	9.9	1.62	0.0	0.7	1.2	0.5	0.309	0.213
8	1.83	62.8	9.6	1.62	0.0	0.7	1.2	0.5	0.293	0.194
					Chenango					
1	0.18	23.5	63.8	1.29	0.15	6.5	0.0	5.6	0.255	0.127
2	0.30	21.1	67.8	1.30	0.15	1.9	0.0	1.3	0.221	0.068
3	0.51	26.4	63.3	1.31	0.45	1.9	0.0	1.3	0.138	0.039
4	0.76	41.5	46.7	1.29	0.60	2.5	0.0	1.3	0.100	0.028
5	1.03	68.0	20.9	1.30	0.65	4.5	0.0	3.3	0.081	0.018
6	1.30	68.0	20.9	1.30	0.65	4.5	0.0	3.3	0.081	0.018
7	1.57	68.0	20.9	1.30	0.65	4.5	0.0	3.3	0.081	0.018
8	1.83	68.0	20.9	1.30	0.65	4.5	0.0	3.3	0.081	0.018
9	2.29	66.3	16.5	1.25	0.75	0.2	0.0	0.0	0.060	0.020
10	254	89.0	8.0	1.37	0.70	0.1	0.0	0.0	0.042	0.009

Table 6-4. Maize genetic inputs.

Variable[†]	Value	Variable	Value
ALA	43.0	LWS	200.0
ALX	90.0	RDX	2.5
CAA	15.0	SFT	0.0
CAX	0.0	TBS	8.0
GSD	0.8	TOP	25.0
GSR	0.7	WCG	2.0
LWM	100.0		

[†] See crop characteristic definitions in computer program listing.

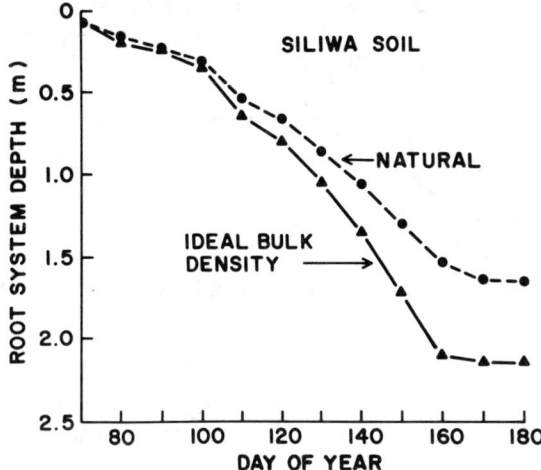

Fig. 6-4. Simulated maize root system depth over time for irrigated and nonirrigated Siliwa soil for natural profile bulk densities and for surface horizon bulk density throughout the profile.

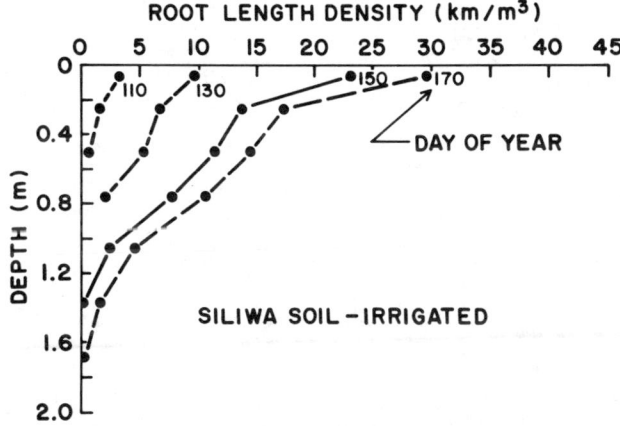

Fig. 6-5. Simulated maize root length densities on four dates for irrigated Siliwa soil.

physiological maturity approaches (Mengel & Barber, 1974). Under well-watered conditions, root length density normally declines exponentially with depth, although root death due to suboptimal soil water can dramatically reduce root length density in surface layers (Allmaras et al., 1975).

The changes in simulated RLV with time and depth on Siliwa soil are shown in Fig. 6-5 (irrigated) and Fig. 6-6 (nonirrigated). For both simulations, root length density declines roughly exponentially with depth and increases with time up to Day of Year 150. Depletion of soil water near the surface in the nonirrigated simulation after Day of Year 150 caused early senescence of root length. Simulated RLV in the top layer of the irrigated soil increased from Day of Year 150 to 170, and then began to decrease due to senescence.

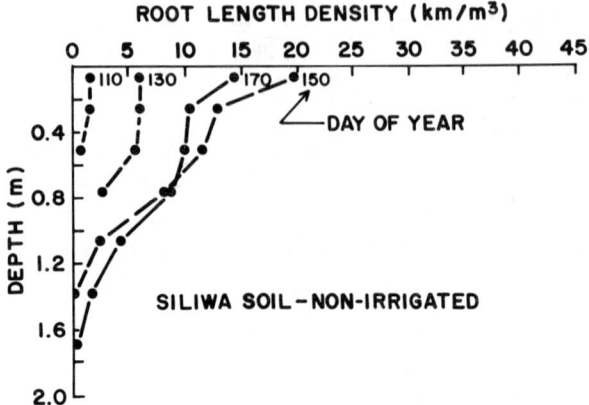

Fig. 6–6. Simulated maize root length densities on five dates for nonirrigated Siliwa soil.

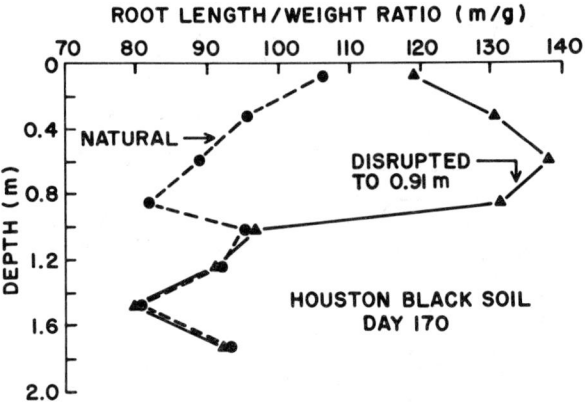

Fig. 6–7. Simulated maize root length/weight ratios at Day of Year 1970 and for soil ripped to 0.91 m for natural Houston Black soil.

Houston Black soil (fine, montmorillonitic, thermic Udic Pellusterts) is typically friable at the surface, but grades into a very plastic, dense clay with low porosity with depth. Burnett and Tackett (1968) reported that profile disruption to 0.61 or 1.22 m increased porosity, soil O_2 contents, root length/weight ratios, and root length density. Simulated profile disruption to 0.91 m increased length/weight ratio (Fig. 6–7) and root length density (Fig. 6–8) in the disrupted zone. The results are consistent with studies indicating that high soil strength causes thicker roots to be produced (Burnett & Tackett, 1968; Glover, 1967; Trouse, 1965). The values of simulated length/weight ratios are within the range reported by Allmaras et al. (1975) for field-grown maize. Considerably smaller values were reported by Follett et al. (1974), but the model could be calibrated to their data by reducing the length/weight ratios at the seedling stage and at the soil surface.

Tillage-induced pans are common in the A2 horizon of Ultisols in the southeastern USA (Campbell et al., 1974; Stitt et al., 1982) where they are

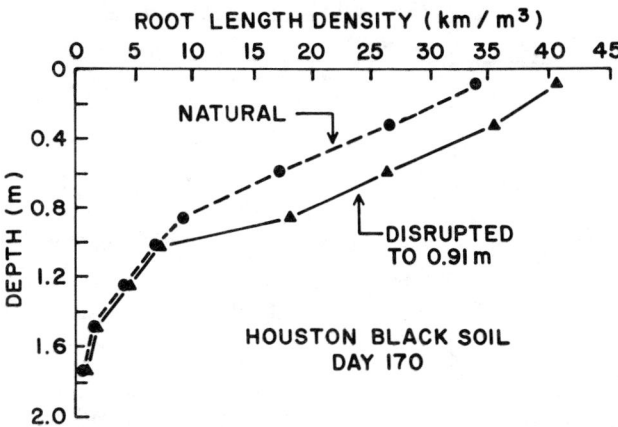

Fig. 6-8. Simulated maize root length densities at Day of Year 1970 and for soil ripped to 0.91 m for natural Houston Black soil.

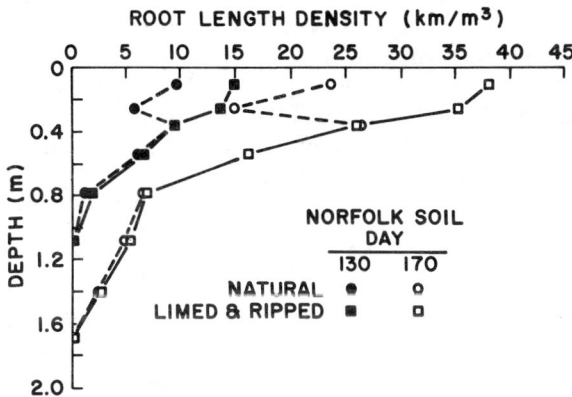

Fig. 6-9. Simulated maize root length densities for Day of Year 110 and 170, and soil limed and ripped to 0.30 m for Norfolk soil.

often associated with low pH and high Al saturation. This complex of factors can inhibit root growth and extraction of subsoil moisture (Taylor & Burnett, 1964). The effects of deep tillage (0.30 m) and liming of a Norfolk soil (fine-loamy, siliceous, thermic Typic Paleudult) with a tillage pan from 0.23 to 0.30 m and 50 to 65% Al saturation to 0.30 m for two dates are shown in Fig. 6-9. The simulated root length density of the limed and tilled soil is approximately twice that of the original soil in the surface 0.30 m.

The Chenango soil (loamy-skeletal, mixed, mesic Typic Dystrochrept) pedon described in Table 3 has coarse fragment contents ranging from 15% in the topsoil to 75% below 1.8 m. The high coarse-fragment content of the subsoil reduces plant-extractable water per unit total soil volume and probably inhibits root growth, although rooting behavior is rarely measured on such soils (Babalola and Lal, 1977; Vine et al., 1981). Simulated root growth in natural Chenango soil and in a soil without coarse fragments, but with

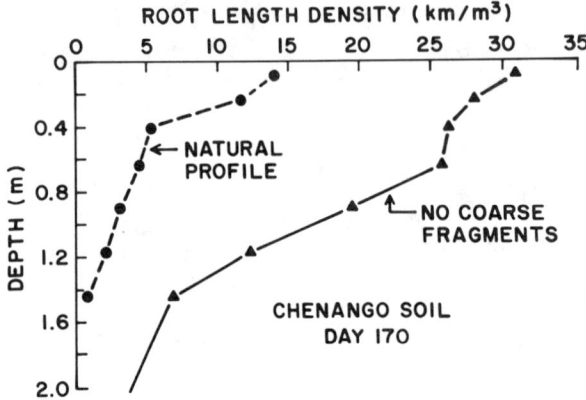

Fig. 6-10. Simulated maize root length densities at Day of Year 1970 and for soil without coarse fragments for natural Chenango soil.

otherwise identical properties, are given in Fig. 6-10. The large coarse-fragment content of the soil reduced root growth throughout the soil profile by two mechanisms: (i) increasing the root growth stress factor for coarse fragments (Eq. [7]); and (ii) increasing the soil strength factor (Eq. [16]). The increase in the soil strength factor resulted from rapid depletion of plant-extractable water due to large amounts of inert coarse fragments. Removal of these coarse fragments resulted in a nearly ideal simulated root development.

V. CONCLUSION

The model described here integrates several of the major factors affecting root system growth and death in soils. It is designed to be a component of simulation models of crop growth and development that provide the following daily inputs: (i) crop growth stage; (ii) dry matter allocation to the root system; (iii) soil layer temperatures; and (iv) soil layer water contents. The root growth model integrates these factors with genetically determined crop rooting characteristics and with soil properties such as texture, bulk density, Al and Ca saturation, and coarse fragment content. It then simulates root system depth, growth, and death as well as weight, length/weight ratio, and root length density by soil layer and by day.

The genetic inputs give the user great flexibility to simulate the root growth of various crops or of crop cultivars with varying root system geometry and/or tolerance to temperature extremes, Al toxicity, Ca deficiency, and poor aeration. Although we have experience with a limited number of crops, we feel that the model is sufficiently flexible to allow simulation of root growth for a variety of soils, climates, and species.

We have used simple algorithms to describe the growth of the root system and its response to soil properties. However, a great deal remains to be learned about root behavior, and we expect users to improve the model by developing more complex and, we hope, more generally applicable algorithms as necessary.

VI. APPENDIX

```
C
C   MAIN     PC VERSION    JUNE, 1987
C
C   ALA = ALUMINUM SATURATION BELOW WHICH ROOT GROWTH IS
C         UNAFFECTED (%)
C   ALS = ALUMINUM SATURATION (%)
C   ALX = ALUMINUM SATURATION ABOVE WHICH ROOT GROWTH IS
C         NEGLIGIBLE (%)
C   ASF = ACTIVE (MINIMUM) LAYER STRESS FACTOR (0-1)
C   ASF1 = MINIMUM OF STRESS FACTORS STP, SCA, SAL, OR SCD
C   ASF2 = THE SQUARE ROOT OF THE MINIMUM OF STRESS FACTORS SST,
C         SAI, OR SCF
C   ASF3 = MINIMUM OF STRESS FACTORS SST, SAL, OR SCF
C   BD  = BULK DENSITY OF THE FINE EARTH FRACTION (MG/M3)
C   BDO = BULK DENSITY BELOW WHICH ROOT GROWTH IS OPTIMUM AT
C         OPTIMUM SOIL WATER CONTENT
C   BDX = BULK DENSITY ABOVE WHICH ROOT GROWTH IS NEGLIGIBLE AT
C         OPTIMUM SOIL WATER CONTENT
C   CA  = EXCHANGEABLE CALCIUM (CMOL/KG)
C   CAA = CALCIUM SATURATION BELOW WHICH ROOT GROWTH IS REDUCED (%)
C   CAS = CALCIUM SATURATION (%)
C   CAX = CALCIUM SATURATION BELOW WHICH ROOT GROWTH IS NEGLIGIBLE
C         (%)
C   CEC = CATION EXCHANGE CAPACITY (CMOL/KG)
C   CLA = CLAY (%)
C   CWP = WATER-FILLED PORE FRACTION AT WHICH AERATION BEGINS TO
C         LIMIT ROOT GROWTH
C   DAC = CUMULATIVE DRY MATTER ACTUALLY INVESTED IN ROOT SYSTEM
C         THROUGH SEASON (KG/HA)
C   DAY = DAY COUNTER USED FOR PRINTING OUTPUT (D)
C   DDI = DAILY DEPTH INCREMENT
C   DLL = DAILY DEATH OF ROOTS IN LAYER (KM)
C   DMC = CUMULATIVE DRY MATTER PARTITIONED TO ROOT SYSTEM (KG/HA)
C   DMD = DRY MATTER ALLOCATION TO ROOT SYSTEM ON A DAY (KG/HA)
C   DPRNT = OUTPUT INTERVAL (D)
C   DRD = POTENTIAL INCREASE IN ROOT DEPTH (M)
C   DWL = DAILY DEATH OF ROOTS IN LAYER (KG/HA)
C   DZ  = THICKNESS OF LAYER PENETRATED BY ROOTING FRONT (M)
C   EAL = EXTRACTABLE ALUMINUM (CMOL/KG)
C   GAL = ACTUAL GROWTH OF ROOT LENGTH ON A DAY (KM/HA)
C   GPL = POTENTIAL ROOT GROWTH IN THE LAYER (KG/HA)
C   GPS = POTENTIAL ROOT SYSTEM GROWTH (KG/HA)
C   GS  = GROWTH STAGE (0-1)
C   GSD = GROWTH STAGE WHEN NORMAL ROOT SENESCENCE BEGINS
C   GSR = GROWTH STAGE WHEN ROOT DEPTH REACHES MAXIMUM
C   GSY = GROWTH STAGE PREVIOUS DAY
C   IJ  = NUMBER OF SOIL LAYERS
C   IR  = NUMBER OF LAYERS WITH ROOTS
C   JDATE = DAY OF YEAR (1-366)
C   LL  = LOWER LIMIT OF PLANT-EXTRACTABLE SOIL WATER FOR FINE
C         EARTH FRACTION (M/M)
C   LT  = LAYER TEMPERATURE (C)
C   LW  = LAYER VOLUMETRIC WATER CONTENT (M/M)
C   LWA = LENGTH/WEIGHT RATIO FOR ROOTS GROWING ON A DAY IN A
C         LAYER (M/G)
C   LWF = LAYER ROOT DISTRIBUTION WEIGHTING FACTOR
C   LWM = NORMAL RATIO OF ROOT LENGTH TO WEIGHT IN PLOW LAYER AT
C         MATURITY (M/G)
```

```
C     LWN = NORMAL LENGTH/WEIGHT RATIO (KM/KG)
C     LWR = LENGTH/WEIGHT RATIO OF ALL ROOTS IN A LAYER (M/G)
C     LWS = NORMAL RATIO OF ROOT LENGTH TO WEIGHT IN SEEDLING (M/G)
C     PD  = PLANTING DEPTH (M)
C     PO  = POROSITY OF THE FINE EARTH FRACTION (%)
C     RD  = ROOT SYSTEM DEPTH (M)
C     RDX = NORMAL MAXIMUM ROOT SYSTEM DEPTH (M)
C     RLL = ROOT LENGTH IN THE LAYER (KM/HA)
C     RLV = ROOT LENGTH DENSITY (CM/CM3)
C     ROK = COARSE FRAGMENTS AS FRACTION OF SOIL VOLUME (0-1)
C     RWL = ROOT WEIGHT IN THE LAYER (KG)
C     SAI = LAYER AERATION STRESS FACTOR (0-1)
C     SAL = ROOT GROWTH STRESS FACTOR FOR ALUMINUM TOXICITY
C     SAN = SAND (%)
C     SBD = ROOT GROWTH STRESS FACTOR FOR EXCESSIVE BULK DENSITY
C     SCA = ROOT GROWTH STRESS FACTOR FOR CALCIUM DEFICIENCY
C     SCD = CODE SET TO 0. TO STOP ROOT GROWTH IF THE HORIZON
C           IS DURIPAN, FRAGIPAN, LITHIC, PARALITHIC, PETROCALCIC,
C           PETROFERRIC, PETROGYPSIC, OR SKELETAL, 1.0 IF OTHER OR
C           NOT KNOWN
C     SCF = ROOT GROWTH STRESS FACTOR FOR EXCESSIVE COARSE FRAGMENTS
C     SFT = FRACTION OF NORMAL ROOT GROWTH WHEN PORE SPACE IS
C           SATURATED (0-1)
C     SIL = SILT (%)
C     SMB = SUM OF BASES (CMOL/KG)
C     SST = LAYER STRENGTH STRESS FACTOR (0-1)
C     STP = LAYER TEMPERATURE STRESS FACTOR (0-1)
C     TBS = BASE TEMPERATURE FOR ROOT GROWTH (C)
C     TOP = OPTIMUM TEMPERATURE FOR ROOT GROWTH (C)
C     TRW = TOTAL ROOT SYSTEM WEIGHT (KG/HA)
C     UL  = DRAINED UPPER LIMIT OF SOIL WATER FOR THE FINE EARTH
C           FRACTION (M/M)
C     WCG = WEIGHTING COEFFICIENT - GEOTROPISM
C     WFL = WEIGHTING FACTOR FOR LAYER (0-1)
C     WFP = FRACTION OF PORE SPACE CONTAINING WATER
C     WFT = WEIGHTING FACTOR, TOTAL
C     Z   = DEPTH TO BOTTOM OF LAYER (M)
C     ZA  = MEAN LAYER DEPTH (M)
C
      CHARACTER*56 TITLE
      CHARACTER*20 PARAM
      CHARACTER*20 WATEMP
      CHARACTER*20 OUTFILE
      CHARACTER*1 SELECT
      REAL LL,LT,LW,LWR,LWM,LWS,LWF
C
      COMMON /BLK1/ CAA,CAX,ALA,ALX,BD(10),SAN(10),SIL(10),ROK(10),
     1    SMB(10),EAL(10),CA(10),CLA(10),ALS(10),BDO(10),BDX(10)
      COMMON /BLK2/ LL(10),UL(10),LT(10),SST(10),LW(10),STP(10),
     1    SCD(10),TBS,TOP,SFT
      COMMON /BLK3/ GAL(10),DWL(10),RLL(10),RWL(10),RLV(10),LWR(10),
     1    RDX,GSR,LWM,LWS,WCG,GSD,GS,DMD,GSY,DMC,DMA,TRW,RD,
     2    IR,DAC
      COMMON /BLK12/ PO(10),CWP(10),SCA(10),SAL(10),SBD(10),SCF(10)
      COMMON /BLK13/ Z(11),ZA(10)
      COMMON /BLK23/ LWF(10),SAI(10),ASF(10),ASF1(10),ASF2(10),ASF3(10)
      COMMON /BLK123/ IJ
C
   10 WRITE (*,'(A/)')' Type in the Input Parameter File Name'
      READ (*,'(A)') PARAM
```

SIMULATION OF ROOT GROWTH

```
      WRITE (*,'(A/)')' Type in the Soil Water/Temperature File Name'
      READ (*,'(A)') WATEMP
      WRITE (*,'(A/)')' Type in the Output File Name'
      READ (*,'(A)') OUTFILE
C
      OPEN(1,FILE=PARAM,ACCESS='SEQUENTIAL',STATUS='OLD')
      OPEN(2,FILE=WATEMP,ACCESS='SEQUENTIAL',STATUS='OLD')
      OPEN(3,FILE=OUTFILE,ACCESS='SEQUENTIAL',STATUS='NEW')
C
C     INITIALIZE VARIABLES
C
      Z(1) = 0.
      IJ = 0
      IR = 0
      RD = 0.
      DAY = 0.
      DMA = 0.
      DAC = 0.
      GSY = 0.
      DMC = 0.
      DO 20 I=1,10
        RWL(I) = 0.
        RLL(I) = 0.
        RLV(I) = 0.
   20 CONTINUE
C
      READ (1,1000) TITLE
      WRITE (3,3000) TITLE
C
C     READ AND WRITE THE GENETIC PARAMETERS
C
      READ (1,1010) RDX,GSR,LWM,LWS,WCG,TBS,TOP,CAA,CAX,ALA,ALX,PD,
     1  GSD,SFT,DPRNT
      WRITE (3,7000)
      WRITE (3,5000) RDX,GSR,LWM,LWS,WCG,TBS,TOP,CAA,CAX,ALA,ALX,PD,
     1  GSD,SFT
C
      RD = PD
      DO 30 I=1,10
        LWR(I) = LWS
   30 CONTINUE
C
C     READ THE FOLLOWING SOIL PROPERTIES FOR UP TO 10 SOIL LAYERS
C
      DO 100 I=1,10
        READ (1,1020,END=110) Z(I+1),BD(I),SAN(I),SIL(I),ROK(I),
     1    SMB(I),EAL(I),CA(I),UL(I),LL(I),SCD(I)
        IJ = IJ+1
  100 CONTINUE
C
C     CALCULATE STATIC SOIL STRESS FACTORS
C
  110 CALL SCALC
C
C     HEADINGS AND OUTPUT FROM SUBROUTINE SCALC
C
      WRITE (3,7010)
      WRITE (3,9000)
      DO 200 I=1,IJ
        WRITE (3,3010) I,Z(I),SAN(I),SIL(I),CLA(I),ROK(I),BD(I),
```

```
      1   PO(I),UL(I),LL(I)
  200 CONTINUE
      WRITE (3,9010)
      DO 210 I=1,IJ
         WRITE (3,3020) I,SMB(I),EAL(I),ALS(I),CA(I),SAL(I),SCA(I),
      1     SBD(I),SCF(I),SCD(I)
  210 CONTINUE
C
C     HEADINGS FOR OUTPUT FROM SUBROUTINE RTDIST
C
      WRITE (3,7020)
      WRITE (3,9020)
C
C     READ DAILY INPUT
C
  300 DAY = DAY+1.
C
C     READ DAY OF THE YEAR, DAILY GROWTH STAGE, ROOT SYSTEM GROWTH
C
      READ (2,1030,END=400) JDATE,GS,DMD
C
C     READ DAILY SOIL TEMPERATURE
C
      READ (2,1040) (LT(I),I=1,IJ)
C
C     READ DAILY SOIL LAYER WATER CONTENT
C
      READ (2,1040) (LW(I),I=1,IJ)
C
C     CALCULATE DYNAMIC SOIL STRESS FACTORS
C
      CALL DCALC
C
C     GROW ROOTS BY LAYER
C
      CALL RTDIST
C
C     OUTPUT FROM SUBROUTINE RTDIST
C
      IF (DAY.EQ.DPRNT) THEN
         DAY = 0.
         WRITE (3,5010) JDATE,GS,RD,TRW,DMA,DMD,DAC,DMC
         DO 310 I=1,IR
            WRITE (3,3030) Z(I+1),RLV(I),GAL(I),DWL(I),RWL(I),
      1        LWR(I),STP(I),SST(I),SAI(I),ASF(I)
  310    CONTINUE
         WRITE (3,3040)
      ENDIF
C
      GO TO 300
C
  400 WRITE (3,9020)
      WRITE (3,3040)
      WRITE (3,3040)
      CLOSE(1)
      CLOSE(2)
      CLOSE(3)
C
      WRITE (*,'(A/)')' Would you like to run another data set? (Y/N)'
      READ (*,'(A)') SELECT
      IF (SELECT.EQ.'Y'.OR.SELECT.EQ.'y') GO TO 10
```

SIMULATION OF ROOT GROWTH

```
C
      STOP
C
 1000 FORMAT (A56)
 1010 FORMAT (9F8.2,/,6F8.2)
 1020 FORMAT (9F8.3,/,2F8.3)
 1030 FORMAT (2X,I3,2(2X,F8.3))
 1040 FORMAT (10F8.3)
 3000 FORMAT (A56,///)
 3010 FORMAT (2X,I3,7X,F4.2,3(4X,F4.1),2(4X,F4.2),4X,F4.2,
     1  2(4X,F4.2))
 3020 FORMAT (2X,I3,3X,4(4X,F4.1),5(4X,F4.2))
 3030 FORMAT (1X,F4.2,1X,F6.2,2X,F6.0,2X,F7.0,2X,F7.1,2X,F8.1,
     1  4(1X,F7.2))
 3040 FORMAT (/)
 5000 FORMAT (2X,'RDX = ',F5.1,' GSR = ',F5.1,' LWM = ',F5.0,
     1  ' LWS = ',F5.0,' WCG = ',F5.1,' TBS = ',F5.0,/,
     2  ' TOP = ',F5.0,' CAA = ',F5.1,' CAX = ',F5.1,
     3  ' ALA = ',F5.0,' ALX = ',F5.0,' PD  = ',F5.2,/,
     4  ' GSD = ',F5.1,' SFT = ',F5.1,//)
 5010 FORMAT (/,1X,I3,2X,'GS =',F6.3,4X,'RD =',F5.2,4X,'TRW =',
     1  F7.1,5X,'DMA =',F7.1,5X,'DMD =',F7.1,/,50X,'DAC =',F7.1,
     2  5X,'DMC =',F7.1,/)
 7000 FORMAT (//,28X,'GENETIC CHARACTERISTICS',//)
 7010 FORMAT (//,29X,'INPUT CHARACTERISTICS')
 7020 FORMAT (//,25X,'DAILY ROOT DISTRIBUTION OUTPUT',//)
 9000 FORMAT (//,2X,'LAYER',6X,'Z',6X,'SAN',5X,'SIL',5X,'CLA',5X,
     1  'ROK',5X,'BD',6X,'PO',6X,'UL',6X,'LL',/)
 9010 FORMAT (//,2X,'LAYER',5X,'SMB',5X,'EAL',5X,'ALS',6X,'CA',5X,
     1  'SAL',5X,'SCA',5X,'SBD',5X,'SCF',5X,'SCD',/)
 9020 FORMAT (1X,'DPTH',3X,'RLV',5X,'GAL',6X,'DWL',6X,'RWL',7X,
     1  'LWR',5X,'STP',5X,'SST',5X,'SAI',5X,'ASF',/)
C
      END
C
C
C*************************************************************
C
C
      SUBROUTINE SCALC
C
C     THIS SUBROUTINE CALCULATES THE STATIC (CONSTANT OVER SEASON)
C     STRESS FACTORS WHICH LIMIT ROOT GROWTH IN PARTICULAR SOIL
C     LAYERS
C
      DIMENSION CEC(10)
C
      COMMON /BLK1/ CAA,CAX,ALA,ALX,BD(10),SAN(10),SIL(10),ROK(10),
     1     SMB(10),EAL(10),CA(10),CLA(10),ALS(10),BDO(10),BDX(10)
      COMMON /BLK12/ PO(10),CWP(10),SCA(10),SAL(10),SBD(10),SCF(10)
      COMMON /BLK13/ Z(11),ZA(10)
      COMMON /BLK123/ IJ
C
      DO 100 I=1,IJ
C
C     CALCULATE PERCENT OF CLAY AND POROSITY OF EACH LAYER
C
      CLA(I) = 100.-SAN(I)-SIL(I)
      PO(I) = (1.-ROK(I))*(1.-BD(I)/2.65)
      CEC(I) = SMB(I)+EAL(I)
C
```

```
      C     CALCULATE DEPTH TO CENTER OF EACH LAYER
      C
            ZA(I) = Z(I)+(Z(I+1)-Z(I))/2.
      C
      C     CALCULATE CRITICAL WATER-FILLED PORE FRACTION
      C
            CWP(I) = .4+0.004*CLA(I)
      C
      C     CALCULATE CALCIUM STRESS FACTOR BY LAYER
      C
            SCA(I) = 1.0
            CAS = CA(I)/(CEC(I)+1.E-20)*100.
            IF (CAS.LT.CAA) SCA(I) = (CAS-CAX)/(CAA-CAX)
      C
      C     CALCULATE ALUMINUM TOXICITY STRESS FACTOR BY LAYER
      C
            ALS(I) = EAL(I)/CEC(I)*100.
            IF (ALS(I).LE.ALA) THEN
              SAL(I) = 1.
            ELSEIF (ALS(I).GE.ALX) THEN
              SAL(I) = 0.
            ELSE
              SAL(I) = (ALX-ALS(I))/(ALX-ALA)
            ENDIF
      C
      C     CALCULATE BULK DENSITY STRESS FACTOR BY LAYER
      C
            BDO(I) = 1.1+0.005*SAN(I)
            BDX(I) = 1.6+0.004*SAN(I)
            IF (BD(I).LE.BDO(I)) THEN
              SBD(I) = 1.
            ELSEIF (BD(I).GE.BDX(I)) THEN
              SBD(I) = 0.
            ELSE
              SBD(I) = (BDX(I)-BD(I))/(BDX(I)-BDO(I))
            ENDIF
      C
      C     CALCULATE COARSE FRAGMENT STRESS FACTOR BY LAYER
      C
            SCF(I) = 1.-ROK(I)
      C
        100 CONTINUE
            RETURN
      C
            END
      C
      C
      C************************************************************************
      C
      C
            SUBROUTINE DCALC
      C
      C     THIS SUBROUTINE CALCULATES DYNAMIC STRESS FACTORS THROUGHOUT
      C     THE SOIL PROFILE ON A DAILY BASIS
      C
            REAL LL,LT,LW,LWF
      C
            COMMON /BLK2/ LL(10),UL(10),LT(10),SST(10),LW(10),STP(10),
           1    SCD(10),TBS,TOP,SFT
            COMMON /BLK12/ PO(10),CWP(10),SCA(10),SAL(10),SBD(10),SCF(10)
```

```
      COMMON /BLK23/ LWF(10),SAI(10),ASF(10),ASF1(10),ASF2(10),ASF3(10)
      COMMON /BLK123/ IJ
C
C     CALCULATE DYNAMIC ROOT GROWTH STRESS FACTORS FOR SOIL LAYERS
C
      DO 100 I=1,IJ
C
C     CALCULATE LAYER STRENGTH FACTORS
C
      IF (LW(I).LT.LL(I)) THEN
        LWF(I) = 0.
      ELSEIF (LW(I).GT.UL(I)) THEN
        LWF(I) = 1.0
      ELSE
        LWF(I) = (LW(I)-LL(I))/(UL(I)-LL(I))
      ENDIF
      SST(I) = SBD(I)*SIN(1.57*LWF(I))
C
C     CALCULATE LAYER AERATION FACTORS
C
      WFP = LW(I)/PO(I)
      SAI(I) = 1.
      IF (WFP.GE.CWP(I)) THEN
        SAI(I) = SFT+(1-WFP)*((1-SFT)/(1-CWP(I)))
      ELSE
        SAI(I) = 1.0
      ENDIF
      IF (SAI(I).LT.0.) SAI(I) = 0.
C
C     CALCULATE LAYER TEMPERATURE FACTORS
C
      IF (LT(I).GE.TBS) THEN
        STP(I) = SIN(1.5707*(LT(I)-TBS)/(TOP-TBS))
      ELSE
        STP(I) = 0.
      ENDIF
C
C     CALCULATE MINIMUM OF STATIC AND DYNAMIC STRESS FACTORS
C
      ASF(I) = MIN(STP(I),SST(I),SAI(I),SCA(I),SAL(I),SCF(I),SCD(I))
      ASF1(I) = MIN(STP(I),SCA(I),SAL(I),SCD(I))
      ASF2(I) = (MIN(SST(I),SAI(I),SCF(I)))**.5
      ASF3(I) = MIN(SST(I),SAL(I),SCF(I))
      IF (ASF(I).LT.0.) ASF(I) = 10E-10
      IF (ASF1(I).LT.0.) ASF1(I) = 10E-10
      IF (ASF2(I).LT.0.) ASF2(I) = 10E-10
      IF (ASF3(I).LT.0.) ASF3(I) = 10E-10
C
  100 CONTINUE
C
      RETURN
C
      END
C
C
C*********************************************************************
C
C
      SUBROUTINE RTDIST
C
```

```
C     THIS SUBROUTINE DISTRIBUTES ROOT GROWTH THROUGHOUT THE SOIL
C     PROFILE IN RESPONSE TO ROOT SYSTEM GROWTH, LAYER DEPTH,
C     STATIC STRESS FACTORS, AND DYNAMIC STRESS FACTORS
C
      DIMENSION WFL(10),DZ(10),GPL(10),DLL(10)
      REAL LWA(10),LWM,LWS,LWF,LWN,LWR
C
      COMMON /BLK3/ GAL(10),DWL(10),RLL(10),RWL(10),RLV(10),LWR(10),
     1    RDX,GSR,LWM,LWS,WCG,GSD,GS,DMD,GSY,DMC,DMA,TRW,RD,
     2    IR,DAC
      COMMON /BLK13/ Z(11),ZA(10)
      COMMON /BLK23/ LWF(10),SAI(10),ASF(10),ASF1(10),ASF2(10),ASF3(10)
      COMMON /BLK123/ IJ
C
C     INITIALIZE VARIABLES
C
      DMA = 0.
      GPS = 0.
      TRW = 0.
      WFT = 0.
C
C     DETERMINE IF ROOT SYSTEM IS STILL GROWING DOWNWARD
C
      IF (GS.GT.GSR) GO TO 300
C
C     DETERMINE POTENTIAL INCREASE IN ROOT DEPTH
C
      DDI = RDX*(GS-GSY)/GSR
      DO 100 I=1,IJ
        IF (RD.LT.Z(I+1)) THEN
          DRD = DDI*MIN(ASF1(I),ASF2(I))
          IF (DRD.GT.(Z(I+1)-RD)) THEN
            DDI = DDI-(Z(I+1)-RD)
            RD = Z(I+1)
          ELSE
            RD = RD+DRD
            GO TO 200
          ENDIF
        ENDIF
  100 CONTINUE
C
C     DETERMINE NUMBER OF LAYERS CURRENTLY CONTAINING ROOTS
C
  200 IF (I.LT.IJ) THEN
        IR = I
      ELSE
        IR = IJ
      ENDIF
C
C
C
  300 DO 400 I=1,IR
C
C     CALCULATE WEIGHTING FACTOR FOR GROWTH HABIT
C
      WFL(I) = (1.-ZA(I)/3.0)**WCG
      DZ(I) = Z(I+1)-Z(I)
      WFT = WFT+WFL(I)
C
C     CALCULATE NORMAL LENGTH/WEIGHT RATIO AS AFFECTED BY GROWTH
```

SIMULATION OF ROOT GROWTH

```
C     STAGE AND LAYER DEPTH
C
      IF (ZA(I).GT.RD) THEN
        LWN = LWS
      ELSE
        LWN = LWS-GS*(LWS-LWM)*(1.-ZA(I)/RD)
      ENDIF
C
C     CALCULATE ACTUAL ROOT LENGTH/WEIGHT RATIOS AS AFFECTED BY
C     ALUMINUM, STRENGTH OR COARSE FRAGMENTS
C
      LWA(I) = LWN/(1.+3.*(1.-ASF3(I)))
C
  400 CONTINUE
C
C     DETERMINE POTENTIAL FOR ROOT GROWTH IN EACH LAYER
C
      DO 500 I=1,IR
        IF (RLV(I).EQ.0.0) THEN
          RLV(I) = 0.01*((RD-Z(I))/DZ(I))
        ENDIF
        WFL(I) = WFL(I)/WFT
        TEMP = (5.0*RLV(I))/(0.025+RLV(I))
        GPL(I) = (TEMP*ASF(I)*WFL(I)*DZ(I)*1E4)/LWA(I)
        GPS = GPS+GPL(I)
  500 CONTINUE
C
C
C
      DO 600 I=1,IR
C
C     DISTRIBUTE ROOT GROWTH BY LAYERS
C
        IF (GPS.GT.DMD) THEN
          GAL(I) = DMD*(GPL(I)/GPS)*LWA(I)
        ELSE
          GAL(I) = GPL(I)*LWA(I)
        ENDIF
        RWL(I) = RWL(I)+(GAL(I)/LWA(I))
        DMA = DMA+GAL(I)/LWA(I)
C
C     CALCULATE ROOT DEATH BY LAYER
C
        DWL(I) = 0.01*RWL(I)*(1+MAX((1.-LWF(I)),(1.-SAI(I))))
        IF (GS.GT.GSD) THEN
          DWL(I) = DWL(I)+RWL(I)*((GS-GSD)/(1.-GSD))**3.
        ENDIF
        RWL(I) = RWL(I)-DWL(I)
        DLL(I) = DWL(I)*LWR(I)
        RLL(I) = RLL(I)+GAL(I)-DLL(I)
        RLV(I) = RLL(I)/(DZ(I)*1E4)
        LWR(I) = RLL(I)/(RWL(I)+10E-10)
        TRW = TRW+RWL(I)
C
  600 CONTINUE
C
      GSY = GS
      DMC = DMC+DMD
      DAC = DAC+DMA
C
```

REFERENCES

Abruna, F., J. Rodriguez, and S. Silva. 1982. Crop response to soil acidity factors in Ultisols and Oxisols in Puerto Rico. VI. Grain sorghum. J. Agric. Univ. P.R. 61:28-38.

Adams, F., and B.L. Moore. 1983. Chemical factors affecting root growth in subsoil horizons of coastal plains soils. Soil Sci. Soc. Am. J. 47:99-102.

Allmaras, R.R., W.W. Nelson, and W.B. Voorhees. 1975. Soybean and corn rooting in southwestern Minnesota: II. Root distributions and related water inflow. Soil Sci. Soc. Am. Proc. 39:771-777.

Anderson, E.L. 1987. Corn root growth and distribution as influenced by tillage and nitrogen fertilization. Agron. J. 79:544-549.

Asady, G.H., A.J.M. Smucker, and M.W. Adams. 1985. Seedling test for the quantitative measurement of root tolerances to compacted soil. Crop Sci. 25:802-806.

Babalola, O., and R. Lal. 1977. Subsoil gravel horizon and maize root growth: 1. Gravel concentrations and bulk density effects. Plant Soil 46:337-346.

Barber, S.A. 1971. Effect of tillage practice on corn (*Zea mays* L.) root distribution and morphology. Agron. J. 63:724-726.

Barber, S.A. 1979. Growth requirements for nutrients in relation to demand at the root surface. p. 5-20. *In* J.L. Harley and R.S. Russell (ed.) The root-soil interface. Academic Press, New York.

Barber, S.A. 1984. Soil nutrient bioavailability. John Wiley & Sons, New York.

Bar-Yosef, B., and J.R. Lambert. 1981. Corn and cotton root growth in response to soil impedance and water potential. Soil Sci. Soc. Am. J. 45:930-935.

Borg, H., D.W. Grimes. 1986. Depth development of roots with time: An empirical description. Trans. ASAE 29:194-197.

Bouldin, D.R. 1979. The influence of subsoil acidity on crop yield potenital. Cornell Univ. Int. Agric. Bull. 34.

Brenes, E., and R.W. Pearson. 1973. Root responses of three gramineae species to soil acidity in an oxisol and an ultisol. Soil Sci. 116:295-302.

Brouwer, R. 1962. Distribution of dry matter in the plant. Neth. J. Agric. Sci. 10:361-376.

Brouwer, R. 1963. Some aspects of the equilibrium between overground and underground plant parts. Meded. Inst. Biol. Scheikd. Onderz. Landbouwgewassan, Wageningen 213:31-39.

Brouwer, R., and C.T. de Wit. 1968. A simulation model of plant growth with special attention to root growth and its consequences. p. 224-242. *In* Proc. 15th Easter School Agric. Sci., Univ. of Nottingham, Nottingham, England.

Burnett, E., and J.L. Tackett. 1968. Effect of soil profile modification on plant root development. Ninth Int. Congr. Soil Sci. Trans., Vol. 3, Paper 34, p. 329-337.

Campbell, R.B., D.E. Reicosky, and C.W. Doty. 1974. Physical properties and tillage of paleudults in the southeastern coastal plains. J. Soil Water Conserv. 29:220-224.

Cannell, R.Q., and M.B. Jackson. 1981. Alleviating aeration stresses. p. 141-192. *In* G.F. Arkin and H.M. Taylor (ed.) Modifying the root environment to reduce crop stresses. ASAE, St. Joseph, MI.

Cassel, D.K. 1982. Tillage effects on soil bulk density and mechanical impedance. p. 45-67. *In* P.W. Unger and D.M. Van Doren (ed.) Predicting tillage effects on soil physical properties and processes. ASA Spec. Publ. 44. ASA, CSSA, and SSSA, Madison, WI.

Cockroft, B., K.P. Barley, and E.L. Greacen. 1969. The penetration of clays by fine probes and root tips. Aust. J. Soil Res. 7:333-348.

Cornish, P.S., H.B. So, and J.R. McWilliam. 1984. Effects of soil bulk density and water regimen on root growth and uptake of phosphorus by rye grass. Aust. J. Agric. Res. 35:631-644.

Cruz, R.T., J.C. O'Toole, M. Dingkuhn, E.B. Yambao, M. Thangaraj, and S.K. DeDatta. 1986. Shoot and root responses to water deficits in rainfed lowland rice. Aust. J. Plant Physiol. 13:567-575.

Derera, N.F., D.R. Marshall, and L.N. Balaam. 1969. Genetic variability in root development in relation to drought tolerance in spring wheats. Exp. Agric. 5:327-337.

Dexter, A.R. 1986. Model experiments on the behaviour of roots at the interface between a tilled seed-bed and a compacted sub-soil. I. Effects of seed-bed aggregate size and sub-soil strength on wheat roots. Plant Soil 95:123-133.

Drew, M.C. 1983. Plant injury and adaptation to oxygen deficiency in the root environment: A review. Plant Soil 75:179-199.

Eavis, B.W. 1972. Soil physical conditions affecting seedling root growth. I. Mechanical impedance, aeration, and moisture availability as influenced by bulk density and moisture levels in a sandy loam soil. Plant Soil 36:613-622.

Evans, H. 1935. The root system of the sugarcane. I. Methods of study. Emp. J. Exp. Agric. 3:351-362.

Evans, H. 1936. The root system of the sugarcane. II. Some typical root-systems. Emp. J. Exp. Agric. 4:208-220.

Fageria, N.K. 1982. Differential aluminum tolerance to rice cultivars in nutrient solution. Pesqui. Agropecu. Bras. 17:1-9.

Farina, M.P.W., M.E. Sumner, C.O. Plank, and W.S. Letzsch. 1980. Exchangeable aluminum and pH as indicators of lime requirement for corn. Soil Sci. Soc. Am. J. 44:1036-1041.

Follett, R.F., R.R. Allmaras, and G.A. Reichman. 1974. Distribution of corn roots in sandy soil with a declining water table. Agron. J. 66:288-292.

Foy, C.D., A.L. Fleming, and G.C. Gerloff. 1972. Differential aluminum tolerance in two snapbean varieties. Agron. J. 64:815-818.

Foy, C.D., H.N. Lafever, J.W. Schwartz, and A.L. Fleming. 1974a. Aluminum tolerance of wheat cultivars related to region of origin. Agron. J. 66:751-758.

Foy, C.D., R.G. Orellana, J.W. Schwartz, and A.L. Fleming. 1974b. Responses of sunflower genotypes to aluminum in acid soil and nutrient solution. Agron. J. 66:293-296.

Gerard, C.J., P. Sexton, and G. Shaw. 1982. Physical factors influencing soil strength and root growth. Agron. J. 74:875-879.

Gerwitz, A., and E.R. Page. 1974. An empirical mathematical model to describe plant root systems. J. Appl. Ecol. 11:773-781.

Glover, J. 1967. The simultaneous growth of sugarcane roots and tops in relation to soil and climate. Proc. S. Afr. Sugar Cane Technol. Assoc. 41:143-159.

Gonzalez-Erico,E., E.J. Kamprath, G.C. Naderman, and W.V. Soares. 1979. Effect of depth of lime incorporation on the growth of corn on an oxisol of central Brazil. Soil Sci. Soc. Am. J. 43:1155-1158.

Goodwin, J.B., F.L.Garagorry, W. Espinosa, L.M. Sans, and L.J. Youngdahl. 1982. Modeling soil-water-plant relationships in cerrado soils of Brazil: The case of maize. Agric. Systems 8:115-127.

Grable,A.R. 1966. Soil aeration and plant growth. Adv. Agron. 18:57-106.

Grable,A.R., and E.G. Siemer. 1968. Effects of bulk density, aggregate size, and soil water suction on oxygen diffusion, redox potentials, and elongation of corn roots. Soil Sci. Soc.Am. Proc. 32:180-186.

Gregory, P.J., M. McGowan, P.V. Biscoe, and B. Hunter. 1978. Water relations of winter wheat. 1. Growth of the root system. J. Agric. Sci. (Cambridge) 91:91-102.

Grossman, R.B., and C.R. Berdanier. 1982. Erosion tolerance for cropland: Application of the soil survey data base. p. 113-130. *In* B.L. Schmidt (ed.) Determinants of soil loss tolerance. ASA, CSSA, and SSSA, Madison, WI.

Hansen, G.K. 1975. A dynamic continuous simulation model of water state and transportation in the soil-plant-atmosphere system. I. The model and its sensitivity. Acata Agric. Scand. 25:129-149.

Hillel, D., and H. Talpaz. 1976. Simulation of root growth and its effect on the pattern of soil water uptake by a nonuniform root system. Soil Sci. 121:307-312.

Hoogenboom, G., and M.G. Huck. 1986. ROOTSIMU V4.0—A dynamic simulation of root growth, water uptake, and biomass partitioning in a soil-plant-atmosphere continuum: Update and documentation. Alabama Agric. Exp. Stn. Agronomy and Soils Dep. Ser. 109.

Howard, D.D., and F. Adams. 1965. Calcium requirement for penetration of subsoils by primary cotton roots. Soil Sci. Soc. Am. Proc. 29:558-562.

Huck, M.G., and D. Hillel. 1983. A model of root growth and water uptake accounting for photosynthesis, respiration, transpiration, and soil hydraulics. p. 273-333. *In* D. Hillel (ed.) Advances in irrigation. Vol. 2. Academic Press, New York.

International Rice Research Institute (IRRI). 1978. IRRI annual report for 1977. Int. Rice Res. Inst., Los Banos, Philippines.

Jackson, M.B., and M.C. Drew. 1984. Effects of flooding on growth and metabolism of herbaceous plants. p. 47-128. *In* T.T. Kozlowski (ed.) Flooding and plant growth. Academic Press, New York.

Jones, C.A. 1983. Effect of soil texture on critical bulk densities for root growth. Soil Sci. Soc. Am. J. 47:1208-1211.

Jones, C.A., and J.R. Kiniry (ed.). 1986. CERES-Maize: A simulation model of maize growth and development. Texas A&M Univ. Press, College Station, TX.

Khalifa, E.M., and S.W. Buol. 1968. Studies of clay skins in a cecil (Typic Hapludult) soil: II. Effect on plant growth and nutrient uptake. Soil Sci. Soc. Am. Proc. 32:102–105.

Kramer, P.J. 1983. Water relations of plants. Academic Press, New York.

Lambers, H. 1983. 'The functional equilibrium', nibbling on the edges of a paradigm. Neth. J. Agric. Sci. 31:305–311.

Lambers, H. 1987. Growth, respiration, exudation, and symbiotic associations: The fate of carbon translocated to the roots. p. 125–145. In P.J. Gregory et al. (ed.) Root development and function. Cambridge Univ. Press, Cambridge, England.

Linn, D.M., and J.W. Doran. 1984. Effect of water-filled pore space on carbon dioxide and nitrous oxide production in tilled and nontilled soils. Soil Sci. Soc. Am. J. 48:1267–1272.

McLean, F.T., and B.E. Gilbert. 1927. The relative aluminum tolerance of crop plants. Soil Sci. 24:163–175.

Mengel, D.B., and S.A. Barber. 1974. Development and distribution of the corn root system under field conditions. Agron. J. 66:341–344.

Monteith, N.H., and C.L. Banath. 1965. The effect of soil strength on sugarcane growth. Trop. Agric. 42:293–296.

Mugwira, L.M., S.U. Patel, and A.L. Fleming. 1980. Aluminum effects on growth and Al, Ca, Mg, K, and P levels in triticale, wheat, and rye. Plant Soil 57:467–470.

Nielsen, N.E., and S.A. Barber. 1978. Differences among genotypes of corn in the kinetics of P uptake. Agron. J. 70:695–698.

O'Toole, J.C., and W.L. Bland. 1987. Genotypic variation in crop plant root systems. Adv. Agron. 41:91–145.

Pavan, M.A., F.T. Bingham, and P.F. Pratt. 1982. Toxicity of aluminum to coffee in ultisols and oxisols amended with $CaCO_3$, $MgCO_3$, and $CaSO_4 \cdot 2H_2O$. Soil Sci. Soc. Am. J. 46:1201–1207.

Pearson, R.W., L.F. Ratliff, and H.M. Taylor. 1970. Effect of soil temperature, strength, and pH on cotton seedling root elongation. Agron. J. 62:243–246.

Porter, J.R., B. Klepper, and R.K. Belfod. 1986. A model (WHTROOT) which synchronizes root growth and development with shoot development for winter wheat. Plant Soil 92:133–145.

Reid, D.A., G.D. Jones, W.H. Armiger, C.D. Foy, E.J. Koch, and T.M. Starling. 1969. Differential aluminum tolerance of winter barley varieties and selections in associated greenhouse and field experiments. Agron. J. 61:218–222.

Retta, A., C.Y. Sullivan, and D.G. Watts. 1984. Relationships of root length to root dry weight in grain sorghum. Sorg. Newsl. 27:146–147.

Reynolds, J.F., and J.H.M. Thornley. 1982. A shoot-root partitioning model. Ann. Bot. 49:585–597.

Ritchey, K.D., J.E. Silva, and U.F. Costa. 1982. Calcium deficiency in clayey B horizons of savanna oxisols. Soil Sci. 133:378–382.

Ritchie, J.T., and E. Burnett. 1971. Dryland evaporative flux in a subhumid climate: II. Plant influences. Agron. J. 63:56–62.

Rose, D.A. 1983. The description of the growth of root systems. Plant Soil 75:405–415.

Simpson, R.J., H. Lambers, and M.J. Dalling. 1982. Translocation of nitrogen in a vegetative wheat plant (*Triticum aestivum*). Physiol. Plant. 56:11–17.

Smucker, A.J.M. 1984. Carbon utilization and losses by plant root systems. p. 27–46. In Roots, nutrient and water influx, and plant growth. ASA, CSSA, and SSSA, Madison, WI.

Soil Survey Staff. 1972. Soil survey laboratory methods and procedures for collecting soil samples. USDA-SCS, Soil Survey Investigations Rep. 1. U.S. Gov. Print. Office, Washington, DC.

Stitt, R.E., D.K. Cassel, S.B. Weed, and L.A. Nelson. 1982. Mechanical impedance of tillage pans in Atlantic Coastal Plains soils and relationships with soil physical, chemical, and mineralogical properties. Soil Sci. Soc. Am. J. 46:100–106.

Stone, J.A., T.C. Kaspar, and H.M. Taylor. 1983. Predicting soybean rooting depth as a function of temperature. Agron. J. 75:1050–1054.

Stone, J.A., and H.M. Taylor. 1983. Temperature and the development of the taproot and lateral roots of four indeterminate soybean cultivars. Agron. J. 75:613–618.

Tackett, J.L., and R.W. Pearson. 1964. Effect of carbon dioxide on cotton seedling root penetration of compacted soil cores. Soil Sci. Soc. Am. Proc. 28:741–743.

Taylor, H.M. 1983. A program to increase plant available water through rooting modification. p. 463-472. *In* Root ecology and its practical application. Int. Symp., Gumpenstein. 27-29 Sept. 1982. Budndesanstalt fur alpenlandische Landwirtschalf, A-8952 Irdning.

Taylor, H.M., and E. Burnett. 1964. Influence of soil strength on the root growth habits of plants. Soil Sci. 89:174-180.

Taylor, H.M., and H.R. Gardner. 1963. Penetration of cotton seedling taproots as influenced by bulk density, moisture content and strength of soil. Soil Sci. 96:153-156.

Taylor, H.M., and B. Klepper. 1974. Water relations of cotton. I. Root growth and water use as related to top growth and soil water content. Agron. J. 66:584-588.

Taylor, H.M., G. M. Roberson, and J.J. Parker, Jr. 1966. Soil strength—root penetration relations for medium to coarse-textured soil materials. Soil Sci. 102:18-22.

Thornley, J.H.M. 1972a. A model to describe the partitioning of photosynthate during vegetative plant growth. Ann. Bot. 36:419-430.

Thornley, J.H.M. 1972b. A balanced quantitative model for root: shoot ratios in vegetative plants. Ann. But. 36:431-441.

Troughton, A. 1980. Environmental effects upon root-shoot relationships. p. 25-41. *In* D.N. Sen (ed.) Environment and root behaviour. Geobios Int., Jodhpur, India.

Trouse, A.C., Jr. 1965. Effects of soil compression on the development of sugar-cane roots. Proc. Int. Soc. Sugar-Cane Technol. 12:137-152.

Venkatraman, T.S., and R. Thomas. 1922. Sugarcane and root systems. Studies in development and anatomy. Agric. J. India 17:381-388.

Vepraskas, M.J., and G.S. Miner. 1986. Effects of subsoiling and mechanical impedance on root growth. Soil Sci. Soc. Am. J. 50:423-427.

Vine, P.N., R. Lal, and D. Payne. 1981. The influence of sands and gravels on root growth of maize seedlings. Soil Sci. 131:124-129.

Voorhees, W.B., R.R. Allmaras, and C.E. Johnson. 1981. Alleviating temperature stress. p. 217-266. *In* G.F. Arkin and H.M. Taylor (ed.) Modifying the root environment to reduce crop stresses. ASAE, St. Joseph, MI.

Voorhees, W.B., D.A. Farrell, and W.E. Larson. 1975. Soil strength and aeration effects on root elongation. Soil Sci. Soc. Am. Proc. 39:948-953.

Wang, J., J.D. Hesketh, and J.T. Woolley. 1986. Preexisting channels and soybean root patterns. Soil Sci. 141:432-437.

Weaver, J.E. 1958. Classification of root systems of forbs of grassland and a consideration of their significance. Ecology 39:393-401.

Zimmerman, R.P., and L.T. Kardos. 1961. Effect of bulk density on root growth. Soil Sci. 91:280-288.

7 Predicting Canopy Light-Use Efficiency from Leaf Characteristics

J. M. NORMAN
University of Wisconsin
Madison, Wisconsin

T. J. ARKEBAUER
University of Nebraska
Lincoln, Nebraska

Vegetation depends on light energy from the sun to convert CO_2 from the air to essential life-sustaining carbon compounds. Light has long been recognized as an essential factor in photosynthesis, which provides biochemical energy and carbon necessary for plant growth. The light-gathering system consists mainly of an aerial array of leaves, with a light trapping effectiveness that depends on many plant and environmental factors. The operation of this light-gathering system is a key determinant of plant productivity. A clearer understanding of it will help us to predict the growth of crops with more reliability.

The accumulated growth of any plant depends on the total carbon fixed by photosynthesis and the fraction of that carbon that can be converted to dry matter (DM). Although nutrients other than carbon also are essential to tissue growth, this chapter will consider only carbon. Only a portion of carbon fixed by photosynthesis eventually appears as standing DM. Some of the carbon is lost through plant respiration in two ways: (i) synthesis of compounds that form the final DM; and (ii) maintenance of the living complex in a functioning condition. These two forms of respiration are usually referred to as growth and maintenance. The distinction between them is somewhat arbitrary, so separation of the two components is difficult experimentally.

A simplistic view of the disposition of carbon in a growing plant can be considered as a combination of three processes: (i) photosynthetic carbon fixation; (ii) maintenance respiration; and (iii) growth respiration. Although considerable uncertainty exists in the knowledge of all three processes, the work of Penning deVries et al. (1974) has increased our confidence in growth respiration estimates so that carbon fixation and maintenance respiration represent greater uncertainty.

In this chapter, we will consider photosynthetic carbon fixation and its efficiency in terms of the mass of CO_2 fixed per unit of absorbed photosyn-

---

Copyright © 1991 ASA-CSSA-SSSA, 677 S. Segoe Rd., Madison, WI 53711, USA. *Modeling Plant and Soil Systems*—Agronomy Monograph no. 31.

thetically active radiation; called the photosynthetic light-use efficiency (LUE_p). Alternatively, the mass of DM produced per unit of absorbed photosynthetically active radiation (PAR) could be termed the dry matter light-use efficiency (LUE_{dm}). Clearly, LUE_{dm} involves maintenance and growth respiration, which may not depend on light directly, and photosynthesis, which does depend on light interception directly.

The concept of a relatively constant LUE_{dm} has great potential for simplifying the prediction of plant productivity. Since incoming solar radiation (SR) or PAR are relatively easy to measure, the simple product of intercepted or absorbed PAR and the appropriate canopy LUE_{dm} could provide an estimate of DM increment. This concept was used by Monteith (1977) to study the effect of climate on crop production in Britain. Monteith's approach was expanded by Charles-Edwards (1981), who carried out a very similar analysis. Numerous investigators have used this approach and measured the canopy LUE_{dm} for various crops. Monteith (1977) suggested a seasonal canopy LUE_{dm} value of 1.4 g DM(MJ ISR)$^{-1}$ (ISR represents intercepted solar radiation) based on above-ground data from apple (*Pyrus* L.) barley (*Hordeum* L.), sugar beet [*Beta vulgaris* (L.) var. Cicla], and potato (*Solanum tuberosum* L.). Gallagher and Biscoe (1978) found a seasonal value of 3.0 g DM(MJ APAR)$^{-1}$ (APAR represents absorbed PAR) for wheat (*Triticum aestivum* L.) and barley in Britain, including roots and tops. This is approximately equivalent to the value used by Monteith (1981) of 4.3 g CO_2 (MJ ISR)$^{-1}$, as well as the value of 1.4 g DM (MJ ISR)$^{-1}$ reported by Monteith (1977). Unsworth et al. (1984) measured a value of about 1.2 g DM(MJ ISR)$^{-1}$ on soybean [*Glycine max* (L.) Merr.] at various moderate levels of ozone treatment. Muchow and Coates (1987) determined seasonal sorghum LUE between 2.1 and 2.4 g DM(MJ IPAR)$^{-1}$ for intercepted PAR on a seasonally integrated basis. Charles-Edwards (1981) has summarized some values of canopy LUE from the literature and simply reported them as g DM MJ^{-1}: Rice (*Oryza sativa* L.), 4.2; corn (*Zea mays* L.), 3.4; sweet potato (*Ipomoea batatas* Lam.), 3.1; kale (*Brassica oleracea* var. Accephala), 2.7; sunflower (*Helianthus annuus* L.), 2.6; cotton (*Gossypium hirsutum* L.), 2.5; clover, 1.6; and soybean, 1.3. Unfortunately these values are difficult to interpret since the form of the radiation (solar or PAR) is not indicated.

The values of LUE reported in the literature vary by more than a factor of three. Either LUE is not very constant across crops, and thus of more limited usefulness than we might desire, or the values reported do not have a common basis. In fact both of these conditions may apply. The main objective of this chapter is to investigate the basis for a canopy LUE and relate various ways of expressing the numerical values.

I. CANOPY LIGHT-USE EFFICIENCY DEFINITIONS

Throughout the literature many definitions are used for canopy LUE; no standard form is apparent. A definition of canopy LUE involves three aspects: (i) the time interval involved—instantaneous, hourly, daily, weekly, or seasonal; (ii) the form of the carbon—dry matter above ground

(DMAG), total plant dry matter including roots (DMT), or net CO_2 uptake by the plant top (CO_2); and (iii) characterization of the radiation—solar radiation intercepted (ISR) or absorbed (ASR), or photosynthetically active radiation intercepted (IPAR) or absorbed (APAR). Intercepted radiation is (I in MJ m^{-2}) the radiation transmitted (T in MJ m^{-2}) to the bottom of the canopy subtracted from the incident radiation.

$$\text{IPAR} = I - T \quad [1]$$

Absorbed radiation is the difference between net radiation (PAR or SR) above the canopy (downward [I] minus upward [RC]) and the net radiation below the canopy (radiation transmitted [T] minus radiation reflected from the soil [RS]).

$$\text{APAR} = (I - \text{RC}) - (T - \text{RS}) \quad [2]$$

The units in Eq. [1] and [2] are MJ m^{-2} in the appropriate wave band (PAR or SR). In the PAR waveband, 1 MJ m^{-2} is 4.6 mol photon m^{-2} s^{-1} if the sun is the source of radiation.

The form of the fixed carbon affects the numerical values of canopy LUE. The conversion between g CO_2 and g DM depends on the composition of the plant because the amount of carbon required to build carbohydrates, proteins, and lipids is different. Furthermore, the amount of these constituents in various plants is different. Appropriate conversion factors can be obtained from Penning deVries et al. (1974) or McDermitt and Loomis (1981). When DM is used to express the carbon fixed, the roots are included only some of the time. Since the fraction of the total plant DM that is in roots may vary from 10% for some agronomic crops to 80% for some native prairie grasses, this is not a minor inconsistency. In addition, roots may lose significant amounts of carbon by sloughing, exudation, and respiration. Furthermore, with annuals, the amount of DM standing is known to be derived from carbon fixed in a given growing season. With perennials, the standing biomass may have been accumulated over several years and the contribution of a given season is difficult to quantify; especially if roots are a major component of carry-over between growing seasons. These uncertainties in respiration and root losses make it essential to obtain good estimates of the carbon fixed by photosynthesis if we are to advance our knowledge of the carbon budget.

Muchow and Coates (1987) discuss the likely effect of including root DM and intercepted vs. absorbed solar and photosynthetically active radiation in cereals. The various definitions of LUE can result in numerical values that differ by more than a factor of two for a given canopy and condition. This chapter contains a discussion of LUE and some interpretation of the concept.

II. MODEL DESCRIPTIONS

This chapter describes the photosynthetic and respiration response functions of individual leaves in various parts of a canopy and then integrates their various contributions to obtain the canopy net photosynthetic rate over some appropriate time interval. This approach requires that we first describe the dependence of leaf photosynthesis and respiration on various factors, such as light, temperature, water status, leaf spectral properties, and position in the canopy. Canopy carbon exchange rates are estimated from a model titled Cupid by combining the equations that describe leaf carbon exchange rates with a characterization of canopy architecture and incident radiation above the canopy. A description of canopy architecture includes the vertical distribution of stem and leaf area, leaf angle distributions, canopy height, and the horizontal distribution of foliage, such as random or clumped. Estimates of canopy photosynthetic LUE can be obtained from Cupid using various definitions.

The stand DM LUE can be obtained by combining the carbon fixed in photosynthesis with estimates of growth and maintenance respiration.

A. Leaf Model

The model that we use to predict leaf net photosynthetic rate is based on the work of von Caemmerer and Farquhar (1981). The leaf model equations used here are very similar to CULEAF, which is described with program listing in Norman (1986), except that the equation for describing the dependence of assimilation rate on temperature was obtained from Schoolfield et al. (1981). The basis of this model is the relation between net assimilation rate (A, μmol m^{-2} s^{-1}) and internal CO_2 concentration (C_i, μmol mol^{-1}). Von Caemmerer and Farquhar (1981) describe how C_i is calculated. The strength of this approach is that limitations to CO_2 uptake that occur in the diffusion path (primarily stomata) can be distinguished from limitations caused by the biochemistry of photosynthesis. Example data for A vs. C_i for soybean (C3) and sorghum [*Sorghum bicolor* (L.) Moench] (C4) leaves, which were obtained in the field under a range of incident light conditions, are given in Fig. 7-1. The leaf photosynthesis submodel of the Cupid model requires mathematical equations that capture the essential characteristics of the dependence of leaf assimilation rate on all the major environmental factors. Not only is incident light important, as shown in Fig. 7-1, but effects of leaf temperature, air vapor pressure, and plant water status are important as well. Measurements of these leaf responses on plants grown in controlled environments are usually not appropriate for estimating canopy photosynthesis of field-grown crops. Therefore, leaf gas-exchange measurements are made in the field with portable gas-exchange instruments.

The leaf response functions for photosynthesis, which are used in this paper, are typical of C4 (corn or sorghum) and C3 crops (soybean or wheat), and are shown in Fig. 7-2. The dependence of assimilation rates on internal CO_2 concentration and light from this model are shown in Fig. 7-2a for corn. Since the von Caemmerer and Farquhar (1981) leaf photosynthesis model is appropriate for C3 plants, the parameter values cannot be interpreted in the same way for a C4 plant. However, the parameters of the von Caemmerer and Farquhar (1981) model were adjusted to fit corn photosyn-

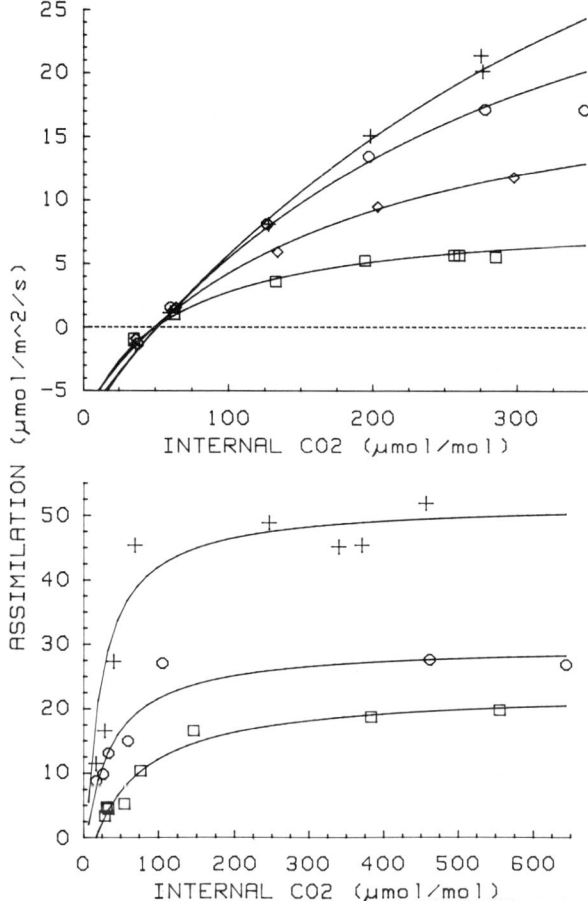

Fig. 7-1. The measured dependence of photosynthetic rate on internal CO_2 concentration for sorghum and soybean for a range of photosynthetically active radiation flux densities. In the upper graph (soybean), the symbols represent the following: + 1000, ○ 430, ◇ 220, and □ 120 μmol photons m^{-2} s^{-1}. In the lower graph (sorghum), the symbols represent the following: + 1900, ○ 900, and □ 400 μmol photons m^{-2} s^{-1}.

thetic and respiratory responses based on our own field measurements and those of Chmora and Oya (1967), Edmeades and Daynard (1979), and Vietor et al. (1977). Figure 7-2b contains a light-photosynthesis response for upper sunlit leaves of C3 and C4 crop types. The C3 parameters are essentially those of von Caemmerer and Farquhar (1981), except for the temperature response mentioned earlier in this section. Also, the maximum rate of electron transfer was increased over the value they used, so maximum rates of photosynthesis in full sunlight are about 50% greater.

The variation of leaf photosynthetic characteristics with depth in the canopy is an important factor to consider in the prediction of canopy photosynthesis from leaf photosynthesis, and enters into the Cupid model through

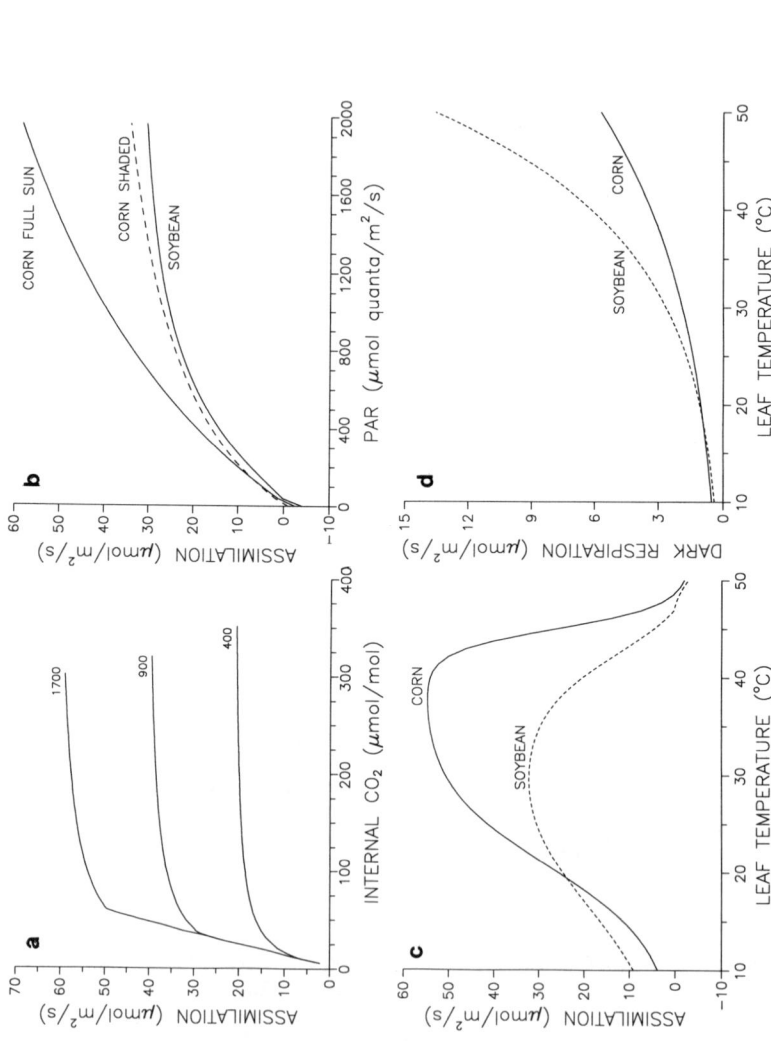

Fig. 7-2. The dependence of leaf assimilation rate on (a) internal CO_2 concentration for corn at several light levels (b) light for upper (solid line) and lower (dashed line) corn leaves and an upper wheat or soybean leaf (solid line); (c) temperature for corn (solid line) and wheat or soybean (dashed line); and (d) the dependence of dark respiration on temperature for corn (solid line) and wheat or soybean (dashed line).

the effects on a leaf's previous light history. The effect of the previous light history of a leaf, which affects the maximum electron transfer rate in full sunlight, the maximum rate of carboxylation and dark respiration, is related to the leaf area index above the particular leaf. Leaves lower in the canopy are less productive when exposed to the sun and more productive in the shade than the uppermost leaves (Fig. 7-2b). Typical leaf photosynthetic rate dependence on temperature is shown in Fig. 7-2c for C4 and C3 plants. This type of response curve is difficult to obtain and probably changes as a plant adapts to varying temperature environments (deWit, 1978). Measurements taken by Chmora and Oya (1967) on corn indicate moderate leaf photosynthetic rates at leaf temperatures above 50 °C. This is higher than upper limits measured in the field in Nebraska, but high temperature and leaf water stress are difficult to separate under field conditions. The simulations done for this chapter did not result in leaf temperatures above 35 °C, so uncertainties in the temperature curve above 40 °C do not affect these results. Curves such as those shown in Fig. 7-2c are difficult to take too seriously for predicting canopy photosynthesis from leaf photosynthesis. The curves may even vary with time of day (Chmora & Oya, 1967). They typically are obtained for a single leaf that will be experiencing a different temperature than the other nonenclosed leaves on the plant. Such leaves tend to show a more marked temperature dependence because of exposure to conditions that they have not had time to adapt to. deWit (1978) recognized this and, as part of his corn model, assumed that corn photosynthetic rate was independent of leaf temperature above 13 °C and linearly decreased to zero between 13 and 8 °C. This response is quite different from those of Chmora and Oya (1967) and Fig. 7-2c. In this chapter, we predict canopy LUE using the temperature dependence in Fig. 7-2c and compare LUE with the results that we obtain by assuming the photosynthetic response to temperature, similar to the method used by deWit (1978).

Stomatal conductance is calculated from leaf photosynthesis, leaf-to-air vapor pressure deficit, and leaf water potential by a method similar to Norman (1986). An unstressed stomatal conductance is calculated from the leaf photosynthetic rate (from the von Caemmerer & Farquhar [1981] model) by assuming C_i/C_a is constant. This ratio is known to vary by species and genotype, but we use 0.4 for C4 crops and 0.72 for C3 crops in this study. High leaf-to-air vapor pressure deficits and low leaf water potentials can result in stomatal closure. Leaf water potential does not result in stomatal closure because all simulations in this chapter are done with adequate soil moisture. The dependence of stomatal conductance on vapor-pressure deficit is controversial for agricultural crops because its magnitude may depend on growing conditions. Controlled-environment-grown plants tend to have a greater dependence on vapor pressure deficit than field grown plants. The complications associated with predicting the effect of vapor pressure deficit on stomatal conductance are formidable (Losch & Tenhunen, 1981; Schulze et al.,1987). Because of this uncertainty, C_i/C_a values listed above were chosen based on midday measurements from corn and soybean leaves when the vapor pressure deficit was approximately 2 to 3 kPa. Choosing C_i/C_a in this way implicitly incorporates a vapor pressure deficit effect in the estimation of stomatal conductance from photosynthesis, and accomplishes reasonable agreement between predicted and measured conductances. A clever approach has been suggested by Ball et al. (1986), but it remains to be determined whether it reduces uncertainty in stomatal conductance prediction under field conditions.

The leaf dark respiration rate also depends on temperature and, in this study, we assume a rate of 2.0 μmol m^{-2} s^{-1} at 25°C for corn or sorghum with a Q_{10} of about 1.8, and 1.1 μmol m^{-2} s^{-1} at 25°C for soybean or wheat (Fig. 7-2d).

Leaf reflectance and transmittance are characteristics that are essential for the prediction of radiation interception or absorption. Since LUE has been expressed in terms of both SR and PAR, we will consider radiation extinction for both these wavelength bands. This is best accomplished by obtaining leaf spectral properties in visible and near-infrared (NIR) wavelength bands, and solving the extinction and scattering equations separately for each of these bands. The light extinction distributions in the PAR are used directly for predicting photosynthesis and stomatal conductance of leaves and the canopy. The radiation penetration results from the PAR and NIR wavelength bands are added together to obtain the ASR or ISR for energy balance considerations. Table 7-1 contains some leaf spectral properties measured in the field on intact plants (Walter-Shea, 1987).

B. Canopy Model

The collective effect of all the leaves must be obtained to predict the canopy light use. The light, temperature, humidity, and wind environments appropriate for each leaf are obtained from a combined solution of the leaf energy budget for leaves in various angle classes and layers, and vertical profile equations for radiation and turbulent transfer (Norman, 1979; Norman & Campbell, 1983). An iterative solution technique is used to solve the combination of vertical profile equations for convective exchange (Norman & Campbell, 1983) and the leaf energy budget equations (Norman, 1979). The fluxes of CO_2, water vapor, energy, and radiation from leaves in various leaf angle classes are totaled to obtain the layer source or sink necessary for the solution of the vertical profile equations. The temperature, humidity, light, and wind from the solution to the vertical profile equations are necessary to the solution of the leaf energy budget equations. Of course, solution of a leaf energy budget equation requires a stomatal conductance for each angle class. Because photosynthetic rate is required to calculate stomatal conductance, the Farquhar-von Caemmerer (1982) model becomes a part of this iteration loop.

This methodology incorporates aerodynamic resistances above and within the canopy, and boundary layer resistances of individual leaves. Because of these additional resistances in the leaf-to-atmosphere water vapor path, stomatal conductance may exhibit minor control over transpiration until severe closure occurs, especially in canopies shorter than 1 or 2 m.

Table 7-1. Leaf reflectance and transmittance for corn and soybean in visible (PAR) and near-infrared (NIR) wavelength bands from field measurements on intact plants using an integrating sphere.

	Leaf reflectance		Leaf transmittance	
	PAR	NIR	PAR	NIR
	%			
Corn	9	38	4	45
Soybean	9	42	4	42

The Cupid model, which is used for the simulations reported in this chapter, incorporates many plant, atmospheric, and soil processes with considerable detail in an attempt to provide a defendable integration from the leaf level to the canopy level (Norman, 1979, 1986; Norman & Campbell, 1983). However, some processes that may affect canopy LUE have not been considered. For example, the fraction of light absorbed by leaves may depend on the incidence angle of the direct beam on the leaf (Walter-Shea, 1987). Because the model does not account for the dependence of light absorption on incident angle, it may slightly overestimate photosynthesis of some leaves. We also have ignored the fluctuating light conditions within the canopy. High frequency light fluctuations from wind effects may increase photosynthetic rates and low frequency fluctuations from sun angle shifts may decrease photosynthetic rates because of stomatal and photosynthetic-induction delays. Furthermore, direct effects of wind on stomata and the leaf angle distribution have not been considered, nor have effects of leaf movement. In fact, the only leaf angle distribution used for these simulations is the spherical distribution, which is reasonable for corn and vegetative stages of soybean growth.

The radiation extinction equations described by Norman (1979), which are used for random leaf positioning, can be modified for the clumping effect of rows by using the clumping factor defined by Nilson (1971). This effect improves light interception estimates for canopies of partial cover, but appears to be minor in crops such as corn or soybean as full cover is approached.

C. Respiration Considerations

The instantaneous net CO_2 taken up by a plant is the difference between CO_2 fixed by photosynthesis and that respired by growth and maintenance respiration processes. The net photosynthetic rate measured with leaf chambers in the field provides us with an estimate of the net CO_2 exchange for leaves. However, other organs enter into carbon exchange considerations, such as leaf sheaths, roots, ears, and tassels. The distinction between photosynthesis and respiration processes is clear in terms of substrates, but becomes blurred in terms of biochemical energy supply since both processes may be occurring in the same leaf cells. Since respiration in the light may be different from respiration in the dark (Sharp et al., 1984; Brooks & Farquhar, 1985), the approach of deWit (1978) seems desirable. deWit's approach uses CO_2 exchange measurements during day and night hours to estimate net carbon exchange of leaves, and considers the growth and maintenance requirements of nonphotosynthesizing organs separately. Another approach would estimate gross photosynthesis of leaves and then consider leaf growth and maintenance requirements by methods similar to those used for nonphotosynthesizing organs, but perhaps with different coefficients (Baker et al., 1972). Measurements of stand photosynthesis with canopy enclosures often are confounded by soil CO_2 evolution. Therefore, measurements in the light are followed immediately with measurements in a darkened chamber to obtain an estimate of gross canopy photosynthetic rate. This is valid to the extent that respiration in the dark is equal to respiration in the light, which may be true for C4 plants like corn. Obviously, a perfect separation of photosynthesis and respiration gas exchanges is very difficult. The best methods may depend on how one plans to use measurements to verify any particular model.

Considering the detailed carbon budget for an entire season, as McCree (1988) did for sorghum, is beyond the scope of this chapter; therefore, let us consider a corn plant just prior to tasseling so that vegetative components dominate. Based on data for corn from Foth (1962), deWit (1978), Yao (1980), and Righes (1980), a typical stand just prior to tasseling in the USA Corn Belt might have the characteristics shown in Table 7-2. The final seasonal grain yield for this typical corn stand might be 1 kg m^{-2} (175 bu acre^{-1}).

For the purpose of this chapter, we will assume that corn leaves are not growing, which is valid just prior to tasseling. We further assume that the net photosynthesis and nighttime respiration relations given in Fig. 7-2 will represent the net carbon source provided by a leaf. Thus, the carbon cost of phloem loading, various carbon transformations, and maintenance respiration of leaves during the day are accommodated in the net photosynthesis relations. Maintenance respiration of leaves at night is accounted for by the dark respiration vs. temperature relation. At 25 °C the daily dark respiration for a leaf respiration rate of 2.0 μmol m^{-2} s^{-1} assumed for corn compares with the daily maintenance coefficient of 0.03 g glucose (g dry wt.)$^{-1}$ d^{-1}, which is provided by McCree (1988) for sorghum leaves, for the stand characteristics summarized in Table 7-2.

The maintenance requirements of leaf sheaths are assumed to be offset by their photosynthesis, the maintenance respiration of the stem is assumed half that of leaves (McCree, 1988), and the maintenance respiration of roots is assumed to be equal to that of leaves because of sloughing, exudation, and ion exchange (Amthor, 1984; Lambers, 1985, 1987). Much uncertainty exists about carbon lost in the root system. Measurements by Andre et al. (1978) of root respiration and exudation are consistent with this assumption of a maintenance coefficient for roots about equal to that of leaves. McCree (1988) assumed that roots had one-third the maintenance cost of leaves, which seems low. Of course these root maintenance costs decrease during grain fill. Thus, at 25 °C the maintenance respiration of roots and stems just before tasseling would be 0.03 and 0.015 g glucose (g dry wt.)$^{-1}$ d^{-1} respectively

Table 7-2. Typical characteristics of a corn stand just before tasseling in the Corn Belt, USA.

Characteristic	Value and unit of measurement
Planting density	6 plants m^{-2}
Height	2.0 m
Row spacing	0.76 m
Leaf area index (LAI)	3.5
Leaf dry weight/leaf area	77 g m^{-2}
Dry weight/ground area:	
Total vegetative	900 g m^{-2}
Leaves	270 g m^{-2}
Sheaths	135 g m^{-2}
Roots	90 g m^{-2}
Stems	405 g m^{-2}

with a Q_{10} of 2.0. These maintenance coefficients for leaves, roots, and stems in units of g glucose (g dry wt.)$^{-1}$ d^{-1} of 0.03, 0.03, and 0.015, respectively, are referred to as Case A in Table 7-7, and represent large coefficients for maintenance respiration. A second example representing lower maintenance requirements uses a nighttime leaf dark respiration rate of 1.4 μmol CO_2 m^{-2} s^{-1}, which agrees well with gas exchange measurements from Nebraska by the authors. From simulations, this leaf respiration rate agrees with a daily maintenance respiration coefficient of 0.02 g glucose (g dry wt.)$^{-1}$ d^{-1}. We assume a root maintenance coefficient one-third that of leaves (McCree, 1988) and a stem maintenance coefficient 1/10 that of leaves. Thus, the lower maintenance respiration case (Case B in Table 7-7) uses maintenance coefficients for leaves, stems, and roots in units of g glucose (g dry wt.)$^{-1}$ d^{-1} of 0.02, 0.0067, and 0.002.

The growth respiration can be computed from the work of Penning deVries based on biochemical pathways using a consistent composition typical of corn (Vertregt & Penning deVries, 1987). The cost of converting glucose to DM, termed the production value inverse (PVI) by Vertregt and Penning deVries (1987), permits an estimate of the DM increment that could be realized from a given amount of net photosynthesis that has had maintenance costs subtracted. For corn, the PVI is 1.42 g glucose (g dry wt.) (Vertregt & Penning deVries, 1987).

III. MODEL COMPARISONS WITH MEASUREMENTS

The radiation model that is used in this chapter is based on several assumptions including random leaf positioning in the horizontal, azimuthal symmetry, and a well-defined leaf inclination distribution, such as spherical, conical or vertical. This kind of model has been compared with measurements from several crops of full cover and shown to perform well in both PAR (Norman et al., 1971; Norman, 1988) and NIR (Norman et al., 1971) wavelength bands.

During the summer of 1981, measurements of intercepted PAR and canopy photosynthesis were made in corn stands of varying planting densities near

O'Neill, NE through a combined effort by Joe Ritchie (Michigan State Univ., East Lansing, MI), Dan Knievel (Penn State Univ., University Park, PA), Don Reicosky (Univ. of Minnesota, Morris, MN), and John Norman (Univ. of Nebraska, Lincoln, NE) (1981, unpublished data). The planting densities varied from 2.3 to 7.4 plants m^2, the leaf area index (LAI) varied from about 1.3 to 3.7, and the date of measurement was 4 August. The measured LUE was 6.9 g CO_2 (MJ IPAR)$^{-1}$. This LUE is in good agreement with the results of the Cupid model, which averaged 6.8 g CO_2 (MJ IPAR)$^{-1}$ over 17 instantaneous periods that correspond to chamber measurements at the three LAIs of 1.3, 2.2, and 3.7.

IV. REVIEW OF LIGHT-USE EFFICIENCY MEASUREMENTS

Measurements of PAR intercepted by a canopy and canopy photosynthesis can be used to estimaste short-term (minutes to hours) canopy LUE. Care must be taken to eliminate soil respiration. Chambers can also be used to estimate CO_2 LUE over longer periods of days, weeks, or even months. Measurements of IPAR and DM over longer periods can be used to estimate DM LUE. We will limit this discussion to canopies of full cover (LAI > 3).

Jones et al. (1986) measured intercepted PAR and canopy (including soil) carbon exchange rates on corn at 50% silking and 6 d later. Assuming a soil respiration rate of 4 μmol CO_2 m^{-2} s^{-1}, at solar noon the LUE was 8.1 and 5.7 g CO_2 (MJ IPAR)$^{-1}$ respectively. These values approach predictions from Cupid presented in the next section.

Jones et al. (1985a) measured soybean canopy LUEs in outdoor controlled-environment chambers at Growth Stages R4 and R5 about 60 to 80 d after planting. The average daytime LUEs were about 6.0 g CO_2 (MJ PAR)$^{-1}$ for the LAI of 3.3. Jones et al. (1985b) determined a seasonal LUE of about 5.0 g CO_2 (MJ IPAR)$^{-1}$ on soybean. Jones et al. (1987, unpublished data) provided unpublished data on tomato for plants grown in outdoor chambers at two nighttime temperatures. The LUE was 6.2 g CO_2 (MJ IPAR)$^{-1}$ for a 20 °C night temperature and 5.6 g CO_2 (MJ IPAR)$^{-1}$ for a 12 °C night temperature.

The LUE was measured for a wheat canopy near midday about 40 to 70 d after sowing when the LAI was between 3.5 and 5.0 (Puckridge & Ratkowsky, 1971). After correcting for soil respiration, the LUE = 4.2 g CO_2 (MJ IPAR)$^{-1}$ for an incident PAR of 1600 μmol photons m^{-1} s^{-1}. At incident PAR of 640 and 320 μmol photons m^{-2} s^{-1}, the LUE was 6.3 and 7.1 g CO_2 (MJ IPAR)$^{-1}$. These values of LUE are similar to those reported for soybean. Spiertz and van deHaar (1978) also reported a midday LUE of 4.7 g CO_2 (MJ IPAR)$^{-1}$ for wheat about 1 wk after anthesis (23 June), with LAI $\approx$ 4 and incident PAR = 1800 μmol photons m^{-2} s^{-1}. Their LUE was 6.9 and 7.1 g CO_2 (MJ IPAR)$^{-1}$ at 700 and 350 μmol photons m^{-2} s^{-1} incident PAR. The daily LUE from Spiertz and van deHaar (1978) was 4.7 g CO_2 (MJ IPAR)$^{-1}$ for 23 June, which had a total incident solar radiation of 22 MJ m^{-2}. The measurements made by Spiertz and van de-

Haar (1978) during the week surrounding 23 June can be used to estimate an average LUE of 2.2 g DMAG (MJ IPAR)$^{-1}$; this DM value is 47% of the daily CO_2 LUE.

Estimates of DM LUE vary between 2.9 (Williams et al., 1965), 3.2 (Yao, 1980), 3.8 (Sivakumar & Virmani, 1984), and 4.4 g DMAG (MJ IPAR)$^{-1}$ (Griffin, 1980) for corn grown under field conditions. This is a wide range of values and a clearer understanding of the reason for this range would be most useful. Light-use efficiency estimates for sorghum are similar; 2.4 (Muchow & Coates, 1986), 2.9 (Sivakumar & Virmani, 1984), and 3.0 g DMAG (MJ IPAR)$^{-1}$ (Steiner, 1986).

Hesketh and Baker (1967) suggested many years ago that canopy LUE may be a relatively constant quantity based on literature data for corn and cotton. The data on corn that they cited from Baker and Musgrave (1964) indicated 3.0 g CO_2 (MJ ISR)$^{-1}$. However, the in-canopy light sensor was not a solarimeter and, even though they tried to correct for wavelength peculiarities, the LUE probably is more appropriately expressed as 6.4 g CO_2 (MJ IPAR)$^{-1}$. Because of the light sensor used, Baker and Musgrave's (1964) LUE should be between the model values for IPAR (~7 g CO_2 MJ^{-1}) and ISR (~4 g CO_2 MJ^{-1}).

V. INTERPRETATIONS FROM THE CUPID MODEL

The plant-environment model Cupid provides a means for studying the characteristics of crop LUE. Using hourly solar radiation and weather data, along with various soil and plant characteristics (Norman & Campbell, 1983), hourly canopy photosynthesis and radiation penetration are predicted with Cupid.

The simulations from Cupid are based on 8 d of hourly solar radiation and weather data collected over corn at Garden City, KS by J. Steiner and E.T. Kanemasu of Kansas State Univ. (1981, unpublished data). They also measured the LAI (2.8) and crop height (2 m). Table 7-3 summarizes the

Table 7-3. Daily summary of solar radiation, air temperature, air vapor pressure, and mean wind speed for the 8 d used in simulations.

Day no.	Solar radiation	Air temperature			Vapor pressure	Wind speed
		Mean	Maximum	Minimum		
	MJ		°C		KPa	m s^{-1}
201	29.7	27.5	35.0	18.2	2.25	2.0
202	29.6	28.0	37.5	21.2	2.32	3.2
203	14.1	22.1	26.1	18.4	2.24	2.6
204	23.4	26.6	33.4	19.9	2.32	2.0
205	19.5	27.6	35.5	22.3	2.24	2.5
206	21.0	23.7	29.9	17.8	2.27	1.7
207	15.3	21.9	25.5	19.2	2.39	2.3
208	16.9	21.8	27.8	17.8	2.20	2.2

solar radiation and weather data for the 8 d. These 8 d represent a wide range of radiation and temperature conditions typical of the corn belt in the central USA.

The LUEs of a C4 crop (typical of corn) and a C3 crop (typical of soybean) were calculated hour-by-hour with Cupid for each of the 8 d. The corn canopy was assumed to have a spherical leaf-angle distribution. We will use the symbolism.

$$Q(CO_2, IPAR) = LUE \qquad [3]$$

where the quantities in parentheses refer to the basis for the efficiency definition. The values for $Q(CO_2, IPAR)$ include daytime photosynthesis plus nighttime dark respiration. This is the sum of measurements made over the canopy with a leaf gas exchange system. The hourly values of $Q(CO_2, IPAR)$ varied from about 6 to 10 g CO_2 (MJ IPAR)$^{-1}$ for corn and 4 to 12 g CO_2 (MJ IPAR)$^{-1}$ for soybean (Fig. 7-3). The upper values for soybean, which occur only in the early morning hours, are close to the photochemical efficiency of about 12 g (MJ APAR)$^{-1}$. Clearly, many factors can affect both light penetration and canopy photosynthesis on an hourly basis. One factor that appears to be causing variation in canopy LUE is the fraction of radiation above the canopy in the form of direct beam (Fig. 7-3). When the radiation above the canopy is mainly diffuse, the canopy LUE

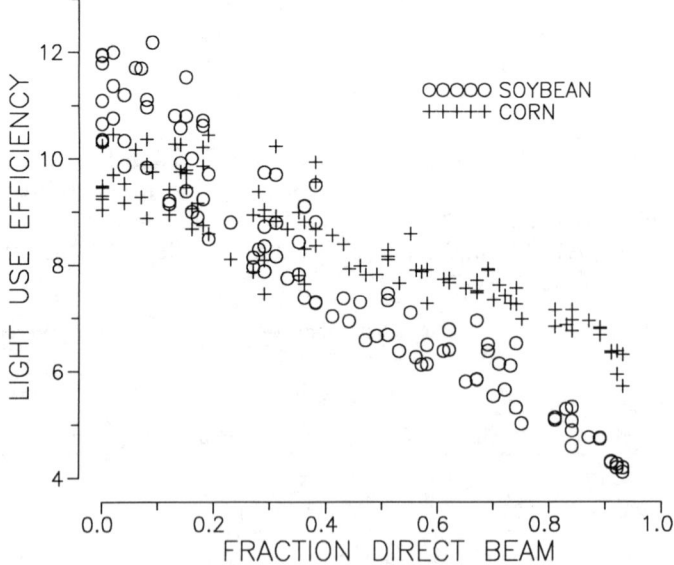

Fig. 7-3. The dependence of predicted light-use efficiency [g CO_2 (MJ IPAR)$^{-1}$] on the fraction of photosynthetically active radiation above the canopy that is from a direct beam. The canopy light-use efficiency is based on intercepted PAR and results are for C3 (o) and C4 (+) canopies.

is larger. This occurs because diffuse light is more uniformly and efficiently distributed over a canopy of leaves that may saturate at high light intensities. When the radiation incident on a canopy is diffuse, the illumination levels usually are low, so that the quantity of intercepted radiation is low even though the LUE may be high. The daily LUE is calculated from the ratio of daily net photosynthesis and daily IPAR. Obviously, the daily LUE should not be calculated from an average of the hourly LUEs. The results in Fig. 7-3 incorporate many factors besides the fraction of beam radiation above the canopy (e.g., temperature, sun zenith angle, vapor pressure deficit, and magnitude of the incident radiation). However, from various tests the dominant factor is the fraction of diffuse radiation.

The daily LUEs do not vary as much as the hourly values do. The average of the eight daily LUEs for soybean are about 85% those of corn (Table 7-4). On overcast days the LUE of corn and soybean are similar, and soybean may have greater efficiency on cool days. On clear, hot days the LUE of soybean is about three-quarters that of corn. Apparently the occurrence of two clear, three overcast, and three partly cloudy days resulted in greater LUEs for corn than soybean. However, if leaf characteristics did not change, we would expect corn to be more efficient in predominantly clear, warm climates and soybean more efficient in cloudy, cool climates. Both C3 and C4 plant types adapt to their radiation environments, so such simple-minded statements may not hold. To simulate adaptation, Cupid simulations assumed that photosynthetic rate was independent of temperature (dark respiration remained temperature dependent) and near the maximum rate of 30 °C for C3 and 35 °C for C4 canopies (Table 7-5). With this assumption, LUEs increased about 5 to 10%.

The conversion factors between absorbed or intercepted, PAR or SR can be confusing. Values for all four combinations are included in Table 7-4. Gallagher and Biscoe (1978) estimated the fraction of PAR absorbed from the fraction of solar radiation transmitted. Their Eq. [A.3] yields absorbed PAR fractions that are smaller than those obtained from the Cupid model, assuming the ISR is given. If LAI is varied from 1.5 to 5, under a range of incident radiation conditions, the APAR fraction of Gallagher and

Table 7-4. Predicted mean and standard deviation (in parentheses) of eight daily light-use efficiencies for typical C3 and C4 canopies at several leaf area indices using measured weather data summarized in Table 2.

Canopy	LAI	$Q(CO_2, IPA)$†	$Q(CO_2, APAR)$‡	$Q(CO_2, ISR)$§	$Q(CO_2, ASR)$¶
C4	1	7.6 (0.4)	7.7 (0.4)	4.7 (0.2)	5.8 (0.3)
(Corn)	3	7.5 (0.5)	7.8 (0.5)	4.2 (0.3)	5.3 (0.4)
	5	7.2 (0.5)	7.5 (0.5)	3.7 (0.2)	4.8 (0.3)
C3	1	6.5 (0.8)	6.6 (0.9)	4.1 (0.5)	5.0 (0.7)
(Soybean	3	6.5 (1.0)	6.7 (1.0)	3.6 (0.5)	4.6 (0.7)
or wheat)	5	6.2 (1.0)	6.5 (1.1)	3.2 (0.5)	4.2 (0.7)

† IPAR represents intercepted photosynthetically active radiation.
‡ APAR represents absorbed photosynthetically active radiation.
§ ISA represents intercepted solar radiation.
¶ ASR represents absorbed solar radiation.

Table 7-5. Mean and standard deviation (in parentheses) of eight daily light-use efficiencies predicted from Cupid for typical C3 and C4 canopies assuming that leaf photosynthetic rate is independent of temperature. The same input data were used as in Table 4.

Canopy	LAI	Q(CO_2, IPAR)†
C4	1	8.2 (0.5)
(Corn)	3	8.0 (0.6)
	5	7.7 (0.6)
C3	1	6.4 (0.7)
(Soybean	3	6.3 (0.8)
or wheat)	5	6.0 (0.8)

† IPAR represents intercepted photosynthetically active radiation.

Table 7-6. The ratios of various combinations of absorbed (A) or intercepted (I), photosynthetically active radiation (PAR) or solar radiation (SR).

Ratio	Leaf area index		
	1.5	2.8	5.0
IPAR/APAR	1.03	1.04	1.05
IPAR/ISR	0.60	0.56	0.51
IPAR/ASR	0.75	0.71	0.66

† Light-use efficiency in g CO_2 (MJ IPAR)$^{-1}$ would be multiplied by the number in the table to correct it to another radiation base. For example 8.0 g CO_2 (MJ IPAR)$^{-1}$ is equivalent to 4.8 g CO_2 (MJ ISR)$^{-1}$ at a leaf area index of 1.5.

Biscoe (1978) varies from 0.67 to 0.97, assuming a spherical leaf-angle distribution and a mean sun incidence angle of 45° in their Eq. [A.4]. Results from Cupid vary from 0.64 to 0.95; thus the Gallagher and Biscoe (1978) Eq. [4] appears to be quite reasonable if ISR is calculated from exponential extinction. If the ISR values calculated from Cupid are used in the Gallagher and Biscoe (1978) Eq. [A.3], the resultant predicted values of APAR vary from 0.53 to 0.83. Thus, results from Cupid suggest that APAR values estimated using their Eq. [A.3] and measured values of ISR with solarimeters should be increased by a factor of 1.25 for a LAI of about 1.5 to a factor of 1.15 for a LAI of about 5. The ratios of various combinations of absorbed or intercepted, PAR or SR are shown in Table 7-6.

The values of LUE in Tables 7-4 and 7-5 are in reasonable agreement with literature values discussed earlier based on CO_2 exchange. However, establishing a common basis for comparison is not easy because of some uncertainty in how maintenance and growth respiration should be accommodated. Because of these difficulties, we will only convert the CO_2 LUE to DM for the corn characteristics given in Table 7-2. The various components of the carbon balance of this corn canopy near tasseling are given in Table 7-7. The final DM LUE of 2.5 g DM MJ^{-1} for Case A is lower than values measured for corn and summarized earlier. The value of 3.2 g DM MJ^{-1} for Case B with lower maintenance respiration values is more typical of measured LUEs, which typically vary between 3 and 4 g DM MJ^{-1}.

Table 7-7. Mean daily carbon exchanges for the corn canopy described in Table 7-2.

	Canopy carbon exchange	
	Case A	Case B
	g CO_2 m^{-2} d^{-1}	
Photosynthesis		
Gross	73.7 (14.0)	73.7 (14.0)
Gross minus leaf respiration	61.9 (12.0)	65.8 (12.7)
Net (Gross minus total maintenance)	49.6 (11.0)	63.8 (12.5)
Maintenance respiration		
Stem	8.8 (1.4)	1.2 (0.2)
Leaf	11.8 (1.9)	7.9 (1.3)
Root	3.5 (0.2)	0.8 (0.1)
Grain	0	0
Total	24.1 (2.4)	9.9 (1.3)
Growth respiration	10.0 (2.1)	12.0 (2.3)
Dry matter increment (g DM m^{-2} d^{-1})	22.6 (5.1)	29.2 (5.7)
IPAR† (MJ m^{-2} d^{-1})	9.0 (2.4)	9.0 (2.4)
Light-use efficiency (CO_2, PAR)	6.9 g CO_2 MJ^{-1}	7.1 g CO_2 MJ^{-1}
Light-use efficiency (DM, PAR)	2.5 g DM MJ^{-1}	3.2 g DM MJ^{-1}

† IPAR represents intercepted photosynthetically active radiation.

VI. SUMMARY

The plant-environment model Cupid is useful in studying canopy LUE by combining knowledge of leaf physiological and radiative properties with canopy architecture. Hourly canopy LUEs vary more than daily values, especially for C3 crops. Measured and modeled short-term (<1 h) canopy LUEs agree well at values near 6.5 g CO_2 (MJ IPAR)$^{-1}$ for corn. A modeled value of canopy LUE over a week, based on DM, is sensitive to choice of maintenance respiration coefficients. A 25% difference resulted between two reasonable choices of coefficients from the literature.

Cupid has been used to predict conversion factors among intercepted or absorbed, SR or PAR. These may be useful for resolving the many ways that canopy LUEs are expressed in the literature.

The canopy LUE can vary with environmental conditions (radiation and temperature), leaf area index, and maintenance and growth respiration. Maintenance respiration coefficients are difficult to measure and are known to vary among cultivars of a single species. These variations in maintenance respiration are not likely to be accommodated indirectly using only a LUE to estimate DM increment from intercepted light. Dry matter LUE coefficients should be used with caution in simple models because of the potential for systematic errors. Furthermore, use of such coefficients may obscure possibilities for increased productivity by reductions in maintenance respiration.

REFERENCES

Amthor, J.S. 1984. The role of maintenance respiration in plant growth. Plant Cell Environ. 7:561–569.

Andre, M., D. Massimino, and A. Daguenet. 1978. Daily patterns under the life cycle of a maize crop. II. Mineral nutrition, root respiration and root excretion. Physiol. Plant 44:197–204.

Baker, D.N., J.D. Hesketh, and W.G. Duncan. 1972. Simulation of growth and yield in cotton: I. Gross photosynthesis, respiration and growth. Crop Sci. 12:431-439.

Baker, D., and R. Musgrave. 1964. Photosynthesis under field conditions. V. Further plant chamber studies of the effects of light on corn. Crop Sci. 4:127-131.

Ball, J.T., I.E. Woodrow, and J.A. Berry. 1986. A model predicting stomatal conductance and its contribution to the control of photosynthesis under different environmental conditions. p. 221-224. *In* J. Biggins (ed.) Progress in photosynthesis research, Vol. 4. Martinus Nijhoff Publ., Dordrecht, Netherlands.

Brooks, A., and G.D. Farquhar. 1985. Effect of temperature on the CO_2/O_2 specificity of ribulose-1,5-bisphosphate carboxylase/oxygenase and the rate of respiration in the light. Planta 165:397-406.

Charles-Edwards, D.A. 1981. Physiological determinants of crop growth. Academic Press, New York.

Chmora, S.N., and V.M. Oya. 1967. Photosynthesis in leaves as a function of temperature. Sov. Plant Physiol. (Engl. Transl.) 14:513-519.

deWit, C.T. 1978. Simulation of assimilation, respiration and transpiration of crops. John Wiley & Sons, New York.

Edmeades, G.O., and T.B. Daynard. 1979. The relationship between yield and photosynthesis at flowering in individual maize plants. Can. J. Plant Sci. 59:585-601.

Farquhar, G.D., and S. von Caemmerer. 1982. Modeling of photosynthetic response to environmental conditions. p. 549-588. *In* O.L. Lange et al. (ed.) Physiological plant ecology II. Encyclopedia of plant physiology. New ser. vol. 12B. Springer-Verlag, Berlin.

Foth, H.D. 1962. Root and top growth of corn. Agron. J. 54:49-52.

Gallagher, J.N., and P.V. Biscoe. 1978. Radiation absorption, growth and yield of cereals. J. Agric. Sci. (Cambridge) 91:47-60.

Griffin, J.L. 1980. Quantification of the effects of water stress on corn growth and yield. M.S. thesis. Univ. of Missouri, Columbia.

Hesketh, J., and D. Baker. 1967. Light and carbon assimilation by plant communities. Crop Sci. 7:285-293.

Jones, J.W., B. Zur, and J.M. Bennett. 1986. Interactive effects of water and nitrogen stresses on carbon and water vapor exchange of corn canopies. Agric. For. Meteorol. 38:113-126.

Jones, P., L.H. Allen, Jr., and J.W. Jones. 1985a. Responses of soybean canopy photosynthesis and transpiration to whole-day temperature changes in different CO_2 environments. Agron. J. 77:242-249.

Jones, P., J.W. Jones, and L.H. Allen, Jr. 1985b. Seasonal carbon and water balances of soybeans grown under stress treatments in sunlit chambers. Trans. ASAE 28:2021-2028.

Lambers, H. 1985. Respiration in intact plants and tissues: Its regulation and dependence on environmental factors, metabolism and invaded organisms. p. 418-473. *In* R. Douce and D.A. Day (ed.) Encyclopedia of plant physiology, Vol. 18. Springer-Verlag, Berlin, West Germany.

Lambers, H. 1987. Growth, respiration, exudation and symbiotic associations: The fate of carbon translocated to the roots. p. 125-145. *In* P.J. Gregory, J.V. Lake, and D.A. Rose (ed.) Root development and function. Cambridge Univ. Press, Cambridge, England.

Losch, R., and J.D. Tenhunen. 1981. Stomatal responses to humidity-phenomenon and mechanism. p. 137-161. *In* P.G. Jarvis and T.A. Mansfield (ed.) Stomatal physiology. Cambridge Univ. Press, Cambridge, England.

McCree, K.J. 1988. Sensitivity of sorghum grain yield to ontogenetic changes in respiration coefficients. Crop Sci. 28:114-120.

McDermitt, D.K., and R.S. Loomis. 1981. Elemental composition of biomass and its relation to energy content, growth efficiency and growth yield. Ann. Bot. 48:275-290.

Monteith, J.L. 1977. Climate and the efficiency of crop production in Britain. Philos. Trans. R. Soc. London B. 281:277-294.

Monteith, J.L. 1981. Climate variation and the growth of crops. Q.J.R. Meteorol. Soc. 107:749-774.

Muchow, R.C., and D.B. Coates. 1986. An analysis of the environmental limitations to yield of irrigated grain sorghum during the dry season in tropical Australia using a radiation interception model. Aust. J. Agric. Res. 37:135-148.

Nilson, T. 1971. A theoretical analysis of the frequency of gaps in plant stands. Agric. Meteorol. 8:25-38.

Norman, J.M. 1979. Modeling the complete crop canopy. p. 249–277. *In* B.J. Barfield and J. Gerber (ed.) Modification of the aerial environment of crops. Am. Soc. Agric. Eng., St. Joseph, MI.

Norman, J.M. 1986. Instrumentation use in a comprehensive description of plant-environment interactions. p. 149–307. *In* W. Gensler (ed.) Advanced agricultural instrumentation. Martinus Nijhof Publ., Dordrecht, Netherlands.

Norman, J.M. 1988. Synthesis of canopy processes. p. 161–175. *In* G. Russell, B. Marshall, and P.G. Jarvis (ed.) Plant canopies: Their growth, form and function. Cambridge Univ. Press, Cambridge, England.

Norman, J.M., and G.S. Campbell. 1983. Application of a plant-environment model to problems in irrigation. p. 155–188. *In* D.I. Hillel (ed.) Advances in irrigation. Academic Press, New York.

Norman, J.M., E.E. Miller, and C.B. Tanner. 1971. Light intensity and sunfleck size distributions in plant canopies. Agron. J. 63:743–748.

Penning deVries, F.W.T., A.H.M. Brunsting, and H.H. van Laar. 1974. Products, requirements and efficiency of biosynthesis: a quantitative approach. J. Theor. Biol. 45:339–377.

Puckridge, D.W., and D.A. Ratkowski. 1971. Photosynthesis of wheat under field conditions: IV. The influence of density and leaf area index on the response to radiation. Aust. J. Agric. Res. 22:11–20.

Righes, A.A. 1980. Water uptake and root distribution of soybeans, grain sorghum and corn. M.S. thesis. Iowa State Univ., Ames.

Schoolfield, R.M., P.J.H. Sharp, and C.E. Magnuson. 1981. Non-linear regression of biological temperature-dependent rate models based on absolute reaction-rate theory. J. Theor. Biol. 88:719–731.

Schulze, E.D., N.C. Turner, T. Gollan, and A. Shackel. 1987. Stomatal responses to air humidity and to soil drought. p. 311–321. *In* E. Zeiger, G.D. Farquhar, and I.R. Cowan (ed.) Stomatal function. Stanford Univ. Press, Stanford, CA.

Sharp, R.E., M.A. Matthews, and J.S. Boyer. 1984. Kok effect and the quantum yield of photosynthesis. Plant Physiol. 75:95–101.

Sivakumar, M.V.K., and S.M. Virmani. 1984. Crop productivity in relation to interception of photosynthetically active radiation. Agric. For. Meteorol. 31:131–141.

Spiertz, J.H.J., and H. van DeHaar. 1978. Differences in grain growth, crop photosynthesis and distribution of assimilates between semi-dwarf and a standard cultivar of wheat. Neth. J. Agric. Sci. 26:233–249.

Steiner, J.L. 1986. Dryland grain sorghum water use, light interception and growth responses to planting geometry. Agron. J. 78:720–726.

Unsworth, M.H., V.M. Lesser, and A.S. Heagle. 1984. Radiation interception and the growth of soybeans exposed to ozone in open-top field chambers. J. Appl. Ecol. 21:1059–1079.

Vertregt, N., and F.W.T. Penning deVries. 1987. A rapid method for determining the efficiency of biosynthesis of plant biomass. J. Theor. Biol. 128:109–119.

Vietor, D.M., R.P. Ariyanayagam, and R.B. Musgrave. 1977. Photosynthetic selection of *Zea mays* L. I. Plant age and leaf position effects and a relationship between leaf and canopy rates. Crop Sci. 17:567–573.

von Caemmerer, S., and G.D. Farquhar. 1981. Some relationships between the biochemistry of photosynthesis and the gas exchange of leaves. Planta 153:367–387.

Walter-Shea, E.A. 1987. Laboratory and field measurements of leaf spectral properties and canopy architecture and their effects on canopy reflectance. Ph.D. diss. Univ. of Nebraska, Lincoln (Diss. Abstr. 87-17268).

Williams, W.A., R.S. Loomis, and C.R. Lepley. 1965. Vegetative growth of corn as affected by population density. I. Productivity in relation to interception of solar radiation. Crop Sci. 5:211–215.

Yao, N.R. 1980. Vegetative and reproductive development of corn at four spring planting dates. M.S. thesis. Penn State Univ., University Park.

8 Soil Spatial Variability

ESHEL BRESLER

The Volcani Center
Bet Dagan, Israel

The traditional approach to modeling infiltration and redistribution (Ch. 9), and solute transport (Ch. 15 and 16) is to derive macroscopic laws from small soil scales and to apply them to large field scales. Applications of such models to actual field conditions were supported by the assumption that the field can be regarded as a homogeneous or equivalent-to-homogeneous medium characterized by equivalent, or effective, soil properties. Such properties are measured by sampling across a few locations in the particular field and subsequently determined by an averaging procedure. This approach does not consider the large spatial variability associated with many field situations.

A rational modeling of field variability is the stochastic approach. Soil properties and, therefore, the model's outcome (e.g., water content and salt concentration distributions) are subjected to uncertainty and are regarded as random variables that can be defined in terms of their statistical moments. The stochastic models are employed to predict the statistical moments of the variables such as crop yield, concentration, water content, and their dependence on the moments of soil properties. These statistical moments exhaust the information the modeler would need in any conceivable application in agriculture. Moreover, in spatially variable fields stochastic modeling may realistically represent the field phenomena, and may provide the main statistical moments, mean and variance, by employing simplified models.

This chapter does not attempt to cover the literature of spatial variability models, but to show illustrations of a few models of water flow and solute transport encountered in some applications for agriculture. These include: (i) nonsteady infiltration caused by application of a given constant water flux at the surface, with a given salt concentration, into soil of initially constant water content and salt concentration; (ii) subsequent (postinfiltration) redistribution and salt transport under these conditions; and (iii) dispersive (and piston-type) salt transport under steady gravitational flow conditions. The details of the models are given in Dagan and Bresler (1979, 1983), and Bresler and Dagan (1979, 1981, 1983a, b).

I. MODELING INFILTRATION AND REDISTRIBUTION

In the particular problem of infiltration and redistribution (Ch. 9), water is applied for a certain time on the soil surface and the solution of Eq. [1]

Copyright © 1991 ASA-CSSA-SSSA, 677 S. Segoe Rd., Madison, WI 53711, USA. *Modeling Plant and Soil Systems*—Agronomy Monograph no. 31.

describes the flow of water in the soil profile as a function of time (t) and depth (z). The solution of the problem for a homogeneous soil column (Ch. 9, 15, and 16) is quite difficult because of the strong nonlinearity of the differential equation (Eq. [1]), and the time required for accurate measurements of the soil hydraulic properties in Eq. [6] and [7]. These difficulties are generally not amplified in a field of spatially variable properties because our interest is in a few statistical averages across the horizontal plane. Paradoxically, the estimation of the average across the horizontal plane may be even simpler than solving the flow problem in a homogeneous column.

A. Theoretical Approach

Infiltration and redistribution can be predicted for a wide variety of conditions from the one-dimensional Richard's Equation

$$\frac{\partial \theta}{\partial t} + \frac{\partial}{\partial z}\left(K \frac{d\psi}{d\theta} \frac{\partial \theta}{\partial z}\right) + \frac{\partial K}{\partial z} = 0 \qquad [1]$$

supplemented by the boundary and initial conditions, an example of which is

$$\theta = \theta_n, \quad t = 0, \quad z > 0 \qquad [2]$$

$$q = q_0, \quad 0 < t < t_i, \quad z = 0 \text{ for } \theta(0, t) < \theta_s \qquad [3]$$

$$q = 0, \quad t > t_i, \quad z = 0 \qquad [4]$$

where the soil lies beneath the horizontal plane z,y, with z as a vertical coordinate directed downward, q is the vertical water flux, θ is the soil moisture, ψ is the suction head, K is the hydraulic conductivity, θ_n is the constant initial moisture content, q_0 is the given constant flux on the surface, and t_i is the infiltration time. If $\theta(0, t)$ reaches the saturated value θ_s for $t_p < t < t_i$, Eq. [3] is replaced by

$$\theta = \theta_s \quad t_p < t < t_i \quad z = 0 \qquad [5]$$

where t_p is defined as ponding time (i.e., the time at which ponding occurs and persists until t_i. Equation [5] neglects the effect of water accumulation on the surface and the resulting positive head or runoff.

The nonhysteretic type of $K(\psi)$ relations of Brooks and Corey (1964) is the one adopted here to characterize a K and θ function by a given set of parameters: K_s, ψ_w, θ_s, θ_r, η, and β. Hence,

$$\frac{K(\psi)}{K_s} = \left(\frac{\psi_w}{\psi}\right)^\eta \qquad [6]$$

$$S = \frac{\theta - \theta_r}{\theta_s - \theta_r} = \left(\frac{\psi_w}{\psi}\right)^\beta, \quad \psi \leq \psi_w \leq 0 \quad [7]$$

where K_s is the hydraulic conductivity at saturation, ψ_w is the air entry value, S is the reduced water content (degree of saturation), θ_s (saturated water content) and θ_r (residual θ) are water contents for saturation and for $K \to 0$, respectively, and η and β are constant, empirical coefficients ($\eta \sim 2.0$–3.5, $\beta \sim 0.25$–0.5, $\beta/\eta \sim 0.1$–0.5). Thus, the soil hydraulic properties are characterized by the six constants θ_s, θ_r, ψ_w, K_s, η, and β, which are all spatially variable. A considerable simplification is achieved by assuming that only K_s is spatially and randomly variable, while the other five parameters (θ_r, θ_s, ψ_w, η, β) are considered constant (for the justification of this assumption see Russo and Bresler, 1982).

Under these conditions, the space variability is expressed mathematically with the aid of the joint probability density function (PDF) $f_k(K_{s1}, K_{s2}, \ldots, K_{sn})$ of K_s at a set of n points of coordinates $r_i(x_i, y_i)$, ($i = 1, 2, \ldots, n$). Since one wishes to apply this to only one field, the space variability in this single realization can be used to derive f_k using some statistical hypotheses. The statistical structure of K_s is determined, up to second-order moments, by the probability density function $f_k(K_s)$, irrespective of the position, and by the autocovariance.

Since θ (or S), ψ, and K depend on K_s, via Eq. [6] and [7], they are random variables as well. Furthermore, Eq. [1] is a stochastic partial differential equation, subjected to the deterministic boundary and initial conditions of Eq. [2] to [5] (incorporation of randomness in the latter equations is possible).

The solution of an infiltration and redistribution problem in a spatially variable field can now be stated as follows: determine the statistical structure of the random variables θ (or S) and the related variables ψ and K which satisfy Eq. [1], the relationships in Eq. [6] and [7], and conditions of Eq. [2] to [5] for given stationary K_s. If the scope is limited to a few statistical moments of the dependent variables only, namely to their unconditional expectation $E(S)$ and variance, σ_S^2, the solution to the problem is stated in more precise terms as

$$E[S(z, t)] = \int_0^\infty S(z, t; K_s) f_k(K_s) \, dK_s \quad [8]$$

$$\sigma_S^2(z, t) = \int_0^\infty \{S(z, t; K_s) - E[S(z, t)]\}^2 f_k(K_s) \, dK_s \quad [9]$$

Once the relationship between S and K_s and solutions of Eq. [1]–[7] are determined, the moments of interest are obtained by a numerical integration (one quadrature) over the PDF of K_s. It should be noted that:

1. The correlation structure of K_s, expressed by its autocovariance, does not enter into the computation of Eq. [8] and [9].
2. The numerical evaluation of the integrals in Eq. [8] and [9] can be carried out by replacing them by sums over equal classes of the cumulative PDF: P, where P is defined by

$$P_k(K_s) = \int_0^{K_s} f_k(K_s) \, dK_s \qquad [10]$$

3. Under the stationarity hypotheses adopted here, the average saturation in the field in the plane at depth z and at time t, is

$$E[S(z, t)] \cong \iint S(x, y; z, t) \, dx \, dy \qquad [11]$$

and similarity for σ_S^2.

Many studies of the PDF of K_s (Dagan & Bresler, 1979; Bresler & Dagan, 1979, 1981) have assumed that K_s is related to scaling parameter δ,

$$\delta = (K_s/K_s^*)^{1/2} \qquad [12]$$

which is lognormally distributed in most field situations so that

$$Y = \ln \delta \qquad [13]$$

and its PDF f_Y is given by

$$f_Y(Y) = \frac{1}{(2\pi)^{1/2} \sigma_Y} \exp\left\{\frac{[Y - E(Y)]^2}{2\sigma_Y^2}\right\} \qquad [14]$$

The PDF of K_s Eq. [14] is defined by the three parameters $E(Y)$, σ_Y, and K_s^* (defined by $K_s^* = [\int \sqrt{K_s} f(K_s) dK_s]^2$). Once they are known, moments like Eq. [8] and [9] can be evaluated by using the transformations

$$K_s = K_s^* \exp(2Y) \qquad [15]$$

$$f_k(K_s) \, dK_s = f_Y(Y) \, dY \qquad [16]$$

Therefore, once the deterministic parameters θ_s, θ_r, ψ_w, β, η, $E(Y)$, σ_Y, and K_s^* (characterizing the field), and q_o, t_i, and θ_n (characterizing the flow) are given, the calculation of the average (Eq. [8]) and variance (Eq. [9]) can be accomplished by first solving Eq. [1] to [7] to determine $S(z, t; K_s)$ for arbitrary K_s and, subsequently, integrating over Y with the aid of Eq. [15] and [16].

The solution for θ (in terms of S) of Eq. [1], subject to Eq. [2] to [5] for given K_s, is complicated, but can be carried out by numerical methods (see Ch. 10). However, one can obtain a good estimate of $E(S)$ and σ_S^2 for a spatially variable field by using a simple approximate solution to Eq. [1] to [5].

The simplest approach applicable to the constant surface flux conditions (Eq. [3]) is the one based on the motion of water content front, the Green and Ampt approach following the approach of Mein and Larson (1973). Under these conditions, the flux q and the water content θ are step functions along vertical lines, with constant values depending on soil properties, for profiles shallower than the wetting front ($z < L$) and with values equal to the initial ones ($z > L$). Hence, at a given time t and at depth z, the magnitude of $q(z, t)$, $\theta(z, t)$, $S(z, t)$, the other related flow variables, and the wetting depth L may vary considerably in the horizontal plane due to the spatial variability of the soil hydraulic properties.

The basic approach is to replace the differential Eq. [1] with its integrated expression between the soil surface and the front. The complete development of the approximate front model from basic principles for both infiltration and redistribution processes is described in detail by Dagan and Bresler (1983) and will not be reported here.

B. Computations of Statistical Moments of the Flow Variables

To calculate the mean and variance of the flow variable S (or θ) the integrals of Eq. [8] and [9] have to be evaluated. Although for some specific flow conditions closed-form analytical evaluations are possible, numerical computations have to be performed for the more general flow conditions considered here. The Monte Carlo method (Dagan, 1989) can be used to obtain the appropriate PDF of K_s with known values of the variance σ_Y and the expectation $m_Y = E(Y)$ of Y. An alternative methods, which has been adopted here, divides the cumulative PDF of Y,

$$P(Y) = \int_{-\infty}^{Y} f(Y) \, dY \qquad [17]$$

into N equal classes ($i = 1, 2, \ldots, N$), to calculate K_s for each class from

$$K_s^i = K_s^* \exp(2Y_i) \qquad [18]$$

and to evaluate numerically the integrals in Eq. [8] and [9] with computed values of $S(z, t; K_s^i)$. This method uses less computing time than the conventional Monte Carlo simulation.

The computation of K_s^i for $i = 1, 2, \ldots, N$ is carried out by the following methods:

1. The area $P(Z)$ under the standard normal density curve $f(Z)$ for $Z:N[0, 1]$ is divided into N equal classes, and the value of Z_i in each class i is recorded.
2. Using the inferred values of $\hat{\sigma}_Y$ and m_Y, the values of Y_i for class of soil are obtained from

$$Y_i = Z_i \sigma_Y + m_Y \qquad [19]$$

These Y_i values are then substituted into Eq. [18] and K_s^i for $i = 1, 2, \ldots, N$ are obtained.

After establishing N values of Y or their equivalent K_s, Eq. [1] to [5] can be solved numerically (Ch. 9) or approximately (Dagan & Bresler, 1983) to obtain the flow variable. The computed $S(x, t; Y)$ values are then substituted into equations similar to Eq. [8] and [9]. These are used to calculate the statistical moments of S by numerical evaluation of the appropriate integrals with the aid of the eight parameters characterizing the field (m_Y, σ_Y, K_s^*, η, β, θ_s, θ_r, ψ_w) and the three parameters characterizing the flow (q_0, t_i, θ_n).

C. Model's Results: An Example For Water Flow

To demonstrate the stochastic model for infiltration and redistribution, the values of the distribution parameters m_Y and σ_Y for Panoche [fine-loamy, mixed (calcareous), thermic Typic Torriorthent] soil of Warrick et al. (1977) have been adopted. For the normal distribution of $Y = \ln \delta$, the values of the statistical distribution parameters are

$$m_Y = -0.616, \qquad \sigma_Y = 1.16 \qquad [20]$$

The functions $K(\psi)$ (Eq. [6]), and $S(\psi)$ or $\theta(\psi)$ (Eq. [7]) have been adopted with values of $\eta = 2.59$, $\beta = 0.36$, $\theta_r = 0.05$ or 0.01, $\theta_s = 0.43$, $K_s^* = 0.22$ cm/h, and $\psi_w = 15$ cm (Bresler & Dagan, 1983a). With these values of the eight deterministic parameters (m_Y, σ_Y, K_s^*, θ_s, θ_r, β, η, and ψ_w) characterizing the field, the desired moments of the flow variables $S(z, t)$ are computed for various combinations of the initial and boundary parameters (q_0, t_i, θ_n) characterizing the flow.

The numerical computations (Ch. 9) were carried out with 90 depth increments. The size of each increment was constant with depth, but changed according to the value of K_s and ranged from 0.05 to 2 cm for $q_0 = 0.5$ and from 0.1 to 4 cm for $q_0 = 6.5$. The minimum value of Δt was chosen to be 0.02 h.

Computations of mean and coefficient of variation (CV) of S (Eq. [7]), for $\theta_n = 0.1$, $q_0 = 0.5$ cm/h, and $t_i = 1$ d at $t = 1$ d and at $t = 6$ d (after 5 d of redistribution), are demonstrated in Fig. 8-1. Also, given in Fig. 8-1 (1-A and 1-B) are deterministic water content profiles (S) calculated with deterministic K_s corresponding to its arithmetic average 0.93 cm/h. This demonstrates the capability of the approximate model to simulate deterministic field conditions during infiltration and redistribution processes. The agreement between numerical and approximate deterministic models is reasonably good, but not always precise. Since the numerical models have been validated under laboratory as well as field conditions (see Ch. 9 and 11), the approximate piston-type solution for water content (and S) may not be too accurate if a particular deterministic value is taken for K_s (and the other parameters). The picture is improved markedly when the statistical moments of S for the entire field are of interest, as demonstrated in Fig. 8-1 (A-2, B-2, A-3, and B-3), which depicts the profiles of the expectation (Eq. [8]) and of the variance (Eq. [9]) in terms of CV. The agreement between the numerical, and piston-type approximation is quite good for the two first central statistical moments, but not as good for CV as for $E(S)$.

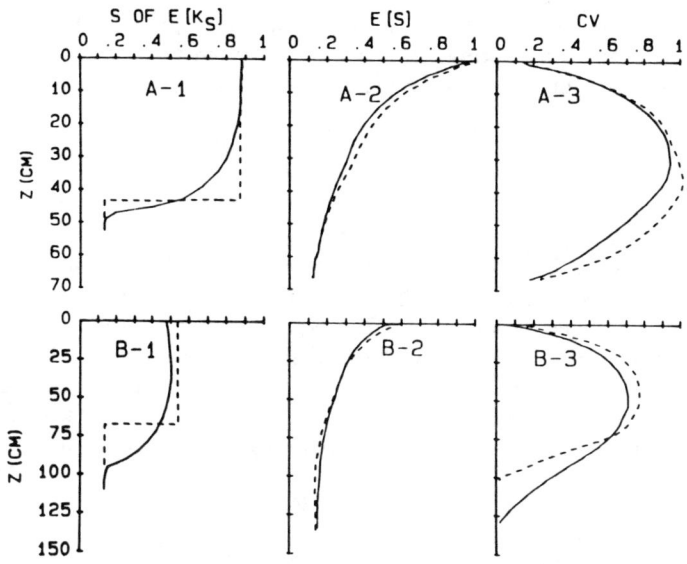

Fig. 8-1. Computer profiles of degree of saturation (S) as a function of depth (z) for $\theta_n = 0.1$, $q_0 = 0.5$ cm/h, after: (A) $t_i = 1$ d of infiltration; (B) $t - t_i = 5$ d of redistribution. The accompanying numbers: (1) represents computations using deterministic value of $E(K_s)$; (2) represents expected value of $S[E(S)]$; and (3) represents coefficient of variation of S (CV). Solid lines were computed with the numerical model (similar to Ch. 9); dashed lines with the approximate model (Bresler & Dagan, 1983a).

II. MODELING SOLUTE TRANSPORT

The most common approach of calculating solute concentration as a function of time and space is to model mass balance by a partial differential equation and, subsequently, to solve the equation for various initial and boundary conditions. The traditional deterministic–numerical modeling approach (Ch. 16) regards the field as a homogeneous unit and applies numerical solutions of the governing dispersion-convection equation for saturated-unsaturated soil valid for laboratory columns (or presuming that the field is a homogeneous porous medium). The use of the traditional approach may not be useful for spatially variable fields, which unlike small laboratory soil columns display variation in their hydraulic properties. Instead, the actual heterogeneous field can be approximated as a collection of vertical homogeneous columns differing in hydraulic properties and, as a result, solute transport will differ from profile to profile depending on the local properties. In practice, one is neither interested nor able to measure the solute concentration at each point of the field, but the representation for the entire field is of interest. Since the hydraulic properties (K_s) are regarded as random, water content (θ), flow velocity ($V = q/\theta$), and, therefore, solute concentration are also random variables that can be characterized by their PDF. Hence, a rational approach for modeling solute transport in spatial variable soils is the stochastic one (as in the water flow modeling case in Section I).

The macroscopic mass balance equation, which governs noninteracting transport in one-dimensional vertical flow through a homogeneous soil lying in the horizontal x, y plane beneath $z = 0$, is

$$\frac{\partial C}{\partial t} + V \frac{\partial C}{\partial z} = \frac{\partial}{\partial z}\left(D \frac{\partial C}{\partial z}\right) \quad [21]$$

where C is the solute concentration and D is the hydrodynamic dispersion coefficient. It will be assumed that $\theta = \theta(z, t)$ and the pore velocity $V(z, t) = q/\theta$ are given solutions of the flow equation (see Section I) independent of C. The concentration C is made dimensionless with respect to the difference between the initial concentration in the soil and that of the water applied to the soil surface. Hence, without loss of generality, C varies between zero and unity. It is also assumed (Saffman, 1959, 1960) that

$$D = \lambda V \quad [22]$$

in which λ is the dispersivity for longitudinal dispersion. Equations [21] and [22] are generally accepted, with some reservation, as valid for unsaturated flows (Bresler, 1973; Segol, 1977) with values of λ that vary in the range of 0.1 to 10 cm for most soils (Biggar & Nielsen, 1976).

The stochastic approximate model of water flow in a spatially variable field, which has been developed in Section I with application to the particular problem involving infiltration and redistribution, is applied here to model the problem of inert solute transport under the same water flow conditions. Solute concentration prior to infiltration (C_n) is assumed uniform while water with uniform concentration $C_o > C_n$ is applied at the surface during infiltration ($0 < t < t_i$). The aim is to compute the expectation $E(C)$ of a dimensionless concentration $C = (c - C_n)/(C_o - C_n)$, where C is the solute concentration in the soil solution, and the variance σ_C^2 as a function of z and t under these conditions (These are quantities of interest in many applications to leaching- or pollution-related problems). For this purpose a simplified model of salt transport (Bresler & Dagan, 1983b) can be used to compute mean concentration and its variance in a spatially variable field.

Equation [21] is subjected to the initial and boundary conditions

$$C = 0, \quad t = 0, \quad z > 0 \quad [23]$$

$$C = 1, \quad 0 < t < t_i, \quad z = 0 \quad [24]$$

$$\partial C/\partial z = 0, \quad t_i < t, \quad z = 0 \quad [25]$$

Adopting the approximate solution for the water flow developed in Section I recall that the actual water content profile is replaced by one of uniform $\bar{\theta}$ or $\bar{S} = [(\bar{\theta} - \theta_r)/\theta_s - \theta_r)]$ extending from the soil surface $z = 0$ to the equivalent front $z = L$, while ahead of the front ($z > L$) the water content is the initial θ_n.

SOIL SPATIAL VARIABILITY

For infiltration with constant flux q_O applied on the surface and entering at $z = 0$, and for redistribution as well, $\overline{V} = \overline{q}/\overline{\theta}$ is also constant with depth in the profile and depends on t soltely.

Hence, the equations satisfied by the concentration V, replacing Eq. [21] under piston water-flow conditions, become

$$\frac{\partial C}{\partial t} + \overline{V}(t)\frac{\partial C}{\partial z} = \frac{\partial}{\partial z}\left(D\frac{\partial C}{\partial z}\right), \quad 0 < z < L \quad [26]$$

To simplify the problem further let

$$\overline{V}(t) = dL/dt \quad [27]$$

and using Eq. [22], Eq. [26] becomes

$$\frac{\partial C}{\partial t} + \frac{dL}{dt}\frac{\partial C}{\partial z} = \lambda\frac{dL}{dt}\frac{\partial^2 C}{\partial z^2}, \quad 0 < z < L \quad [28]$$

To simplify matters, it is assumed that the condition $\partial C/\partial z = 0$ applies for a region that is sufficiently far from $z = L$ to justify adopting for C the solution pertaining to a vertical column of indefinite length. Furthermore, except for exceedingly short times, L is sufficiently large to allow the approximate solution for $C(z, t; Y)$ of Bresler and Dagan (1983b),

$$C(z, t; Y) = \mathrm{erf}\left[\frac{L(t; Y) - z}{(2/26)^{1/2} L(t; Y)}\right]; L < 13\,\lambda_{\max} \quad [29]$$

$$C(z, t; Y) = \mathrm{erf}\left[\frac{L(t; Y) - z}{2\lambda_{\max}^{1/2}(L - 6.5)^{1/2}}\right]; L > 13\,\lambda_{\max} \quad [30]$$

It is assumed, somewhat arbitrarily, that for $L > 13\,\lambda_{\max}$ the dispersivity becomes constant and equal to $\lambda_{\max}$.

The derivation of the expectation $E(C)$ and the variance σ_C^2 is straightforward by substituting Eq. [29] and [30], with values of $L(t; Y)$ determined explicitly during infiltration and redistribution in Section I

$$E[C(z, t)] = \int_{-\infty}^{\infty} C(z, t; Y) f_Y(Y)\, dY \quad [31]$$

and

$$\sigma_C^2(z, t) = \int_{-\infty}^{\infty} [C - E(C)]^2 f_Y(Y)\, dY \quad [32]$$

for f_Y normal (Eq. [4]). Note that one can also determine $f_C(z, t; C)$, the PDF of C, and P_c, the cumulative f_c, by using the relationships

$$f_C \, dC = f_Y \, dY; \quad P_c = \int f_c \, dC = \int_{-\infty}^{Y(z,t;C)} f_Y \, dY \qquad [33]$$

which, in terms of spatial distribution, permits one to evaluate the fractional area of the entire field for which C is smaller than a given number.

Note that the integration over $f_Y \, dY$ has also been performed in Section I when computing the various statistical moments of S at different z's and t's (Eq. [8] and [16]) and it is simple to incorporate C in the same integration process.

A. Computations of the Statistical Moments of Concentration

With values of m_Y, σ_Y, and six deterministic parameters, K_s^*, θ_s, θ_r, β, η, and ψ_w (characterizing the field), the desired moments of $C(z, t)$ are computed for various combinations of the initial and boundary parameters q_0, t_i, and θ_n, characterizing the flow (see Section I).

The mean $E[C]$ and the variance σ_C^2 have been computed by two methods: (i) by using the numerical "exact" solution (Ch. 9) of the water flow problem (Section I) and, subsequently, the numerical solution by finite differences as described by Bresler (1973); and (ii) by employing the approximate model of water flow to calculate $L(t; Y)$, which is substituted to Eq. [29] or [30] to obtain $C(z, t; Y)$.

The computation procedure for the approximate model involves the following steps:

1. Set $i = 1$, define t, and calculate $L_i(t)$ by the approximate water flow model (Dagan & Bresler, 1983).
2. Define and list several values of z.
3. Compare $L_i(t)$ with the input values of z. If $L_i(t) \geq z$, compute $C_i(z, t) = C(z, t; Y_i)$ from Eq. [29] or [30] for any z listed in Step 2 above. Otherwise, $C_i(z, t) = 0$.
4. Set $i = 2, 3, \ldots, N$ and repeat Steps 1 to 3.
5. Once $C_i(z, t)$ has been determined for the whole set of N values, two moments of C (averages and variances) are computed from

$$E[C(z, t)] = \frac{1}{N} \sum_{i=1}^{N} C_i(z, t) \qquad [34]$$

$$\sigma_C^2(z, t) = \frac{1}{N} \sum_{i=1}^{N} \{C_i(z, t) - E[C(z, t)]\}^2 \qquad [35]$$

6. Change t and repeat Steps 1 to 5.

For the numerical model, finite difference equations to approximate the partial differential equation (Eq. [21]), the initial conditions (Eq. [23]), and the boundary conditions (Eq. [24] and [25]) are first formulated and then solved numerically [for details, see Bresler (1973) or Ch. 16]. The soil moisture retention curves that are adapted for the numerical computations are those of Eq. [7], with the same values of β, ψ, θ_s, and θ_r, as for the approximate model. The hydraulic conductivity function Eq. [6] is also calculated as in the approximate model with the same deterministic η, K_s^*, m_Y, and σ_Y, and the same values of K_s^i, $i = 1, 2, \ldots, N$. The computed results of $C_i(z, t, Y_i)$ $i = 1, 2, \ldots, N$, are substituted into Eq. [34] and [35] to calculate the two central statistical moments of C. The numerical computations were carried out with 90 depth increments. The size of each increment was constant with depth, but changed according to the value of K_s^i, as discussed in section I.

B. Model's Results: An Example For Solute Transport

The results demonstrated in Fig. 8-2 are again derived from the Panoche soil, as in Fig. 8-1, with $m_Y = -0.616$, $\sigma_Y = 1.16$ $\lambda = \lambda_{max} = 3$ cm, $K_s^* = 0.22$, $\theta_s = 0.43$, $\theta_r = 0.05$, $\beta = 0.36$, $\eta = 2.59$, $\psi_w = 15$ cm, $\theta_n = 0.1$, $q_0 = 0.5$ cm/h, and $t_i = 24$ h.

The capability of the approximate model to simulate deterministic field conditions during infiltration and redistribution is demonstrated in Fig. 8-2 (A-1 and B-1), respectively. It can be seen that the agreement is reasonable, but not always precise. This suggests that the approximate piston-type solution for water flow and the corresponding approximate solution for $C(z, t)$

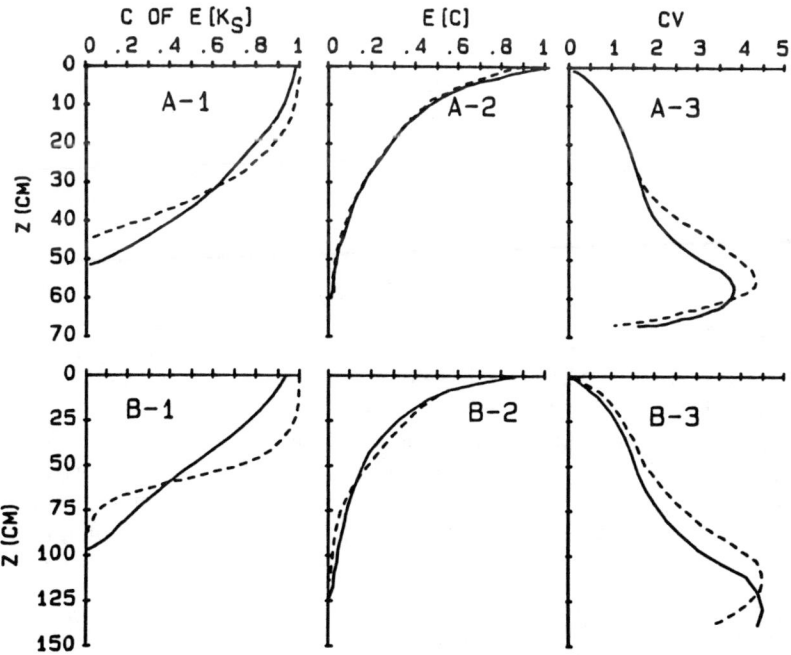

Fig. 8-2. Computed profiles of concentration (C) as a function of depth (z) for $\lambda = 3$ cm. (A) $t_i = 1$ d of infiltration; (B) $t - t_i = 5$ d of redistribution. The accompanying numbers: (1) represents computations using deterministic value of $E(K_s)$; (2) represents expected value of S [$E(S)$]; and (3) represents coefficient of variation of S (CV).

are not too accurate if a particular deterministic value is taken for K_S and the other parameters. The picture is drastically improved, however, when the statistical moments of C for the entire field are of interest. To illustrate this point, the distribution of the first central moment $E(C)$ is given in Fig. 8-2 (A-2 and B-2) and that of the second moment σ_C^2 in Fig. 8-2 (A-3 and B-3) in terms of CV of C. The agreement between the two methods of computation is very good for the expected values and good for the CV of both S (Fig. 8-1) and C (Fig. 8-2). The results given in Fig. 8-2 also emphasize the difference between the shapes of the $E[C]$ profile from numerical computations and/or approximate computations, and the deterministic profile of the concentration C.

C. Stochastic Solute Transport in Steady Gravitational Flow

Steady gravitational flow conditions can be considered acceptable after a certain irrigation time, needed for the wetting front to move away from the soil surface, has elapsed and the irrigation rate q_0 does not change with time. These conditions may be particularly applicable to a leaching-field situation in which, for instance, a saline soil is leached for a sufficiently long time. Under these conditions, Eq. [21] is solved for steady uniform flow (q, θ, and therefore, V are constant and do not depend on z and t, but vary in the x, y plane). For boundary and initial conditions

$$C = 1 \quad z = 0, \quad t > 0$$
$$C = 0 \quad z > 0, \quad t = 0 \quad [36]$$

the concentration profile of a noninteracting solute is described by the solution of Eq. [21] subject to Eq. [36] as

$$C = (0.5) \left[1 - \text{erf} \frac{z - Vt}{2(Dt)^{1/2}} \right] \quad [37]$$

where C again is the dimensionless concentration. The approximate given by Eq. [37] is valid for $t \gg D/V^2$.

Whereas V and D are constant for a homogeneous column, in a heterogeneous field they vary with x and y. Therefore in Eq. [37] C is a random function of the random variables V and D (because of the random nature of V and θ) and it depends deterministically upon z and t. The main aim here, therefore, is to evaluate $f(z, t; C)$ the PDF of concentration for fixed z and t.

To relate the variables V and D in Eq. [27] to the spatially variable soil properties, the stochastic steady flow velocity V is determined from Eq. [6], [7], [13], [14], and [15]. The randomness of C is also associated with the boundary and initial conditions in addition to its dependence on the variability of the soil variables. For this example, which simulates the practical case of leaching with solution that has been applied on the soil surface, at a rate q_0 (Eq. [3]), q_0 in turn is also a random variable characterized by a PDF $f(q_0)$. A simple, realistic approximation to the random variation of q_0 with x,y is the uniform (rectangular) PDF:

$$f(q_0) = 0 \text{ for } q_0 < \bar{q}_0 - d_{q_0} \text{ and } q_0 > \bar{q} + d_{q_0}$$

$$f(q_0) = \frac{1}{2d_{q_0}} \text{ for } \bar{q}_0 - d_{q_0} \leq q_0 \leq \bar{q}_0 + d_{q_0} \quad [38]$$

The PDF $f(q_0)$ is defined by the two parameters $\bar{q}_0$ and d_{q_0}; so that the PDF of q_0 is constant in the interval of width $2d_{q_0}$ around the average $\bar{q}_0$. In the case of rainfall over the field, d_{q_0} is very small, but for many irrigation methods or rainfall over orchards $d_{q_0}/\bar{q}_0$ may be quite large. Note that this spatial variability is completely independent of soil variability, but dependent on nonuniformity of water application.

During steady gravitational leaching at any point in the field, one of two situations will occur: (i) if $q_0 \geq K_s$ then ponding takes place in that portion of the field, the flow is saturated, and, with neglect of additional head on the surface (i.e., for unit gradient) Darcy's law yields

$$V = K_s/\theta_s; \quad [39]$$

and (ii) for $q_0 < K_s$ the soil is unsaturated and the velocity is given by

$$V = K(\theta)/\theta \quad \text{and} \quad K(\theta) = q_0 \quad [40]$$

Taking Eq. [22] as the representative expression for $D(V)$, the soil parameter λ (the dispersivity) fits lognormal distribution (Biggar & Nielsen, 1976) so that

$$\mu = \ln \lambda \quad [41]$$

and $f(\mu)$ is given by Eq. [14] after replacing Y by μ with the statistical parameters σ_μ and $m_\mu = E(\mu)$.

Now C is a function of V and λ (Eq. [22] and [37]), which, in turn, are functions of Y, q_0, and μ (Eq. [6], [7], [13], [14], [15], [38], [39], [40], and [41]). The PDF of C can be written in a general manner.

$$f(C) \, dC = f(Y, q_0, \mu) \, dY \, dq_0 \, d\mu \quad [42]$$

Furthermore, it is assumed here, for the sake of simplicity, that Y, q_0, and μ are independent random variables. Hence Eq. [41] becomes

$$f(C) \, dC = f(Y) \, f(q_0) \, f(\mu) \, dY \, dq_0 \, d\mu \quad [43]$$

with $f(Y)$ (Eq. [14]), $f(q_0)$ [Eq. [38]), and $f(\mu)$ similar to Eq. [14].

The computation of $P(z, t; C) = \int_0^C f(z, t; C) \, dC$ (see details in Bresler & Dagan, 1981) is carried out as follows:

1. Given z, t, C, and $\lambda = \exp(\mu)$, the velocity V is obtained from

$$V = [z + 2\phi^2\lambda - 2\phi(\lambda z + \phi^2\lambda^2)^{1/2}]/t \qquad [44]$$

where

$$\phi(C) = \text{erf}^{-1}(1 - 2C) \qquad [45]$$

2. To find out which one of the two relationships Eq. [39] or [40] holds, V is compared with q_0/θ_s, which is the value separating unsaturated flow (Eq. [40]) from ponding (Eq. [39]). If $V < q_0/\theta_s$ ponding takes place and Eq. [39] is valid. Then, by Eq. [15] and [13]

$$V = \frac{K_s}{\theta_s} = \frac{K_s^* \exp(2Y)}{\theta_s} \qquad [46]$$

or

$$Y = (1/2)\ln(V\theta_s/K_s^*) \qquad [47]$$

If $V > q_0\theta_s$, the flow is unsaturated, Eq. [40] is valid, and by Eq. [6] [7], [14], and [12],

$$V = \frac{q_0}{\theta} = q_0/[\theta_r + (q_0/K_s^*)^{\beta/\eta}(\theta_s - \theta_r)\exp(-2\beta Y/\eta)] \qquad [48]$$

or

$$Y = \frac{1}{2\beta/\eta} \ln \frac{(q_0/V) - \theta_r}{(q_0/K_s^*)^{\beta/\eta}(\theta_s - \theta_r)} \qquad [49]$$

Hence, for given θ_r, θ_s, K_s^*, and β/η, $Y(z, t; C, q_0, \mu)$ is determined from Eq. [47] or [49].

3. This value of Y is substituted in Eq. [50] with $q_0 = \bar{q}_0$ or in Eq. [51] with $\lambda = \bar{\lambda}$:

$$P(z, t; C) = \frac{1}{2(2\pi)^{1/2} \sigma_\mu} \int_{-\infty}^{\infty} \left\{ \exp\left[-\frac{(\mu - m_\mu)^2}{2\sigma_\mu^2}\right] \right.$$

$$\left. \times \left[1 + \text{erf} \frac{Y(z, t; C, \mu, \bar{q}_0) - m_Y}{2^{1/2} \sigma_Y}\right] \right\} d\mu \qquad [50]$$

$$P(z, t; C) = \frac{1}{4d_{q_0}} \int_{\bar{q}-d_{q_0}}^{\bar{q}_0+d_{q_0}}$$

$$\left[1 + \text{erf} \frac{Y(z, t; C, \bar{\lambda}, q_0) - m_Y}{2^{1/2} \sigma_Y}\right] dq_0 \qquad [51]$$

4. The cumulative probability P is now determined by a numerical quadrature (numerical integration) with negligible truncation error.
5. Once $C(z, t; Y)$ has been determined numerically for a set of Y values, various moments of C are calculated from Eq. [31] and [32], using Eq. [50] and [51].

D. Illustration of Results

To illustrate the capability of the model, $P(z, t; C)$ computations from Eq. [50] or [51], which provide the most detailed information, are presented in Fig. 8-3 for nine values of t. For these computations, $z = 20$ cm, $m_Y = -0.16$, $\sigma_Y = 0.57$, $\sigma_\mu = 0$, $\bar{\lambda} = 2.9$ cm, $q_0 = 2.78$ cm/h, $d_{q_0} = 0$, $\beta/\eta = 0.33$, $\theta_r \approx 0$, $\theta_s = 0.41$, and $K_s^* = 13.4$ cm/h.

The curves $P(z, t; C)$ in Fig. 8-3 have a physical interpretation; they represent the area of the field relative to the total area, which is at concentration smaller than C at depth $z = 20$ cm and time t (the number labeling the curves). The application of curves like those in Fig. 8-3 is immediate if one wishes to know, for example, what proportion of the field has a concentration larger than $C = 0.5$ at depth 20 cm and after 2 h ($t = 2$) of leaching with water with $C = 1$.

Computations of Fig. 8-3 are compared in Fig. 8-4 with measured chloride concentration data at 24 locations in the Bet Dagan, Israel field. Measurements that were made by Naor (1989) fit reasonably well with the computed lines.

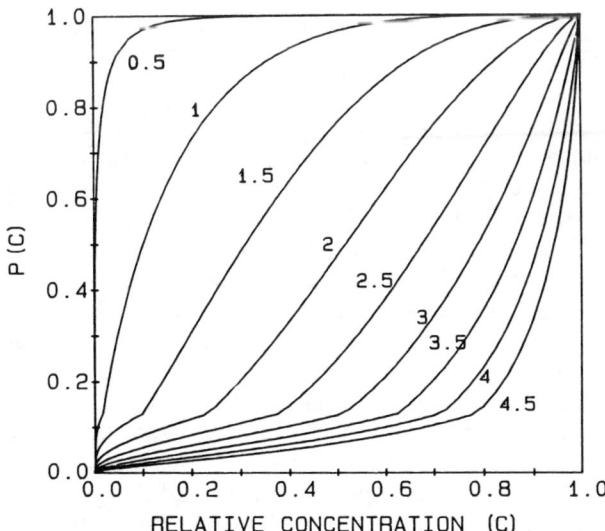

Fig. 8-3. Cumulative probability $P(z, t; C)$ as a function of concentration C, for nine various times (t) in hours (the numbers labeling the lines) and fixed $z = 20$ cm computed with $\bar{\lambda} = 2.98$ cm, $\sigma_\mu = 0.0$, $m_Y = -0.16$, $\sigma_Y = 0.56$, $q_0 = 2.78$, and $d_{q_0} = 0$.

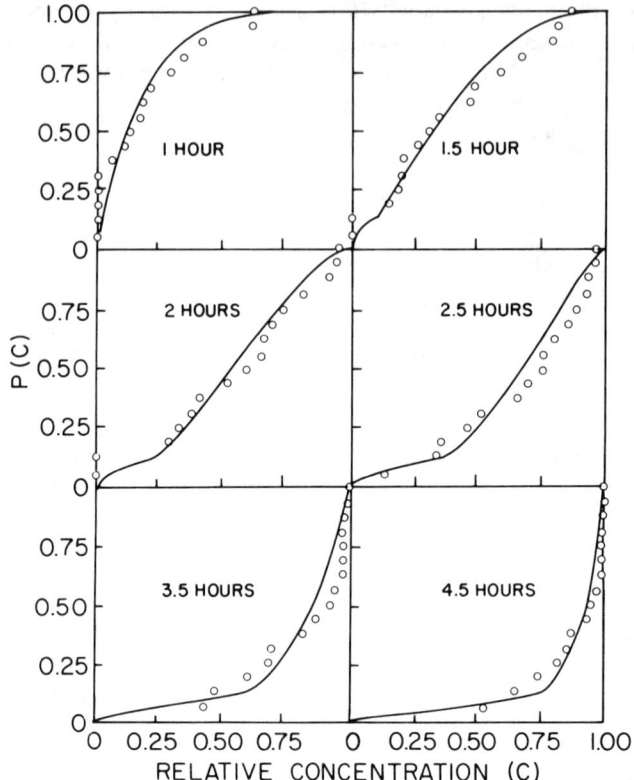

Fig. 8-4. Computed cumulative probability $P(z, t; C)$ of Fig. 8-3 (solid lines) in comparison with measured data (circles).

After evaluating $C(z, t; Y)$ the average concentration distribution $\overline{C}(z, t)$, is computed from Eq. [31] by one quadrature over the integral. The $\overline{C}(z, t)$ results for dispersive flow ($\bar{\lambda} = 2.9$ cm as in Fig. 8-3) and piston displacement ($\bar{\lambda} = 0$) are illustrated in Fig. 8-5, as a function of time for $z = 20$ cm, by the solid and dashed lines, respectively. Also given in Fig. 8-5 (by the blank squares) are the measured chloride concentration data averaged across 24 locations in a field at Bet Dagan (Naor, 1989). The dashed-dotted curve in Fig. 8-5, representing deterministic $C(z, t)$ results, is calculated from Eq. [37] with deterministic average values of Y. This curve is given for orientation purposes of comparison with traditional deterministic models.

The data of Fig. 8-4 and 8-5 may be used for model validation, as they compare between measured data of chloride concentration and the model's results. Figure 8-5 demonstrates the superiority of the stochastic model over the deterministic one, for this and similar situations (with $\sigma_Y - 0.57$, $Y = -0.16$, and $K_s^* = 13.8$ cm/h of chloride), a stochastic model, which includes dispersive transport, represents better average field concentration than does a stochastic model with a piston-type displacement.

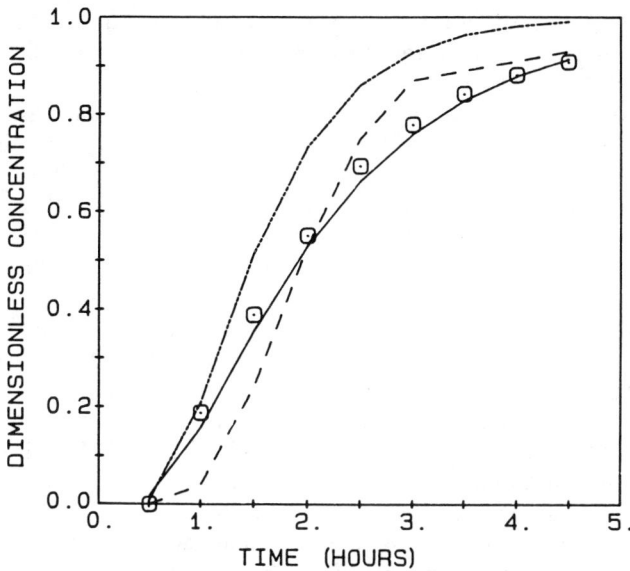

Fig. 8-5. Average concentration profile as a function of time computed from Eq. [31] for dispersive flow (solid line) and for piston flow (dashed line). Experimental data averaged across 24 locations in the field at Bet Dagan, Israel are represented by the squares. The dashed-dotted line represents $C(z = 20, t)$ data computed from Eq. [37] with deterministic (average) values for V and $\lambda = D/V$.

III. CRITICAL ASSUMPTIONS

The traditional deterministic modeling approach regards the field as homogeneous laboratory columns. However, large-scale natural fields may display a wide variation in their properties, so that considering the field as a unique, uniform column is highly questionable. The soil properties are considered to be regionalized variables that vary in an irregular manner in space and, since these properties are determined at a finite (and not too large) number of points in the field, their distribution throughout the field is uncertain. Hence, stochastic modeling, in which soil properties are regarded as random variables characterized by their PDF rather than by deterministic values, is a rational framework to deal with variability and uncertainty.

In this stochastic modeling approach, the statistical structure of concentration and water content as a function of space and time is determined for given, in a statistical sense, soil properties, and initial and boundary conditions. The problem is not a formidable one because it is not necessary to know the pointwise distribution of salt and water, but rather some gross parameters, like average and variance, for the entire field. Consequently, simplified models of the flow and transport mechanisms can be employed since minute details are lost in the process of averaging. The simplified approaches, such as those demonstrated here, are not as general as the deterministic models. The latter (Ch. 9–18) can be used to predict general flow situations, and the responses to various initial and boundary conditions. The results are physically sound and may give insight into the problem. Thus, both approaches are needed.

The models assume that, in spite of soil heterogeneity and eventual nonuniform water application, the flow is vertical. This assumption is bound to be quite accurate, for the upper layer comprises the agricultural soil, the thickness of which is on the order of one to a few meters. Furthermore, the length scale characterizing the variability in the horizontal plane is much larger then the root zone depth and, henceforth, lateral flows are negligibly small compared with the vertical flows. Another far-reaching simplification is the neglect of heterogeneity in the vertical direction as compared with variability in the horizontal plan. This assumption can be justified by the contrast between the horizontal heterogeneity scale and the depth, the latter being much smaller. In a heterogeneous field and because of uncertainty, the variation in soil properties is viewed as random and the actual field is regarded as a realization of an ensemble of fields of the same PDF of the variables of interest. It is assumed that these PDF are statistically stationary, and under appropriate ergodic assumptions, the ensemble and space averaging in the given realization over the horizontal plane are equivalent.

IV. RESEARCH NEEDS

The modeling of water flow and solute transport in heterogeneous formations can be divided into two approaches. In the first, the stationary PDF of the flow and transport variables serves as input, without making direct use of the field measurements, which serve only for the inference of the PDF. Such an approach is justified when the measurements are not available and the statistical structure is inferred by some qualitative considerations, or if one is interested in some space averages of the random function, which by ergodic arguments is equal to ensemble averages. This is the approach that is described in this chapter and suggests that simplified models may provide a quite accurate prediction of statistical moments of concentration and water content distribution in heterogeneous fields. Hence, refinements of such models do not seem to have an appreciable effect on the prediction of statistical moments in spatially variable fields, which are more sensitive to the statistical parameters characterizing soil hydraulic and transport parameters than to model accuracy. These conclusions, although based on a limited body of information, may serve as guidelines for future research, as they point on the factors that affect the flow and transport phenomena in heterogeneous fields when the first approach is taken. With this approach, and in spite of the apparent complexity, modeling of additional factors, such as chemical reaction, chemicals decay, adsorption and exclusion, soil structure changes, and the influence of root uptake, seems feasible and should be considered in future research.

The second approach (which has not been considered here and should also be considered in future research), known also as the geostatistical approach, makes explicit use of the measurements in order to reduce the uncertainty of the predicted values of the random function at any point in the field. The widely used method in this context is kriging (Dagan, 1989), which leads to estimates of the expectation and covariance values conditioned on measurements. An alternative and similar approach, which relies on the concept of conditional probability of the classical statistics has also been suggested (Feinerman et al., 1986). Both methods, in which the ensuing values of both statistical moments expectation and variance vary throughout the field space, are recommended for future research.

V. SUMMARY

Conceptual models for water flow and solute transport in saturated-unsaturated soils for heterogeneous fields are described. These models are applied to compute water content and concentration distributions under infiltration and redistribution flow conditions. Unlike laboratory columns or hypothetical homogeneous fields, the properties of a heterogeneous field vary at random. Consequently, solute concentration and water content should be regarded as random variables that can be defined in statistical terms. Mean, variances, and frequency distribution of concentration and water content have been computed for vertical steady and nonsteady flows caused by positive recharge (during infiltration) or zero recharge (during redistribution) on the soil surface. These computations are made for a field in which the hydraulic conductivity and the recharge magnitude vary at random in the horizontal plane. Three mechanisms responsible for the transport are examined: (i) convection by vertical velocity, which changes in the horizontal plane because of variability of soil properties; (ii) recharge applied on the surface; and (iii) pore scale dispersivity. The models are capable of determining detailed statistical information about the distribution in the field and, specifically, the probability function representing the area of the field that at depth z and time t is at concentration less than a given concentration C.

It has been shown that the average concentration of an inert solute and water content profiles have shapes and evolve with time in manners that differ from the profiles predicted by the usual complimentary error function erfc and Richard's profiles. Furthermore, approximate solutions of transport and flow can be used as input for the statistical averaging procedure because approximate models, although generally lacking accuracy for a particular set of soil water properties, lead to accurate values of the expectations and variances for the entire field. This happens because the model errors cancel in the averaging process. The models suggested here depict, quite accurately, the measured statistics of probability function and the expectation of concentration distribution.

This chapter provides an introduction to the stochastic modeling approach of flow and transport in spatially variable soils, and to the concepts that are important to modelers. It emphasizes that spatial variability consideration may allow simpler models to be used.

VI. LISTINGS

The details of the use of the models are provided in two program listings. A listing of the input and output files, of each program, is also provided.

The input file listing of Program 1 consists of 15 lines of data. This is the file as configured for computing the results of Fig. 8-1 and 8-2. The input file listing of Program 2 consists of seven lines. This is the file as configured for computing the results given in Fig. 8-3, 8-4, and 8-5.

The output files are generally listed in an abbreviated form. In some cases the full title is given. Only a small part of the output file listing of Program 1 is provided as an example of output for the data given in Fig. 8-1 and 8-2. The output file listing for Program 2 is the one from which the computed results for Fig. 8-3, 8-4, and 8-5 were taken.

```
**************************************************************************
C              Program N0. 1
C              ------------
C              Eshel Bresler
C
C              PROGRAM  LISTING
c              ---------------
c
C---MODEL FOR COMPUTING EXPECTATION, VARIANCE, AND SKEWNESS OF
C---THETA AND C
C---DURING INFILTRATION AND REDISTRIBUTION USING PISTON FLOW
C---APPROXIMATION.
C----KS is number of K-sat data
C----KZ is number of depth nodes
C----KT is number of time output
C----KRD is number of time output before redistribution calculations
C----KRK is number of increments in Runge-Kutte calculations
C----KKOD is index for computation. Generally use zero.
C----ETA,BETA,TETR,TETAS are the cresponding hydraulic parameters
C----UM=1-(1-BETA)/ETA, HE is hw, QU is recharge rate,TETAI is initial teta
C----DIS is dispersivity
C----SKS is average K-sat
C----SIGY is sigma of Y, SMEAN is mean of Y
C----SKMEAN is K-*, PORDIS is pore scale dispersivity
C----TOUT is output time, Z is output depth
C----ZK(I) is a table of equal increments from standard normal table
C----SK(I) is a table of KS K-sat data
C----Q(I) is the flux (q) as a function of time
C----AVERS is average of S, AVERC is average of C, etc.
C----SDS is standard deviation of S, SDC is S.D. of C , etc.
C----CVS is coefficient of variation of S, CVC is CV of C, etc.
C----VARS is variance of S, VARC is variance of C, etc.
C----SKES is skewness of S, SKEC is skewness of C, etc.
      IMPLICIT  REAL*8(A-H,O-Z)
      DIMENSION SKD(100),TD(100),TN(100),DL(100),V(100),S(100)
      DIMENSION VD(100),Z(100),ZL(100),SC(100),TT(100),CC(100),SK1(100)
      DIMENSION S4(100),V1(100),T1(100),VD1(100),S1(100),SK2(100),PS(99)
```

SOIL SPATIAL VARIABILITY

```
      DIMENSION SK(50),TOUT(50),SKK(100),TDP(100),DKS(100),CZ(41,18)
      DIMENSION ZK(50),G(50),Q(50),TZ(99)
      READ *,KS,KZ,KT,KRD,KRK,KKOD
      PRINT 136
      PRINT 94,KS,KZ,KT,KRD,KRK,KKOD
      READ *,ETA,BETA,UM,HE,QU,TETAS,TETAI,DMAX,DIS,SKS
      TDISP=DIS
      READ *,TETAR,SIGY,SMEAN,SKMEAN,PORDIS
      AK=SKMEAN*DEXP(2.*SMEAN+2.*SIGY**2)
      PRINT 102
      PRINT 101,ETA,BETA,UM,HE,QU,TETAS,TETAI,DMAX,DIS,SKS
      PRINT 1020
      PRINT 101,TETAR,SIGY,SMEAN,SKMEAN,AK,PORDIS
      READ 97,(TOUT(M),M=1,KT)
      PRINT 121
      PRINT 96,(TOUT(M),M=1,KT)
      TSTI=TETAS-TETAR
      SI=(TETAI-TETAR)/TSTI
      AA=1./(((1.-SI)*(1.-SI**((ETA-1.)/BETA)))
      READ 149,(ZK(I),I=1,KS)
      CALL ASHER(SK,KS,ZK,SIGY,SMEAN,SKMEAN,KKOD)
      READ 97,(Z(J),J=1,KZ)
      PRINT 123
      PRINT 96,(Z(J),J=1,KZ)
      DO 67 J=1,KZ
   67 TZ(J)=Z(J)
      PRINT 120
      PRINT 140
      PRINT 100,(SK(I),I=1,6)
      PRINT 141
      PRINT 100,(SK(I),I=7,12)
      PRINT 142
      PRINT 100,(SK(I),I=13,18)
      PRINT 143
      PRINT 100,(SK(I),I=19,20)
      DO 61 M=1,KT
      KC=0
      SUML=0.
      SUMV=0.
      DO 62 J=1,KZ
      SUMS=0.
      SUMPS=0.
      SUMG=0.
      SUMQ=0.
      SUMY=0.
      SUMC=0.
      SUMK=0.
      DO 63 I=1,KS
      FKL=SK(I)*SI**(ETA/BETA)/QU
      QL=FKL
      BB=QL/SK(I)
      IF(M.GT.KRD) GO TO 3
      IF(KC.GT.0) GO TO 57
      SKS=SK(I)
```

```
      DKS(I) = SKS/QU
      DPELTA = 1.-SI
      TDP(I) = DKS(I)/(1.-DKS(I))*DPELTA
      TN(I) = TOUT(M)
      TD(I) = TN(I)*QU*(ETA-1.)/(HE*TSTI)
      TDOUT = (TOUT(M)*QU*(ETA-1.))/(HE*TSTI)
      IF(DKS(I).GT.1.) GO TO 40
      IF(TDOUT.GT.TDP(I)) GO TO 12
   40 VD(I) = TD(I)
      TJ = TD(I)
      DKP = DKS(I)
      CALL FILT (TJ,DKP,UM,SKX,SI,ETA,KKOD)
      SKD(I) = SKX
      SK1(I) = SKD(I)
      SKK(I) = SKD(I)*QU
      S(I) = (SKD(I)/DKS(I))**(BETA/ETA)
      S4(I) = S(I)
      DL(I) = VD(I)/S(I)
      V(I) = TD(I)*HE*TSTI/(ETA-1.)
      ZL(I) = V(I)/(S(I)*TSTI + (TETAR-TETAI))
      IF(M.LT.KRD) GO TO 57
      V1(I) = V(I)
      T1(I) = TD(I)
      SK2(I) = SKD(I)
      VD1(I) = VD(I)
      S1(I) = S(I)
      GO TO 57
   12 S(I) = 1.
      S4(I) = S(I)
      VDP = TDP(I)
      SKD(I) = DKS(I)
      SK1(I) = SKD(I)
      SKK(I) = SKD(I)*QU
      TN(I) = TOUT(M)
      TD(I) = TN(I)*QU*(ETA-1.)/(HE*TSTI)
      TI = TD(I)-TDP(I)
      DKP = DKS(I)
      CALL   POND(TI,VDP,DKP,VX,BB,AA,KKOD)
      VD(I) = VX
      DL(I) = VD(I)
      V(I) = VD(I)*HE*TSTI/(ETA-1.)
      ZL(I) = V(I)/(S(I)*TSTI + TETAR-TETAI)
      IF(M.LT.KRD) GO TO 57
      V1(I) = V(I)
      T1(I) = TD(I)
      SK2(I) = SKD(I)
      VD1(I) = VD(I)
      S1(I) = S(I)
      GO TO 57
    3 T2 = T1(I)
      TN(I) = TOUT(M)
      TD(I) = TN(I)*QU*(ETA-1.)/(HE*TSTI)
      ALFA = ETA/(BETA*VD1(I))*(1. + (0.5*SK2(I)/DKS(I))**((BETA-1.)/ETA)/
     1VD1(I))
```

SOIL SPATIAL VARIABILITY

```
      SC(I) = S1(I)/((1.+ALFA*S1(I)**(ETA/BETA)*(TD(I)-T1(I))*DKS(I))**
     1(BETA/ETA))
      SKDI1 = SK2(I)
      B1 = ETA/(BETA*VD1(I))
      B2 = (BETA-1.)/ETA
      DELTA = -QL/SKDI1-SI/VD1(I)*(DKS(I)/SKDI1)**(1./ETA)-FKL**
     1(1.-1./ETA)/VD1(I)*(SKDI1**(BETA/ETA-1.))/(DKS(I)**B2)+SI/VD1(I)*
     2FKL**(1.-1./ETA)*DKS(I)**(1./ETA)/(SKDI1)
      TOSEF = (1.-SI*SKDI1**(-BETA/ETA)*DKS(I)**(BETA/ETA))
      FHK = (-B1*SKDI1**2*(1.+1./VD1(I)*(SKDI1/DKS(I))**B2+DELTA))*
     1TOSEF
   54 H = (TD(I)-T2)/KRK
   53 DO 5 I4 = 1,KRK
      IF(SKDI1.LE.0.) GO TO 41
      DELTA = -QL/SKDI1-SI/VD1(I)*(DKS(I)/SKDI1)**(1./ETA)-FKL**
     1(1.-1./ETA)/VD1(I)*(SKDI1**(BETA/ETA-1.))/(DKS(I)**B2)+SI/VD1(I)*
     2FKL**(1.-1./ETA)*DKS(I)**(1./ETA)/(SKDI1)
      TOSEF = (1.-SI*SKDI1**(-BETA/ETA)*DKS(I)**(BETA/ETA))
      IF(KKOD.NE.0) DELTA = 0.
      IF(KKOD.NE.0) TOSEF = 1.
      H1 = H*(-B1*SKDI1**2*(1.+1./VD1(I)*(SKDI1/DKS(I))**B2+DELTA))*
     1TOSEF
      SK1(I) = SKD(I)
      SKDI1 = SKDI1+H1
    5 CONTINUE
      SKD(I) = SKDI1
      SKK(I) = SKD(I)*QU
   21 S(I) = (SKDI1/DKS(I))**(BETA/ETA)
      S4(I) = S(I)
      DL(I) = VD1(I)/S(I)
   41 ZL(I) = VD1(I)*HE*TSTI/((ETA 1.)*S(I)*TSTI+TETAR-TETAI)
   57 IF(ZL(I)-Z(J)) 51,52,52
   52 DD = ZL(I)
      DIS = TDISP
      ZL1 = (DD-Z(J))/(DSQRT(2.D+00/26.D+00)*DD)
      IF(DD.LT.13.*DIS) GO TO 44
      ZL9 = DIS*DD-6.5
      IF(ZL9.LT.0.) ZL9 = 0.
      ZL1 = (DD-Z(J))/(2.*DSQRT(DIS*DD-6.5))
   44 ZL5 = (DD-Z(J))/(2.*DSQRT(DIS*DD))+DSQRT(DD/DIS)
      ZL6 = (2.*DD-Z(J))/DIS
      CALL ERRFT (ZL1,ZL5,ZL6,CXY)
      CC(I) = CXY
      CZ(I,J) = CC(I)
      Q(I) = SKK(I)*G(I)
      GO TO 6
   51 CC(I) = 0.
      CZ(I,J) = CC(I)
      S4(I) = SI
      SKK(I) = SI**(ETA/BETA)*SK(I)
      SK1(I) = SKK(I)/QU
      G(I) = 1.
      Q(I) = FKL*QU
    6 SUMS = SUMS+S4(I)
```

```
      PS(I) = S4(I)**(-1./BETA)*HE
      SUMPS = SUMPS + PS(I)
      SUMC = SUMC + CC(I)
      SUMK = SUMK + SK1(I)
      SUMG = SUMG + G(I)
      SUMQ = SUMQ + Q(I)
      SUMY = SUMY + DLOG(SKK(I))
      IF(KC.GT.0) GO TO 63
      SUMV = SUMV + V(I)
      SUML = SUML + ZL(I)
   63 CONTINUE
      KC = KC + 1
    7 AVERS = SUMS/KS
      AVERC = SUMC/KS
      AVERPS = SUMPS/KS
      AVERQ = SUMQ/KS
      AVERG = SUMG/KS
      AVERK = SUMK/KS
      AVERY = SUMY/KS
      REFK = (AVERQ/AVERG)/AK
      IF(REFK.LT.1.E-35) GO TO 65
      ALREFK = DLOG(REFK)
      ALRPS = DLOG(HE/AVERPS)
   65 ALAVRS = DLOG(AVERS)
      IF(AVERS.LT.1.E-05.AND.AVERC.LT.1.E-05.AND.AVERK.LT.1.E-05) GO TO
     1 62
      PRINT 1320
      PRINT 100,TOUT(M),Z(J)
      PRINT 132
      PRINT 100,AVERS,AVERC,AVERK,AVERY,AVERQ,AVERG
      SUM1S = 0.
      SUM1C = 0.
      SUM1K = 0.
      SUM1Y = 0.
      SUM2S = 0.
      SUM2C = 0.
      SUM2K = 0.
      SUM2Y = 0.
      DO 1 K = 1,KS
      SUM1S = SUM1S + (S4(K)-AVERS)**2
      SUM1C = SUM1C + (CC(K)-AVERC)**2
      SUM1K = SUM1K + (SK1(K)-AVERK)**2
      SUM1Y = SUM1Y + (DLOG(SKK(K))-AVERY)**2
      SUM2S = SUM2S + (S4(K)-AVERS)**3
      SUM2C = SUM2C + (CC(K)-AVERC)**3
      SUM2K = SUM2K + (SK1(K)-AVERK + 1.E-16)**3
      SUM2Y = SUM2Y + (DLOG(SKK(K))-AVERY)**3
    1 CONTINUE
      VARS = SUM1S/KS
      VARC = SUM1C/KS
      VARK = SUM1K/KS
      VARY = SUM1Y/KS
      SKES = SUM2S/KS
      SKEC = SUM2C/KS
```

```
      SKEK=SUM2K/KS
      SKEY=SUM2Y/KS
      SDS=DSQRT(VARS)
      SDC=DSQRT(VARC)
      CVS=SDS/AVERS
      IF(AVERC.LT.1.E-29) GO TO 64
      CVC=SDC/AVERC
   64 PRINT 154
      PRINT 100,SDS,SDC,CVS,CVC
      PRINT 1330
      PRINT 100,SKEK,SKEY
      PRINT 133
      PRINT 100,VARS,VARC,VARK,VARY,SKES,SKEC
      VARSAV=(DSQRT(VARS))/AVERS
   62 CONTINUE
      AVERV=SUMV/KS
      AVERL=SUML/KS
      SUM1V=0.
      SUM1L=0.
      SUM2V=0.
      SUM2L=0.
      DO 2 K=1,KS
      SUM1V=SUM1V+(V(K)-AVERV)**2
      SUM1L=SUM1L+(ZL(K)-AVERL)**2
      SUM2V=SUM2V+(V(K)-AVERV)**3
      SUM2L=SUM2L+(ZL(K)-AVERL)**3
    2 CONTINUE
      VARV=SUM1V/KS
      VARL=SUM1L/KS
      SKEV=SUM2V/KS
      SKEL=SUM2L/KS
      PRINT 135
      PRINT 100,AVERV,AVERL,VARV,VARL,SKEV,SKEL
   61 CONTINUE
   92 FORMAT(1E8.2,9F8.3)
   93 FORMAT (1X,1E8.2,15F8.3)
   94 FORMAT (6I6)
   97 FORMAT (10F8.3)
   95 FORMAT (11E12.4)
   96 FORMAT (10F8.3)
   98 FORMAT (8E10.4)
  100 FORMAT (10E13.4)
  101 FORMAT (14F8.3,E12.4,1F6.2)
  102 FORMAT ('  ETA     BETA    UM      HE      QU     TETAS    TETA
     1I    DMAX    DIS    SKS ')
 1020 FORMAT ('  TETAR   SIGY   SMEAN  SKMEAN    AK    PORDIS')
  120 FORMAT ('          20 VALUES OF K AT SATURATION ')
  121 FORMAT ('          TOUT ')
  123 FORMAT ('          Z ')
  132 FORMAT ('       AVERS    AVERC    AVERK    AVERY
     1 AVERQ    AVERG    ')
 1320 FORMAT ('       TOUT      Z ')
  133 FORMAT ('       VARS     VARC     VARK     VARY
     1 SKES     SKEC        ')
```

```
      1330 FORMAT ( '   SKEK       SKEY   ' )
      135 FORMAT ( '   AVERV      AVERL      VARV        VARL
     1SKEV       SKEL ' )
      136 FORMAT ( '   KS    KZ    KT   KRD   KRK   KKOD      ' )
      140 FORMAT(' NO  1        2          3          4
     15       6 ')
      141 FORMAT(' NO  7        8          9         10
     111      12')
      142 FORMAT(' NO 13       14         15         16
     117      18')
      143 FORMAT(' NO 19       20      ')
      149 FORMAT (7E10.3)
      154 FORMAT ( '   SDS        SDC        CVS        CVC ' )
      158 FORMAT ( '            Z ' )
      STOP
      END
      SUBROUTINE ERRFT (ZL1,ZL5,ZL6,CXY)
      IMPLICIT REAL*8(A-H,O-Z)
      TX=1./(1.+0.3275911*ZL1)
      TERFX=0.25482959*TX-0.284496736*TX*TX+1.421413741*TX**3
     1-1.453152027*TX**4+1.061405429*TX**5
      TY=1./(1.+0.3275911*ZL5)
      TERFY=0.25482959*TY-0.284496736*TY*TY+1.421413741*TY**3
     1-1.453152027*TY**4+1.061405429*TY**5
      IF(ZL1.GT.10.) ZL1=10.
      CERF=1.-DEXP(-ZL1*ZL1)*TERFX
      IF(ZL5.GT.6) GO TO 1
      CERFC=(DEXP(-ZL5*ZL5)*TERFY)*DEXP(ZL6)
    1 CERFC=0.
    2 CXY=CERF+CERFC
      RETURN
      END
      SUBROUTINE POND (T,VDP,DKS,VX,B,A,KKOD)
      IMPLICIT REAL*8(A-H,O-Z)
      DIMENSION X(100)
      X(1)=200.
      DO 1 I=1,15
      KI=I+1
      IF(KKOD.EQ.0) GO TO 3
      F=X(I)-VDP-DLOG((1.+X(I))/(1.+VDP))-DKS*T
      DF=1.-1./(1.+X(I))
      GO TO 4
    3 F=X(I)-VDP-1./(A*(1.-B))*DLOG((1.+A*(1.-B)*X(I))/(1.+A*(1.-B)
     1*VDP))-T*DKS*(1.-B)
      DF=1.-1./(A*(1.-B))*1./(1.+A*(1.-B)*X(I))/(1.+A*(1.-B)*VDP)*
     1(A*1.-B)/(1.+A*(1.-B)*VDP)
    4 X(I+1)=X(I)-F/DF
      IF(DABS((X(I+1)-X(I))/X(I+1)).LT.0.05) GO TO 2
    1 CONTINUE
    2 VX=X(KI)
      100 FORMAT (10E13.4)
      101 FORMAT ( '             X ' )
      RETURN
      END
```

```
      SUBROUTINE FILT (T,DKS,UM,SKX,SI,ETA,KKOD)
      IMPLICIT REAL*8(A-H,O-Z)
      DIMENSION X(50),D2F(100),F(100)
      X(1)=1.E-14
      DO 1 I=1,15
      KI=I+1
      IF(KKOD.EQ.0) GO TO 3
      DELTA=0.
      DDELTA=0.
      GO TO 4
    3 DELTA=-SI/T*(X(I)/DKS)**(1.-1./ETA)
      DDELTA=-SI/T*(1./DKS)**(1.-1./ETA)*(1.-1./ETA)*X(I)**(1./ETA)
    4 F(I)=X(I)/DKS+1./T*(X(I)/DKS)**UM-1./DKS+DELTA
      DF=1./DKS+(1./T)*(1./DKS)**UM*UM*X(I)**(UM-1.)+DDELTA
      D2F(I)=(1./T)*(1./DKS)**UM*UM*(UM-1.)*X(I)**(UM-2.)
      X(I+1)=X(I)-F(I)/DF
      IF(DABS((X(I+1)-X(I))/X(I+1)).LT.0.05) GO TO 2
    1 CONTINUE
    2 SKX=X(KI)
  100 FORMAT (10E13.4)
  101 FORMAT ( '          X ' )
  102 FORMAT ( '          D2F ' )
      RETURN
      END
      SUBROUTINE ASHER(SK,KS,ZK,SIGY,SMEAN,SKMEAN,KKOD)
      IMPLICIT REAL*8(A-H,O-Z)
      DIMENSION ZK(50),Y(50),SK(50)
      DO 1 I=1,KS
      Y(I)=ZK(I)*SIGY+SMEAN
    1 SK(I)=SKMEAN*DEXP(2.*Y(I))
      RETURN
      END
```

Input File Listing

```
  20   18    5    3  200    0
 2.592  0.36  0.753 15.   0.50   0.43   0.1  120.0  3.0 0.93
 0.05  1.16  -0.616 0.22  0.1
  5.   13.   24.   48.   138.
-1.950E 00-1.433E 00-1.175E 00-9.250E-01-7.500E-01-6.000E-01-4.500E-01
-3.200E-01-2.000E-01-6.250E-02 6.250E-02 2.000E-01 3.200E-01 4.500E-01
 6.000E-01 7.500E-01 9.250E-01 1.175E 00 1.433E 00 1.950E 00
  0.5  1.   2.   4.   8.   14.   20.   28.   36.   44.
  55.  68.  80.  95.  105.  115.  125.  140.
-2.2 -1.775 -1.525 -1.35 -1.213 -1.075 -0.9885
-0.8775 -0.8 -0.7 -0.625 -0.55 -0.4825 -0.4075
-0.35 -0.2875 -0.225 -0.1575 -0.1 -0.03 0.03
 0.1 0.1575 0.225 0.2875 0.35 0.4075 0.4825
 0.55 0.625 0.7 0.8 0.8775 0.9885 1.075
 1.213 1.35 1.525 1.775 2.2
```

Output File Listing (Prog. N0. 1)

```
KS   KZ   KT   KRD  KRK  KKOD
20   18   5    3    200  0
ETA     BETA    UM      HE      QU      TETAS   TETAI   DMAX    DIS     SKS
2.592   0.360   0.753   15.000  0.500   0.430   0.100   120.000 3.000   0.930
TETAR   SIGY    SMEAN   SKMEAN  AK      PORDIS
0.050   1.160   -0.616  0.220   0.947   0.100
        TOUT
5.000   13.000  24.000  48.000  138.000
        Z
0.500   1.000   2.000   4.000   8.000   14.000  20.000  28.000  36.000  44.000
55.000  68.000  80.000  95.000  105.000 115.000 125.000 140.000
        20 VALUES OF K AT SATURATION
NO 1            2           3           4           5           6
0.6960E-03   0.2310E-02   0.4202E-02   0.7505E-02   0.1126E-01   0.1595E-01
NO 7            8           9           1011        12
0.2259E-01   0.3055E-01   0.4035E-01   0.5551E-01   0.7419E-01   0.1021E+00
NO 13           14          15          1617        18
0.1348E+00   0.1823E+00   0.2582E+00   0.3656E+00   0.5487E+00   0.9801E+00
NO 19           20
0.1783E+01   0.5917E+01
    TOUT        Z
0.2400E+02   0.5000E+00
    AVERS       AVERC       AVERK       AVERY       AVERQ       AVERG
0.9652E+00   0.9988E+00   0.2779E+00   -0.3033E+01  0.1389E+00   0.1000E+01
    SDS         SDC         CVS         CVC
0.8375E-01   0.2682E-02   0.8677E-01   0.2685E-02
    SKEK        SKEY
0.1753E-01   -0.3966E+01
    VARS        VARC        VARK        VARY        SKES        SKEC

0.7014E-02   0.7194E-05   0.8609E-01   0.3388E+01   -0.1486E-02  -0.4442E-07
    TOUT        Z
0.2400E+02   0.1050E+03
    AVERS       AVERC       AVERK       AVERY       AVERQ       AVERG
0.1316E+00   0.0000E+00   0.4796E-06   -0.1735E+02  0.2398E-06   0.1000E+01
    SDS         SDC         CVS         CVC
0.3469E-17   0.0000E+00   0.2637E-16   0.4359E+01
    SKEK        SKEY
0.5924E-17   0.6484E-14
    VARS        VARC        VARK        VARY        SKES        SKEC

0.1204E-34   0.0000E+00   0.1416E-11   0.5040E+01   0.0000E+00   0.0000E+00
```

SOIL SPATIAL VARIABILITY

```
C                   Program N0. 2
C                   --------------
C                   Eshel Bresler
C                   PROGRAM LISTING
C                   ----------------
C
C
C  MODEL FOR COMPUTING PROBABILITY FUNCTION OF CONC. AND
C  EXPECTATION
C  -------------------------------------------------------
C
C---- N is number of times field averages are computed
C---- N3 is number of c and phi data in table
C---- KK is number of increments in the numerical integration
C---- KCOD = 1 if dispersivity is stochastic, otherwise R is stochastic
C---- SKST is K*, YM is E[Y], SY is SD of Y, AL is E[log(dispersivity)]
C---- SL is SD of log(dispersivity), DD is average dispersivity
C---- BU is lower integration limit for stochastic dispersivity
C---- TS is saturated theta, TR is residual theta, DP is difusivity
C---- B is beta/etha, Z is (constant) soil depth, TT is (constant) time
C---- R is (average recharge)/K*, S is dr/(average recharge)
      DIMENSION X(150),C(150),CD(150),XI(103,6),CI(103,6),IX(103,6),
     1IY(103,6),ISEMEL(3,6),ZERO(2),SCAL(2),COT(14),CTOX(4),CTOY(4),
     2NP(6),LSEG(6),XJ(103,6),CTOZ(4),SE(103,6),DERC(150),DERCR(150)
      DIMENSION SUMLL(150),SUMRR(150),VVV(150),CAVL(150),tz(150)
      DIMENSION CAV(150),ZZ(150),CSTAR(20),CSTAR1(20),A9(150),A10(150)
      DIMENSION CC(25),PHI(25),PC(150),PCR(150),CAV1(150),CAV2(150)
      DIMENSION A1(50),A2(50),A3(50),A4(50),A5(50),A6(50),A7(50),A8(50)
      READ *,N,N3,K,KK,KCOD
      PRINT 38
      PRINT 37,N,N3,K,KK,KCOD
      READ *,SKST,YM,AL,SL,DD,SY,BU
      PRINT 36
      PRINT 35,SKST,YM,AL,SL,DD,SY,BU
      READ *,TS,TR,DP,B,Z,TT,R,S
      PRINT 105
      PRINT 101,TS,TR,DP,B,Z,TT,R,S
      READ *, (CC(IL),IL=1,N3)
      READ *, (PHI(IL),IL=1,N3)
      PRINT 102
      PRINT 109, (CC(IL),IL=1,N3)
      PRINT 103
      PRINT 109, (PHI(IL),IL=1,N3)
      KKK=KK+1
      PI=3.1416
      FKK=KK
      M=0
      HC=1.
      CH=HC
      F=SQRT(2.)*SY
      P=(2.*B*SY*SY)/(1.-B)
      RB=SKST*R
      S=1.E-20
      L=0
```

```
      ZZ(1)=0.
      CAV(1)=0.
      tz(1)=0.0
      x(1)=0.0
      DO 3 I=2,N
      ZZ(I)=Z
      tz(i)=tz(i-1)+0.5
      x(i)=z*ts/(tz(i)*skst)
      RBAR=SKST*R
      DR=S*RBAR
      FR=0.5/DR
      DELR=2.*DR/FKK
      DO 31 IL=1,N3
      SUMY=0.
      SUML=0.
      SUMPC=0.
      SUMLL(IL)=0.
      SUMRR(IL)=0.
      SUMPC1=0.
      SUMR=0.
      RR=RBAR-DR-DELR
      YL=BU
      Y2=YL
      DELL=-2.*YL/KK
      VV=SKST*X(I)/TS+SKST*X(I)/(ZZ(I)*TS)*(2.*PHI(IL)**2*DD-2.*PHI(IL)
     1*SQRT(DD*ZZ(I)+PHI(IL)**2*DD**2+DP*ZZ(I)*TS/(X(I)*SKST)))
      VVV(IL)=VV
      IF(VV.LE.0.) GO TO 43
      DO 25 LI=1,KKK
      RR=RR+DELR
      YL=YL+DELL
      DL=EXP(YL)
      SUMR=SUMR+RR
      IF(VV.LT.RR/TS) GO TO 39
      YY=-0.5/B*ALOG((RR/VV-TR)/((RR/SKST)**B*(TS-TR)))
      GO TO 40
   39 YY=0.5*ALOG(VV*TS/SKST)
   40 TETA=TR+(R/SKST)**B*(TS-TR)
      tt=tz(i)
      VT1=-VV*TT/(2.*SQRT(DD*VV*TT+DP*TT))
      VT2=(ZZ(I)-VV*TT)/(2.*SQRT(DD*VV*TT+DP*TT))
      VT5=VT1
      VT6=VT2
      IF(VT5.GT.12.5) VT5=12.5
      IF(VT6.GT.12.5) VT6=12.5
      EXPVT1=EXP(-VT5**2)
      EXPVT2=EXP(-VT6**2)
      CSTAR(IL)=0.5/ZZ(I)*(-VT1*(1.-ERF(VT1))+1./SQRT(PI)*EXP(VT1)
     1+VT2*(1.-ERF(VT2))-1./SQRT(PI)*EXP(VT2))*2.0*SQRT(DD*VV*TT+DP
     2*TT)
      SUMPC=(ERF((YY-YM)/F)+1)+SUMPC
      VF=SKST*X(I)/TS+SKST*X(I)/(ZZ(I)*TS)*(2.*PHI(IL)**2*DL-2.*PHI(IL)
     1*SQRT(DL*ZZ(I)+PHI(IL)**2*DL**2+DP*ZZ(I)*TS/(X(I)*SKST)))
      IF(VF.LE.0.) GO TO 43
```

```
      IF(VF.LT.RB/TS) GO TO 33
      YF=-0.5/B*ALOG((RB/VF-TR)/((RB/SKST)**B*(TS-TR)))
      GO TO 34
33    YF=0.5*ALOG(VF*TS/SKST)
34    SUMPC1=SUMPC1+EXP(-(YL-AL)**2/(2.*SL*SL))*(ERF((YF-YM)/F)+1.)
      tt=tz(i)
      VT3=-VF*TT/(2.*SQRT(DL*VF*TT+DP*TT))
      VT4=(ZZ(I)-VF*TT)/(2.*SQRT(DL*VF*TT+DP*TT))
      VT7=VT3
      VT8=VT4
      IF(VT7.GT.12.5) VT7=12.5
      IF(VT8.GT.12.5) VT8=12.5
      EXPVT3=EXP(-VT7**2)
      EXPVT4=EXP(-VT8**2)
      CSTAR1(IL)=0.5/ZZ(I)*(-VT3*(1.-ERF(VT3))+1./SQRT(PI)*EXP(VT3)
     1+VT4*(1.-ERF(VT4))-1./SQRT(PI)*EXP(VT4))*2.0*SQRT(DL*VF*TT+DP
     2*TT)
      IF(KCOD.EQ.1) GO TO 25
      IF(IL.LT.N3) GO TO 25
27    Y1=BU
      DEL1=-2.*Y1/KK
      Y2=Y2+DELL
      D2=EXP(Y2)
      FL=1./(SQRT(2.*PI)*SL)*EXP(-(Y2-AL)**2/(2.*SL*SL))
      tt=tz(i)
      DO 24 LJ=1,KKK
      Y1=Y1+DEL1
      V=SKST/TS*EXP(2.*Y1)
      U=V
      IF(RR.LT.TS*V) V=(RR*EXP(2.*B*Y1))/(TS*(RR/SKST)**B)
      IF(RB.LT.TS*V) U=(RB*EXP(2.*B*Y1))/(TS*(RB/SKST)**B)
      FY=1./(SQRT(2.*PI)*SY)*EXP(-(Y1-YM)**2/(2.*SY*SY))
      SUMY=SUMY+DEL1*(1.-ERF((Z-V*TT)/(2.*SQRT(DD*V*TT+DP*TT))))*FY/FKK
      SUML=SUML+DEL1*(1.-ERF((Z-U*TT)/(2.*SQRT(D2*U*TT+DP*TT))))*FY*FL
     1*DELL
      SUMLL(IL)=SUMLL(IL)+(DEL1*(1.-ERF((Z-U*TT)/(2.*SQRT(D2*U*TT+DP*TT
     1))))*FY*FL)*CSTAR1(IL)*DELL
      SUMRR(IL)=SUMRR(IL)+(DEL1*(1.-ERF((Z-V*TT)/(2.*SQRT(DD*V*TT+DP*TT
     1))))*FY*FR)*CSTAR(IL)*DELR
24    CONTINUE
      tt=tz(i)
25    CONTINUE
      GO TO 46
43    SUMPC=0.
      CSTAR(IL)=0.
      SUMPC1=0.
      CSTAR1(IL)=0.
46    IF(IL.LT.N3) GO TO 47
      CBAR=0.5*SUMY
      CBARL=0.5*SUML
47    PC(IL)=0.5/(SQRT(2.*PI)*SL)*SUMPC1*DELL
      PCR(IL)=SUMPC*0.5/FKK
      IF(PCR(IL).GT.1.0)PCR(IL)=1.0
      IF(PCR(IL).LT.0.0)PCR(IL)=0.0
```

```
   31 CONTINUE
      print 150,tt,Z,DR
      PRINT 106
      PRINT 11,(PCR(IL),IL=1,N3)
      SUMC1=(PCR(1)*CC(1)+(1.-PCR(N3))*(1.-CC(N3)))*0.5
      SUSTRR=(PCR(1)*CSTAR(1)+(1.-PCR(N3))*(1.-CSTAR(N3)))*0.5
      SUMC=0.5*(PC(1)*CC(1)+(1.-PC(N3))*(1.-CC(N3)))
      SUSTRL=0.5*(PC(1)*CSTAR1(1)+(1.-PC(N3))*(1.-CSTAR1(N3)))
      SUMC1R=SUMC1*(0.5*CC(1)+0.5*(1.+CC(N3)))
      SUMCL=SUMC*(0.5*CC(1)+0.5*(1.+CC(N3)))
      DO 41 IL=2,N3
      DELC=CC(IL)-CC(IL-1)
      DELSTR=CSTAR(IL)-CSTAR(IL-1)
      DELSTL=CSTAR1(IL)-CSTAR1(IL-1)
      SUMC=SUMC+(PC(IL-1)+PC(IL))*0.5*DELC
      DERC(IL)=(PC(IL)-PC(IL-1))/DELC
      DERCR(IL)=(PCR(IL)-PCR(IL-1))/DELC
      SUMC1=SUMC1+(PCR(IL-1)+PCR(IL))*0.5*DELC
      CCIL=(CC(IL)+CC(IL-1))*0.5
      SUMC1R=SUMC1R+CCIL*(PCR(IL-1)+PCR(IL))*0.5*DELC
      SUSTRR=SUSTRR+(PCR(IL)+PCR(IL-1))*0.5*DELSTR
      SUSTRL=SUSTRL+(PC(IL)+PC(IL-1))*0.5*DELSTL
      SUMCL=SUMCL+CCIL*((PC(IL-1)+PC(IL))*0.5*DELC)
   41 CONTINUE
      CBAR1=1.-SUMC
      CBAR2=1.-SUMC1
      CBRSTR=1.-SUSTRR
      CBRSTL=1.-SUSTRL
      CAV1(I)=CBAR1
      SIGMR=1.-CBAR2**2-2.*SUMC1R
      SIGML=1.-CBAR1**2-2.*SUMCL
      SKEWR=(1.-CBAR2)**2-2.*SUMC1R+SUMC1*CBAR2
      SKEWL=(1.-CBAR1)**2-2.*SUMCL+SUMC*CBAR1
      CAV2(I)=CBAR2
      CAV(I)=CBAR
   30 IF(X(I).GE.R*(1.-S)) GO TO 6
      G=(0.5*ALOG(X(I))-YM)/(F)
      C(I)=0.5*(1.-ERF(G))
      GO TO 7
    6 IF(X(I).GT.R*(1.+S))GO TO 8
      G=(0.5*ALOG(X(I))-YM)/F
      H=(0.5/B*ALOG(X(I)*R**(B-1)*(1.-S)**(B-1))-YM)/F
      C(I)=1.-0.5*(1+ERF(G))+(1.-S)/(4.*S)*(ERF(H)-ERF(G))-(X(I)**(1./(1
     1.-B))/(4.*S*R))*EXP(-2.*B*YM/(1.-B))*EXP(2.*B*B*SY*SY/((1-B)**2))*
     2(ERF(H+P/F)-ERF(G+P/F))
      GO TO 7
    8 Q=(0.5/B*ALOG(X(I)*R**(B-1)*(1.+S)**(B-1))-YM)/F
      H=(0.5/B*ALOG(X(I)*R**(B-1)*(1.-S)**(B-1))-YM)/F
      C(I)=1.-0.5*(1.+ERF(Q))+(1.-S)/(4.*S)*(ERF(H)-ERF(Q))-(X(I)**(1./(
     11.-B))/(4.*S*R))*EXP(-2.*B*YM/(1.-B))*EXP(2.*B*B*SY*SY/((1-B)**2))
     2*(ERF(H+P/F)-ERF(Q+P/F))
    7 IF(L.GT.0) GO TO 3
      IF(C(I).GT.0.010) GO TO 3
      L=L+1
```

SOIL SPATIAL VARIABILITY

```
      3 CONTINUE
        PRINT 151
        PRINT 11,(TZ(I),I=2,N)
        PRINT 120
        PRINT 11, (CAV(I),I=2,N)
        PRINT 123
        PRINT 11,(X(I),I=1,N)
        PRINT 124
        PRINT 11, (ZZ(I),I=1,N)
      5 CONTINUE
      1 CONTINUE
     11 FORMAT(5E12.4)
     35 FORMAT (10F8.4)
     36 FORMAT ( '   SKST    YM    AL    SL    DD    SY    BU')
     37 FORMAT (5I4)
     38 FORMAT ( '  N  N3  K  KK  KCOD ' )
    101 FORMAT (8(2X,F6.3))
    102 FORMAT ( '          CC ' )
    103 FORMAT ( '          PHI' )
    105 FORMAT ( '   TS    TR    DP    B    Z    TT    R
      1    S')
    106 FORMAT ( '      P(C) FOR  STOCHASTIC RECHARGE ' )
    107 FORMAT ( '      P(C) FOR  STOCHASTIC  LAMDA ' )
    108 FORMAT (10F8.5)
    109 FORMAT (3X,5F8.5)
    120 FORMAT ( '   AVERAGE C VS. TIME OR XI ' )
    123 FORMAT ( '   XI(I)=Z*TS/(Time*K*)       ' )
    124 FORMAT ( '   ZZ(I)=Z IN  CM ' )
    151 FORMAT ( '   TZ(I)=TIME IN HOURS')
    150 FORMAT ( ' TIME=',F5.2,4X,'Z=',F6.2,5X,'DR=',E8.2)
    999 FORMAT (5I6,50X)
   1000 FORMAT (6E12.4,8X)
   1001 FORMAT (3E12.4,44X)
   1002 FORMAT (9F8.4,8X)
        STOP
        END
         function erF(z)
         REAL*8 Z,ZB,T,A1,A2,A3,A4,A5,A6,ZZ
        zb=1.
        if(z.lt.0.) zb=-1.
        Zz=Dabs(z)
        t=1/(1+ZZ*.3275911)
        a1=.254829592
        a2=-.284496736
        a3=1.421413741
        a4=-1.453152027
        a5=1.061405429
        a6=-Zz**2
        func=1-(a1*t+a2*t**2+a3*t**3+a4*t**4+a5*t**5)*Dexp(a6)
        if(zb.lt.0.) go to 30
        erf=func
        go to 50
     30 erf=-func
     50 return
        end
```

Input File Listing (no. 2)

```
 10  19   1   21   0
13.4  -0.160  0.0000  1.e-10  2.89  0.57  -3.5  24.    0.33
0.41   0.00   0.002   0.33   20.0   24    0.20750   1.0E-20
0.0023  .0113  .025   .05    .1    .2    .3    .4    .45   .475
 .6    .7    .8    .9    .95   .975  .985  .995  .9975
2.    1.98   1.39   1.165  .91   .6    .375  .18   .09   .045
-0.18  -0.375 -0.6  -0.91  -1.165 -1.388 -1.54 -1.83 -1.99
```

Output File Listing (no. 2.)

```
  N  N3   K   KK  KCOD
 10  19   1   21   0
  SKST    YM      AL      SL      DD      SY      BU
 13.4000 -0.1600  0.0000  0.0000  0.0000  0.5700 -3.5000
    TS     TR     DP      B       Z      TT      R       S
  0.410  0.000  0.002   0.330  20.000  24.000   0.207   0.000
           CC
  0.00230 0.01130 0.02500 0.05000 0.10000
  0.20000 0.30000 0.40000 0.45000 0.47500
  0.60000 0.70000 0.80000 0.90000 0.95000
  0.97500 0.98500 0.99500 0.99750
           PHI
  2.00000 1.98000 1.39000 1.16500 0.91000
  0.60000 0.37500 0.18000 0.09000 0.04500
 -0.18000-0.37500-0.60000-0.91000-1.16500
 -1.38800-1.54000-1.83000-1.99000
TIME=  0.50    Z= 20.00    DR=0.28E-19
   P(C) FOR STOCHASTIC RECHARGE
 0.1000E+01  0.1000E+01  0.1000E+01  0.1000E+01  0.1000E+01
 0.1000E+01  0.1000E+01  0.1000E+01  0.1000E+01  0.1000E+01
 0.1000E+01  0.1000E+01  0.1000E+01  0.1000E+01  0.1000E+01
 0.1000E+01  0.1000E+01  0.1000E+01  0.1000E+01
TIME=  1.00    Z= 20.00    DR=0.28E-19
   P(C) FOR STOCHASTIC RECHARGE
 0.1000E+01  0.1000E+01  0.1000E+01  0.1000E+01  0.1000E+01
 0.1000E+01  0.1000E+01  0.1000E+01  0.1000E+01  0.1000E+01
 0.1000E+01  0.1000E+01  0.1000E+01  0.1000E+01  0.1000E+01
 0.1000E+01  0.1000E+01  0.1000E+01  0.1000E+01
TIME=  1.50    Z= 20.00    DR=0.28E-19
   P(C) FOR STOCHASTIC RECHARGE
 0.7839E+00  0.7840E+00  0.7869E+00  0.7879E+00  0.7892E+00
 0.7907E+00  0.7918E+00  0.7927E+00  0.7931E+00  0.7933E+00
 0.7944E+00  0.7953E+00  0.7964E+00  0.7979E+00  0.7991E+00
 0.8001E+00  0.8008E+00  0.8022E+00  0.8029E+00
TIME=  2.00    Z= 20.00    DR=0.28E-19
   P(C) FOR STOCHASTIC RECHARGE
 0.4820E+00  0.4821E+00  0.4863E+00  0.4879E+00  0.4897E+00
 0.4918E+00  0.4934E+00  0.4948E+00  0.4954E+00  0.4957E+00
 0.4973E+00  0.4987E+00  0.5002E+00  0.5024E+00  0.5042E+00
 0.5057E+00  0.5068E+00  0.5088E+00  0.5099E+00
TIME=  2.50    Z= 20.00    DR=0.28E-19
```

P(C) FOR STOCHASTIC RECHARGE
0.2543E+00 0.2544E+00 0.2581E+00 0.2595E+00 0.2611E+00
0.2631E+00 0.2645E+00 0.2657E+00 0.2663E+00 0.2666E+00
0.2680E+00 0.2692E+00 0.2707E+00 0.2726E+00 0.2743E+00
0.2757E+00 0.2766E+00 0.2785E+00 0.2795E+00
TIME= 3.00   Z= 20.00   DR=0.28E-19
P(C) FOR STOCHASTIC RECHARGE
0.1359E+00 0.1360E+00 0.1369E+00 0.1372E+00 0.1376E+00
0.1381E+00 0.1384E+00 0.1387E+00 0.1389E+00 0.1389E+00
0.1393E+00 0.1396E+00 0.1399E+00 0.1404E+00 0.1408E+00
0.1411E+00 0.1414E+00 0.1418E+00 0.1420E+00
TIME= 3.50   Z= 20.00   DR=0.28E-19
P(C) FOR STOCHASTIC RECHARGE
0.1081E+00 0.1081E+00 0.1089E+00 0.1092E+00 0.1096E+00
0.1100E+00 0.1103E+00 0.1106E+00 0.1107E+00 0.1108E+00
0.1111E+00 0.1114E+00 0.1117E+00 0.1121E+00 0.1125E+00
0.1128E+00 0.1130E+00 0.1134E+00 0.1137E+00
TIME= 4.00   Z= 20.00   DR=0.28E-19
P(C) FOR STOCHASTIC RECHARGE
0.8746E-01 0.8749E-01 0.8825E-01 0.8854E-01 0.8887E-01
0.8926E-01 0.8955E-01 0.8981E-01 0.8992E-01 0.8998E-01
0.9027E-01 0.9052E-01 0.9081E-01 0.9121E-01 0.9154E-01
0.9183E-01 0.9203E-01 0.9240E-01 0.9261E-01
TIME= 4.50   Z= 20.00   DR=0.28E-19
P(C) FOR STOCHASTIC RECHARGE
0.7187E-01 0.7189E-01 0.7259E-01 0.7285E-01 0.7315E-01
0.7352E-01 0.7378E-01 0.7401E-01 0.7412E-01 0.7417E-01
0.7444E-01 0.7467E-01 0.7494E-01 0.7530E-01 0.7561E-01
0.7587E-01 0.7605E-01 0.7639E-01 0.7658E-01
TZ(I)=TIME IN HOURS
0.5000E+00 0.1000E+01 0.1500E+01 0.2000E+01 0.2500E+01
0.3000E+01 0.3500E+01 0.4000E+01 0.4500E+01
AVERAGE C VS. TIME OR XI
0.6372E-04 0.2041E-01 0.1990E+00 0.6508E+00 0.8553E+00
0.8553E+00 0.9770E+00 0.9770E+00 0.9770E+00
XI(I)=Z*TS/(Time*K*)
0.0000E+00 0.1224E+01 0.6119E+00 0.4080E+00 0.3060E+00
0.2448E+00 0.2040E+00 0.1748E+00 0.1530E+00 0.1360E+00
ZZ(I)=Z IN CM
0.0000E+00 0.2000E+02 0.2000E+02 0.2000E+02 0.2000E+02
0.2000E+02 0.2000E+02 0.2000E+02 0.2000E+02 0.2000E+02

REFERENCES

Biggar, J.W., and D.R. Nielsen. 1976. Spatial variability of leaching characteristics of a field soil. Water Resour. Res. 12:78–84.

Bresler, E. 1973. Simultaneous transport of solute and water under transient unsaturated flow conditions. Water Resour. Res. 9:975–986.

Bresler, E., and G. Dagan. 1979. Solute dispersion in unsaturated heterogeneous soil at field scale: II. Applications. Soil Sci. Soc. Am. J. 43:467–472.

Bresler, E., and G. Dagan. 1981. Convection and pore scale dispersive solute transport in unsaturated heterogeneous fields. Water Resour. Res. 17:1683–1693.

Bresler, E., and G. Dagan. 1983a. Unsaturated flow in spatially variable fields. 2. Application of water flow models to various fields. Water Resour. Res. 19:421–428.

Bresler, E., and G. Dagan. 1983b. Unsaturated flow in spatially variable fields. 3. Solute transport models and their application to two fields. Water Resour. Res. 19:4293435.

Brooks, R.H., and A.T. Corey. 1964. Hydraulic properties of porous media. Hydrology Paper No. 3, Colorado State Univ., Fort Collins.

Dagan, G., and E. Bresler. 1979. Solute dispersion in unsaturated heterogeneous soil at field scale: I. Theory. Soil Sci. Soc. Am. J. 43:461–467.

Dagan, G., and E. Bresler. 1983. Unsaturated flow in spatially variable fields. 1. Derivation of models of infiltration and redistribution. Water Resour. Res. 19:413–420.

Dagan, G. 1989. Flow and transport in porous formation. Springer-Verlag, Berlin.

Feinerman, E., E. Bresler, and G. Dagan. 1986. Statistical inference of spatial random functions. Water Resour. Res. 6:935–942.

Mein, R.G., and C.L. Larsen. 1973. Modeling infiltration during a steady rain. Water Resour. Res. 9:384–394.

Naor, A. 1989. Distribution of interactive solutes in a spatially variable field. (Hebrew with English summary.) Ph.D. diss. The Hebrew University of Jerusalem, Jerusalem, Israel.

Russo, D., and E. Bresler. 1982. A univariate versus a multivariate parameter distribution in a stochastic-conceptual analysis of unsaturated flow. Water Resour. Res. 18:483–488.

Saffman, P.G. 1959. A theory of dispersion in a porous medium. J. Fluid Mech. 6:321–349.

Saffman, P.G. 1960. Dispersion due to molecular diffusion and macroscopic mixing in flow through a network of capillaries. J. Fluid Mech. 7:194–208.

Segol, G. 1977. A three-dimensional Galerkin finite element model for the analysis of a contaminant transport in saturated-unsaturated porous media. p. 2123–2144. *In* W.G. Grey et al. (ed.) Finite element in water resources. Penetech Press, London.

Warrick, A.W., G.J. Mullen, and D.R. Nielsen. 1977. Scaling field measured soil hydraulic properties using a similar media concept. Water Resour. Res. 13:355–362.

9 Infiltration and Redistribution

R. J. HANKS

Utah State University
Logan, Utah

Infiltration was one of the first processes that soil scientists attempted to model. Klute (1952) was the first to recognize that it is necessary to account for the large change in soil properties (hydraulic conductivity and matric potential) with water content if the distinct wetting front associated with infiltration is predicted. Philip (1955, 1957) further improved numerical procedures for predicting infiltration. Numerous other investigators have added methdology and information in recent years (see Klute, 1973, and Freeze, 1969 for review). This chapter will not attempt to exhaustively cover the literature of infiltration prediction models, but will try to show an illustration of one model, similar to many others, that will allow reasonable predictions of the infiltration and redistribution process. Illustrating a model that demonstrates the influence of many factors on infiltration, such as initial water content, time, application rate, and soil hydraulic properties, was thought to be important.

I. INFILTRATION EQUATIONS

One should recognize that there may be many instances when the illustrated model is more complicated than needed and where simple equations are sufficient. Two of the most useful simple equations are:

$$I = K t^a \qquad [1]$$

and

$$I = S t^{1/2} + A t \qquad [2]$$

where I is cumulative infiltration, t is time, a and K are constants, S is the sorptivity, and A is closely related to the saturated hydraulic conductivity for many situations. Equation [2] (Philip, 1957) is often preferred because the value of S and A can be estimated from basic soil properties. Nevertheless, Eq. [1] and [2] are restricted to certain situations; ponding of water on the surface, uniform initial water content, and no change of soil properties with time.

Copyright © 1991 ASA-CSSA-SSSA, 677 S. Segoe Rd., Madison, WI 53711, USA. *Modeling Plant and Soil Systems*—Agronomy Monograph no. 31.

II. GENERAL EQUATION FOR UNSATURATED FLOW

Infiltration of water into the soil can be predicted for a wide variety of situations if the general flow equation for vertical flow is

$$\frac{\partial \psi_m}{\partial t} C_\theta = \frac{\partial}{\partial z}\left(K_\theta \frac{\partial \psi_m}{\partial z} + K_\theta\right) \qquad [3]$$

where θ is volumetric water content, t is time, z is soil depth, K_θ is hydraulic conductivity, C_θ is the water capacity ($\partial\theta/\partial\psi_m$), and ψ_m is soil matric potential. Hydraulic conductivity and matric potential are both strongly dependent on θ as shown in Table 9-1 (and also in the input and output listings), which is an example of the soil input data needed to solve an infiltration problem. Over the range of saturation to air dry, θ of a soil may vary from about 0.50 to about 0.02, whereas ψ_m may vary from 0 to about $-1.0\,E+4$ KPa, and K_θ may vary from about 50 to 5 $E-7$ mm/h. Thus, as the soil water content dries from saturation to air dry, both ψ_m and K_θ change over several orders of magnitude.

III. NUMERICAL MODEL

The solution of Eq. [3] presents difficulties because ψ_m and K_θ change so much as water content changes. Many analytical solutions have been given for restricted conditions such as ponded surface conditions, uniform initial water content, semi-infinite soil depth, and specific relations of $\theta - K_\theta$ and $\theta - \psi_m$. However, using numerical methods, it is possible to solve Eq. [3] in general without these restrictions. This method will be presented herein.

A numerical approximation of Eq. [3] is:

$$\left(\frac{h_i^j - h_i^j}{\Delta t}\right) C_i^{j+1/2} = K_{i-1/2}^{j+1/2}\left(\frac{\beta(h_{i-1}^j - h_i^j) + \alpha(h_{i-1}^{j+1} - h_i^{j+1}) + \Delta z}{\Delta z^2}\right)$$

$$- K_{i+1/2}^{j+1/2}\left(\frac{\beta(h_i^j - h_{i+1}^j) + \alpha(h_i^{j+1} - h_{i+1}^{j+1}) + \Delta z}{\Delta z^2}\right) \qquad [4]$$

where i is a depth subscript, j is a time superscript, $h = \psi_m$ at position (or depth i) and time j, Δt is the finite time increment ($t^{j+1} - t^j$), Δz is the finite depth increment ($z_{i+1} - z_i$), $C_i^{j+1/2} = C_\theta$ at depth i averaged over Δt, $K_{i-1/2}^{j+1/2} = K_\theta$ averaged across depth i and $i-1$, and $K_{i+1/2}^{j+1/2} = K_\theta$ averaged across depth i and $i+1$ at Δt. The coefficients α and β are used to allow for different approximations of ψ_m across the time interval, subject to the restriction that $\alpha + \beta = 1$. Note that the values of the ψ_m, h^j, constitute the initial (known) values, and h^{j+1} is not known, but will be computed at the end of the time interval.

Table 9-1. Hydraulic conductivity (K_θ) and matric potential (ψ_m) as related to water content (θ) for Sarpy loam and Millville silt loam.

Water content	Millville silt loam		Sarpy loam	
	ψ_m	K_θ	ψ_m	K_θ
m³/m³	K Pa	mm/h	K Pa	mm/h
0.00	−2.5E+6	1.0E−10	−2.5E+5	1.0E−10
0.02	−4.0E+4	1.0E−9	−3.9E+4	3.2E−10
0.04	−6.4E+3	1.6E−8	−6.0E+3	1.1E−9
0.06	−2.5E+3	1.0E−7	−8.8E+2	6.5E−8
0.08	−9.5E+2	2.8E−7	−1.3E+2	4.1E−6
0.10	−3.8E+2	1.9E−6	−6.6E+1	6.7E−5
0.12	−1.5E+2	9.1E−6	−3.3E+1	1.1E−3
0.14	−7.7E+1	5.9E−5	−2.6E+1	3.9E−3
0.16	−3.9E+1	4.5E−4	−2.1E+1	1.4E−2
0.18	−3.1E+1	1.8E−3	−1.7E+1	3.1E−2
0.20	−2.4E+1	5.2E−3	−1.3E+1	7.2E−2
0.22	−2.0E+1	2.0E−2	−1.0E+1	1.4E−1
0.24	−1.7E+1	3.0E−2	−7.8E+0	2.8E−1
0.26	−1.4E+1	4.5E−2	−6.4E+0	5.1E−1
0.28	−1.2E+1	7.8E−2	−5.3E+0	9.3E−1
0.30	−1.0E+1	1.3E−1	−4.3E+0	1.6E+0
0.32	−8.4E+0	2.1E−1	−3.5E+0	2.7E+0
0.34	−6.9E+0	3.4E−1	−2.5E+0	4.7E+0
0.36	−5.6E+0	5.1E−1	−1.8E+0	8.0E+0
0.38	−4.2E+0	8.3E−1	−7.3E−1	1.9E+1
0.40	−3.2E+0	1.3E−0	−3.0E−1	4.6E+1
0.42	−2.2E+0	2.2E−0	0	5.0E+1
0.44	−1.2E+0	7.0E−0		
0.46	−3.5E−1	9.3E−0		
0.48	0.0	1.2E+1		

If β is set at 1 and $\alpha = 0$, then the computation can be made with known values of h^j on the right-hand side of Eq. [4], but the size of Δz and Δt increments must be small, and much computation is required. However, the numerical solution is quite simple. An equation is written for each depth increment and values of h^{j+1} are found for each depth. These values are taken as the new initial conditions and the process is repeated.

However, if $\alpha = \frac{1}{2}$ to 1 and $\beta = \frac{1}{2}$ to 0, there is no restriction on the size of the Δt and Δz increments. A set of n equations with n unknowns results (n is the number of depth increments), which calls for a procedure to solve simultaneous equations. Fortunately, a very efficient method is available; the tri-diagonal matrix solution. Using this scheme decreases computation time, and places no restrictions on the time and depth increments. This solution is shown in statements 1840 to 2170 in the program listing (see Appendix).

A. Estimating Hydraulic Conductivity and Water Capacity

A difficulty still exists, however, because the values of $K_{i-1/2}^{j+1/2}$ and $K_{i+1/2}^{j+1/2}$, as well as $C_i^{j+1/2}$, are assumed constant during the time interval. This necessitates that the time increment be small, which may increase computer time. This problem is most serious during infiltration when large differences in water content occur because of the sharp wetting front.

There is also the difficulty of estimating K_θ between two depths with quite different water contents. We have tried many schemes and believe the one described below provides the best results.

The method used to estimate conductivity involves, first, developing a table of $\Sigma\,(D\,\Delta\theta)$ vs. θ from data of $K - \theta$ and $\psi_m - \theta$, where $D = K_\theta(\Delta\psi_m/\Delta\theta)$, (diffusivity) as shown in the output listing. The conductivity (K_θ) between depth increments i and $i-1$ is then computed as

$$K_{i-1/2}^{j+1/2} = D_{i-1/2}^{j+1/2} \frac{(\theta_{i-1}^j - \theta_i^j)}{(h_{i-1}^j - h_i^j)} \qquad [5]$$

where

$$D_{i-1/2}^{j+1/2} = \frac{\left(\sum_{\theta_L}^{\theta_{i-1}^j}(D\Delta\theta) - \sum_{\theta_L}^{\theta_i^j}(D\Delta\theta)\right)}{(\theta_{i-1}^j - \theta_i^j)} \qquad [6]$$

where θ_L is the lowest water content in the water content table, as shown in statements 1160 to 1780 of the program listing (see Appendix).

It is also necessary to estimate the water capacity $C_i^{j+1/2}$. Since the value required applies to depth z_i, it was taken as

$$C_i^{j+1/2} = \frac{\Delta\theta}{\Delta\psi_m} \qquad [7]$$

from tables of θ vs. ψ_m, as shown in statements 1790 to 1810 in program listing (see Appendix).

B. Boundary Conditions

The next problem concerns the boundary conditions. The boundary conditions, of course, determine what type of process is involved. A common boundary condition used for infiltration is surface ponding causing saturation at the soil surface after water is applied

$$\theta_{z=0}^{t\geq 0} = \theta \text{ (saturation)} \quad \text{or} \quad h_{z=0}^{t\geq 0} = h \text{ (saturation)} \qquad [8]$$

as when sudden flooding occurs in basin irrigation.

However, the above boundary condition is not all that is needed. Most field situations involve water applications to the soil in which all of the water is absorbed and saturation is not immediately reached at the surface. This commonly occurs with rainfall and sprinkler irrigation. Thus, a more realistic boundary condition is the flux boundary condition, where water is applied to the soil surface at a given intensity (rainfall or sprinkler rate). The soil absorbs all of the water if the flux is small. However, if the flux is large, then the soil absorbs water until the soil surface saturates, after which only part of the applied water is absorbed by the soil. The difference between applied and infiltrated water can then appear as runoff.

This is difficult to model because the boundary condition may change from a flux to a ponded condition. The model described herein used an iterative procedure to find the soil surface conditions that are appropriate according to

$$Jw = Ja, \text{ provided } \theta^t_{z=0} < \theta \text{ (saturation)} \qquad [9]$$

where Jw is the surface and Ja is the applied flux of water. The scheme for setting the top boundary conditions are accomplished in statements 1410 to 1750 of the program listing (see Appendix).

There is also a boundary at the bottom of the finite column that must be considered for any solution. For many conditions, like infiltration, the lower boundary condition can be assumed to be no flow

$$Jw \text{ (bottom)} = 0.0. \qquad [10]$$

Bottom boundary conditions are determined by the value of the parameter TAA. The model provides for these conditions:

1. No flow (Eq. [10], TAA = 1) and
2. Constant ψ_m (TAA = 0).

C. Time Steps

One problem with using numerical methods is the size of the time steps. Smaller time steps generally lead to more accurate solutions but, of course, take more computer time. The model presented herein provides for variable time steps through parameter CONQ, which is the absolute amount of total water content change in the entire profile for each time step. The value 0.035 has been found by trial and error to be reasonable. The computation is shown in statements 3000 to 3070 of the program listing (see Appendix). Thus, if infiltration is occurring, time steps will be small because there is a relatively large amount of water entering the soil. During redistribution, the time steps will be larger because the water content change within the soil is slower. Typically, during an infiltration-redistribution event, the time steps will change from 0.0024 to 2.4 h. There are situations when this time step is too large, so the computations are corrected. Program listing statements 2280 to 2310 and 2690 to 2720 provide procedures in which the time steps are decreased and the computations repeated, if clearly inaccurate estimates are made (see Appendix).

IV. MODEL VERIFICATIONS

The models used to compute infiltration and redistribution have been validated under several different conditions. Hanks and Bowers (1962) showed that this model computed infiltration correctly for a specified infiltration problem solved by Philip (1955). Hanks et al. (1969) showed that the computed results agree well with measured results for infiltration and distribution, as well as evaporation and drainage from a soil. The model shown has also been used as part of a more complicated model discussed in detail in Chapter 11 and validated over an entire season under field conditions with good results (Nimah & Hanks, 1973a, b).

V. EXAMPLES

A. Example 1

The model has been used to simulate several situations. The first is a uniform initial water content and a high rate of water applied (300 mm/h) at the soil surface for 1 h for Sarpy loam (mixed mesic Typic Udipsamments). The resulting soil water content profiles at various times are shown in Fig. 9-1. With these initial and boundary conditions, the infiltrated water was 80 mm and runoff was 220 mm. Redistribution was then computed until 50 h, with no flow at the top and bottom boundaries. There was a sharp wetting front at about 27 cm at the end of water application. The surface water content was at saturation (42%) and decreased to 31% at 25 cm. At 20 h, after redistribution for 19 h, the wetting front had penetrated to about 50 cm and the water content behind the wetting front was about 23%. At 50 h the wetting front had penetrated to about 65 cm and the water content behind the wetting front was about 20%.

B. Example 2

Figure 9-2 shows the simulation where the water application was lower, at 30 mm/h, so no runoff occurred. It was necessary to increase the time of infiltration to attain the same infiltration as for the conditions presented in Fig. 9-1. During water applicataion, the water content behind the wetting front did not saturate, but reached about 37%, and the wetting front penetrated deeper than at the end of wetting for Fig. 9-1. However, at 20 and 50 h, after redistribution, the water content profiles are essentially the same as for the ponded situation (Fig. 9-1). The main factor to consider for long redistribution time is the amount of water that gets into the soil.

C. Example 3

Figure 9-3 illustrates the results for ponded infiltration for Millville silt loam (coarse silty carbonic mesic Typic Haploceroll). Because of the differ-

INFILTRATION AND REDISTRIBUTION

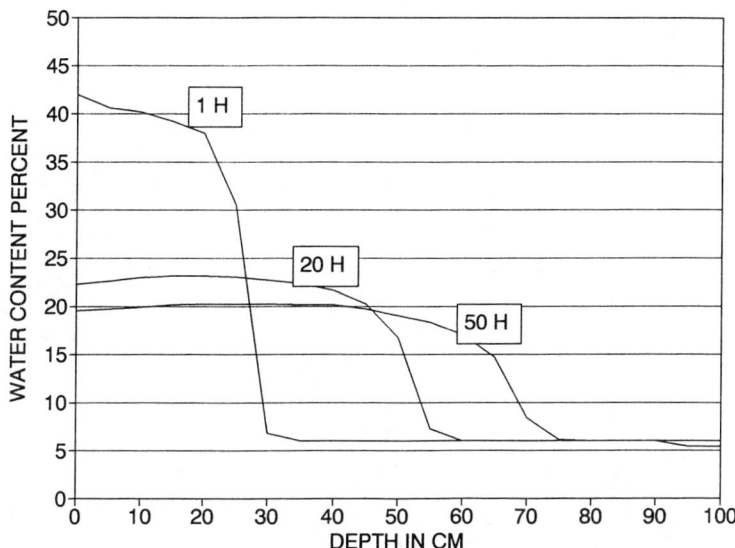

Fig. 9-1. Simulated water content during infiltration for 1 h and subsequent redistribution for Sarpy loam, with an initial water content of 6% and a high application rate (300 mm/h) causing runoff.

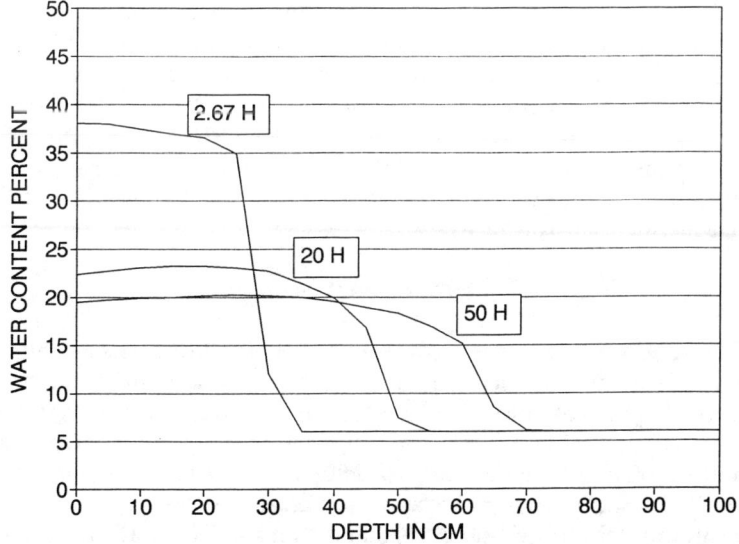

Fig. 9-2. Simulated water content during infiltration for 2.67 h and subsequent redistribution for Sarpy loam, with an initial water content of 6% and a low application rate (30 mm/h) with no runoff.

ence in soil properties, it took 3.34 h to infiltrate about 80 cm of water for the Millville soil, whereas it took only 1 h for the Sarpy loam, with the same water application rate (300 mm/h).

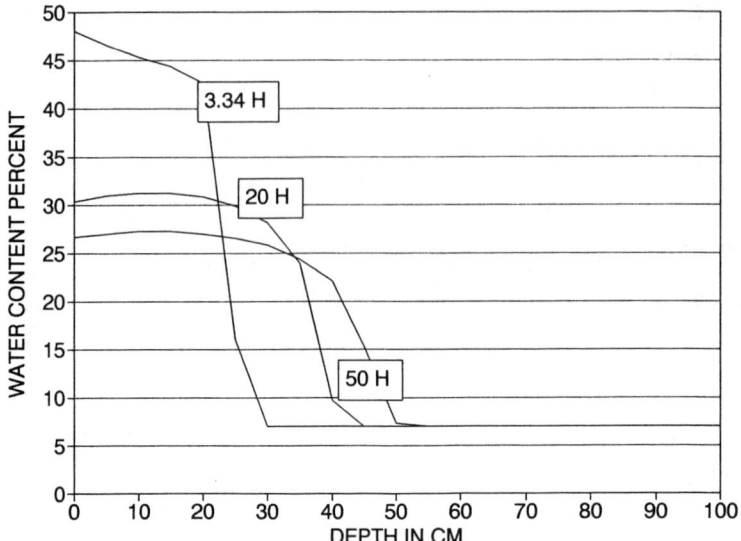

Fig. 9-3. Simulated water content during infiltration for 3.34 h and subsequent redistribution for Millville silt loam, with an initial water content of 7% and a high application rate (300 mm/h) causing runoff.

At the end of infiltration, the water content behind the wetting front was higher for the Millville soil than for the Sarpy, but the depth of penetration was less. After 20 h, the water content behind the wetting front was about 31% and the depth of penetration was about 35 cm. After 50 h, the water content behind the wetting front was about 27% and the depth of wetting was about 45 cm. Thus the water content behind the wetting front was higher for the Millville than for the Sarpy, but the depth of wetting was lower for Millville than for Sarpy.

D. Example 4

Figure 9-4 presents water content profiles for Millville silt loam with no runoff. The water application rate was lowered to 10 mm/h, and the time of infiltration increased to 8 h to get the same amount of water infiltrated as with ponding. At the end of water application, the water content behind the wetting front was about 43%, which was lower than the ponded case (Fig. 9-3), but the depth of penetration was greater. However, the water content profiles after redistribution at 20 and 50 h was essentially the same for the ponded and nonponded cases. Thus again, the important factor for redistribution is the amount of water that actually got into the soil.

E. Sample 5

Figure 9-5 illustrates another situation similar to Fig. 9-3, except with a higher initial water content (14% compared with 7%). The amount of water

INFILTRATION AND REDISTRIBUTION

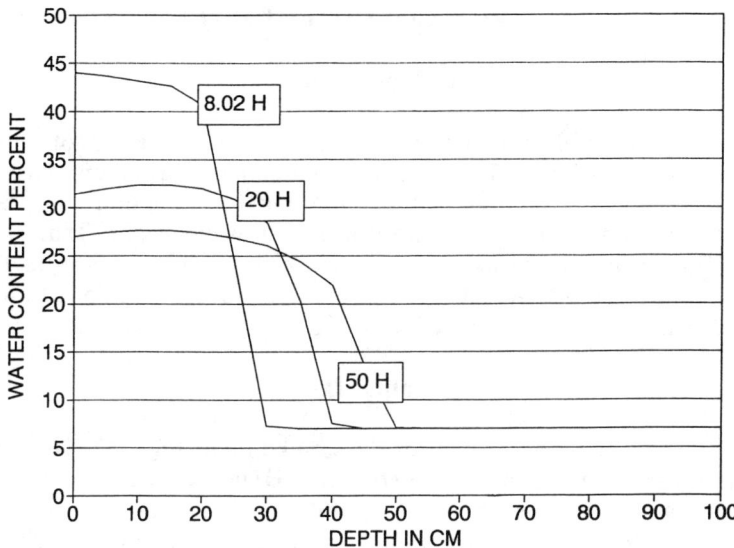

Fig. 9-4. Simulated water content during infiltration for 8.02 h and subsequent redistribution for Millville silt loam with an initial water content of 7% and a low application rate (15 mm/h) with no runoff.

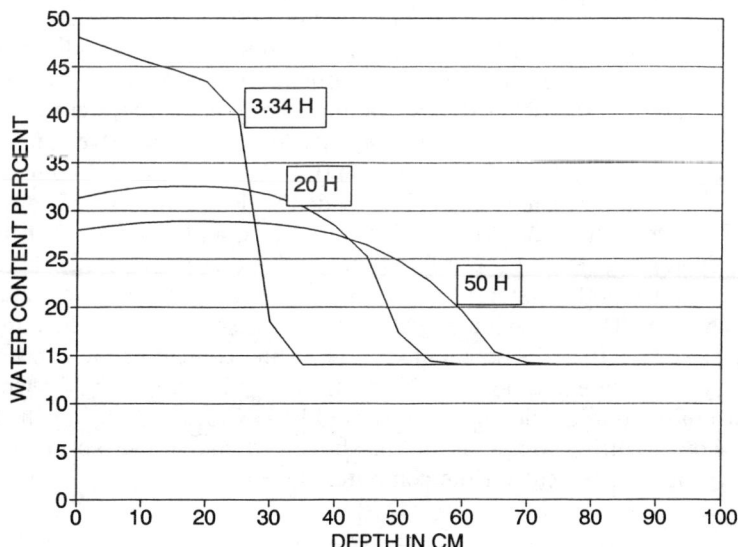

Fig. 9-5. Simulated water content during infiltration for a short time (1 h) and subsequent redistribution for Millville silt loam, with an initial water content of 14% and a high application rate (300 mm/h) causing runoff.

infiltrated in the same time was 78 mm; slightly less than for the low initial water content. The depth of penetration was greater for the high compared with the low initial water content. The position of the wetting front is less distinct when initial water content is high compared with low.

VI. INPUT AND OUTPUT DATA

A. Input Data

The input data file, shown after the program listing, consists of 13 lines of data, which is followed by comments containing information about the data. The file shown is configured for computing the simulation of Fig. 9-3. To change the boundary condition, information is changed in Line 3, the surface flux array. To change soils, the information on Lines 5 through 11 and 13 needs to be changed. To change initial conditions (as done for Fig. 9-5), the water content in Lines 10 and 11 needs to be changed.

B. Output Data

The output file is also listed for the simulation of Fig. 9-3. The information down to and including the constants HDRY etc., is a printout of the input data or is computed from input data information. Note that the parameter HDRY and the other parameters are defined in remark statements (REM) in the beginning of the program listing (see Appendix). The column headed by DIFFUSIV is a summation function, Eq. [6], and is not diffusivity. The column WAT CONT is generated within the program and placed in Array T as water content with equal increments. The TIME column is the cumulative time. The CWF column is the cumulative water added, computed from totaling the change of water content. The cumulative irrigation and rain is given in the IRR + RAIN column. The cumulative water flow out of the bottom of the profile is given by CUMB and cumulative runoff is provided under RUNOFF. The cumulative flux at the soil surface, computed from a summation of Darcy's law, is provided under CUMS. If the computation is made with small errors, CUMS should equal CWF, which it does for the example given. The WATBAL column is a check to see if there is a water balance in the system; thus, a test of the accuracy of the computations. The WFDD is the value of water flux at the soil surface, which must be equal to or less than the potential flux. The water content profile is printed out every time there is a boundary condition time listed (Array V), which is shown in the output file under TIME and FLUX (3, 20, 50 h) for the example shown. During computation, the above variables are also printed out for every time increment on the computer screen.

VII. CRITICAL ASSUMPTIONS

As with all models that are simplifications of more complicated processes, many critical assumptions are made that the user should be aware of if the models are to be useful. Unfortunately, the list of assumptions is so large it is seldom listed separately from the discussion.

The most critical assumptions found using this model are those that involve soil properties. As discussed herein, the soil properties, the $\psi_m - \theta$

and the $K - \theta$ relations, are assumed to be unique. Thus, hysteresis is not accounted for nor are changes in these relations that might occur under normal conditions. These changes can be taken into account, but that makes the model much more complicated. However, the changes in soil surface properties for many soils are too great to be ignored if there is a desire to estimate real field situations. Oliveira et al. (1987) measured the changes in soil surface properties during a season for a Millville silt loam under sprinkler irrigation and rainfall. They found significant reduction in soil surface $K - \theta$ values, especially near saturation. The reduction was less with a surface straw mulch, but was still significant. As a consequence, runoff during the season, with the same intensity of sprinkler irrigation, increased. In earlier papers we showed that simulated infiltration was greatly influenced by both the $K - \theta$ and the $\psi - \theta$ relations near saturation, but were little influenced by the data for these relations under dry situations (Hanks & Bowers, 1963).

Another critical problem is the assumption of no hysteresis effects during redistribution. As shown by Hanks et al. (1969), simulations in which hysteresis in the soil water relations were considered, indicated a significant effect on the water content profile for the field capacity condition.

Because the soil is assumed to be uniform, this model assumes soil cracking is not significant, which may be a critical assumption for many soils.

Nevertheless, these simplifying assumptions may still be useful under many field situations. The model does assume water balance and provides good predictions of field situations over long-time periods, if field data are used as initial conditions and water is added at such a slow rate that it all enters the soil. Thus, even though the soil surface properties may have changed with time, all of the water may still have entered the soil. Models do help to organize a very complicated situation, but should always be checked or verified in the field. Thus, a good motto might be "Modelers or users of models should keep one foot in the field." Conversely, the field scientist should be well acquainted with the models used and the assumptions involved to draw any conclusions from simulations.

VIII. OTHER MODELS

There are many other models that could be used to simulate infiltration and redistribution. An example, considerably simpler than the model described herein, is the model of Campbell (1985). If a mathematical relation between θ and ψ_m and between θ and K_θ is assumed, considerable simplification is possible. Thus, all of the table look-up program steps used in this chapter could be replaced by equations, and thus, fewer program steps. Other techniques could be used, as discussed by Campbell (1985) to allow for longer time steps.

IX. SUMMARY

A model is described that can be used to estimate the infiltration of water into a uniform soil with known initial water content profile and with different intensities of water added to the soil surface. All of the water may be infiltrated or some runoff may occur. Once water is added, the model allows for computation of water redistribution within the soil profile. When compared with field data, the model provides good approximations for many conditions. However, the model is quite sensitive to soil properties of matric potential and hydraulic conductivity versus water content relations. If these relations change with time and rainfall, or water additions have a high intensity, the model predictions will be in error, unless modified to account for these changes. The user is cautioned to "Always keep one foot in the field" when using this or any other model.

X. RESEARCH NEEDS

There are many research needs related to this model and many others. In particular, there is need to find more useful procedures for estimating the soil properties needed for the model and to account for the changes that occur under many situations in the field. More study needs to be done to consider the effects of hysteresis, especially as related to the field capacity of the soil.

XI. APPENDIX

PROGRAM LISTING

```
10 REM MODEL FOR COMPUTING INFILTRATION,
*  REDISTRIBUTION AND EVAPORATION
20 REM DETT initial and smallest time increment.
30 REM CONQ largest water content change allowed.
40 REM WATH saturated water content.
50 REM HWET matric head of saturated water content.
60 REM WATL the lowest water content-air dry.
70 REM HDRY the matric head corresponding to WATL.
80 REM TAA used to tell condition of the bottom
*  boundary-if TAA=1 flux=0, TAA=0 cons. mat head.
100 REM TT=1.0 for Lassonen = 0.5 for Cr-Nich.
110 REM IER twice number of surfaces flux changes.
120 REM TIME cumulative time at start.
130 REM CUMT time at end to stop.
140 REM DELW water content increment-T(I) array.
150 REM K number of soil depth increments.
160 REM NB number of depth increments to compute
    flux-usually=K.
170 REM DD(I) depth array giving depth from top.
180 REM B(I) hydraulic conductivity vs depth.
```

INFILTRATION AND REDISTRIBUTION

```
190 REM C(I) water capacity vs depth.
200 REM W(I) ending soil water content vs depth.
210 REM Y(I) begin soil water content vs depth.
220 REM H(I) ending matric head vs depth.
230 REM G(I) begin matric head vs depth.
240 REM V(I) surface boundary potential flux
*   followed by time; + for irrigation or rain and
*   - for ET.
260 REM T(I) array of equal spaced water content.
270 REM P(I) matric head array vs T(I).
280 REM E(I) hydraulic conduct. array vs T(I).
290 REM F(I) temporary array.
300 REM D(I) array to sum diffusivity x DELW.
310 DIM B(25), C(25), F(25), DD(25), V(50), P(55)
320 DIM E(55),W(25),D(55),T(55),H(25),G(25),Y(25)
330 OPEN "SOWATD.DAT" FOR INPUT AS #1
340 OPEN "SOWAT.OUT" FOR OUTPUT AS #2
350 DEFINT I-N
360 INPUT #1, LAB$
370 INPUT #1, K, IER, NB, ND, KI, KCPMAX
380 KK = K + 1
390 IR = INT(IER / 2)
400 FOR I = 1 TO IER: INPUT #1, V(I): NEXT I
410 FOR I = 1 TO KK: INPUT #1, DD(I): NEXT I
420 FOR I = 1 TO ND: INPUT #1, P(I): NEXT I
430 FOR I = 1 TO ND: INPUT #1, E(I): NEXT I
440 PRINT #2, LAB$, DATE$, TIME$
450 PRINT #2, "   K   IER   NB   ND   KCPMAX"
460 PRINT #2, USING " ####"; K, IER, NB, ND,
*   KCPMAX
470 T(1) = 0
480 FOR I = 1 TO KK: INPUT #1, W(I): NEXT I
490 INPUT #1, DETT, CONQ, TAA, TIME, TT, CUMT
500 IF TAA < 1 THEN ITAA = 0
510 IF TAA >= 1 THEN ITAA = 1
520 INPUT #1, HDRY, HWET, WATL, WATH, HLOW, DELW
530 D(1) = (E(1) * (P(2) - P(1)))
540 FOR I = 2 TO ND
550   D(I) = E(I) * (P(I) - P(I - 1)) + D(I - 1)
560   T(I) = DELW + T(I - 1)
570 NEXT I
580 PRINT #2, "WAT CONT  MAT HEAD  CONDUCTI
*   DIFFUSIV  WAT CONT  MAT HEAD  CONDUCTI
*   DIFFUSIV"
590 NE = ND / 2
600 FOR I = 1 TO NE
610   J = NE + I
620   PRINT #2, USING "#.##^^^^  "; T(I), P(I),
*   E(I), D(I), T(J), P(J), E(J), D(J)
```

```
630 NEXT I
640 KC = 1
660 CWFLX = 0
670 DELT = DETT
680 TM = 1 - TT
690 TBB = 1 - TAA
700 YMAX = WATH
710 RUNOF = 0
720 CUMS = 0
740 RPI = 0
750 CUMB = 0
760 CUMM = 0
770 IRDF = 0
780 EVAP = 0
790 SIR = 0
800 PIT = 0
810 J = INT((W(1) - T(1)) / DELW) + 1
820 H(1) = (P(J + 1) - P(J)) * (W(1) - T(J)) /
*  DELW + P(J)
830 G(1) = H(1)
840 C(1) = DELW / (P(J + 1) - P(J))
850 FOR I = 2 TO K
860   PIT = W(I) *(DD(I + 1) - DD(I - 1))/2+ PIT
870 NEXT I
880 FOR I = 2 TO KK
890   J = INT((W(I) - T(1)) / DELW) + 1
900   H(I) = (P(J + 1) - P(J)) * (W(I) - T(J)) /
*  DELW + P(J)
910   C(I) = DELW / (P(J + 1) - P(J))
920   G(I) = H(I)
930 NEXT I
940 PRINT #2, "  DEPTH    C(I)  W-DEPTH    H-DEPTH "
950 FOR I = 1 TO KK
960    IF I = 1 THEN 980
970    PRINT #2, USING " ###.  #.##^^^^  #.###
*  #.###^^^^  #.### "; DD(I), C(I), W(I), H(I)
980    Y(I) = W(I)
990 NEXT I
1000 PRINT #2, "   TIME      FLUX "
1010 FOR I = 2 TO IER STEP 2
1020   PRINT #2, USING "###.####  "; V(I), V(I - 1)
1030 NEXT I
1040 WFDD = V(1)
1050 EOR = V(1)
1060 PRINT #2, "   DETT     CONQ      TAA     TIME
*  TT      CUMT    DELW"
1070 PRINT #2, USING "###.#### "; DETT, CONQ, TAA,
*  TIME, TT, CUMT, DELW
1080 PRINT #2, "   HDRY     HWET     WATL
```

INFILTRATION AND REDISTRIBUTION

```
*      WATH       HLOW       HHI"
1090 PRINT #2, USING "#.###^^^^ "; HDRY, HWET,
*    WATL, WATH, HLOW, HHI
1100 KCK = 1
1150 BOT = WATL:TOP=WATH: WKP=W(1): HKP=H(1)
1160 REM  COMP OF COND.(B) AND WATER CAPAC (C)
1170 IF EOR > 0 THEN 1210
1180 W(1) = WATL
1190 H(1) = HDRY
1200 GOTO 1230
1210 W(1) = WATH
1220 H(1) = HWET
1230 TWW = (W(1) + Y(1)) * .5
1240 IF (TWW > WATH) THEN TWW = WATH
1250 J = INT((TWW - T(1)) / DELW) + 1
1260 BB = (TWW - T(J)) / DELW
1270 DIFFA = (D(J + 1) - D(J)) * BB + D(J)
1280 HI = (P(J + 1) - P(J)) * BB + P(J)
1290 FOR I = 1 TO K
1300   TW = (W(I + 1) + Y(I + 1)) * .5
1310   J = INT((TW - T(1)) / DELW) + 1
1320   BB = (TW - T(J)) / DELW
1330   DIFFB = (D(J + 1) - D(J)) * BB + D(J)
1340   GI = (P(J + 1) - P(J)) * BB + P(J)
1350   IF ABS(HI - GI) < .0001 THEN 1740
1390   B(I) = (DIFFA - DIFFB) / (HI - GI)
1400   IF I > 1 THEN 1760
1410   ER = (B(1) * (H(1) * TT - H(2) * TT - C(2) *
*    TM + G(1) * TM + DD(2))) / DD(2)
1420   IF ABS(ER) > ABS(EOR) THEN 1440
1430   IF H(1) = HWET OR H(1) = HDRY THEN 1760
1440   IF ABS(1.1 * EOR - ER) - ABS(.1 * EOR) <= 0
*  THEN 1470
1450   IF KCK = 1 THEN 1510
1460   IF KCK < 12 THEN 1550
1470   H(1) = (EOR * DD(2) / B(1) + H(2) * TT -
*  G(1) * TM + G(2) * TM - DD(2)) / TT
1480   IF H(1) < HDRY THEN H(1) = HDRY
1490   IF H(1) > HWET THEN H(1) = HWET
1500   GOTO 1760
1510   H(1) = HKP
1520   W(1) = WKP
1530   KCK = KCK + 1
1540   GOTO 1230
1550   KCK = KCK + 1
1560   IF ER = EOR THEN 1760
1565   IF ER > EOR THEN 1610
1570   IF W(1) = WATH THEN 1760
1580   BOT = W(1)
```

```
1590  W(1) = (W(1) + TOP) * .5
1600  GOTO 1640
1610  IF W(1) = WATL THEN 1760
1620  TOP = W(1)
1630  W(1) = (W(1) + BOT) * .5
1640  J = INT((W(1) - T(1)) / DELW) + 1
1650  BB = (W(1) - T(J)) / DELW
1660  REM IF ABS(EOR)<.000001 THEN 1780
1670  H(1) = (P(J + 1) - P(J)) * BB + P(J)
1680  TWW = (W(1) + Y(1)) * .5
1690  J = INT((TWW - T(1)) / DELW) + 1
1700  BB = (TWW - T(J)) / DELW
1710  DIFFA = (D(J + 1) - D(J)) * BB + D(J)
1720  HI = (P(J + 1) - P(J)) * BB + P(J)
1730  GOTO 1380
1740  B(I) = (D(J + 1) - D(J)) / (P(J + 1) - P(J))
1750  IF I = 1 THEN 1410
1760  TWW = TW
1770  HI = GI
1780  DIFFA = DIFFB
1790  TW = (W(I + 1) + Y(I + 1)) * .5
1800  J = INT((TW - T(1)) / DELW) + 1
1810  C(I + 1) = DELW / (P(J + 1) - P(J))
1820 NEXT I
1830 KCK = 1: KCP = 0
1840 REM*WATER FLOW TRIDIAGONAL MATRIX SOLUTION
1850 FOR I = 2 TO K
1860  POT = (DD(I + 1) - DD(I - 1)) / (2 * DELT)
1870  DLXA = (DD(I) - DD(I - 1))
1880  DLXB = (DD(I + 1) - DD(I))
1890  AA = B(I - 1) / DLXA
1900  CC = B(I) / DLXB
1910  BB = C(I) * POT / TT + CC + AA
1920  DA = (C(I) * POT * G(I) + CC * (TM * (G(I +
*  1) - G(I)) - DLXB) + AA * (TM * (G(I - 1) -
*  G(I)) + DLXA)) / TT
1930  IF I > 2 THEN 2000
1940  IF H(1) >= HWET OR H(1) <= HDRY THEN DA = DA
*  + AA * H(1): GOTO 1970
1950  DA = DA - (AA * (TM * (G(I - 1) - G(I)) +
*  DLXA)) / TT + EOR / TT
1960  BB = BB - AA
1970  F(I) = DA / BB
1980  E(I) = CC / BB
1990  GOTO 2030
2000  IF I >= K THEN 2040
2010  E(I) = CC / (BB - AA * E(I - 1))
2020  F(I) = (DA + AA * F(I - 1)) / (BB - AA * E(I
*  - 1))
```

```
2030 NEXT I
2040 IF ITAA = 0 THEN EX = (G(I) - G(I + 1) +
*    DLXB) * B(I) / DLXB ELSE EX = 0
2050 BB = BB - CC
2060 DA = DA + CC * ((G(I) - G(I + 1)) * TM +
*    DLXB) / TT - EX / TT
2070 H(I) = (DA + AA * F(I - 1)) / (BB - AA * E(I
*    - 1))
2080 FOR I = K - 1 TO 2 STEP -1
2090   H(I) = E(I) * H(I + 1) + F(I)
2100 NEXT I
2110 IF ITAA = 0 THEN 2140
2120 H(KK) = H(K) + DD(KK) - DD(K)
2130 G(KK) = G(K) + DD(KK) - DD(K)
2140 IF TAA = 2 AND H(KK) >= HWET THEN ITAA = 0
*    ELSE ITAA = 1
2150 FOR I = 2 TO K
2160   IF H(I) > HWET THEN 2280
2170 NEXT I
2180 REM Compute new water content vs matric head.
2190 IF H(1) <= HDRY OR H(1) >= HWET THEN 2220
2200 WFDD = EOR
2210 GOTO 2260
2220 WFDD = (B(1) * (H(1) * TT - H(2) *TT-G(2) *
*    * TM + G(1) * TM + DD(2))) / DD(2)
2230 IF H(1) >= HWET THEN W(1) = WATH
2240 IF H(1) <= HDRY THEN W(1) = WATL
2250 GOTO 2550
2260 HI = (EOR * DD(2) / B(1) + H(2) * TT - G(1) *
*    TM + G(2) * TM - DD(2)) / TT
2270 IF HI > HDRY AND HI < HWET THEN 2360
2280 IF KCP >= KCPMAX THEN 2320
2290 KCP = KCP + 1
2300 DELT = DELT * .5
2310 GOTO 1850
2320 IF HI < HDRY THEN H(1) = HDRY
2330 IF HI > HWET THEN H(1) = HWET
2340 WFDD = (B(1) * (H(1) * TT - H(2) * TT -G(2)*
*    TM + G(1) * TM + DD(2))) / DD(2)
2350 GOTO 2230
2360 H(1) = HI
2370 I = 1
2380 IF ABS(H(I) - G(I)) < .0001 THEN 2540
2390 NHI = ND
2400 NLO = 1
2410 J = INT(ND / 2)
2420 IF H(I) = P(J) THEN 2510 ELSE IF H(I) > P(J)
*    THEN 2450
2430 NHI = J
```

```
2440 GOTO 2460
2450 NLO = J
2460 JT = J
2470 J = INT((NHI - NLO) / 2) + NLO
2480 IF J <> JT THEN 2420
2490 IF H(I) >= P(J) THEN 2510
2500 J = J - 1
2510 WAT = (H(I) - P(J)) * DELW / (P(J + 1) -
*  P(J)) + T(J)
2520 W(I) = WAT
2530 GOTO 2550
2540 W(I) = Y(I)
2550 FOR I = 2 TO KK
2560   W(I) = C(I) * (H(I) - G(I)) + Y(I)
2570   IF W(I) > WATH THEN W(I) = WATH
2580   IF W(I) < WATL THEN W(I) = WATL
2590 NEXT I
2600 SUM3 = 0
2610 SUM2 = 0
2620 SUM1 = 0
2630 FOR I = 2 TO K
2640   SUM1 = W(I) + SUM1
2650   SUM2 = Y(I) + SUM2
2660   IF ABS(SUM1 - SUM2) <= ABS(SUM3) THEN 2680
2670   SUM3 = SUM1 - SUM2
2680 NEXT I
2690 IF ABS(SUM3) <= ABS(CONQ) THEN 2730
2700 IF DELT <= DETT * .1 THEN 2730
2710 DELT = .5 * DELT
2720 GOTO 1850
2730 SUM1 = 0
2740 SUM2 = 0
2750 WFUU = B(NB) * ((H(NB) - H(NB + 1)) * TT +
*  (G(NB) - G(NB + 1)) * TM + DD(NB + 1) - DD(NB))
*  / (DD(NB + 1) - DD(NB))
2760 FOR I = 2 TO K
2770   SUM1 = W(I) * (DD(I + 1) - DD(I - 1)) / 2 +
*  SUM1
2780   SUM2 = Y(I) * (DD(I + 1) - DD(I - 1)) / 2 +
*  SUM2
2790 NEXT I
2800 CWF = SUM1 - PIT
2810 WFRDD = (SUM1 - SUM2) / DELT
2820 CUMS = WFDD * DELT + CUMS
2830 IF EOR > 0 THEN SIR = EOR * DELT + SIR
2840 IF EOR < 0 THEN EVAP = WFDD * DELT + EVAP
2850 IF EOR > 0 THEN RPI = RPI + WFDD * DELT
2860 IF EOR > 0 THEN RUNOF = (EOR - WFDD) * DELT +
*  RUNOF
```

INFILTRATION AND REDISTRIBUTION

```
2870 CUMB = WFUU * DELT + CUMB
2880 CUMET = CUMET + ET * DELT
2890 HRFLUX = WFUU
2900 CTRAN = CTRAN + ETPL * DELT
2910 CWFLX = (SUM1 - SUM2)
2920 KB = K - 1
2930 TIME = TIME + DELT
2940 WATBAL = SIR + EVAP - RUNOF - CUMB - CWF
2950 PRINT USING "####.#### "; TIME, CWF, SIR,
*  CUMB, RUNOF, CUMS, WATBAL, EOR
2960 REM ---CHANGE DELT HERE
2970 IF ABS(SUM3 - 0) > .0001 THEN 3000
2980 DELT = 3 * DELT
2990 GOTO 3090
3000 TW = ABS(CONQ * DELT / SUM3)
3010 IF TW >= .1 * DETT THEN 3040
3020 TW = .1 * DETT
3030 GOTO 3060
3040 IF TW <= 1000 * DETT THEN 3060
3050 TW = 1000 * DETT
3060 IF TW > 2 * DELT THEN 2980
3070 DELT = TW
3080 REM  SEE IF EVAP OR RAIN ETC. HAS CHANGED
3090 IF IDELT = 1 THEN DELT = DELT1
3100 IDELT = 0
3110 IF DELT < DETT THEN DELT = DETT
3120 IF DELT > 6 THEN DELT = 6
3130 IF TIME - V(KC + 1) < 0 THEN 3250
3150 PRINT #2, "    TIME       CWF      IRR+RAIN
 CUMB      RUNOFF     CUMS
*  WATBAL      EOR"
3160 PRINT #2, USING " #.###^^^^"; TIME, CWF, SIR,
*  CUMB, RUNOF, CUMS, WATBAL, EOR
3170 PRINT #2, "  WATER CONTENT VS DEPTH"
3180 FOR I = 1 TO KK: PRINT #2, USING " #.### ";
*  W(I); : NEXT I
3190 PRINT #2,
3200 EOR = V(KC + 2)
3210 IR = INT((KC + 2) / 2)
3220 KC = KC + 2
3230 DELT = DETT
3240 GOTO 3270
3250 IF (TIME + DELT) <= V(KC + 1) THEN 3270
3260 DELT = V(KC + 1) - TIME
3270 IF DELT < DETT THEN DELT = DETT
3290 IF TIME >= CUMT THEN 3490
3300 Y(1) = (W(1) + Y(1)) * .5
3310 J = INT((Y(1) - T(1)) / DELW) + 1
3320 BB = (Y(1) - T(J)) / DELW
```

```
3330 IF ABS(EOR - 0) < .0001 THEN 3350
3340 G(1) = (P(J + 1) - P(J)) * BB + P(J)
3350 FOR I = 2 TO KK
3360   J = INT((W(I) - T(1)) / DELW) + 1
3370   BB = (W(I) - T(J)) / DELW
3380   G(I) = (P(J + 1) - P(J)) * BB + P(J)
3390   TW = (W(I) - Y(I)) + W(I)
3400   IF TW > WATH THEN 3440
3410   IF TW >= WATL THEN 3450
3420   TW = WATL
3430   GOTO 3450
3440   TW = WATH
3450   Y(I) = W(I)
3460   W(I) = TW
3470 NEXT I
3480 GOTO 1150
3490 END
```

* Denotes a continuation of the previous line if the first character in a line is *.

INPUT FILE
SOWATD.DAT FOR MONOGRAPH-MILLVILLE-INFILTRATION
* AND REDISTRIBUTION
```
20,10,20,26,0,3
30,3,0,20,0,30,0,40,0,50
0,5,10,15,20,25,30,35,40,45,50,55,60,65,70,75,80,8
*  5,90,95,100
-2.50E+06,-4.00E+05,-6.40E+04,-2.47E+04,-9.50E+03,
*  -3.77E+03,-1.50E+03,-7.65E+02
-3.90E+02,-3.06E+02,-2.40E+02,-2.02E+02,-1.70E+02,
*  -1.43E+02,-1.20E+02,-1.00E+02
-8.40E+01,-6.86E+01,-5.60E+01,-4.23E+01,-3.20E+01,
*  -1.96E+01,-1.20E+01,-3.46E+00
0,10000
1.0E-11,1.0E-10,1.6E-09,1.0E-08,2.8E-08,1.9E-07,
*  9.1E-07,5.9E-06
4.50E-05,1.8E-04,5.2E-04,2.0E-03,3.0E-03,4.5E-03,
*  7.8E-03,1.3E-02
2.1E-02,3.4E-02,5.1E-02,8.3E-02,1.3E-01,2.2E-01,
*  7.0E-01,9.3E-01
1.2E+00,1.2E+00
.07,.07,.07,.07,.07,.07,.07,.07,.07,.07,.07,.07,
*  .07,.07,.07,.07,.07,.07,.07,.07,.07
.0024,.038,1.,0.,1.,50
-4.0E5,0,.02,.48,-.15E5,0.02
```

CONTENTS OF FILE

Line 1-- Label
Line 2-- K,IER,NB,ND,KI,KCPMAX
Line 3-- V Array   upper boundary condition
Line 4-- DD Array   depth increments
Line 5-8 P Array   Matric potential-water content
* starting at 0.0
Line 9-12 E Array   Hydraulic conductivity-water
* content starting at 0.0
Line 13 W Array   Initial water content
* corresponding to depth array
Line 14--DETT,CONQ,TAA,TIME,TT,CUMT
Line 15--HDRY,HWET,WATL,WATH,HLOW,DELW

NOTE THAT DEPTHS ARE IN CM
MATRIC POTENTIAL IS IN CM (HEAD)
HYDRAULIC CONDUCTIVITY IS IN CM/HR
TIME IS IN HOURS
WATER CONTENTS ARE VOLUME FRACTIONS

* Denotes a continuation of the previous line if
the first character in a line is *.

                  OUTPUT FILE
SOWATD.DAT FOR MONOGRAPH-MILLVILLE-INFILTRATION
* AND REDISTRIBUTION   01-10-1990   06:38:45
   K   IER   NB   ND   KCPMAX
   20   10   20   26   3
WAT CONT MAT HEAD CONDUCTI DIFFUSIV
* WAT CONT MAT HEAD CONDUCTI DIFFUSIV
0.00E+00 -.25E+07 0.10E-10 0.24E-04 0.26E+00
* -.14E+03 0.45E-02 0.34E+00
0.20E-01 -.13E+06 0.10E-09 0.26E-03 0.28E+00
* -.12E+03 0.78E-02 0.51E+00
0.40E-01 -.64E+05 0.30E-09 0.28E-03 0.30E+00
* -.10E+03 0.13E-01 0.76E+00
0.60E-01 -.29E+05 0.28E-08 0.38E-03 0.32E+00
* -.84E+02 0.21E-01 0.11E+01
0.80E-01 -.95E+04 0.22E-07 0.80E-03 0.34E+00
* -.69E+02 0.34E-01 0.16E+01
0.10E+00 -.48E+04 0.19E-06 0.17E-02 0.36E+00
* -.56E+02 0.53E-01 0.23E+01
0.12E+00 -.27E+04 0.91E-06 0.36E-02 0.38E+00
* -.44E+02 0.83E-01 0.33E+01
0.14E+00 -.72E+03 0.59E-05 0.15E-01 0.40E+00

```
*  -.32E+02   0.13E+00   0.48E+01
 0.16E+00  -.39E+03   0.45E-04   0.30E-01   0.42E+00
*  -.22E+02   0.22E+00   0.71E+01
 0.18E+00  -.30E+03   0.18E-03   0.47E-01   0.44E+00
*  -.12E+02   0.70E+00   0.14E+02
 0.20E+00  -.24E+03   0.52E-03   0.77E-01   0.46E+00
*  -.34E+01   0.93E+00   0.22E+02
 0.22E+00  -.20E+03   0.19E-02   0.15E+00   0.48E+00
*   0.00E+00   0.12E+01   0.26E+02
 0.24E+00  -.17E+03   0.24E-02   0.23E+00   0.50E+00
*   0.10E+05   0.12E+01   0.12E+05
  DEPTH    C(I)    W-DEPTH    H-DEPTH
   5.    0.10E-05    0.070    -.191E+05
  10.    0.10E-05    0.070    -.191E+05
  15.    0.10E-05    0.070    -.191E+05
  20.    0.10E-05    0.070    -.191E+05
..................................
..................................
  95.    0.10E-05    0.070    -.191E+05
 100.    0.10E-05    0.070    -.191E+05
   TIME      FLUX
   3.0000    30.0000
  20.0000     0.0000
  30.0000     0.0000
  40.0000     0.0000
  50.0000     0.0000
   DETT      CONQ     TAA      TIME      TT
*  CUMT      DELW
  0.0024    0.0380   1.0000   0.0000   1.0000
*  50.0000   0.0200
   HDRY      HWET     WATL     WATH     HLOW
*  HHI
 -.130E+06  0.000E+00  0.200E-01  0.480E+00  -.150E+05
*  0.000E+00
   TIME      CWF    IRR+RAIN   CUMB     RUNOFF
*  CUMS     WATBAL    EOR
 0.300E+01  0.755E+01  0.900E+02  0.000E+00  0.825E+02
*  0.755E+01  -.391E-04   0.300E+02
   WATER CONTENT VS DEPTH
 0.480  0.463  0.450  0.437  0.405  0.105  0.070
*  0.070  0.070  0.070  0.070  0.070  0.070  0.070
*  0.070  0.070  0.070  0.070  0.070  0.070  0.070
   TIME      CWF    IRR+RAIN   CUMB     RUNOFF
*  CUMS     WATBAL    EOR
 0.200E+02  0.755E+01  0.900E+02  0.000E+00  0.825E+02
*  0.755E+01  -.362E-04   0.000E+00
   WATER CONTENT VS DEPTH
 0.300  0.306  0.309  0.308  0.303  0.293  0.274
*  0.201  0.077  0.070  0.070  0.070  0.070  0.070
```

```
*  0.070    0.070    0.070    0.070    0.070    0.070    0.070
   TIME       CWF      IRR+RAIN      CUMB       RUNOFF
*  CUMS       WATBAL       EOR
 0.300E+02 0.755E+01 0.900E+02 0.000E+00 0.825E+02
*  0.755E+01 -.343E-04 0.000E+00
   WATER CONTENT VS DEPTH
 0.283    0.288    0.291    0.290    0.286    0.280    0.265
*  0.237    0.132    0.071    0.070    0.070    0.070    0.070
*  0.070    0.070    0.070    0.070    0.070    0.070    0.070
   TIME       CWF      IRR+RAIN      CUMB       RUNOFF
*  CUMS       WATBAL       EOR
 0.400E+02 0.755E+01 0.900E+02 0.000E+00 0.825E+02
*  0.755E+01 -.343E-04 0.000E+00
   WATER CONTENT VS DEPTH
 0.272    0.277    0.279    0.279    0.276    0.269    0.259
*  0.234    0.186    0.081    0.070    0.070    0.070    0.070
*  0.070    0.070    0.070    0.070    0.070    0.070    0.070
   TIME       CWF      IRR+RAIN      CUMB       RUNOFF
*  CUMS       WATBAL       EOR
 0.500E+02 0.755E+01 0.900E+02 0.000E+00 0.825E+02
*  0.755E+01 -.343E-04 0.000E+00
   WATER CONTENT VS DEPTH
 0.264    0.268    0.270    0.270    0.267    0.262    0.252
*  0.234    0.207    0.109    0.070    0.070    0.070    0.070
*  0.070    0.070    0.070    0.070    0.070    0.070    0.070
```

* Denotes a continuation of the previous line if the first character in a line is *.

REFERENCES

Campbell, G.S. 1985. Soil physics with basic-transport models for soil plant systems. Elsevier, Amsterdam, The Netherlands.

Freeze, R.A. 1969. The mechanism of natural ground water recharge and discharge. I. One dimensional vertical, unsteady, unsaturated flow above a recharging or discharging ground water flow system. Water Resour. Res. 5:153-171.

Hanks. R.J., and S.A. Bowers. 1962. Numerical solution of the moisture flow equation for infiltration into layered soils. Soil Sci. Soc. Am. Proc. 26:530-534.

Hanks, R.J., and S.A. Bowers. 1963. Influence of variations in the diffusivity-moisture content relation on infiltration. Soil Sci. Soc. Am. Proc. 27:263-265.

Hanks, R.J., A. Klute, and E. Bresler. 1969. A numeric method for estimating infiltration redistribution drainage and evaporation of water from soil. Water Resour. Res. 5:1064-1069.

Klute, A. 1952. A numerical method for solving the flow equation for water in unsaturated materials. Soil Sci. 73:105-116.

Klute, A. 1973. Soil water flow theory and its application in field situations. p. 9-35. *In* Field soil water regime. ASA Spec. Publ. 5, ASA, CSSA, and SSSA, Madison, WI.

Nimah, M.N., and R.J. Hanks. 1973a. Model for estimating soil water and atmospheric interrelations: I. Description and sensitivity. Soil Sci. Soc. Am. Proc. 37:522-527.

Nimah, M.N., and R.J. Hanks. 1973b. Model for estimating soil water and atmospheric interrelations: II. Field test of the model. Soil Sci. Soc. Am. Proc. 37:528-532.

Oliveira, C.A.S., R.J. Hanks, and U. Shani. 1987. Infiltration and runoff as affected by pitting, mulching and sprinkler irrigation. Irrig. Sci. 8:49-64.

Philip, J.R. 1955. Numerical solution of equations of the diffusion type with diffusivity concentration dependent. Trans. Faraday Soc. 51:885-892.

Philip, J.R. 1957. The theory of infiltration: 4. Sorptivity and algebraic infiltration equations. Soil Sci. 84:257-264.

10 Drainage

R. WAYNE SKAGGS
North Carolina State University
Raleigh, North Carolina

This chapter describes a water management simulation model that was developed to characterize drainage and water table control practices in poorly drained soils. Drainage of water from the soil profile is an important hydrologic component in most agricultrual soils. Natural drainage processes are sufficient for crops in many agricultural lands. However, artificial drainage is required for efficient production on a large percentage of the world's most fertile soils. For example, as of 1987 over 44 million ha of agricultural land in the USA benefited from drainage. Between 40 and 70% of the cropland in Delaware, Florida, Indiana, Louisiana, Michigan, North Carolina, and Ohio require artificial drainage, and over 0.5 million ha of land are drained in each of 17 states (Pavelis, 1987).

Drainage is needed to provide trafficable conditions for seedbed preparation and planting in the spring, to ensure a suitable environment for plant growth during the growing season, and to permit harvest in the fall. At the same time, excessive drainage is undesirable because it reduces soil water available to growing plants and leaches fertilizer nutrients, carrying them to receiving streams where they act as pollutants. In some cases, water table control or subirrigation can be used to maintain a relatively high water table during the growing season, thereby supplying irrigation water for crop growth as well as preventing excessive drainage.

The design and operation of agricultural drainage and associated water management systems should depend on soil properties, site parameters, crop factors, and climatological factors. Further, the design of one component should depend on the other components. For example, a field with good surface drainage will require less intensive subsurface drainage than it would if surface drainage is poor. This has been clearly demonstrated in both field studies of crop response (Schwab et al., 1974) and by theoretical methods (Skaggs, 1974). The relative importance of water management components varies with climate; so in humid regions, a well-designed drainage system may be critical in some years yet provide essentially no benefits in others. Thus, methods for designing and evaluating multicomponent water management systems should allow designers to identify sequences of weather conditions that are critical to crop production and to describe the performance of the system during those periods.

Copyright © 1991 ASA-CSSA-SSSA, 677 S. Segoe Rd., Madison, WI 53711, USA. *Modeling Plant and Soil Systems*—Agronomy Monograph no. 31.

Objective methods of describing drainage processes have evolved from steady state solutions or models, such as those developed by Dr. S.B. Hooghoudt in The Netherlands and Dr. Don Kirkham in the USA in the 1940s and 1950s (Kirkham, 1957; Luthin, 1978; van Schilfgaarde, 1974), to comprehensive computer simulation models now available. The model described herein is named DRAINMOD. The model is a computer simulation program that characterizes the response of the soil water regime to various combinations of surface and subsurface water management. The DRAINMOD model predicts the response of the water table and the soil water above the water table to other hydrologic components, such as infiltration and evapotranspiration (ET), as well as to surface and subsurface drainage, and the use of water table control or subirrigation practices. Surface irrigation can also be considered. The model has also been used to determine hydraulic loading capacities of wastewater disposal sites. Climatological data are used in the model to simulate the performance of a given water management system across several years. In this way, optimum water management systems can be designed on a probabilistic basis, as initially proposed for subsurface drainage by van Schilfgaarde (1965), and subsequently used by Young and Ligon (1972) and Wiser et al. (1974).

I. DRAINMOD

A. Problem Definition

A schematic of the type of water management system considered is provided in Fig. 10-1. The soil is nearly flat and has an impermeable layer at a relatively shallow depth. Subsurface drainage is provided by drain tubes or parallel ditches at a distance, d, above the impermeable layer and which are spaced a distance, L, apart. When rainfall occurs, water infiltrates at the surface and percolates through the profile, raising the water table and increasing the subsurface drainage rate. If the rainfall rate is greater than the capacity of the soil to infiltrate, water begins to collect on the surface. When good surface drainage is provided so that the surface is smooth and on grade, most of the surface water will be available for runoff. However, if surface drainage is poor, a certain amount of water must be stored in depressions before runoff can begin. After rainfall ceases, infiltration continues until the water stored in the surface depressions is infiltrated into the soil. Thus, poor surface drainage effectively lengthens the infiltration event for a given storm, permitting more water to infiltrate and a larger rise in the water table than would occur if depression storage did not exist.

The rate that water is drained from the profile depends on the hydraulic conductivity of the soil, the relationship between soil water content and the pressure head, the drain depth and spacing, the effective profile depth, and the depth of water in the drains. When the water level is raised in the drainage ditches for purposes of controlled drainage or subirrigation, drainage rates will decrease and may become negative. Under these conditions water

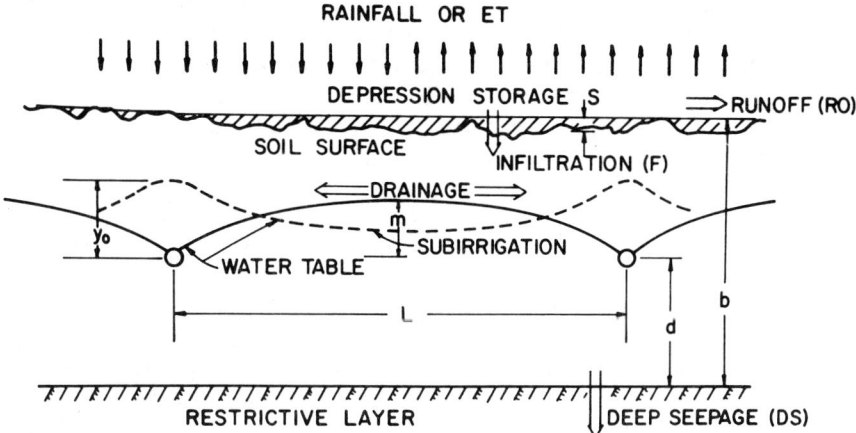

Fig. 10-1. Schematic of water management with drain tubes that may be used for drainage or subirrigation.

will move from the drains into the soil profile, yielding the shape shown by the broken curve in Fig. 10-1. A high water table reduces the amount of storage available for infiltrating rainfall and may result in frequent conditions of excessive soil water if the system is not properly designed and managed (Skaggs, 1974). Water may also be removed from the profile by ET and by deep seepage, both of which must be considered in the calculations if the soil water regime is to be modeled successfully.

B. Background, Philosophy, and Governing Equations

The first version of DRAINMOD was written in the early 1970s (Skaggs, 1975b). Results of further developments, including field experiments designed to test the validity of the model, were provided by Skaggs (1978). Methods to estimate the effects of water-management system design on crop yields have been incorporated in DRAINMOD (Skaggs et al., 1981; Hardjoamidjojo & Skaggs, 1982). The model has been accepted by the USDA Soil Conservation Service (SCS) for design and evaluation of drainage and subirrigation systems in humid regions. It is available to all states on the SCS computer system. A cooperative effort in the North Central states resulted in the development of a microcomputer version of the model (North Central Comuter Inst., 1985). User manuals and program documentation for DRAINMOD are available from the SCS (1985). A new microcomputer version of the program, Version 4.0, is available from the author.

Two important criteria were adopted at the outset of the model development process. First, the model should be capable of describing water movement and storage in the profile to characterize the soil water regime and drainage rate with time. Secondly, the model should be developed so that the computer time necessary to simulate long term processes and input data requirements are not prohibitive. The movement of water in soil is a complex process and it would have been easy to become so involved with reaching exact solutions to every possible situation that the final answer would have never been obtained. The guiding principle in the model development was, therefore, to assemble the linkages between various components of the system and to allow the specifics to be incorporated as subroutines so that they may be readily modified as better methods are developed.

The rates of infiltration, drainage, ET, and the distribution of soil water in the profile can be computed by obtaining numerical solutions to the two- or three-dimensional Richards Equation for saturated and unsaturated flow (Freeze, 1971). Other approaches include the use of numerical methods to solve the Richards Equation for one-dimensional vertical flow in combination with approximate analytical methods for flow in the horizontal direction. Such an approach was employed in a forerunner to the DRAINMOD program (Skaggs, 1974). Numerical solutions were used to calculate infiltration, surface runoff, and the subsequent distribution of soil water. The drainage rate was calculated by the Hooghoudt Equation (van Schilfgaarde, 1974) and applied as a bottom boundary condition in the numerical solutions. A similar approach was used by Feddes et al. (1978) and by Smith (1985), who have developed efficient numerical schemes for solving the Richards Equation for vertical flow. Karvonen (1988) used this approach to develop a model to predict the effect of drainage on soil moisture, soil temperature, and crop yield. Solutions to the Richards Equation are used in Ch. 11 herein to characterize ET. Numerical solutions may require prohibitive amounts of computer time for long-term simulations. However, computer capabilities have increased tremendously since the model discussed in this chapter was first developed and numerical approaches are now feasible. It is still important to simplify the model inputs and minimize the computer resources necessary, if the model is to be applied routinely for design and analysis. Approximate methods were used to characterize the water movement processes in the DRAINMOD model.

The model is based on a water balance for a section of soil of unit surface area that extends from the impermeable layer to the surface and is located midway between adjacent drains. By assuming the shape of the water table between drains, the water balance can just as easily be conducted for the entire profile (Smith, 1985). However, it should be realized that the shape changes with time, drain spacing, and the relative rates of ET and drainage. This problem was solved by McCarthy (1990), who modified DRAINMOD to compute the water balance based on the average water table depth between two drains. The water balance for a time increment of Δt may be expressed as

$$\Delta V_a = D + \text{ET} + \text{DS} - F \qquad [1]$$

where ΔV_a is the change in the air volume or water free pore space (cm) in the section, D is drainage (cm) from (or subirrigation into) the section, ET is evapotranspiration (cm), DS is deep seepage (cm), and F is infiltration (cm) entering the section in Δt.

The terms on the right-hand side of Eq. [1] are computed as functions of the water table elevation, soil water content, soil properties, site and drainage system parameters, crop and stage of growth, and atmospheric conditions. The amount of runoff and storage on the surface is computed from a water balance at the soil surface for each time increment, which may be written as

$$P = F + \Delta S + \text{RO} \qquad [2]$$

where P is the precipitation (cm), ΔS is the change in volume of water stored on the surface (cm), and RO is runoff (cm) during Δt. The basic time increment used in Eq. [1] and [2] is 1 h. However, when rainfall does not occur and drainage and ET rates are slow, such that the water table position moves slowly with time, Eq. [1] is based on Δt of 1 d. Conversely, time increments ≤ 0.1 h are used to compute F when rainfall rates exceed the infiltration capacity.

C. Model Components

1. Precipitation

Precipitation records are one of the major inputs to DRAINMOD. The accuracy of the model's prediction for infiltration, runoff, and surface storage is dependent on the complete description of rainfall. Therefore, a short time increment for rainfall input data will allow better estimates of these model components than will less frequent data. A basic time increment of 1 h was selected for use in the model because of the availability of hourly rainfall data. While data for shorter time increments are available for a few locations, hourly rainfall data are readily available for many locations in the USA. Records for 77 stations in the humid regions of the USA are available on SCS files. Hourly data from several additional stations are currently being added to these files. Models for generating weather data artificially have been developed (e.g., Richardson & Wright, 1984; Robbins & Skaggs, 1986) and may be used when recorded data are unavailable.

2. Infiltration

Water infiltration at the soil surface is a complex process that has been studied extensively during the past two decades. Models for quantifying the infiltration process are presented in Ch. 9. Infiltration is affected by: (i) soil factors, such as hydraulic conductivity, the soil water characteristic, initial water content, surface compaction, depth of profile, and water table depth; (ii) plant factors, such as extent of cover and depth of root zone; and (iii) rainfall factors such as intensity, duration, and time distribution.

One method for characterizing the infiltration process involves the solution of the nonlinear partial differential equation first derived by Richards (1931) (c.f. Eq. [3] in Ch. 9) for transient unsaturated flow under rainfall or ponded surface conditions. Although the Richards Equation provides a rather comprehensive method of determining the effects of many interactive factors on infiltration, input and computational requirements prohibit its use in DRAINMOD. Nevertheless, these solutions can be used to evaluate approximate methods and, in some cases, to determine parameter values required in these methods. The reader is referred to Ch. 9 for a discussion of numerical solutions to the Richards Equation.

Approximate equations for predicting infiltration have been proposed by Green and Ampt (1911), Horton (1939), Philip (1957), and Holtan et al. (1967), among others. The Green-Ampt equation was chosen to characterize the infiltration component in DRAINMOD. The Green-Ampt Equation was originally derived by assuming deep homogeneous profiles with a uniform initial water content. The equation may be written as

$$f = K_s + K_s M_d S_f F \qquad [3]$$

where f is the infiltration rate, F is accumulative infiltration, K_s is the hydraulic conductivity of the transmission zone, M_d is the difference between final and initial volumetric water contents ($M_d = \theta_o - \theta_i$), and S_f is the effective suction at the wetting front. For a given soil with a given initial water content Eq. [3] may be written as

$$f = A/F + B \qquad [4]$$

where A and B are parameters that depend on soil properties, such as initial water content and distribution, and surface conditions, such as cover, crusting, etc.

In addition to the uniform profiles for which it was originally derived, the Green-Ampt equation has been used with good results for profiles that become denser with depth (Childs & Bybordi, 1969) and for soils with partially sealed surfaces (Hillel & Gardner, 1969). Bouwer (1969) showed that it may also be used for nonuniform initial water contents. Resistance to air movement may be quite significant for shallow water tables where air may be trapped between the water table and the advancing wetting front (McWhorter, 1971, 1976). Morel-Seytoux and Khanji (1974) showed that the Green-Ampt equation retained its original form when the effects of air movement were considered for deep soils. The equation parameters were simply modified to include effects of air movement.

Mein and Larson (1973) used the Green-Ampt equation to predict infiltration from steady rainfall. Their results were in good agreement with rates obtained from solutions to the Richards Equation for a wide variety of soil types and application rates. Mein and Larson's results imply that, for uniform deep soils with constant initial water contents, the infiltration rate may be expressed in terms of cumulative infiltration (F) alone, regardless of the application rate. Investigations by Smith (1972) and Reeves and Miller (1975) tend to support this conclusion. These results are extremely important for modeling efforts of the type discussed herein. If the infiltration relationship is independent of application rate, the only input parameters required are those pertaining to the necessary range of initial conditions.

The model requires input for infiltration in the form of a table that quantifies relationships between A and B and water table depth. The parameters A and B in Eq. [4] may be determined by using regression methods to fit the equation to observed infiltration data. The resultant parameter values will reflect the effects of air movement, as well as other factors that would have otherwise been neglected. Infiltration predictions based on such measurements will usually be more reliable than if the predictions are obtained from basic soil property measurements. When infiltration data are not available, which is usually the case, infiltration parameters can be estimated from soil water characteristic data (Bouwer, 1969; Mein & Larson, 1973) or from soil texture (Rawls et al., 1983).

Although it is assumed in the present version of the model that the A and B matrix is constant, it is possible to allow it to vary with time or to be dependent on events that affect surface cover, compaction, etc. Shirmohammadi and Skaggs (1984) found that the infiltration characteristics of a soil changed significantly over the growing season.

3. Surface Drainage

Surface drainage is characterized by the average depth of depression storage that must be satisfied before runoff can begin. In most cases, it is assumed that depression storage is evenly distributed over the field. Depression storage may be further broken down into: (i) a microcomponent representing storage in small depressions, due to surface structure and cover; and (ii) a macrocomponent, which is due to larger surface depressions and which may be altered by land forming, grading, etc. A field study conducted by Gayle and Skaggs (1978) showed that the micro-storage component varies from about 0.1 cm for soil surfaces that have been smoothed by weathering to several centimeters for rough plowed land. Macro-storage values for eastern North Carolina fields varied from nearly zero for fields that have been land formed and smoothed or that are naturally on grade, to >3 cm for fields with numerous pot holes and depressions or which have inadequate surface outlets. Surface storage could be considered as a time-dependent function or dependent on other events such as rainfall and the time sequence of tillage operations. Therefore, the variation in the micro-storage component during the year can be simulated. However, it is assumed to be constant in the present version of the model.

A value for the maximum storage depth is required as input to the model. When the surface storage depth, as determined by Eq. [2], exceeds this value, the additional excess is allotted to surface runoff. The model assumes that water available for runoff moves immediately from the surface to the outlet. The amount of runoff is part of the output on a daily, monthly, or yearly basis as specified.

4. Subsurface Drainage

The rate of subsurface water movement into drain tubes or ditches depends on the hydraulic conductivity of the soil, drain spacing and depth, profile depth, and water table elevation. Water moves toward drains in both the saturated and unsaturated zones, and can best be quantified by solving the Richards Equation for two-dimensional flow. Solutions have been obtained for several drainage boundary conditions (Skaggs & Tang, 1976; Fipps & Skaggs, 1986). Input and computational requirements prohibit the use of these numerical methods in DRAINMOD, as was the case for infiltration discussed previously. However, numerical solutions provide a very useful means of evaluating approximate methods of computing drainage flux.

The method used in DRAINMOD to calculate drainage rates is based on the assumption that lateral water movement occurs mainly in the saturated region. Water movement in the saturated zone can be characterized for most drainage situations by making the Dupuit-Forchheimer (D-F) assumptions and solving the resulting Boussinesq Equation (Luthin, 1978):

$$\frac{\partial(sh)}{\partial t} = K \frac{\partial}{\partial x} \left[h \frac{\partial h}{\partial x} \right] + e \qquad [5]$$

where h is the height of the water table above the datum, K is saturated hydraulic conductivity, s is drainable porosity, which may vary with h, e is the rate that water is added vertically due to rainfall, or that is lost (negative values of e) by ET or deep seepage, t is time, and x is the horizontal distance from the origin. Equation [5] is nonlinear, but can be rather easily solved using either finite-difference or finite-element methods (Moody, 1967; Skaggs, 1975a). Parsons (1987) used finite-element solutions to Eq. [5] as the basis for a simulation model for watershed scale drainage to a single canal. While this approach provides a good description of the drainage process and allows consideration of lateral variation in soil properties, surface elevations, and crop species between drains, it requires much more computational time and input information than other alternatives.

Several methods for simplifying Eq. [5] and obtaining analytical solutions are discussed by van Schilfgaarde (1974). However, the method proposed by Bouwer and van Schilfgaarde (1963) was selected for use in the present version of DRAINMOD. This equation may be written as

$$q = \frac{4 K_e m (2d_e + m)}{CL^2} \qquad [6]$$

where q is the flux in cm/h, m is the midpoint water table height above the drain (Fig. 10-2), K_e is the equivalent lateral hydraulic conductivity (cm/h), d_e is the equivalent depth from the drains to the impermeable layer (cm), L is the distance between drains (cm), and C is the ratio of the average flux to the flux at a point midway between drains. Solutions based on a water balance at the midpoint and on a drainage rate given by Eq. [6] are compared with numerical solutions of Eq. [5] in Fig. 10-3. While good agreement was obtained when constant $C = 1.0$ was used, almost exact agreement

DRAINAGE

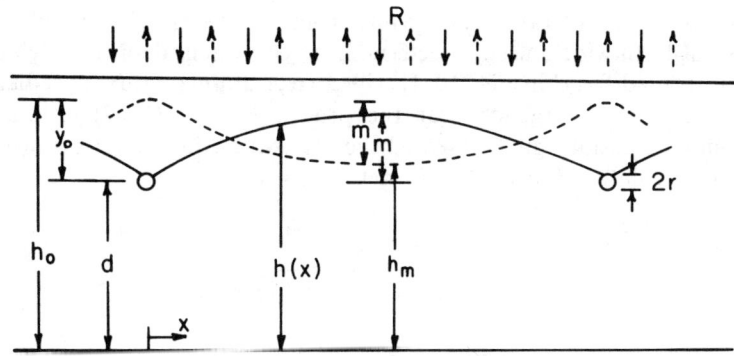

Fig. 10-2. Schematic of water table shape for drainage (———) and subirrigation from (---) parallel drain tubes.

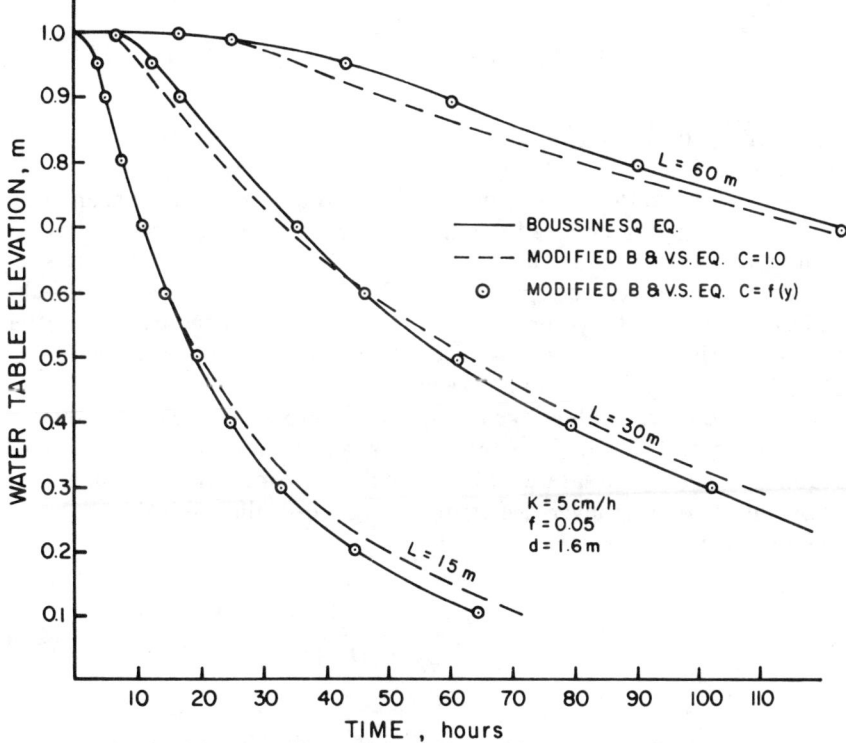

Fig. 10-3. Midpoint water table drawdown as predicted by solutions to the Boussinesq equation and by a simplified model that uses a modification of the Bouwer and van Schilfgaarde (1963) equation (B & VS Eq.).

with the Boussinesq solutions was found when C was allowed to vary with the water table elevation, y. The present version of the model uses $C = 1.0$, however.

Hooghoudt (van Schilfgaarde, 1974) characterized flow to cylindrical drains by considering radial flow in the region near the drains and applying the D-F assumptions to the region away from the drains. The Hooghoudt analysis has been widely used to determine an equivalent depth, d_e, which, when substituted for d in Fig. 10-1, will correct drainage fluxes predicted by Eq. [6] for convergence near the drains. Moody (1967) examined Hooghoudt's solutions and presented the following equations from which d_e can be calculated: For $0 < d/L < 0.3$

$$d_e = \frac{d}{1 + d/L\,[(8/\pi) \ln (d/r) - \alpha]} \quad [7]$$

in which

$$\alpha = 3.55 - 1.6d/L + 2(d/L)^2 \quad [8]$$

and for $d/L > 0.3$

$$d_e = \frac{L\pi}{8\,[\ln (L/r) - 1.15]} \quad [9]$$

in which r = drain-tube radius. Usually α can be approximated as $\alpha = 3.4$ with negligible error for design purposes (van Schilfgaarde, 1974).

For actual, rather than completely open drain tubes, there is an additional loss of hydraulic head due to convergence as water approaches the finite number of openings in the tube. The effect of various opening sizes and configurations can be approximated by defining an effective drain-tube radius, r_e, such that a completely open drain tube with radius r_e will offer the same resistance to inflow as an actual tube with radius r (Skaggs, 1978; Dierichx, 1980; Mohammad & Skaggs, 1983).

Most soils are not homogeneous, but are layered. Since the subsurface water movement to a drain is primarily in the lateral direction, the model requires, as input, the saturated lateral hydraulic conductivity of each layer above the restricting layer. Referring to Fig. 10-4, the equivalent conductivity is calculated as

$$K_e = \frac{K_1 d_1 + K_2 D_2 + K_3 D_3 + K_4 D_4}{d_1 + D_2 + D_3 + D_4} \quad [10]$$

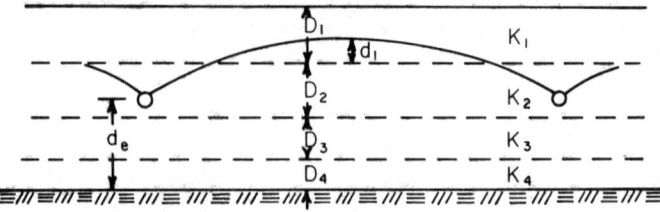

Fig. 10-4. Equivalent lateral hydraulic conductivity is determined for soil profiles with up to five layers.

Because the thickness of the saturated zone in the upper layer, d_1, is dependent on the water table position, K_e is determined prior to every flux calculation using the updated value of d_1. If the water table is below Layer 1, $d_1 = 0$, and a similarly defined d_2 is substituted for D_2 in Eq. [10].

The methods discussed above for predicting drainage flux assume a curved (elliptical) water table completely below the soil surface, except at the midpoint, where it may be coincident with the surface. However, in some cases, the water table may rise to completely inundate the surface, with ponded water remaining there for relatively long periods of time. Then the D-F assumptions will not hold because the streamlines will be concentrated near the drains with most of the water entering the soil surface in that vicinity. Kirkham (1957) showed that, in one case, more than 95% of the flow entered the surface in a region bounded by $\pm \frac{1}{4}$ of the drain spacing. At inundation, the drainage flux is calculated using an equation derived by Kirkham (1957) as

$$q = \frac{4\pi K_e (\delta + b - r_e)}{GL} \quad [11]$$

where

$$G = 2 \ln \left\{ \frac{\tan [\pi(2b - r)/4h]}{\tan (\pi r/4h)} \right\} + 2 \sum_{m=1}^{\infty}$$

$$\ln \left\{ \frac{\cosh (\pi m L/4h) + \cos (\pi r/2h)}{\cosh (\pi m L/4h) - \cos (\pi r/2h)} \right.$$

$$\left. \frac{\cosh (\pi m L/4h) - \cos [\pi(2d - r)/2h]}{\cosh (\pi m L/4h) + \cos [\pi(2d - r)/2h]} \right\}, \quad [12]$$

h is the actual depth of the profile (cm), δ is the depth of the water on the surface (cm), b is the depth to the drain (cm), d is the actual depth from the drain to the impermeable layer (cm), and L is the drain spacing (cm) as shown in Fig. 10-5.

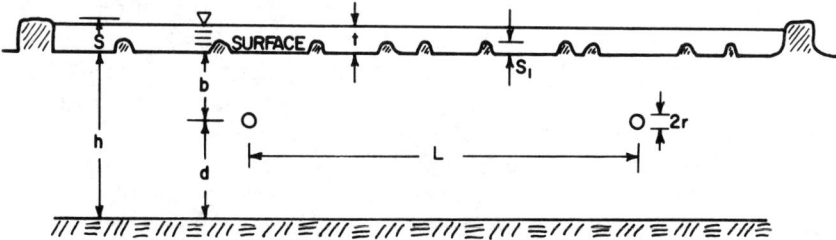

Fig. 10-5. Schematic of drainage from a ponded surface. Water will move over the surface to the vicinity of the drains until the ponded depth becomes $< S_1$. The maximum depressional storage is S.

Equation [10] is used after the water table rises to the surface for as long as surface water can move freely toward the drains. This is determined by the micro-storage component illustrated as S_1 in Fig. 10-5. When the rows are oriented perpendicular to the drain-tube direction, water may move along the furrows to the region above the drains, but still remain in lower depressional areas. When the ponded depth becomes $< S_1$, water can no longer move freely over the surface, the depth of water ponded over the drains will decrease more rapidly than that near the midpoint, and Eq. [6] is used to estimate drainage flux. Both S and S_1 (Fig. 10-5) are model inputs.

Use of Eq. [6] or [11] assumes that drainage is limited by the rate of soil water movement to the lateral drains and not by the hydraulic capacity of the drain tubes or the outlet. Usually, the sizes of the drain tubes are chosen to provide a design flow capacity, which is called the drainage coefficient, DC. Typically the DC, which depends on the size and slope of the drain tubes, may be 1 to 2 cm/d (about 3/8 to 3/4 in./d), depending on the geographic location and crops to be grown. The drainage capacity for a given slope and size of drain, a required DRAINMOD input, can be obtained from the SCS National Engineering Handbook (SCS, 1985; Section 16; Fig. 4-36) or by direct use of the Manning Equation. When the calculated flux exceeds the DC, q is set equal to the DC.

5. Subirrigation

When subirrigation is used, water is raised in the drainage outlet to maintain a pressure head (y_o) (refer to the broken curve in Fig. 10-2). Then a relationship corresponding to Eq. [6] for flux is

$$q = [4K_e (2 h_o m + m^2)]/L^2 \qquad [13]$$

where $h_o = y_o + d_e$ is the equivalent water table elevation at the drain and m is defined as $m = h_m - h_o$, with h_m as the equivalent water table elevation midway between the drains. For subirrigation, $h_o > h_m$, and both m and q are negative. Convergence losses at the drain are treated in the same manner as in drainage by using the equivalent depth to the impermeable layer, d_e, rather than the actual depth, d, to define h_o in Eq. [13]. Equation [13] was derived by making the D-F assumptions and solving the resulting flow equation for steady evaporation from the field surface at rate q. The magnitude of q increases as m becomes more negative (i.e., as h_m becomes smaller) until the water table at the midpoint reaches the equivalent depth of the impermeable layer ($h_m = 0$). For deeper midpoint water table depths, which can occur because the actual depth to the impermeable layer is deeper than the equivalent depth, Eq. [13] predicts a decrease in the magnitude of q. Ernst (1975) observed that this is inconsistent with the physics of flow, since the maximum subirrigation rate should occur when the midpoint water table reaches the impermeable layer. He derived an equation similar to Eq. [13] to correct these deficiencies. The equation may be written in the present notation as

$$q = \frac{4K_e m \left(2h_o + \dfrac{h_o}{D_o} m\right)}{L^2} \qquad [14]$$

where $D_o = y_o + d$, and d is the distance from the rain to the impermeable layer. Equation [14] is used in DRAINMOD to predict subirrigation flux.

For controlled drainage, a weir is set at a given elevation in the drainage outlet. The actual water level in the drain is not fixed as it is with subirrigation, but is dependent on the size of the outlet, previous drainage, etc. If the water table elevation in the field is higher than the water level in the drain, drainage will occur and the water level in the drain will increase. If it rises to the weir level, additional drainage water will spill over the weir and leave the system. When the water table in the field is lower than that in the drain, water will move into the field at a rate given by Eq. [14]. This raises the water table in the field or supplies ET demands while reducing the water level in the drain. The water level in the outlet during subirrigation or controlled drainage is computed at each time increment by a subroutine named YDITCH. This subroutine uses the geometry of the outlet, weir setting, and drainage or subirrigation flux to determine the water level in the outlet at each time step.

For conventional drainage and surface irrigation, the weir should be set at or below the drain depth. For controlled drainage or subirrigation, the weir should be set above the drain depth to provide control for the desired water table. Presently, as part of the input to the model, the weir depth for each month is specified, along with the day of the month that a new weir setting is to take effect. For subirrigation, the water level in the outlet is assumed to be maintained at the weir elevation.

D. Evapotranspiration

The ET component of the model accounts for water loss from the system by evaporation from the soil surface or by transpiration from the plants. This subject is covered in detail in Ch. 11. The determination of ET is a two step process in DRAINMOD. First, the daily potential ET (PET) is calculated in terms of atmospheric data and is distributed on an hourly basis. The PET represents the maximum amount of water that will leave the soil system by ET when there is a sufficient supply of soil water. For any hour in which rainfall occurs, hourly PET is set equal to zero. After PET is calculated, checks are made to determine if ET is limited by the soil water conditions. If not, ET is set equal to PET; otherwise ET is set equal to the smaller amount that can be supplied from the soil system, discussed below.

Potential ET depends on climatological factors, which include net radiation, temperature, humidity, and wind velocity. Methods for predicting PET in humid regions were reviewed by McGuinness and Borden (1972) and Jensen (1974). Among the most reliable methods are the ones developed by Penman (1948, 1956) and van Bavel (1966). These methods require net radiation, relative humidity, temperature, and wind speed as input data. Additional methods that could be used include, among others, those by Jensen et al. (1963), and Stevens and Stewart (1963). Daily PET may be determined by any methods and input directly as model data.

The most reliable methods for calculating PET require daily solar or net radiation as input data; these data are available for only a few locations. Because we are interested in conducting simulations in many locations throughout the USA, it is desirable to estimate ET based on readily available input data. Therefore, the empirical method developed by Thornthwaite (1948) was selected as an option for predicting PET. When this option is used, PET is computed from recorded daily maximum and minimum temperature values. The heat index must be determined from long-term temperature records and entered separately, along with the latitude of the site. Adjustments for day length and number of days in the month are made in the program based on latitude and date.

The PET obtained can be adjusted through a correction factor entered for each month. The correction factor is normally determined as the ratio of monthly PET as determined by the most reliable source available (i.e., lake evaporation, pan data, or one of the prediction methods such as the Penman Method) to the long-term average monthly PET predicted by the Thornthwaite Method (1948). When PET values are input directly, a correction factor of 1.0 is used. The daily PET is evenly distributed between 0600 h and 1800 h.

Once the PET is determined, a check is made to determine if soil water conditions are limiting. As long as the soil water content in the root zone is above θ_c, the lower limit water content (usually taken as the wilting point), ET is set equal to PET. Roots are assumed to be concentrated within an effective root zone depth, which is a model input dependent on the crop and time after planting, as discussed in a later section. The effect of the crop species on ET is reflected by θ_c and the effective root depth function.

In addition to infiltration from rainfall or irrigation, water may enter the root zone by upward flux from the water table. Figure 10–6 shows the relationship between the maximum steady rate of upward water movement and the water table depth for seven North Carolina soils (Skaggs, 1978). Relationships such as these are read as inputs to the model and used to calculate upward flux based on current water table depth. When upward flux is not sufficient to meet ET demand, the deficit is supplied by water stored in the root zone. For convenience, this water is assumed to be removed from layers of soil, starting at the surface. As water is removed, a dry zone is created. When the dry zone depth becomes equal to the effective root depth, ET is limited by soil water conditions and is set equal to the rate of upward water movement. The depleted soil water storage in the dry zone is accounted for separately from the rest of the unsaturated zone. It is accounted for on a day-to-day, hour-to-hour basis and is assumed to be the first volume filled when rainfall or irrigation occurs.

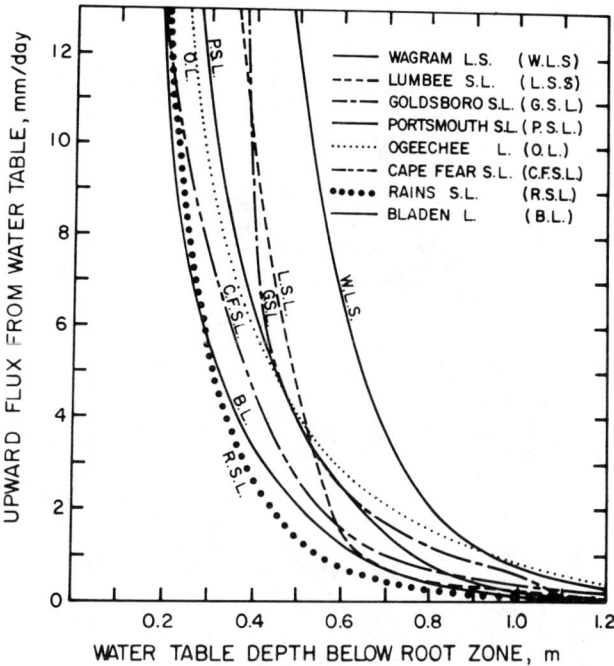

Fig. 10-6. Effect of the water table depth on steady upward flux from the water table for several North Carolina soils including: Wagram (loamy, siliceous, thermic Arenic Kandiudult). Lumbee (fine-loamy over sandy or sandy-skeletal, siliceous, thermic Typic Ochraquult); Goldsboro (fine-loamy, siliceous, thermic Aquic Paleudult). Portsmouth (fine-loamy over sandy or sandy-skeletal, mixed, thermic Typic Umbraquult); Ogeechee (fine-loamy, siliceous, thermic Typic Ochraquult); Cape Fear (clayey, mixed, thermic Typic Umbraquult); Rains (fine-loamy, siliceous, thermic Typic Paleaquult); and Bladen (clayey, mixed, thermic Typic Albaquult).

E. Soil Water Distribution

Knowledge of the soil water distribution in the soil profile is needed to evaluate individual components such as drainage and ET. These depend on the position of the water table and the soil water distribution in the unsaturated zone. The water table depth is a key variable that is determined at the end of every water balance calculation. The soil water content below the water table is assumed to be essentially saturated; actually it is slightly less than the saturated value due to residual trapped air in soils with fluctuating water tables. In some earlier models, the water content in the unsaturated zone was assumed to be constant and equal to the difference between the saturated value and the drainable porosity. However, except for the region close to the drains, the pressure head distribution above the water table during drainage may be assumed nearly hydrostatic for most field scale drainage systems (Skaggs & Tang, 1976). The soil water distribution under these conditions is the same as in a column of soil drained to equilibrium with a static water table. The water table drawdown is slow in most fields with artificial drains, and the unsaturated zone in a sense keeps up with the saturated zone. This is supported by the results plotted in Fig. 10-7 for pressure head versus

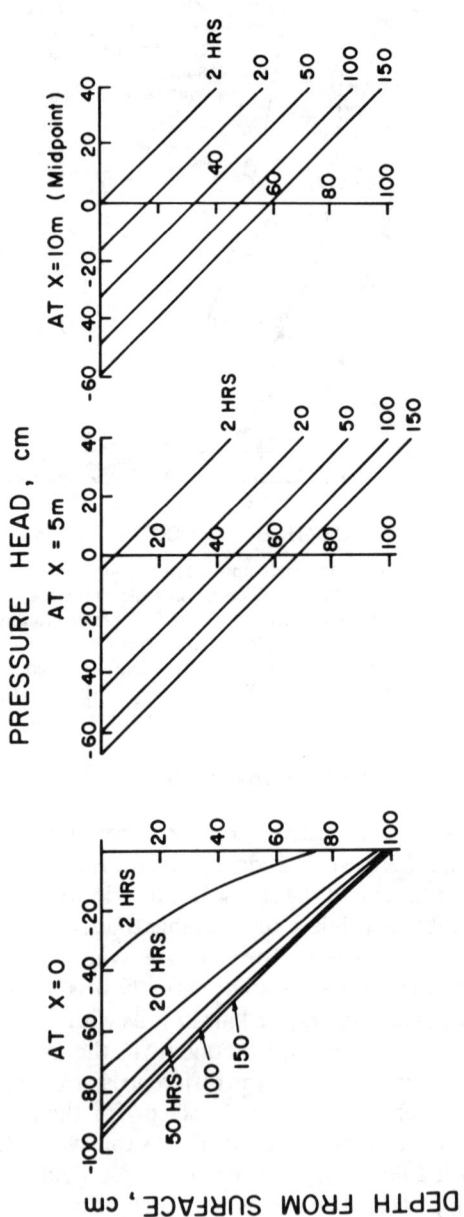

Fig. 10-7. Pressure head distribution with depth at midpoint, quarter point, and next to the drain for various times after drainage begins for a Panoche loam (from Skaggs & Tang, 1976).

depth at three locations between drains spaced 20 m apart in a panoche loam (fine-loamy, mixed, thermic Typic Torriorthents): (i) adjacent to the drain ($x = 0$); (ii) at the quarter point ($x = 5$ m); and (iii) at the midpoint between the drains ($x = 10$ m). These results were obtained by solving the two-dimensional Richards Equation for saturated and unsaturated flow during drainage. The pressure head at the quarter and midpoints increase with depth in a 1:1 fashion, indicating that the unsaturated zone is essentially drained to equilibrium with the water table (located where pressure head = 0) at all times after drainage begins.

The assumption of a hydrostatic condition above the water table during drainage will generally hold for conditions in which the D-F assumptions are valid. This will be true for situations in which the ratio of the drain spacing to profile depth is large, but may cause errors for deep profiles with narrow drain spacings.

Water is also removed from the profile by ET, which results in water table drawdown and changes in the water content of the unsaturated zone. In this case the vertical hydraulic gradient in the unsaturated zone is in the upward direction. However, when the water table is near the surface, the vertical gradient will be small and the water content distribution still close to the equilibrium distribution. Solutions for the water content distribution in a vertical column of soil under simultaneous drainage and evaporation are provided in Fig. 10-8. The water table was initially at the surface of the soil column. Solutions were obtained for various evaporation rates and a drainage rate at the bottom of the column equal to that resulting from drains spaced 30 m apart and 1 m deep.

Results in Fig. 10-8 indicate that the soil water distribution was independent of the evaporation rate, except for the region close to the surface at the high evaporation rate (4.8 mm/d). The distribution for no evaporation is exactly the same as that which would result from the profile draining to equilibrium with a water table 0.7 m deep. Thus the drained-to-equilibrium assumption appears to provide a good approximation of the soil water distribution for this soil for both drainage and evaporation when the water table depth is relatively shallow. Even when the water table is very deep, the soil water distribution for some distance above the water table will be approximately equal to the equilibrium distribution.

The zone directly above the water table is called the wet zone and the water content distribution is assumed to be independent of the means in which water was removed from the profile. Thus, the air volume, or the volume of water leaving the profile by drainage, ET, and deep seepage, may be plotted as a function of water table depth as shown in Fig. 10-9. Assuming hysteresis can be neglected, Fig. 10-9 would allow the water table depth to be determined simply from the volume of water that enters or is removed from the profile over an arbitrary period of time.

The maximum water table depth for which the approximation of a drained-to-equilibrium water content distribution will hold depends on the hydraulic conductivity functions of the profile layers and the ET rate. The maximum depth will increase with the hydraulic conductivity of the soil and

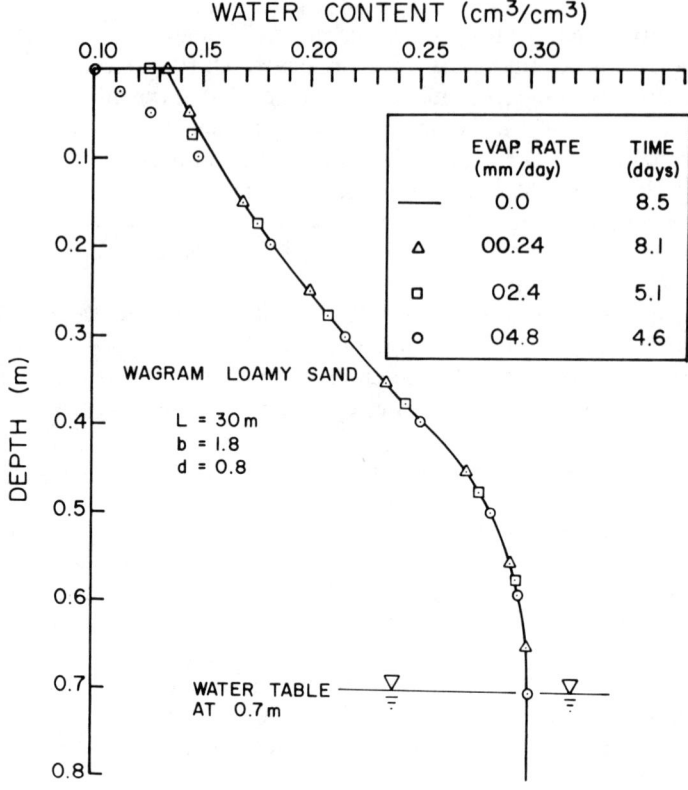

Fig. 10-8. Soil water distribution for a water table depth of 0.7 m for various drainage and evaporation rates.

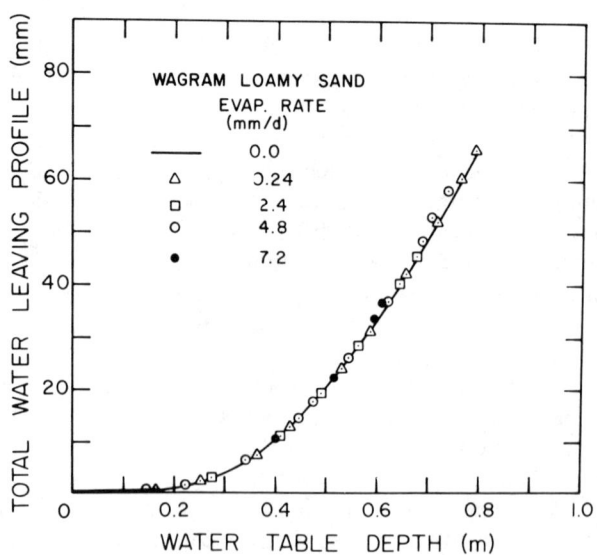

Fig. 10-9. Volume of water leaving profile (cm^3/cm^3) by drainage and evaporation versus water table depth. Solutions for five evaporation rates are provided.

decrease with the ET rate. Because the unsaturated hydraulic conductivity decreases rapidly with water content, large upward gradients may develop near the surface or near the bottom of the root zone when the soil water distribution departs from the equilibrium profile. At this point, the upward flux cannot be sustained for much deeper water table depths and additional water, necessary to supply the ET demand, would be extracted from storage in the root zone creating a dry zone as discussed in the ET section. This is shown schematically in Fig. 10-10.

For purposes of calculation in DRAINMOD, the soil water is assumed to be distributed in two zones: (i) a wet zone extending from the water table up to the root zone and possibly through the root zone to the surface; and (ii) a dry zone. The water content distribution in the wet zone is assumed to be that of a drained-to-equilibrium profile. When the maximum rate of upward water movement, determined as a function of the water table depth, is not sufficient to supply the ET demand, water is removed from the root zone storage creating a dry zone as discussed in the ET section.

To save computational time and reduce input requirements, a number of approximations were made concerning the soil water distribution above the water table as discussed above. These assumptions may cause errors during dry periods, particularly in soils with deep water tables and low conductivity in the subsurface layers. Deep water tables may result from vertical seepage into an underlying aquifer or from deep subsurface drains. For such conditions, the soil water at the top of the wet zone just beneath the root zone may be depleted by slow upward movement and by roots extending beyond

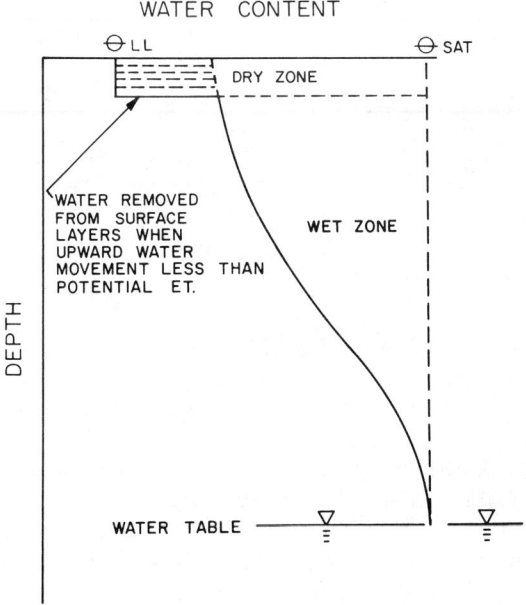

Fig. 10-10. Schematic of the soil water distribution when a dry zone is created at the surface.

the assumed depth of the concentrated root mass. Such conditions may cause the water content at the top of the wet zone to significantly depart from the drained-to-equilibrium distribution. However, this will not cause a problem for wet conditions and for most shallow water table soils for which the model was derived. Rogers (1985) found that predictions of drainage and ET in DRAINMOD could be improved for a deep sandy soil by modifying the methods for calculating the dry zone.

F. Rooting Depth

The effective rooting depth is used in the model to define the zone from which water can be removed, as necessary, to supply ET demands. Since the simulation process is usually continuous for several years, an effective depth is defined for all periods. When the soil is fallow the effective depth is defined as the depth of a thin layer (2–5 cm deep) that will dry out at the surface. When a second crop or a cover crop is grown, its respective rooting depth function is also included. The rooting depth function is input as a table of effective rooting depth versus date. The rooting depth for days other than those listed in the table is obtained by interpolation.

This method of treating the rooting depth is, at best, an approximation. The depth and distribution of plant roots is affected by many factors in addition to crop species and date after planting. These factors include physical barriers such as hardpans and plow pans, chemical barriers, fertilizer distribution, tillage treatments, and others as reviewed in detail by Allmaras et al. (1973) and Danielson (1967). Soil water is one of the most important factors influencing root growth and distribution. This includes both depth and fluctuation of the water table as well as the distribution of soil water during dry periods. Since the purpose of the model is to predict the water table position and soil water content, a model that includes the complex plant growth processes would be required to accurately characterize the change of the root zone with time. Such models have been developed for very specific situations, but their use is limited by input-data and computational requirements. However, the variation of root zone depths with time after planting may be approximated for some crops from experimental data reported in the literature.

II. FIELD EVALUATION OF THE MODEL

Field experiments were conducted over a 5-yr period at three locations in the North Carolina Coastal Plains to test the reliability of the model. Three soil types and five different drainage system designs were included in the experiment from which 21 site-years of data were obtained. Subirrigation was practiced on all of the sites during some years. Rainfall intensity and water table elevations were measured continuously at each site and the observed day-end water table elevations were compared with predicted values. Effective lateral hydraulic conductivity values were measured in the field using both auger hole and water table drawdown methods. Numerous other field and laboratory measurements were made for each soil to determine input soil property and site parameter data.

In general, comparison of predicted and measured water table elevations were in good agreement. The average absolute deviation between predicted and observed water table depths for 21 site-years of data (approximately 7400 pairs of daily predicted and measured values) was 10.3 cm. This agreement was not the result of data fitting. Input parameters were determined independently as noted above. Results are described in detail by Skaggs (1982).

The DRAINMOD model has also been tested using 8 yr of field data from Ohio (Skaggs et al., 1981). Predicted and measured outflow volumes were compared for field plots with surface drainage alone, subsurface drainage alone, and for plots with both surface and subsurface drainage. Further testing was conducted for a tight commerce (fine-silty, mixed, nonacid, thermic Aeric Fluvaquent) silt loam soil in Louisiana (Gayle et al., 1985) and for a Moen (fine-loamy, mixed Typic Argiboroll) silt loam soil in Belgium (Hendrosusanto et al., 1987). Agreement between predicted and observed values was good in all field studies. It was concluded, in each case, that the model was reliable for use in design and evaluation of agricultural water management systems. The model also performed well for irrigated California soils when deep and lateral seepage losses were considered (Chang et al., 1983).

III. WATER MANAGEMENT SYSTEM OBJECTIVES

Agricultural water management systems may be installed to satisfy a variety of objectives. In most cases, the overall objective is to maximize yields, profits, or return on the investment. The relationship between water management system design, crop yields and profit is complex. It varies from year to year with both weather and economic conditions. Traditional objectives of water management systems include eliminating water-related factors that limit crop production or reducing those factors to an acceptable level. The acceptable level for a given situation may also depend on the landowner. Some owners prefer to operate at a greater level of risk than others, so an acceptable level of drainage protection may be less for one owner than for another.

Water-related factors that limit production on poorly drained soils include: (i) stresses caused by excessive soil water conditions during the growing season; (ii) stresses due to deficient soil water conditions; and (iii) the lack of trafficable or workable soil conditions, which cause delays in planting, harvesting, and other field operations. In irrigated arid regions, additional stresses may be caused by salinity and drainage is needed to reduce those stresses. The DRAINMOD model was initially developed for humid regions. The following objective functions are routinely computed for use in evaluating the performance of the water management system design that is simulated:

1. Number of working days: This is used to characterize the ability of the water management system to ensure trafficable conditions during specified periods.
2. The Sum of Excess Water (SEW) at depths less than a specified depth. The SEW provides a measure of excessive soil water conditions during the growing season, which could be detrimental to plant growth.
3. Number of dry days during growing season: This quantifies the number of days during which deficient soil water conditions exist.
4. Irrigation volume: This is the amount of irrigation water that can be applied in a specified time interval. This objective function is used when a drainage system is designed for land disposal of wastewater.

The DRAINMOD model is used to simulate the performance of a given system design for a period of climatological record, usually several years. The objective functions are evaluated and provided as part of the output for each simulation. Daily, monthly, and yearly summaries may be obtained. Through multiple simulations, the most economical system can be chosen that will satisfy specified limits on the objective functions. In order to integrate the effects of the various objective functions on crop yields, the model was modified to use stress-day index methods to estimate yields (Skaggs et al., 1982). Modifications for the yield version of DRAINMOD are discussed in a subsequent section. First, the four objective functions are explained in more detail.

A. Working Day

A day is counted as a working day if: (i) the air volume (drained volume) in the profile exceeds some limiting value (AMIN); (ii) the rainfall occurring that day is less than a maximum allowable value (ROUTA); and (iii) a minimum number of days (ROUTT) have elapsed since that amount of rainfall occurred. The beginning and ending dates of two time periods are specified; usually one for spring planting and one for fall harvest. The model accumulates and prints out the number of working days during the specified periods for each year.

B. Sum of Excess Water

The concept of SEW_{30} was discussed by Wesseling (1974) and Bouwer (1974). It was originally defined by Sieben (1964) to evaluate the influence of high fluctuating water tables during the winter on cereal crops. It is used herein to quantify excessive soil water conditions (cm-d) during the growing season and is defined as

$$\text{SEW} = \sum_{i=1}^{n} (\text{SEWX} - y_i) \qquad [15]$$

where y_i is the water table depth on Day i, with $i = 1$ as the first and n as the number of days in the growing season. The term SEWX is a specified depth defined as the threshold for excess water, usually taken to be 30 cm for corn. Negative terms inside the summation are neglected. In some cases, the water table depth may vary significantly during the day, thus SEW is calculated in DRAINMOD on an hourly interval to better define the time that the water table remains in the root zone.

Although the SEW concept has a number of weaknesses, it still provides a convenient method of approximating the quality of drainage. Sieben found that cereal yields decreased for SEW_{30} values greater than 100 to 200 cm-d.

C. Dry Days

A dry day is defined as a day in which ET is limited by soil water conditions. When the water table is at a shallow depth, water removed from the root zone by ET is replenished by upward movement from the wetter zones near the water table. After the water table is drawn down to a certain depth, the ET demand can no longer be sustained by upward movement alone and the root zone water will be depleted. Evapotranspiration will continue at a rate governed by atmospheric conditions until the soil water content in the root zone reaches some lower limit as discussed previously. When this condition occurs, ET will be limited to the rate that water can move upward to the root zone from the vicinity of the water table. Days when ET is less than the potential ET, because of the limiting soil water conditions, are presumed detrimental to optimum crop production and are counted as dry days.

The three parameters, working days, SEW, and dry days may be used to evaluate the performance of a proposed drainage or subirrigation system. Ideally, the design should ensure a given number of working days during the season, SEW values below a given maximum to prevent crop damage by excessive soil water, and a minimum number of dry days to prevent crop losses due to deficient soil water conditions.

D. Wastewater Irrigation Volume

The DRAINMOD model can be used to evaluate hydraulic loading limitations of land treatment of wastewater. The objective function for evaluating a system design and irrigation scheme, in this case, is the amount of wastewater that can be applied per unit area. This function may be evaluated on an annual basis to determine the size of the required system, and on a month-to-month basis to assess the wastewater storage capacity that may be required during wet months.

Water application to the surface may be scheduled at a specified interval (INTDAY) during a given period. The model determines whether the profile has sufficient drained pore space for each scheduled wastewater application. If not, the application is skipped until the next scheduled irrigation. Applications may also be rained out if rainfall occurs during the time of scheduled irrigation. The amount and rate that water is applied may be varied from month to month. In addition to the amount of wastewater irrigation, the number of skips and postponements, the consecutive days when irrigation cannot be applied, are outputs of the model.

IV. CROP YIELD

In most cases, the principle objective of designing a water management system is to increase yields and profits. The objective functions discussed in the previous section influence yield, but it is often difficult to determine their relative importance for a specific application. To evaluate the effects of combinations of stresses, approximate methods were incorporated into a yield version of DRAINMOD. This version can be used to predict the relative yield on a year-by-year basis. The relative yield is the ratio (expressed as a percent) of the actual yield to the potential yield, where the potential yield is the long-term average yield that would be obtained if all the soil water stresses are eliminated. In this version, the effects of planting date delay, excessively wet conditions and excessively dry conditions are integrated to predict the cumulative effect on yield. This allows the impact of various water management alternatives to be more easily interpreted and, since yields are estimated, economic analysis can be conducted to determine economically optimum designs.

It should be emphasized that the methods presented herein were developed for corn (*Zea mays* L.). Furthermore, all the required inputs are currently available only for corn. Inputs for other crops are being developed and can be approximated for some crops at this time.

Separate models for crop response to excessive and deficient soil-water conditions, as well as to delays in the planting date, have been incorporated into this version of DRAINMOD (Skaggs et al., 1982). The general crop response model may be written as

$$YR = Y/Y_o = YR_w \times YR_d \times YR_P \qquad [16]$$

where $YR_w = Y_w/Y_o$, $YR_d = Y_d/Y_o$, $YR_P = Y_P/Y_o$, and where YR is the relative yield, Y is the yield for a given year, and Y_o is the base yield or potential yield. If only wet stresses occur, Y_w is the yield that would be obtained, Y_d is the yield that would be obtained if only drought stresses occur, and Y_P is the yield that would be obtained if the only reduction is due to a delay in planting date. Individual submodels are used to calculate the relative yields, YR_w, YR_d, and YR_P.

A. Crop Response to Excessive Soil Water Conditions

The model for predicting corn yield response to excessive soil water conditions was originally developed by Hardjoamidjojo et al. (1982). It uses the stress-day index (SDI) method (Hiler, 1969) to quantify the effect of high water tables on corn yield. Reanalysis of the Hardjoamidjojo et al. (1982) data based on improved crop factors results in the following model:

$$YR_w = 100 - 0.71 \times SDI_w \quad \text{for } SDI_w < 141 \quad [17a]$$

$$YR_w = 0 \quad \text{for } SDI_w \geq 141 \quad [17b]$$

The SDI_w is the stress day index for excessively wet conditions and may be expressed as

$$SDI_w = \sum_{j=1}^{N} CS_{w_j} \times SDW_j \quad [18]$$

where N is the number of days in the growing season, CS_{w_j} the crop susceptibility factor for excessive soil water conditions for Day j, and SDW_j is the stress day factor for Day j. The SDW_j is taken to be the same as the daily value of SEW. The units for SDI_w are centimeter-days. Crop susceptibility factors for corn were obtained from studies by Evans et al. (1990) and are shown in Table 10-1. These factors are dependent on the stage of plant development. They were normalized so that the sum of the factors for the five growth stages = 1.0. These values are based on data collected since the original study by Hardjoamidjojo et al. (1982) and were used to determine the parameters in Eq. [17].

B. Yield Response to Deficient Soil Water Conditions

Shaw's model (1978) for corn response to deficient soil water conditions may be written as

$$YR_d = 100 - 1.22 \, SDI_d \quad [19]$$

where SDI_d is the stress-day index for drought conditions. The coefficients of the equation, 100 and 1.22, are data inputs. The SDI_d is the stress-day index for drought conditions and is defined on a cumulative basis as

$$SDI_d = \sum_{j=1}^{N} SD_{d_j} \times CS_{d_j} \quad [20]$$

Table 10-1. Crop (corn) susceptibility factors for excessive soil water conditions.

Growth stage	Days after planting	Crop susceptibility factor (CS_w)
Establishment	0–19	0.20
Early vegetative	20–35	0.22
Late vegetative	36–55	0.32
Flowering	56–70	0.19
Yield formation	71–89	0.08
Ripening	90–	0.02

where SD_{d_j} and CS_{d_j} are the stress-day and crop susceptibility factors, respectively, for growth period j, and N is the number of periods (5 d long in this case) in the growing season. The stress-day factor is defined as

$$SD_{d_j} = \sum_{k=1}^{n_j} (1.0 - AET_k/PET_k) \qquad [21]$$

where AET_k and PET_k are the actual and potential daily ET, respectively, and n_j is the number of days in the jth growing period. The CS_d values were developed by Shaw (1974) for 5-d growing periods relative to the silking stage and are shown in Table 10-2. Whenever $SD_d \geq 4.5$ for two or more consecutive 5-d periods, Shaw multiplied the index for those periods by an additional factor of 1.5 to account for greatly reduced yields found under severe stress periods of more than a few days duration (Shaw, 1976).

Shaw's method is used in DRAINMOD to calculate YR_d, which is printed out along with SDI_d for each year of the simulation.

C. Crop Response to Delayed Planting Date

Corn yields are significantly reduced if the planting date is delayed beyond an optimum period. Inputs to DRAINMOD to calculate YR_P include the date to begin seedbed preparation and planting, the number of days needed to complete this task, and the threshold date beyond which yield reduction occurs if the crop is not yet planted. The model determines whether trafficable conditions exist each day and maintains a running total of suitable working days during the planting period. When enough working days have occurred to complete the operation, the planting date is fixed, the length of planting date delay beyond the optimum is determined, and YR_P is estimated from the following equations:

if PDELAY < DELAY1,

$$YR_P = 100.0 - PDRF \times PDELAY \qquad [20a]$$

Table 10-2. Crop susceptibility factors used to evaluate deficient soil water conditions on corn yield Shaw (1974).

Period†	CS_d	Period	CS_d
−8	0.50	+1	2.00
−7	0.50	+2	1.30
−6	1.00	+3	1.30
−5	1.00	+4	1.30
−4	1.00	+5	1.30
−3	1.00	+6	1.30
−2	1.75	+7	1.20
−1	2.00	+8	1.00
		+9	0.50

† Five-day periods relative to silking where silking occurs in the 10-d period −1 to +1 (e.g., −8 is the period 35 to 40 d before silking, ×7 is 30 to 35 d after silking).

if PDELAY > DELAY1,

$$YR_P = 100.0 - PDRF \times DELAY1$$
$$- PDRF2 \times (PDELAY - DELAY1) \qquad [20b]$$

where DELAY1 is a breakpoint past which the yields decrease at a faster rate, and PDRF and PDRF2 are the slopes before and after the breakpoint. The values of PDRF, PDRF2, and DELAY1 are input into the program. The last day that planting can be completed without reduced yields (JLAST), the required number of working days for seedbed preparation and planting, REQWRK, and the number of days in the growing season, IGROW, are also program inputs.

Seymour (1986) studied corn yield response to planting delay. She reviewed results of studies conducted in other parts of the USA and determined DRAINMOD inputs for corn as shown in Table 10-3.

D. Overall Crop Response Model

The DRAINMOD model was programmed so that the time-varying root depth function and the computation of the stress-day indices are initiated after the required number of working days is satisfied and the planting date established. The stress-day factors are calculated on a daily basis and stored. At the end of the year's simulation, the stress-day factors are multiplied by the appropriate crop susceptibility factors and the stress-day indices determined. The relative yields as affected by excessive soil water conditions, deficient soil water conditions, and delay in planting date are determined and overall relative yields are calculated as discussed.

The basic assumption in the approach outlined above is that the effect of either excessive or deficient soil-water conditions or planting-date delay is independent of the other two factors. Interactive effects between the three factors are neglected. The validity of this approach was tested by Hardjoamidjojo and Skaggs (1982) for a Portsmouth sandy loam (fine-loamy over sandy, mixed, thermic Typic Umbraqualts) in eastern North Carolina.

Table 10-3. Summary of equation parameters for predicting corn yield response to delay in planting for several locations. (After Seymour, 1986.)

Location	Optimum planting date	DELAY1	PDRF	PDRF2
England	30 Apr.	30	0.95	3.6
Illinois	30 Apr.	31	0.81	--
Iowa	15 May	30	0.76	2.1
Kansas	5 May	20	0.62	--
Louisiana	15 Mar.	45	0.72	--
New York	1 May	13	0.89	1.32
North Carolina	10 Apr.	42	0.88	1.62
North Dakota	17 May	38	0.79	--
Ohio	1 May	22	0.60	1.8

V. APPLICATION OF THE YIELD VERSION OF DRAINMOD—AN EXAMPLE

The yield version of DRAINMOD is demonstrated in this section by analyzing the effect of drainage and subirrigation system design for a Portsmouth sandy loam soil. The Portsmouth is a poorly drained soil with a nearly level surface, which normally requires artificial drainage for efficient crop production. Simulations were conducted for the 30-yr period from 1950 through 1979 using weather data for Wilson, NC. The performance of drainage systems with drain spacings varying from 5 to 300 m and surface depressional storage depths of s = 25 mm (poor surface drainage) and s = 2.5 mm (good surface drainage) were simulated. Seedbed preparation was assumed to begin on 15 March (Day of Year 74) and eight trafficable days were required to plant the corn. If 8 d suitable for seedbed preparation occurred before 10 April (Day of Year 100), the corn was assumed planted on Day of Year 100. If the eighth day occurred after Day of Year 100, the corn was assumed planted on that day. It is recognized that the actual length of time required for the planting operation depends on size of operation, equipment and labor available, and many other factors. A period of 8 d was chosen arbitrarily as being typical for an eastern North Carolina operation.

A sample model output is given in Table 10-4. The output is divided into two sections. The first section summarizes inputs in the following order:

1. Weather data
2. Drainage system parameters
3. Soil properties
4. Crop inputs
5. Wastewater irrigation inputs.

The second section contains results of the simulation. Daily, monthly, or annual summaries may be obtained at the option of the user.

Average predicted yields are plotted as a function of drain spacing for the Portsmouth sandy loam in Fig. 10-11. Relationships are plotted for both good and poor surface drainage for a drain depth of 1 m. The maximum yield of 78% was obtained for a drain spacing of about 27 m for good surface drainage. For poor surface drainage the maximum yield was about the same, but occurred at a drain spacing of about 20 m. Higher average yields were not obtained because of deficit soil water conditions, which caused drought stresses during several years. The quality of surface drainage does not have much effect on average predicted yields for drain narrow spacings, which provide good subsurface drainage. Surface drainage is extremely important, however, for wider drain spacings when subsurface drainage is poor. For example, at a 100-m drain spacing the average predicted relative yields were 57 and 45% for good and poor surface drainage, respectively.

The effect of drain depth on the relationship between yield and drain spacing is shown in Fig. 10-12. This soil demonstrates some tendency for overdrainage at drain spacings that are closer together than necessary and for the deeper drain depths. The highest yields were obtained for 10- to 15-m

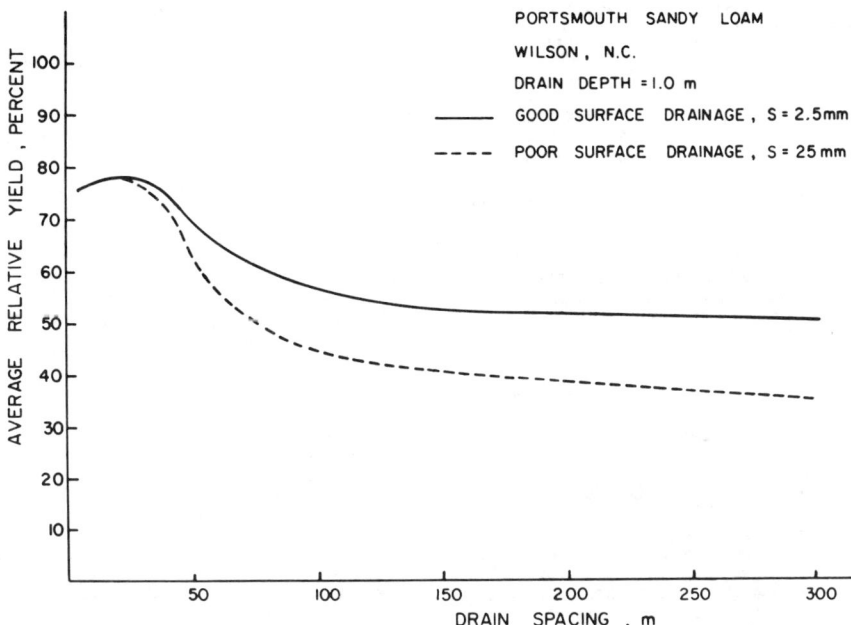

Fig. 10-11. Effect of drain spacing on average relative corn yields for a Portsmouth sandy loam. Average yields for both good and poor surface drainage were obtained for a 30-yr period using climatological record from Wilson, NC.

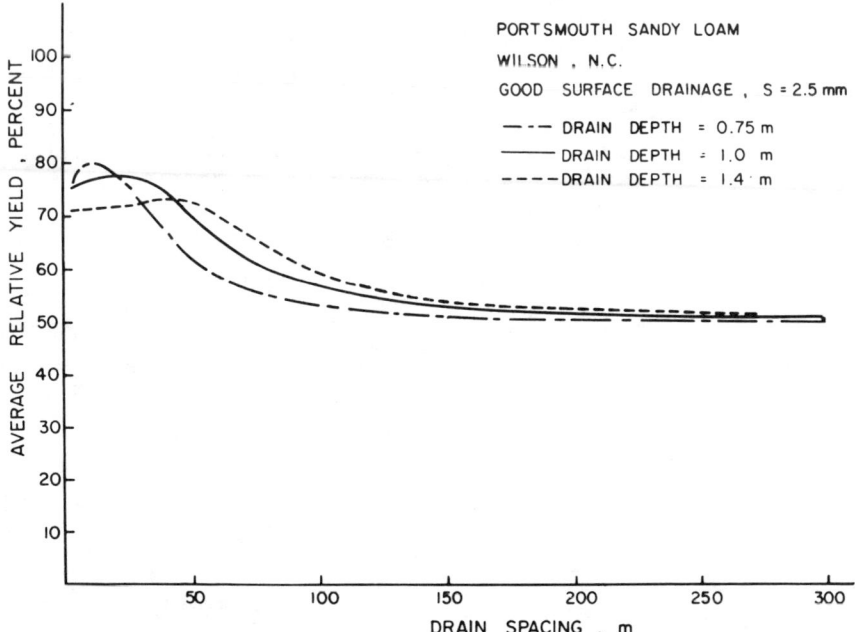

Fig. 10-12. Effect of drain spacing and depth on long term average relative yield for a Portsmouth sandy loam.

Table 10-4. Example of DRAINMOD Printout for Portsmouth Sandy Loam Soil.

```
***************************************************************************
                     ********************
                     *  D R A I N M O D  *
                     ********************

       DATA READ FROM INPUT FILE: D:\DM40\INPUT40\PORTA102.LIS

                           TITLE OF RUN
                           ************

       EXAMPLE SIMULATION FOR PORTSMOUTH SANDY LOAM LOCATED NEAR WILSON, N.C.
       CONVENTIONAL DRAINAGE; DRAIN SPACING =100 M; DEPTH = 100 CM; GOOD SURFACE DRAIN



                               CLIMATE INPUTS
                               ******* ******

          DESCRIPTION                         (VARIABLE)     VALUE    UNIT
       ---------------------------------------------------------------------
       FILE FOR RAINDATA ............ D:\DM40\WEATHER\WILSON.RAI
       FILE FOR TEMPERATURE/PET DATA .. D:\DM40\WEATHER\WILSON.TEM
       RAINFALL STATION NUMBER........................(RAINID)    319476
       TEMPERATURE/PET STATION NUMBER..................(TEMPID)    319476
       STARTING YEAR OF SIMULATION..................(START YEAR)    1950    YEAR
       STARTING MONTH OF SIMULATION................(START MONTH)       1    MONTH
       ENDING YEAR OF SIMULATION......................(END YEAR)    1979    YEAR
       ENDING MONTH OF SIMULATION....................(END MONTH)      12    MONTH
       TEMPERATURE STATION LATITUDE...................(TEMP LAT)   35.47    DEG.MIN
       HEAT INDEX.........................................(HID)   75.00

       ET MULTIPLICATION FACTOR FOR EACH MONTH
          1.00  1.00  1.00  1.00  1.00  1.00  1.00  1.00  1.00  1.00  1.00  1.00

                               DRAINAGE SYSTEM DESIGN
                               **********************

                           *** CONVENTIONAL DRAINAGE ***

          JOB TITLE:

                   EXAMPLE SIMULATION FOR PORTSMOUTH SANDY LOAM LOCATED NEAR WI
                   CONVENTIONAL DRAINAGE; DRAIN SPACING =100 M; DEPTH = 100 CM;
```

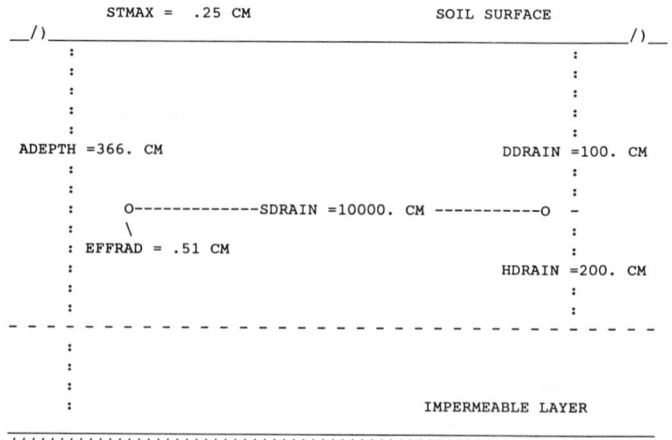

DRAINAGE

Table 10-4. Continued.

DEPTH (CM)	SATURATED HYDRAULIC CONDUCTIVITY (CM/HR)
.0 - 30.0	8.000
30.0 - 300.0	2.000

```
DEPTH TO DRAIN = 100.0 CM
EFFECTIVE DEPTH FROM DRAIN TO IMPERMEABLE LAYER = 200.0 CM
DISTANCE BETWEEN DRAINS = 10000.0 CM
MAXIMUM DEPTH OF SURFACE PONDING =    .25 CM
EFFECTIVE DEPTH TO IMPERMEABLE LAYER =  300.0 CM
DRAINAGE COEFFICIENT(AS LIMITED BY SUBSURFACE OUTLET) =  2.50 CM/DAY
ACTUAL DEPTH FROM SURFACE TO IMPERMEABLE LAYER = 366.0 CM
SURFACE STORAGE THAT MUST BE FILLED BEFORE WATER
    CAN MOVE TO DRAIN =  .25 CM
FACTOR -G- IN KIRKHAM EQ. 2-17 =12.06
```

DEPTH OF WEIR FROM THE SURFACE

DATE	1/ 1	2/ 0	3/ 0	4/ 0	5/ 0	6/ 0
WEIR DEPTH	100.0	100.0	100.0	100.0	100.0	100.0

DATE	7/ 0	8/ 0	9/ 0	10/ 0	11/ 0	12/ 0
WEIR DEPTH	100.0	100.0	100.0	100.0	100.0	100.0

SOIL INPUTS
****

TABLE 1   TABLE 2

DRAINAGE TABLE		SOIL WATER CHARACTERISTIC VS VOID VOLUME VS UPFLUX			
VOID VOLUME (CM)	WATER TABLE DEPTH (CM)	HEAD (CM)	WATER CONTENT (CM/CM)	VOID VOLUME (CM)	UPFLUX (CM/HR)
.0	.0	.0	.3655	.00	1.0000
1.0	34.2	10.0	.3325	.19	.5000
2.0	56.1	20.0	.3270	.49	.2000
3.0	71.2	30.0	.3205	.83	.0625
4.0	84.3	40.0	.3155	1.23	.0306
5.0	95.8	50.0	.3105	1.69	.0142
6.0	106.8	60.0	.3070	2.20	.0112
7.0	117.3	70.0	.3035	2.91	.0073
8.0	127.6	80.0	.3000	3.62	.0035
9.0	137.6	90.0	.2974	4.49	.0024
10.0	147.4	100.0	.2949	5.36	.0012
11.0	157.0	110.0	.2923	6.31	.0007
12.0	166.3	120.0	.2897	7.25	.0001
13.0	175.4	130.0	.2871	8.24	.0000
14.0	184.3	140.0	.2846	9.23	.0000
15.0	192.9	150.0	.2820	10.27	.0000
16.0	201.5	160.0	.2804	11.31	.0000
17.0	210.1	170.0	.2788	12.41	.0000
18.0	218.7	180.0	.2772	13.51	.0000
19.0	227.3	190.0	.2756	14.66	.0000
20.0	235.8	200.0	.2740	15.82	.0000
21.0	244.4	210.0	.2721	16.99	.0000
22.0	253.4	220.0	.2702	18.15	.0000
23.0	262.9	230.0	.2683	19.32	.0000
24.0	272.5	240.0	.2664	20.49	.0000
25.0	282.1	250.0	.2645	21.65	.0000
26.0	291.6	260.0	.2626	22.69	.0000
27.0	301.2	270.0	.2607	23.74	.0000
28.0	310.8	280.0	.2588	24.78	.0000
29.0	320.4	290.0	.2569	25.83	.0000
30.0	329.9	300.0	.2550	26.87	.0000
35.0	377.8	350.0	.2440	32.10	.0000
40.0	425.7	400.0	.2330	37.32	.0000
45.0	473.5	450.0	.2208	42.54	.0000
50.0	521.4	500.0	.2085	47.77	.0000
60.0	617.1	600.0	.1900	58.21	.0000
70.0	712.8	700.0	.1838	68.66	.0000
80.0	808.6	800.0	.1775	79.11	.0000
90.0	904.3	900.0	.1713	89.55	.0000

(continued on next page)

Table 10-4. Continued.

```
                    GREEN AMPT INFILTRATION PARAMETERS
                    W.T.D.        A           B
                    (CM)         (CM)        (CM)
                     .000        .000        .000
                   50.000       1.200        .750
                  100.000       6.500       1.200
                  150.000      10.000       1.500
                  200.000      12.000       1.500
                  500.000      15.000       1.500
                 1000.000      15.000       1.500

                            TRAFFICABILITY
                            **************
                                                  FIRST       SECOND
REQUIREMENTS                                      PERIOD      PERIOD
  -MINIMUM AIR VOLUME IN SOIL (CM):                3.00        3.00
  -MAXIMUM ALLOWABLE DAILY RAINFALL(CM):           1.20        1.20
  -MINIMUM TIME AFTER RAIN BEFORE TILLING CAN CONTINUE: 2.00   2.00

WORKING TIMES
  -DATE TO BEGIN COUNTING WORK DAYS:               3/15       12/31
  -DATE TO STOP COUNTING WORK DAYS:                8/18       12/31
  -FIRST WORK HOUR OF THE DAY:                       8          8
  -LAST WORK HOUR OF THE DAY:                       20         20
                                 CROP
                                 ****

SOIL MOISTURE AT CROP WILTING POINT =    .13

    HIGH WATER STRESS:   BEGIN STRESS PERIOD ON    4/10
                         END STRESS PERIOD ON     12/20
                         CROP IS IN STRESS WHEN WATER TABLE IS ABOVE  30.0 CM

    DROUGHT STRESS:      BEGIN STRESS PERIOD ON    4/10
                         END STRESS PERIOD ON     12/20


                       MO    DAY   ROOTING DEPTH(CM)
                        1     1         3.0
                        4    10         3.0
                        4    28        10.0
                        5    11        15.0
                        5    25        25.0
                        6    14        30.0
                        7    18        30.0
                        9    21        30.0
                       10     1        10.0
                       10    19         3.0
                       12    31         3.0
```

(Example: the following table may be printed for each year)

MONTHLY VOLUMES IN CENTIMETERS FOR YEAR 1965

MONTH	RAIN	INFIL	ET	DRAIN	RUNOFF	DRY DAYS	WRK DAYS	SEW	PUMP
1	4.93	4.11	1.27	2.41	.68	.00	.00	.00	.00
2	8.89	4.04	1.59	2.84	4.98	.00	.00	.00	.00
3	9.70	4.85	2.08	3.02	4.85	.00	.00	.00	.00
4	8.74	7.17	5.51	1.27	1.57	.00	11.89	88.75	.00
5	3.53	3.53	12.09	.49	.00	.00	24.76	11.50	.00
6	16.54	16.54	10.74	.28	.00	4.00	18.62	.00	.00
7	26.70	18.43	13.83	1.87	8.26	.00	3.29	213.06	.00
8	4.06	4.06	14.11	.58	.00	4.00	9.84	31.99	.00
9	3.78	3.78	3.31	.00	.00	23.00	.00	.00	.00
10	4.42	4.42	1.17	.00	.00	27.00	.00	.00	.00
11	2.72	2.72	.57	.00	.00	21.00	.00	.00	.00
12	1.45	1.45	1.35	.00	.00	.00	1.00	.00	.00
TOTALS	95.45	75.10	67.63	12.77	20.35	79.00	69.39	345.30	.00

(continued on next page)

DRAINAGE

Table 10-4. Continued.

(Summary of Predicted Annual Stresses and Yields)

EXAMPLE SIMULATION FOR PORTSMOUTH SANDY LOAM LOCATED NEAR WILSON, N.C.
CONVENTIONAL DRAINAGE; DRAIN SPACING =100 M; DEPTH = 100 CM; GOOD SURFACE
DRAIN
****

```
        ---------RUN STATISTICS ----------       time:  7/ 5/1990 @ 11:22
    input file:    D:\DM40\INPUT40\PORTA102.LIS
    parameters:      free drainage         and yields calculated
                     drain spacing =    10000. cm   drain depth =  100.0 cm
```

	SDI - STRESS		plant	plant	harv.	RELATIVE YIELDS (%)			
	excess	drought	date	delay	date	excess	drought	delay	overall
1950	53.7	.0	115	10.	245	61.8	100.0	91.2	56.3
1951	.0	22.5	124	19.	254	100.0	72.6	83.3	60.5
1952	.0	55.0	111	6.	241	100.0	32.9	94.7	31.1
1953	28.0	33.2	143	38.	273	81.0	59.5	66.6	32.1
1954	.0	63.7	117	12.	247	100.0	22.3	89.4	19.9
1955	.0	5.9	109	4.	239	100.0	92.8	96.5	89.5
1956	14.4	33.2	130	25.	260	91.2	59.5	78.0	42.3
1957	.0	23.7	119	14.	249	100.0	71.1	87.7	62.4
1958	.0	15.3	143	38.	273	100.0	81.3	66.6	54.1
1959	43.2	5.5	130	25.	260	69.6	93.3	78.0	50.6
1960	36.3	23.3	116	11.	246	74.8	71.6	90.3	48.4
1961	16.9	5.1	153	48.	283	89.3	93.8	53.3	44.7
1962	.0	7.7	135	30.	265	100.0	90.6	73.6	66.7
1963	.0	1.5	115	10.	245	100.0	98.2	91.2	89.6
1964	.0	35.5	119	14.	249	100.0	56.7	87.7	49.8
1965	33.8	2.0	113	8.	243	76.7	97.6	93.0	69.6
1966	46.0	.0	110	5.	240	67.5	100.0	95.6	64.5
1967	.0	.0	100	0.	230	100.0	100.0	100.0	100.0
1968	45.0	.8	120	15.	250	68.3	99.0	86.8	58.7
1969	.0	6.3	134	29.	264	100.0	92.3	74.5	68.8
1970	.0	13.4	138	33.	268	100.0	83.6	71.0	59.3
1971	.0	8.7	147	42.	277	100.0	89.4	63.0	56.4
1972	.0	2.2	168	63.	298	100.0	97.3	29.0	28.2
1973	.0	6.3	100	0.	230	100.0	92.3	100.0	92.3
1974	.0	2.7	110	5.	240	100.0	96.8	95.6	92.5
1975	39.1	1.2	126	21.	256	72.7	98.5	81.5	58.4
1976	.0	46.4	115	10.	245	100.0	43.3	91.2	39.5
1977	.0	30.6	114	9.	244	100.0	62.7	92.1	57.7
1978	.0	1.6	143	38.	273	100.0	98.1	66.6	65.3
1979	9.4	14.7	191	86.	321	94.9	82.0	.0	.0
AVG	12.2	15.6	127.	22.	257.	91.6	81.0	78.9	57.0

drain spacings at a depth of 0.75 m. Maximum yields occur at wider drain spacings for deeper drain depths. However, overdrainage during some years causes the maximum yield for the deep drains to be less than that for shallow drains.

The effect of subirrigation was simulated by raising the weir depth in the drainage outlet to within 50 cm of the soil surface. The outlet water levels were raised after planting and held at that elevation until the corn was mature. Lowering the outlet to increase the drainage rates during high rainfall periods would potentially increase yields and allow somewhat wider drain spacings. Use of DRAINMOD to simulate the effectiveness of using feedback from field water-table sensors to control the outlet elevation during subirrigation has been studied (Fouss, 1983; Smith et al., 1985).

Average predicted yields for subirrigation are plotted as a function of drain spacing in Fig. 10-13. Results for drainage alone are also given for comparison. Good surface drainage was assumed for both cases. Average predicted yields for subirrigation increased with a decrease in drain spacing. The maximum was 98% at a drain spacing of 5 m. An economic analysis (Skaggs et al., 1988, unpublished data) showed that maximum annual return to land and management would be obtained for a drain spacing of 20 m where the average predicted yield was 93%.

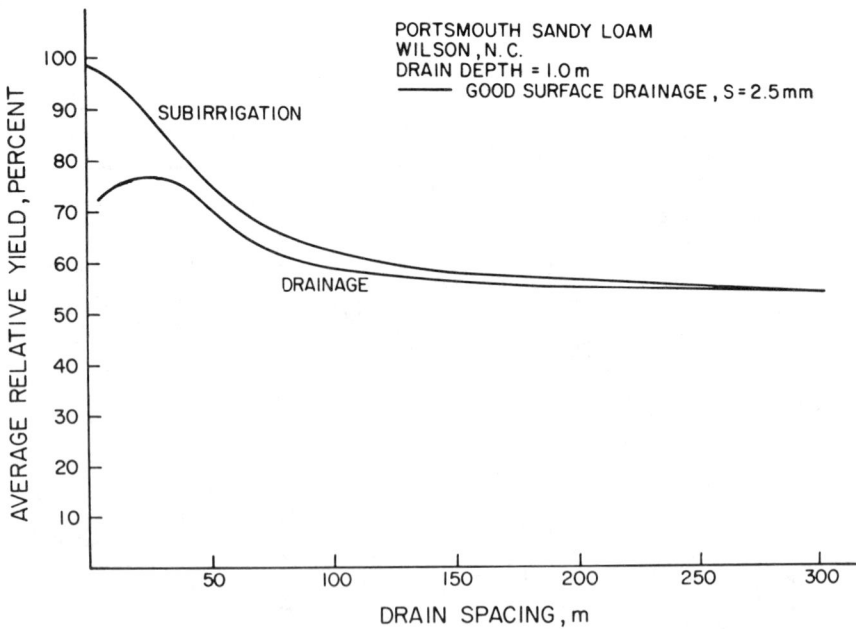

Fig. 10-13. Effect of drain spacing and surface drainage on average relative yields predicted for conventional drainage and subirrigation for a Portsmouth sandy loam.

VI. RESEARCH NEEDS

Simulation models have progressed to the point at which they are applied to the evaluation, design, and management of drainage and water-table control systems. However, present use is only a small fraction of the potential use. Probably the single greatest limitation to the application of drainage simulation models is the lack of a strong data base for soil property inputs. These properties can be measured for research needs, but it is usually not feasible to measure the unsaturated $K(h)$ and $h(\theta)$ functions on an applied scale. Methods for rapidly determining field-effective values for these properties are needed. Soil property data are needed for the dominate soils in regions where simulation models will be applied. They are readily available in some states (e.g., Carlisle et al., 1978), but nonexistent in others. The data base for soil water tensions at the wet end of the spectrum (tensions <0.33 bar) needs to be strengthened if we are to routinely apply simulation models for design and management of drainage and related systems.

Models need to be improved and new algorithms developed for predicting the movement of nutrients and other pollutants from drained lands. Existing models, such as DRAINMOD, can be used to determine the rate and amount of water that leaves the field by surface runoff and subsurface flow to drains. These results can be combined with experimental data to evaluate the effects of drainage system design and operation on pollutant loading (Deal et al., 1986). Process-based models such as these discussed in Ch. 13 to 15 are needed to characterize transformations, uptake, storage, and movement of nutrients in the soil water system. Such methods need to be adapted on a field scale for drained soils.

Existing models compute stresses due to dry conditions and predict yields by adjusting a potential yield. Physiologically based crop simulation models, such as those discussed in Ch. 3 to 5, could be used to better describe the effects of drainage and related water management practices on yields. The combination of these approaches with drainage simulation models would allow the interaction between water management, fertility, and cultural practices to be considered.

One of the important benefits ascribed to subsurface drainage is that it lowers water tables early in the growing season, permitting deep root growth and thereby reducing drought stress later in the year. Current simulation models cannot directly predict this benefit. They are capable of considering time-dependent root depths, which are inputs, but the effect of high water tables or dry conditions on root depth is not computed. Methods are needed to predict the effects of the soil water regime, especially high water tables, on root depth.

The effects of macropores on drainage processes in heavy soils are not considered directly in current simulation models. Research to characterize drainage processes in these soils should be continued. Methods such as those developed by Leeds-Harrison and Jarvis (1986) and by Branswijk (1988) should be incorporated in water management simulation models.

VII. SUMMARY

This chapter describes a water management simulation model for high water-table soils requiring drainage. The DRAINMOD model was developed for design and evaluation of multicomponent water management systems, which may include facilities for subsurface drainage, surface drainage, subirrigation or controlled drainage, and irrigation of wastewater onto land. The model is based on a water balance in the soil profile. It uses climatological data to predict, on a day-to-day, hour-by-hour basis, the response of the water table and the soil water regime above it to various combinations of surface and subsurface water management. By simulating the performance of alternative systems across several years of record, an optimum water management system can be designed. The DRAINMOD model is composed of a number of separate components, incorporated as subroutines, to evaluate the various mechanisms of water movement and storage in the soil profile. These components include methods to evaluate infiltration, subsurface drainage, surface drainage, potential evapotranspiration (PET), actual ET, subirrigation, and the soil water distribution. In order to simplify the model and reduce the number and complexity of inputs, approximate methods are used for each component. The model is constructed so that improved methods can be easily substituted for existing components as they become available.

The DRAINMOD model routinely predicts the following objective functions to quantify the performance of the water management system: (i) number of working days; (ii) sum of excess water (SEW) to indicate stress due to excessively wet conditions; (iii) number of dry days; and (iv) the amount of wastewater that can be irrigated for a given drainage design. The model also estimates annual yields using the stress-day index method. An example of the use of DRAINMOD for designing a drainage and subirrigation system is provided for an eastern North Carolina soil.

REFERENCES

Alessi, J., and J.F. Power. 1975. Response of an early maturing corn hybrid to planting date and population in the northern plains. Agron. J. 67:762-765.

Allmaras, R.R., A.L. Black, and R.W. Rickman. 1973. Tillage, soil environment and root growth. p. 62-86. *In* Proc. Natl. Conserv. Tillage Conf., Des Moines, IA.

Benson, G.O., H.E. Thompson, and J.P. Shroyer. 1978. Late planting of corn and soybeans. Iowa State Univ. Coop. Ext. Serv. Pamphlet 843.

Bouwer, H. 1969. Infiltration of water into nonuniform soil. J. Irrig. Drain. Div. Am. Soc. Civ. Eng. 95(IR4):451-462.

Bouwer, H. 1974. Developing drainage design criteria. *In* J. van Schilfgaarde (ed.) Drainage for agriculture. Agronomy 17:67-90.

Bouwer, H., and J. van Schilfgaarde. 1963. Simplified method of predicting fall of water table in drained land. Trans. ASAE 6(4):288-291.

Branswijk, J.J.B. 1988. Modeling of water balance, cracking and subsidence of clay soils. J. Hydrol. (Amsterdam) 97:199-212.

Bunting, E.S. 1968. The influence of date of sowing on development and yield of maize in England. J. Agric. Sci. (Cambridge) 71:117-125.

Carlisle, V.W., R.E. Caldwell, F. Sodek, III, L.C. Hammond, F.G. Calhoun, M.A. Granger, and H.L. Berland. 1978. Characterization data for selected Florida soils. Univ. Florida Soil Sci. Res. Rep. 78-1.

Chang, A.C., R.W. Skaggs, L.F. Hermsmeier, and W.R. Johnson. 1983. Evaluation of a water management model for irrigated agriculture. Trans. ASAE 26(2):412-418.

Childs, E.C., and M. Bybordi. 1969. The vertical movement of water in stratified porous material. 1. Infiltration. Water Resour. Res. 5(2):466-459.

Danielson, R.E. 1967. Root systems in relation to irrigation. In R.M. Hagan et al. (ed.) Irrigation of agricultural lands. Agronomy 11:390-424.

Deal, S.C., J.W. Gilliam, R.W. Skaggs, and K.D. Konyha. 1986. Prediction of nitrogen and phosphorus losses as related to agricultural drainage system design. Agric. Ecosyst. Environ. 18:37-51.

Dierickx, W. 1980. Electric analog study of the effect of openings and surroundings of various permeabilities on the performance of field drainage pipes. Publ. nr. 77. Meded. Rijksst. Landbouwtech.

Dillon, M.A., and R.E. Gwin, Jr. 1979. How planting date and full-season or early hybrids affect corn yields. Kansas State Univ. Agric. Exp. Stn. Bull. 600.

Ernst, L.F. 1975. Formulae for groundwater flow in areas with subirrigation by means of open conduits with a raised water level. Misc. Reprint 178. Inst. Land Water Manage. Res., Wageningen, The Netherlands.

Evans, R.O., R.W. Skaggs, and R.E. Sneed. 1990. Normalized crop susceptibility factors for corn and soybean to excess water stress. Trans. ASAE 33(4):1153-1161.

Feddes, R.A., P.J. Kowalik, and H. Zaraday. 1978. Simulation of water use and crop yield. PUDOC, Wageningen, The Netherlands.

Fipps, G., and R.W. Skaggs. 1986. Effect of canal seepage on drainage to parallel drains. Trans. ASAE 29(5):1278-1283.

Free, G.R., C.S. Wionkelblech, H.M. Wilson, and C.E. Bay. 1966. Time of planting in a comparison of plow-plant and conventional seedbed preparation for corn. Agron. J. 58:333-336.

Freeze, R.A. 1971. Three dimensional, transient saturated-unsaturated flow in a groundwater basin. Water Resour. Res. 7:347-366.

Fouss, J.L. 1983. Daily weather forecast rainfall probability as an input to a water management simulation model. p. 401-406. In Proc. Natural Resour. Modeling Symp., Pingree Park, CO. 16-21 Oct. 1983. Agric. Res. Service Publ. ARS-30. U.S. Gov. Print. Office.

Gayle, G.A., and R.W. Skaggs. 1978. Surface storage on bedded cultivated lands. Trans. ASAE 21(1):102-104, 109.

Gayle, G., R.W. Skaggs, and C.E. Carter. 1985. Evaluation of a water management model for a Louisiana sugar cane field. J. Am. Soc. Sugar-Cane Technol. 4:18-28.

Green, W.H., and G. Ampt. 1911. Studies of soil physics, part I—the flow of air and water through soils. J. Agric. Sci. 4:1-24.

Hardjoamidjojo, S., and R.W. Skaggs. 1982. Predicting the effects of drainage systems on corn yields. Agric. Water Manage. 5:127-144.

Hardjoamidjojo, S., R.W. Skaggs, and G.O. Schwab. 1982. Corn yield response to excessive soil water conditions. Trans. ASAE 25(4):922-927.

Hendrosusanto, R., J. Feyen, W. Dierickx, and G. Wyseure. 1987. The use of simulation models to evaluate the performance of subsurface drainage systems. p. 67-76. In Proc. Third Int. Drain. Workshop, Columbus, OH, 7-11 Dec. 1987. Ohio State Univ., Columbus, OH.

Henshaw, J.N., R.W. Lockett, J.L. Rabb, J.G. Marshall, and L.W. Sloane. 1975. Response of corn hybrids to planting dates in Louisiana. p. 61-72. Louisiana State Univ. Agric. Mech. College of Agric. Exp. Stn. Rep. of Projects, Louisiana State Univ., Baton Rouge, LA.

Hiler, E.A. 1969. Quantitative evaluation of crop drainage requirements. Trans. ASAE 12(4):499-505.

Hillel, D., and W.R. Gardner. 1969. Steady infiltration into crust topped profiles. Soil Sci. 108:137-142.

Holtan, H.N., C.B. England, and V.O. Shanholtz. 1967. Concepts in hydrologic soil grouping. Trans. ASAE 10(3):407-410.

Hooghoudt, S.B. 1940. Bijdragen tot de kennis van eenige natuurkundige grootheden van den gond, 7, Algemeene beschouwing van het probleem van de detail ontwatering en de infiltratie door middel van parallel loopende drains, greppels, slooten, en kanale. Versl. Landbouwk. Onderzoek. 46:515-707.

Horton, R.E. 1939. Analysis of runoff plot experiments with varying infiltration capacity. Trans. Am. Geophys. Union IV:693-694.

Jensen, M.E. (ed.). 1974. Consumptive use of water and irrigation water requirements. Rep. Tech. Com. on Irrig. Water Requirements. ASCE.

Jensen, M.E., H.R. Haise, and R. Howard. 1963. Estimating evapotranspiration from solar radiation. J. Irrig. Drain. Div. Am. Soc. Civ. Eng. 89(IR4):15-41.

Karvonen, T. 1988. A model for predicting the effect of drainage on soil moisture, soil temperature, and crop yield. Ph.D. diss. Helsinki Univ. of Technol., Helsinki, Finland.

Kirkham, D. 1957. Theory of land drainage. In J.N. Luthin (ed.) Drainage of agricultural lands. Agronomy 7:139-181.

Leeds-Harrison, P.B., and N.J. Jarvis. 1986. Drainage modeling in heavy clay soils. p. 198-220. In J. Saavalainen and P. Vakkilainen (ed.) Proc. Int. Seminar Land Drain. Helsinki, Finland. 9-11 July 1986. Helsinki Univ. of Tech., Helsinki, Finland.

Luthin, J.N. 1978. Drainage engineering. R.F. Krieger Publ. Co., Huntington, NY.

McCarthy, E. 1990. Hydrologic model for a drained forest watershed. Ph.D. diss. North Carolina State Univ., Raleigh.

McGuinness, J.L., and E.F. Borden. 1972. A comparison of lysimeter-derived potential evapotranspiration with computed values. USDA Tech. Bull. 1452.

McWhorter, D.B. 1971. Infiltration affected by flow of air. Colorado State Univ. Hydrology Paper 49.

McWhorter, D.B. 1976. Vertical flow of air and water with a flux boundary condition. Trans. ASAE. 19(2):259-261, 265.

Mein, R.G., and C.L. Larson. 1973. Modeling infiltration during a steady rain. Water Resour. Res. 9(2):384-394.

Mohammad, F.S., and R.W. Skaggs. 1983. Drain opening effects on drain inflow. J. Irrig. Drain. Eng. 109(4):393-404.

Moody, W.T. 1967. Nonlinear differential equation of drainage spacing. J. Irrig. Drain. Div. Am. Soc. Civ. Engr. 92(IR2):1-9.

Morel-Seytoux, H.J., and J. Khanji. 1974. Derivation of an equation of infiltration. Water Resour. Res. 10(4):795-800.

Nolte, B.H., D.M. Byg, and E. Gill. 1976. Timely field operations for corn and soybeans in Ohio. Ohio State Univ. Coop. Ext. Serv. Bull. 605.

Nolte, B.H., and R.D. Duvick. 1985. Economic factors of drainage related to corn production. Ohio State Univ. Coop. Ext. Serv. Natl. Corn Handb.

North Central Computer Institute. 1985. DRAINMOD: Documentation for the water management simulation model. Software J., North Central Computer Inst., Madison, WI.

Parsons, J. 1987. A water management model for agricultural drainage districts. Ph.D. thesis. North Carolina State Univ., Raleigh.

Pavelis, G.A. 1987. Economic survey of farm drainage. p. 110-136. In G.A. Pavelis (ed.) Farm drainage in the United States. USDA-ERS Misc. Publ. 1455:110-136.

Pendleton, J.W., and D.B. Egli. 1969. Potential yield of corn as affected by planting date. Agron. J. 61:70-71.

Penman, H.L. 1948. Natural evaporation from open water, bare soil and grass. Proc. R. Soc. London, A. 193:120-145.

Penman, H.L. 1956. Evaporation—an introductory survey. Neth. J. Agric. Sci. 4:9-29.

Philip, J.R. 1957. The theory of infiltration: 4. Sorptivity and algebraic infiltration equations. Soil Sci. 84:257-264.

Rawls, W.J., D.L. Brakensiek, and N. Miller. 1983. Green-Ampt infiltration parameters from soils data. J. Hydraul. Eng. 109(1):62-69.

Reeves, M., and E.E. Miller. 1975. Estimating infiltration for erratic rainfall. Water Resour. Res. 11(1):102-110.

Richards, L.A. 1931. Capillary conductivity of liquid through porous media. Physics 1:318-333.

Richardson, C.W., and D.A. Wright. 1984. WGEN: a model for generating daily weather variables. USDA-ARS, ARS-8.

Robbins, K.D., and R.W. Skaggs. 1986. Simulated weather parameters for DRAINMOD. ASAE Paper 86-2550, ASAE, St. Joseph, MI.

Rogers, J.S. 1985. Water management model evaluation for shallow sandy soils. Trans. ASAE 28(3):785-790.

Schwab, G.O., N.R. Fausey, and D.W. Michener. 1974. Comparison of drainage methods in a heavy textured soil. Trans. ASAE 17(3):424-425, 428.

Seymour, R.M. 1986. Corn yield response to planting date in eastern North Carolina. M.S. thesis. North Carolina State Univ., Raleigh.

Shaw, R.H. 1974. A weighted moisture stress index for corn in Iowa. Iowa State J. Res. 49(2):101-114.

Shaw, R.H. 1976. Moisture stress effects on corn in Iowa. Iowa State J. Res. 50(4):335-343.

Shaw, R.H. 1978. Calculation of soil moisture and stress conditions in 1976 and 1977. Iowa State J. Res. 53(2):119-127.

Shirmohammadi, A., and R.W. Skaggs. 1984. Effect of soil surface conditions on infiltration for shallow water table soils. Trans. ASAE 27(6):1780-1787.

Sieben, W.H. 1964. Het verban tussen ontwatering en opbrengst bij de jonge zavalgronden in de Noordoostpolder. Van Zee tot Land. 40, Tjeenk Willink V, Zwolle, Netherlands.

Skaggs, R.W. 1974. The effect of surface drainage on water table response to rainfall. Trans. ASAE 17(3):406-411.

Skaggs, R.W. 1975a. Drawdown solutions for simultaneous drainage and evapotranspiration. J. Irrig. Drain. Div. Am. Soc. Civ. Eng. 101(IR4):279-291.

Skaggs, R.W. 1975b. A water management model for high water table soils. ASAE Paper No. 75-2524, ASAE, St. Joseph, MI.

Skaggs, R.W. 1978. A water management model for shallow water table soils. Univ. of North Carolina Water Resour. Res. Inst. Tech. Rep. 134.

Skaggs, R.W. 1982. Field evaluation of a water management simulation model. Trans. ASAE 25(3):666-674.

Skaggs, R.W., N.R. Fausey, and B.N. Nolte. 1981. Water management evaluation for North Central Ohio. Trans. ASAE 24(4):922-928.

Skaggs, R.W., S. Hardjoamidjojo, E.H. Wiser, and E.A. Hiler. 1982. Simulation of crop response to surface and subsurface drainage systems. Trans. ASAE 25(6):1675-1678.

Skaggs, R.W., and Y.K. Tang. 1976. Saturated and unsaturated flow to parallel drains. J. Irrig. Drain. Div.. Am. Soc. Civ. Eng. 102(IR2):221-238.

Smith, R.E. 1972. The infiltration envelope: Results from a theoretical infiltrometer. J. Hydrol. (Amsterdam) 17:1-21.

Smith, R.E. 1985. Simulation of draintile flow in CREAMS2. p. 400. *In* Proc. Natural Resour. Modeling Symp., Pingree Park, CO. 16-21 Oct. 1983. USDA-ARS Publ. ARS-30. U.S. Gov. Print. Office.

Smith, M.C., R.W. Skaggs, and J.E. Parsons. 1985. Subirrigation system control for water use efficiency. Trans. ASAE 28(2):489-496.

Soil Conservation Service (SCS). 1985. DRAINMOD user's manual. Interim Tech. Release. USDA-SCS,

Stevens, J.C., and E.H. Stewart. 1963. A comparison of procedures for computing evaporation and evapotranspiration. p. 123-133. *In* Intl. Assoc. Sci. Hydrol., Intl. Union of Geod. and Geophys. Publ. 62.

Thornthwaite, C.W. 1948. An approach toward a rational classification of climate. Geogr. Rev. 38:55-94.

van Bavel, C.H.M. 1966. Potential evaporation: the combination concept and its experimental verification. Water Resour. Res. 2(3):455-467.

van Schilfgaarde, J. 1965. Transient design of drainage systems. J. Irrig. Drain. Div. Am. Soc. Civ. Eng. 91(IR3):9-22.

van Schilfgaarde, J. 1974. Nonsteady flow to drains. *In* J. van Schilfgaarde (ed.) Drainage for agriculture. Agronomy 17:245-270.

Wesseling, J. 1974. Crop growth and wet soils. *In* J. van Schilfgaarde (ed.) Drainage for agriculture. Agronomy 17:7-32.

Wiser, E.H., R.C. Ward, and D.A. Link. 1974. Optimized design of a subsurface drainage system. Trans. ASAE 17(1):175-178.

Young, T.C., and J.T. Ligon. 1972. Water table and soil moisture probabilities with tile drainage. Trans. ASAE 15(3):448-451.

11 Soil Evaporation and Transpiration

R. J. HANKS

Utah State University
Logan, Utah

The successful modeling of evaporation and transpiration has been accomplished by many authors (see Molz, 1981 for a recent review). Evaporation involves only water loss into the atmosphere at the soil surface. However, transpiration involves water extraction within the soil wherever there are roots and subsequent loss of water through the plant stems and leaves into the atmosphere. Thus, transpiration is a much more complicated process than evaporation because a biological system is involved and water flow through the plant is much different than through the soil. These two processes have almost always been considered together and only evapotranspiration has been considered as a single entity. This is undoubtedly because it is very difficult, under field conditions, to differentiate between the two processes. Modeling these processes has emphasized their differences so that they should be considered separately.

This chapter will treat a model of these processes in quite simple terms, particularly with regard to plant root extraction. Chapter 12 (Campbell) will consider plant root extraction in detail. There are many situations in which it may be sufficient to simply measure or estimate evapotranspiration (ET) using procedures such as those given by Doorenboos and Pruitt (1977). It may even be possible to estimate ET from crop yield measurements using simple models like that of Stewart et al. (1977), which takes advantage of the strong linear relation between ET and yield (especially dry matter yield). Stewart's relation can be written as

$$Y/Ym = 1 - A (1 - ET/ETm) \qquad [1]$$

where ETm is maximum ET, Y is yield, Ym is maximum yield measured under the conditions of ETm, and A is a calibration constant. This equation assumes that there are no other factors limiting yield except water stress. Equation [1] assumes that yield is measured and evolved from a semi-empirical process of curve fitting. Hanks (1983) has discussed this approach in some detail and has shown that it can be justified on a rational basis.

Nevertheless, there is still the problem of estimating ET, and more precisely evaporation (E) and transpiration (T). This process is complicated because both E and T are influenced by many soil, climate, and crop factors, as well as soil and water management practices.

Copyright © 1991 ASA-CSSA-SSSA, 677 S. Segoe Rd., Madison, WI 53711, USA. *Modeling Plant and Soil Systems*—Agronomy Monograph no. 31.

I. MODELING EVAPORATION

Evaporation of water from a bare soil can be estimated using the model of Ch. 9 (Hanks) because the general equation of flow is the same as infiltration and redistribution. The only difference is the boundary conditions; instead of water being applied at the soil surface, water is being removed. Thus, a potential E rate (Ep) is needed to determine how much water could be lost from the soil if water was freely available. If water cannot be moved to the soil surface fast enough to meet this potential demand, then E < Ep. If the Ep was constant with time, then E would be characterized by a constant-rate period and then by a falling-rate period, as discussed widely in general texts on the subject. Figure 11-1 shows such a computation made with the model. The classical constant rate is shown, C, for about the first 24 h, after which the evaporation rate falls. The falling rate period started when the soil surface had dried to the lowest possible value (the air-dry water content), at which time the boundary condition changed from a constant flux to a constant matric head at the soil surface.

However, under natural conditions Ep is not constant, but varies with time during the day. Using the energy balance equation (for a bare soil), this rate can be estimated from

$$Ep = Rn - S + H \qquad [2]$$

where Rn is net radiation, S is soil heat flow, and H is sensible heat exchange with the atmosphere. Thus, Ep is related to soil surface and climatic charac-

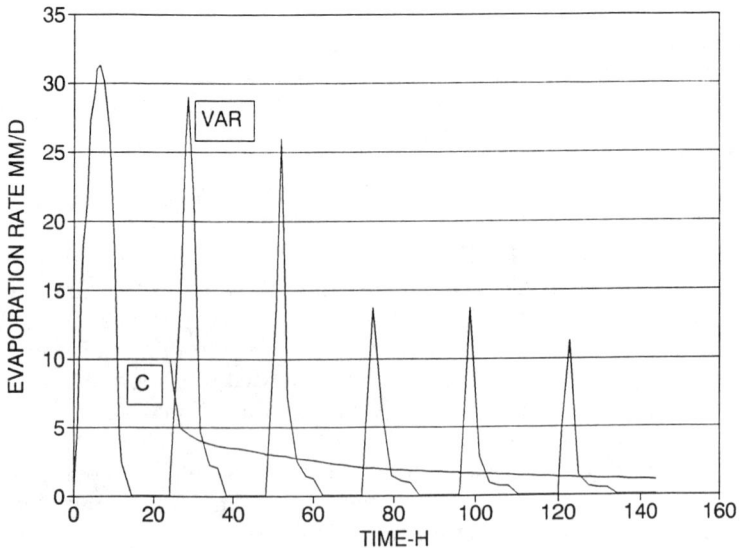

Fig. 11-1. Evaporation rate simulated for a constant rate (C) compared with a variable rate (VAR) during the day and zero rate at night.

teristics because they influence *Rn*. The Ep is also related to soil temperature, since temperature is related to *S*. Climatic variables are also involved because air temperature and wind influence *H*. Climatic factors also influence Ep because precipitation influences soil water content, which in turn influences *Rn* and *S*. The dominant factor in Eq. [2] is *Rn*. A good approximation of Ep can often be obtained from an estimation of *Rn*. However, *Rn* is closely related to solar radiation and changes greatly during a single day. The model described in this chapter incorporates this diurnal variation by a sinusoidal approximation assuming water loss only during daylight hours. The sinusoidal approximation was devised because most climatic data is measured only on a daily basis or longer. Figure 11-1 also shows this E simulation using the model listed in this chapter (see Appendix) for the same initial conditions as for the steady flow. This simulation shows neither a distinct constant rate nor a distinct falling rate. These attributes were overwhelmed by the diurnal variation of Ep. The peak rates do fall from day to day until the fourth day, after which the peak rate tends to level out. However, daily E falls throughout the period, as could be determined from the area under the rate curve for each day. The peak rate was high early in the morning because the soil water flow to the surface occurred during the night, causing the soil surface to rise above the lower air-dry limit. Thus, this simulation, which is certainly more realistic than assuming a constant rate during day and night, is an indication that the classical constant rate and falling rate stages of E apply only to laboratory-type conditions and not to natural field conditions. However, if E had been plotted versus cumulative Ep rather than time, the constant rate and falling rate stages would be apparent.

For many purposes, it may be sufficient to know only how much E occurs over a period of several days. Figure 11-2 shows cumulative E for the

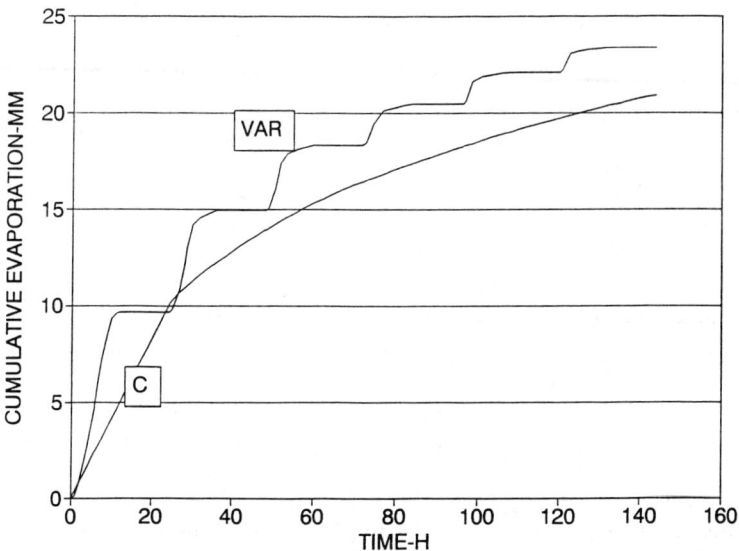

Fig. 11-2. Cumulative evaporation rate simulated for a constant rate (C) compared with a variable rate (VAR) during the day and zero rate at night.

situation demonstrated by Fig. 11-1, which shows that, after 10 d, cumulative E was slightly higher for a varying Ep (23 mm) compared with a constant surface boundary condition for E (21 mm). Thus, for many purposes the constant rate approach may be useful.

The part of the BASIC program that converts a constant input rate into a sinusoidal rate is given in statements 4350 to 4570 of the program listing (see Appendix).

Evaporation is also influenced by plant cover. If plants are present, Ep is changes and water is extracted as T, so soil water content and water flow to the surface is influenced.

II. MODELING TRANSPIRATION

Modeling of T is much more complex than E because plants extract water throughout the soil profile, and because plant roots and aboveground portions of the plant may be growing with time, thus changing the surface boundary condition. The model shown in this chapter takes care of these additional complications in a rather simple, but by no means complete way. The basic equation of one-dimensional water flow, shown in Ch. 9, which is sufficient for infiltration, redistribution, and E, must be modified to account for root extraction as follows (Nimah & Hanks, 1973):

$$\frac{\partial \theta}{\partial t} = \frac{\partial}{\partial x}\left(K_\theta \frac{\partial H}{\partial z}\right) + A(z,t) \qquad [3]$$

where θ is water content, t, is time, z is depth, K_θ is hydraulic conductivity, H is soil hydraulic head (sum of matric and gravity heads), and $A(z,t)$ is the root-extraction term. The root-extraction term is defined as follows:

$$A(z,t) = \frac{[H\text{root} + Rz - h(z,t)] \text{ RDF}(z,t) K_\theta}{\Delta x \, \Delta z} \qquad [4]$$

where Hroot is an effective root water potential at the soil surface, Rz is a correction for the root water potential at other soil depths, $h(z,t)$ is the matric head, RDF(z,t) is a root density function, x is an arbitrary distance from the root surface (assumed 10 mm) to the point in the soil, where $h(z,t)$ is a mean value and Δz is the size of the depth increment.

The root-extraction term is dynamic because it is a biological interface between climatic and soil factors, both of which are dynamic. The simplified RDF(z,t) function used is determined from the relative mass of live roots measured, as discussed by Nimah and Hanks (1973).

The model also allows for a growing plant above- and belowground for the purpose of estimating plant cover. The aboveground growth is estimated in statement 1170 (see Appendix), which serves the purpose of splitting ETp into Ep and Tp. The variable ESTART gives the time in days when the crop may have grown enough for transpiration to be significant. For a crop like corn (*Zea mays* L.) in Utah, this value is about 30 d. The crop then grows as a function of time to reach maximum T by the time designated by ESTOP. Thereafter, Ep is given as ETp $(1 - AK1)$. Typical values of $AK1$ for a crop like corn would be 0.9. If $AK1 = 0.9$ and ETp = 7.0 mm/d, then Tp would be 6.3 mm/d and Ep would be 0.7 mm/d (since ETp = Ep + Tp). Figure 11-3 shows how Ep and Tp vary during the season for these conditions. In the program, the input data of ETp are read into Array V and converted to Array TET (ETp) and V (Ep). Within the program, Tp is computed as the difference between ETp and Ep. If a perennial crop like alfalfa (*Medicago sativa* L.) or grass is used, then ESTART and ESTOP are set to be very small values so the plant cover is constant throughout the period. statement 1300 (see Appendix) prints out the time variation of Ep and ETp.

The program also provides for a root growth function, as shown in statements 1500 to 1750. The values for the fractional amount of roots in each soil layer at maximum growth are input data and contained in Array RDFSAV. This general root pattern is imposed as a function of a given time according to the equation on statement 1560. The variable RDFDAY gives the days to maximum root depth.

Root extraction as a function of depth and time are found (statements 2450 to 2880). The purpose of the section is to find a value of *H*root that matches the interface constraints. The constraints are:

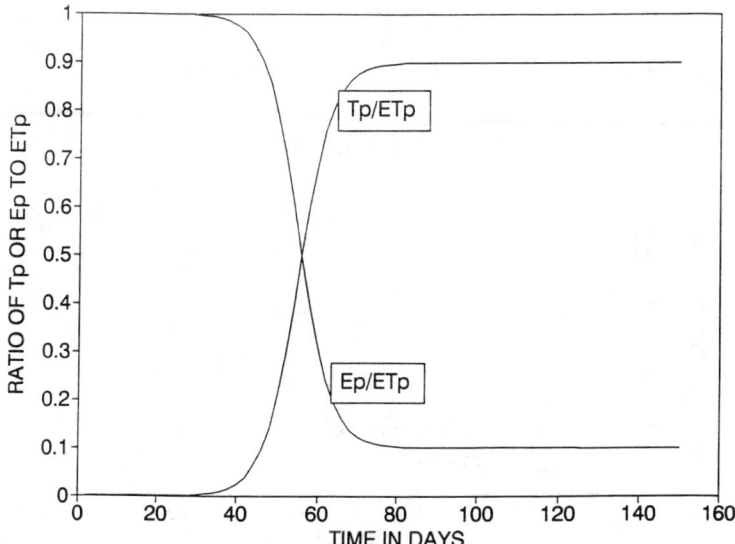

Fig. 11-3. Computation of the ratio of potential evaporation (Ep) and potential transpiration (Tp) to potential evapotranspiration (ETp) as a function of time for an annual crop such as corn.

$$T = Tp \quad \text{if } H\text{root} > H\text{low} \qquad [5]$$

$$T = \text{value provided by Eq. [4] for } H\text{root} = H\text{low} \qquad [6]$$

Thus, this model assumes that plants have some mechanism whereby the root water potential will fall only to some low value and no lower, thus simulating stomatal closure. It is also assumed that water cannot flow out of the plant into the soil. The Hroot value is hunted for, subject to the above constraints, because there are more unknowns than equations.

Climatic factors that influence ETp, and thus Tp, influence the value of Hroot through Eq. [5]. Soil factors influence the Hroot value because Eq. [4] must be solved for each layer and integrated with all other layers subjected to the above constraints. The model finds an extraction value for each depth, which is then used in the solution of the general flow equation (statements 2890 to 3220).

The rest of the program is devoted to finding the new values of water content from the matric potential just computed (statements 3230 to 3640), preparing data for output, and initializing the recently computed values for the next computation.

Figure 11-4 shows a simulation for a crop like alfalfa, starting with a partially wet soil profile as shown in the input file. Cumulative T and E are shown to vary during each day, but for the first 3 d the daily rate of T is essentially constant. Figure 11-5 shows the value of Hroot computed for this same period. The value of Hroot (suction) that was reached on the third

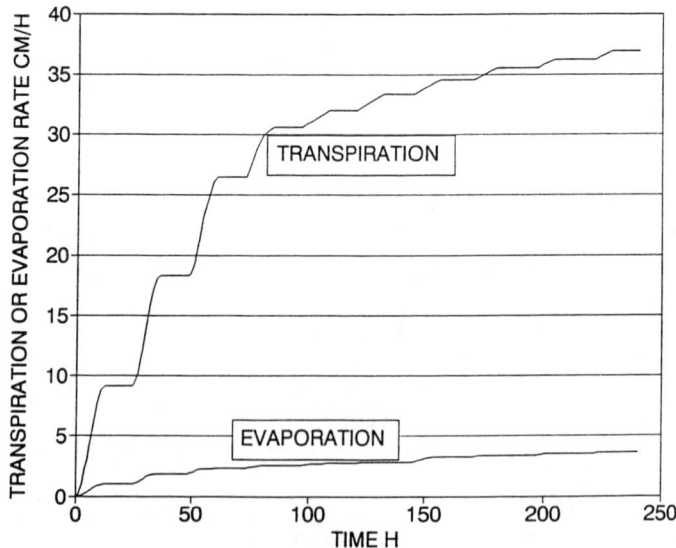

Fig. 11-4. Cumulative evaporation and transpiration of a crop with complete cover (such as alfalfa) as related to time, for conditions of the input file.

day, varied with time during the day, and generally increased. After the third day, T was limited during the day. The nighttime values of *H*root generally increased with time, which would reflect the changes in soil matric suction as the soil dried.

The computations illustrate the difference between E and T. As shown in Fig. 11-1, when the soil was bare and E alone was occurring, E dropped off after 24 h, whereas T did not drop off until about 70 h.

Part of the difficulty with modeling T is related to the boundary conditions. The simplifications used in this model were useful for some situations, but there is no provision for decreasing T as the plant matures or is harvested (alfalfa). Under these conditions, it is often useful to compute Tp and Ep external to the model for the desired situation and include this as input data. Retta and Hanks (1980) also show a simplified method for making these computations for wheat (*Triticum aestivum* L.).

The question of how to go about splitting ETp into Ep and Tp is still open to question. The approximation used in the models is similar to the one described herein is discussed by Childs and Hanks (1975). The crop coefficient curves developed over the years for many crops and various locations (Doorenbos & Pruitt, 1977; Jensen, 1973) are used as base information. An estimation of Tp is then made 5 to 10% below the crop coefficient curve when the crop is actually growing. Figure 11-6 illustrates how this was done for corn by Stewart et al. (1977). While this approach seems somewhat arbitrary, our experience has shown that it works quite well.

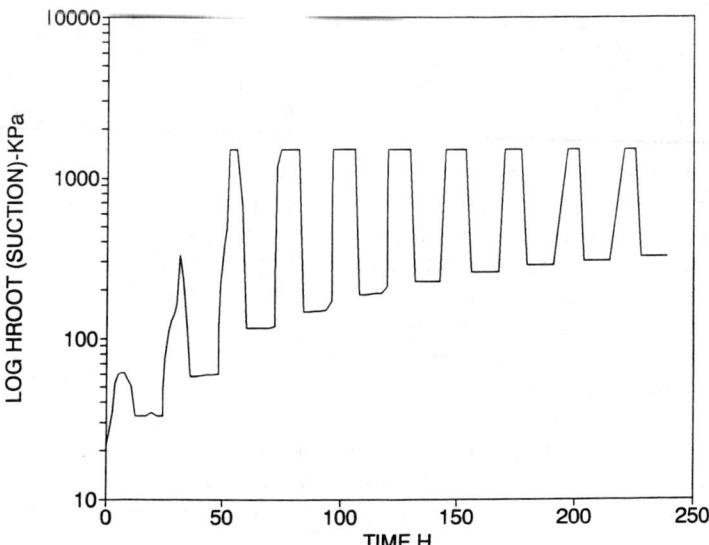

Fig. 11-5. Simulated root water suction of a crop with complete cover (such as alfalfa) as related to time, for conditions of the input file.

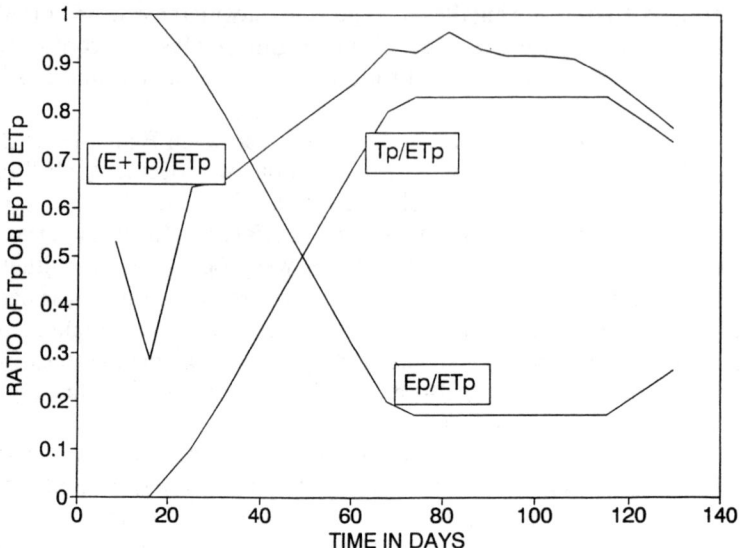

Fig. 11-6. Ratios of potential evaporation (Ep) and transpiration (Tp) to potential evapotranspiration (ETp) deduced from the crop coefficient curves of Pruitt in Davis, CA, but applied to Logan, UT conditions. Data is from Stewart et al. (1977).

III. MODEL VERIFICATION

This model has been verified under several situations. The first attempt was made by Nimah and Hanks (1973) for oat (*Avena sativa* L.) over a short period and alfalfa over an entire season.

Model verification for the oat study shows an interesting value of the models. The field study showed quite different soil water profiles in a lysimeter and in the adjacent field sometime after irrigation. Imposing the boundary and initial conditions of the lysimeters and the adjacent field, a model run indicated considerable water flow upward from the water table into the crop root zone in the field, but nothing in the lysimeter (since there was no flow at the bottom of the lysimeter). This simulation matched the measured conditions well, as shown in Table 11-1. Thus, the model helped to alert us to a condition not previously considered.

Full-season comparisons of measured and computed soil water content were also performed for alfalfa. These data, shown in Fig. 11-7, indicate very good agreement between measured and simulated data across the entire season. However, a more detailed look at the data indicates the best agreement immediately after irrigation and the poorest agreement just before irrigation. There were no significant accumulated errors during the season.

This model has also been tested for various irrigation levels where salinity has also been a problem. If salinity is a problem, further modifica-

SOIL EVAPORATION AND TRANSPIRATION

Table 11-1. Comparison of simulated and measured evapotranspiration (ET) and upward water flow (Nimah & Hanks, 1973).

	Root depth assumed	ET	Upward flow
		mm	
Simulation	300	49	22
	450	58	23
Measured	†	53	21

† No measurements made of root density.

tions are needed, as discussed by Hanks (1984). A comparison of measured and computed ET is shown in Fig. 11-8, where ET was limited by low soil water and/or excess salinity. There is good agreement between measured and simulated ET.

A similar model has also been developed (Hanks, 1974) that uses the same procedures to split ETp into Tp and Ep, but that uses a more simple soil water flow computation. A field capacity concept of water flow is used so computations are made on a daily basis only. This model assumes root extraction from only one or two layers on any one day on the basis of relative water content only. Thus, much less input data is required. This simpler model provides very similar predictions to the more detailed model described herein, provided there is no upward flow, salinity is not a problem, and long-time periods are involved.

The illustrated simulations are only a small part of the information made possible with this model (or similar models). Such information as soil water content, soil water flux, and root extraction as a function of time and depth are computed and could be printed out if necessary. Also, many other types of problems could be simulated, such as water flow upward (or drainage) from a water table. Many of these other problems are illustrated in the references cited.

The details of the use of this model are given in the program listing, and the input and output file listings (see Appendix). The first 13 lines of the input file are the same as needed for infiltration and redistribution so they are discussed in Ch. 9. Line 14 lists the constants associated with E and T. Line 15 gives the data for the root density functions used in Array RDFSAV.

The output file, discussed in Ch. 9, contains most of the data that applies to this chapter. In addition, the output file herein shows the constants, related to E and T, included on Line 14 of the input file, as well as the RDFSAV array. There are also several columns of data different from Ch. 9; TRAN is cumulative transpiration, EVAP is cumulative evaporation, EPOT is instantaneous E rate, TRRAT is instantaneous T rate, and HROOT is the root water potential.

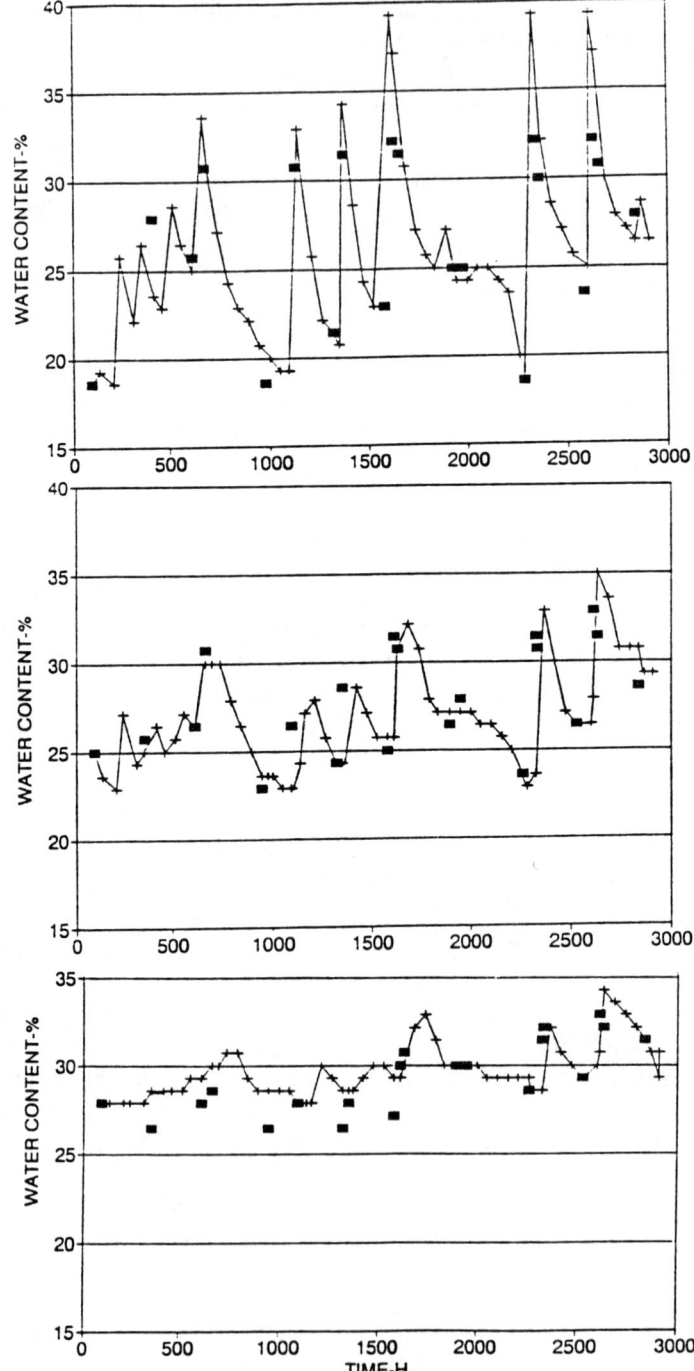

Fig. 11-7. Comparison of predicted, +, and measured, ■, soil water content for a season as predicted by the model (Nimah & Hanks, 1973).

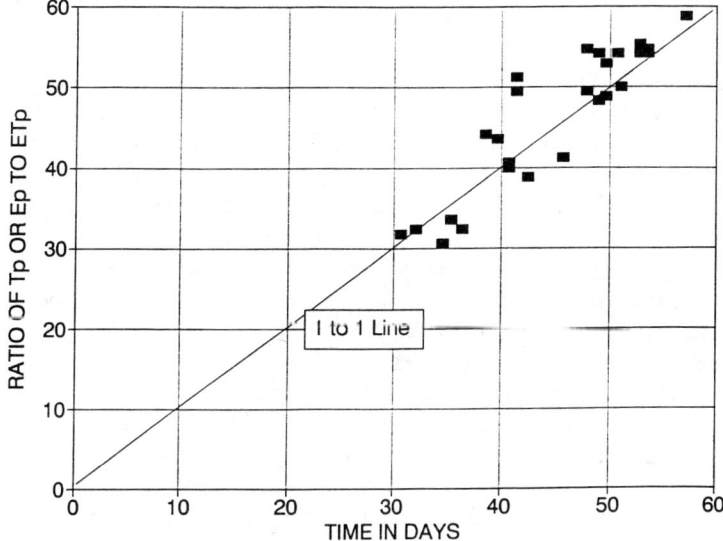

Fig. 11-8. Comparison of predicted and measured evapotranspiration (ET) for various irrigation and salinity levels. Data is from Stewart et al. (1977).

IV. CRITICAL ASSUMPTIONS

This model has many limitations. Other than the probable oversimplification of the root relations, the soil hydraulic properties that are required are the most serious limitations. The model assumes a homogeneous soil in which the basic soil properties change only with water content. Experience has shown that most soils go through very significant changes in soil properties with time upon wetting and drying, especially at the soil surface. This is a particular problem with infiltration and soil E estimates. Future modifications are planned, based on field research, to account for these changes. Another critical assumption is the procedure used to split ETp into Ep and Tp. Experience has shown that some adjustments have to be made when the model is used in a different climatic location.

V. COMPUTER REQUIREMENTS

The program as shown has been tested using GW-BASIC and QUICK-BASIC (Microsoft Corp., Redmond, WA). The programs that are shown required about 1 min to run using QuickBASIC and an IBM compatable computer (AT) with a mathematical coprocessor. Simulation of a complete season of about 120 d takes about 30 min, depending on how many times the boundary conditions change. The program required <20K bytes of memory.

VI. SUMMARY

The model described in this chapter was designed, in addition to that in Ch. 9, to include the effects of root extraction of water and loss of the water by T through the leaves. The model is fairly general in that infiltration and redistribution, as well as E and T, can be simulated. Data needed are basic soil properties such as hydraulic conductivity and matric potential-water-content relations. Crop properties, such as the amount of plant cover, are needed to estimate the potential water loss from the soil as E and the loss of water through the plant as T. Transpiration is also dependent on the root proliferation within the profile. Predictions made with the model have agreed with field measurements for several studies.

Critical assumptions include a homogeneous soil, both in depth and time, and known boundary conditions (especially at the soil surface). Root extent and water adsorption are considered in a very simplified form. The model is conservative in that errors do not increase with time for many situations. The user is cautioned to check model predictions against field-measured data as much as possible. The soils and crop data needed are best measured under field conditions. This model should only be used if the user has enough field experience so that unreasonable results can be detected. Thus, the user of this model is advised to keep "one foot in the field."

VII. APPENDIX

PROGRAM LISTING

```
10 REM --Model of evaporation and transpiration
20 REM  IER twice the No. of surfaces flux changes.
30 REM  K the number of soil depth increments.
40 REM  NB number of depth increments to compute
*  flux-usually=K.
50 REM  DETT initial and smallest time increment.
60 REM  CONQ largest water content change allowed.
70 REM  TAA used to tell the condition of the
*  bottom boundary-if TAA=1 flux=0, if  TAA=0
*  then constant mat head.
90 REM  TIME is cumulative time.
100 REM  TT 1.0 for Lassonen and 0.5 for
*  Crank-Nicholson methods.
110 REM  CUMT time at end- time to stop.
120 REM  DELW water content increment vs T(I)
130 REM  HDRY matric head corresponding to WATL.
140 REM  HWET matric head of sat. water content.
150 REM  WATL lowest water content-air dry.
160 REM  WATH saturated water content.
170 REM  HLOW lowest root water head allowed.
```

SOIL EVAPORATION AND TRANSPIRATION

```
180 REM   HHI highest root water head allowed.
190 REM   RDFDAY time-days-for max root depth.
200 REM   RDFDEL No. hours between recomputing root
*   depths.
210 REM   ESTART time-days-when above ground growth
*   starts.
220 REM   ESTOP time-days-when above ground growth
*   is maximum.
230 REM   AK1 Tpot/ETpot when time > ESTOP.
240 REM   DD(I) depth array from top of profile.
250 REM   B(I) hydraulic conductivity vs depth.
260 REM   C(I) water capacity vs depth.
270 REM   W(I) ending soil water content vs depth.
280 REM   Y(I) begin soil water content vs depth.
290 REM   H(I) ending matric head vd depth.
300 REM   G(I) begin matric head vs depth.
310 REM   V(I) surface boundry potential flux
*   followed by time; + for irrigation or rain
*   - flux for ET.
330 REM   T(I) array of equal spaced water content.
340 REM   P(I) matric head vs T(I).
350 REM   E(I) hydraulic conductivity vs T(I).
360 REM   F(I) temporary array used.
370 REM   D(I) sum of diffusivity x DELW.
380 REM   TET(I) ETpot array.
390 REM   RDFSAV(I) mature root density vs depth.
400 REM   RDF(I) root depth array vs depth.
410 DIM A(25), B(25), C(25), F(25F), DD(25),V(50)
415 DIM V(50), P(55), E(55), W(25), D(55), T(55)
420 DIM H(25), G(25), Y(25), TET(25), RDFSAV(25)
425 DIM ROOT(25), RDF(25)
430 OPEN "evaptran.DAT" FOR INPUT AS #1
440 OPEN "evaptran.OUT" FOR OUTPUT AS #2
450 DEFINT I-N
460 INPUT #1, LAB$
470 INPUT #1, K, IER, NB, ND, KCPMAX
480 KK = K + 1
490 IR = INT(IER / 2)
500 FOR I = 1 TO IER: INPUT #1, V(I): NEXT I
510 FOR I = 1 TO KK: INPUT #1, DD(I): NEXT I
520 FOR I = 1 TO ND: INPUT #1, P(I): NEXT I
530 FOR I = 1 TO ND: INPUT #1, E(I): NEXT I
540 PRINT #2, LAB$, DATE$, TIME$
550 PRINT #2, "   K   IER   NB   ND   KCPMAX"
560 PRINT #2, USING " ####"; K; IER; NB; ND;
*   KCPMAX
570 T(1) = 0
580 FOR I = 1 TO KK: INPUT #1, W(I): NEXT I
590 INPUT #1, DETT, CONQ, TAA, TIME, TT, CUMT
```

```
600 IF TAA < 1 THEN ITAA = 0
610 IF TAA >= 1 THEN ITAA = 1
620 INPUT #1, HDRY, HWET, WATL, WATH, HLOW, DELW
630 INPUT #1, RDFDAY, RDFDEL, ESTART, ESTOP, AK1,
 *  RRES, HHI
640 FOR I = 1 TO KK: INPUT #1, RDFSAV(I): NEXT I
650 D(1) = (E(1) * (P(2) - P(1)))
660 CLOSE 1
670 FOR I = 2 TO ND
680 D(I) = E(I) * (P(I) - P(I - 1)) + D(I - 1)
690 T(I) = DELW + T(I - 1)
700 NEXT I
710 PRINT #2, "WAT CONT  MAT HEAD  CONDUCTI
 *  DIFFUSIV  WAT CONT  MAT HEAD  CONDUCTI
DIFFUSIV"
720 NE = ND / 2
730 FOR I = 1 TO NE
740 J = NE + I
750 PRINT #2, USING "#.##^^^^  "; T(I); P(I);
 *  E(I); D(I); T(J); P(J); E(J); D(J)
760 NEXT I
770 KC = 1
780 AK4 = .5 / RDFDAY
790 DELDAY = RDFDAY * 24 / RDFDEL
800 RDXDAT = DELDAY
810 HROOT = HLOW
820 CWFLX = 0
830 DELT = DETT
840 TM = 1 - TT
850 TBB = 1 - TAA
860 YMAX = WATH
870 RUNOF = 0
880 CUMS = 0
890 RPI = 0
900 CUMB = 0
910 CUMM = 0
920 IRDF = 0
930 EVAP = 0
940 SIR = 0
950 CTRAN = 0
960 PIT = 0
970 PIEB2 = 6.2832
980 J = INT((W(1) - T(1)) / DELW) + 1
990 H(1) = (P(J + 1) - P(J)) * (W(1) - T(J)) /
 *  DELW + P(J)
1000 G(1) = H(1)
1010 C(1) = DELW / (P(J + 1) - P(J))
```

SOIL EVAPORATION AND TRANSPIRATION

```
1020 FOR I = 2 TO K
1030 PIT = W(I) * (DD(I + 1) - DD(I - 1)) / 2 +
*  PIT
1040 NEXT I
1050 FOR I = 2 TO KK
1060 J = INT((W(I) - T(1)) / DELW) + 1
1070 H(I) = (P(J + 1) - P(J)) * (W(I) - T(J)) /
*  DELW + P(J)
1080 C(I) = DELW / (P(J + 1) - P(J))
1090 G(I) = H(I)
1100 NEXT I
1110 PRINT #2, "  DEPTH   C(I)     W(I)     H(I)
*  RDFSAV(I)"
1120 FOR I = 1 TO KK
1130 IF I = 1 THEN 1150
1140 PRINT #2, USING " ###.   #.##^^^^   #.###
*  #.###^^^^  #.###"; DD(I); C(I); W(I); H(I);
*  RDFSAV(I)
1150 Y(I) = W(I)
1160 NEXT I
1170 REM*COVER GROWTH LOOP
1180 AK3 = .5 / (ESTOP - ESTART)
1190 PRINT #2, " TIME END   SOIL FLUX   ET FLUX"
1200 FOR I = 2 TO IER STEP 2
1210 IR = I / 2
1220 IF V(I - 1) < 0 THEN 1250
1230 TET(IR) = 0
1240 GOTO 1300
1250  TET(IR) = V(I - 1)
1260 IF V(I) / 24 < ESTART THEN 1290
1270 V(I - 1) = TET(IR) - TET(IR) * AK1 / (1 +
*  EXP(6 - AK3 * (V(I) - ESTART * 24)))
1280 GOTO 1300
1290  V(I - 1) = TET(IR)
1300 PRINT #2, USING " #.###^^^^ "; V(I); V(I-1);
*  TET(IR)
1310 NEXT I
1360 WFDD = V(1)
1370 EOR = V(1)
1380 PRINT #2, "   DETT    CONQ     TAA    TIME
*  TT      CUMT    DELW"
1390 PRINT #2, USING "###.#### "; DETT; CONQ; TAA;
*  TIME; TT; CUMT; DELW
1400 PRINT #2, "   HDRY    HWET     WATL
*  WATH    HLOW    HHI"
1410 PRINT #2, USING "#.###^^^^ "; HDRY; HWET;
*  WATL; WATH; HLOW; HHI
```

```
1420 PRINT #2, "      RDFDAY      RDFDEL      ESTART
*    ESTOP       AK1         "
1430 PRINT #2, USING "#.###^^^^ "; RDFDAY; RDFDEL;
*    ESTART; ESTOP; AK1
1440 KCK = 1
1450 HROOT = G(2)
1460 PRINT #2, "      TIME        CWF         IRR+RAIN
*    TRAN        EPOT        EVAP        WATBAL      TRRAT
*    HROOT"
1470 VIEW PRINT 1 TO 2
1480 PRINT "      TIME        CWF         IRR+RAIN      TRAN
*    EPOT        EVAP        WATBAL      TRRAT  "
1490 VIEW PRINT 2 TO 24
1500 REM ROOT GROWTH LOOP
1510 IF IRDF = 1 THEN 1770
1520 IF ABS(RDFDEL) < .000001 THEN 1730
1530 IF TIME < RDXDAY THEN 1770
1540 IF TIME > DELDAY * RDFDEL THEN 1730
1550 RDXDAY = DELDAY + RDXDAY
1560 DROOT = DD(KK) / (1 + EXP(6 - AK4 * TIME))
1570 J = 2
1580 FOR I = 2 TO KK
1590 RDF(I) = 0
1600 IF J >= KK THEN 1710
1610 ROOT(J) = DROOT * DD(J) / DD(KK)
1620 IF ROOT(J) >= DD(I) THEN 1670
1630 RDF(I) = RDFSAV(J) * (ROOT(J) - DD(I - 1)) /
*    (ROOT(J) - ROOT(J - 1)) + RDF(I)
1640 IF (ROOT(J - 1) > DD(I - 1)) THEN RDF(I) =
*    RDFSAV(J) * (1 - ((ROOT(J) - DD(I - 1)) /
*    (ROOT(J) - ROOT(J - 1)))) + RDF(I)
1650 J = J + 1
1660 GOTO 1600
1670 RDF(I) = (DD(I) - DD(I - 1)) / (ROOT(J) -
*    ROOT(J - 1)) * RDFSAV(J) + RDF(I)
1680 IF (ROOT(J - 1) > DD(I - 1)) THEN RDF(I) =
*    RDF(I) - (ROOT(J - 1) - DD(I - 1)) / (ROOT(J) -
*    ROOT(J - 1)) * RDFSAV(J)
1690 IF ROOT(J) > DD(I) THEN 1710
1700 J = J + 1
1710 NEXT I
1720 GOTO 1770
1730 FOR I = 1 TO KK
1740 RDF(I) = RDFSAV(I)
1750  NEXT I
1760 IRDF = 1
1770 BOT = WATL
1780 TOP = WATH
```

```
1790 REM*COMP. CONDUCT. (B) AND WATER CAPACITY (C)
1800 HKP = H(1)
1810 WKP = W(1)
1820  IF EOR > 0 THEN 1860
1830  W(1) = WATL
1840  H(1) = HDRY
1850  GOTO 1880
1860  W(1) = WATH
1870  H(1) = HWET
1880  TWW = (W(1) + Y(1)) * .5
1890  IF (TWW > WATH) THEN TWW = WATH
1900  J = INT((TWW - T(1)) / DELW) + 1
1910  BB = (TWW - T(J)) / DELW
1920  DIFFA = (D(J + 1) - D(J)) * BB + D(J)
1930  HI = (P(J + 1) - P(J)) * BB + P(J)
1940  FOR I = 1 TO K
1950  TW = (W(I + 1) + Y(I + 1)) * .5
1960  J = INT((TW - T(1)) / DELW) + 1
1970  BB = (TW - T(J)) / DELW
1980  DIFFB = (D(J + 1) - D(J)) * BB + D(J)
1990  GI = (P(J + 1) - P(J)) * BB + P(J)
2000 IF ABS(HI - GI) < .0001 THEN 2350
2010  B(I) = (DIFFA - DIFFB) / (HI - GI)
2020  IF I > 1 THEN 2370
2030  ER = (B(1) * (H(1) * TT - H(2) * TT - G(2) *
* TM + G(1) * TM + DD(2))) / DD(2)
2040  IF ABS(ER) > ABS(EOR) THEN 2060
2050  IF H(1) = HWET OR H(1) = HDRY THEN 2370
2060  IF ABS(1.1 * EOR - ER) - ABS(.1 * EOR) <= 0
* THEN 2090
2070  IF KCK = 1 THEN 2130
2080  IF KCK < 12 THEN 2170
2090  H(1) = (EOR * DD(2) / B(1) + H(2) * TT -
* G(1) * TM + G(2) * TM - DD(2)) / TT
2100  IF H(1) < HDRY THEN H(1) = HDRY
2110  IF H(1) > HWET THEN H(1) = HWET
2120  GOTO 2370
2130  H(1) = HKP
2140  W(1) = WKP
2150  KCK = KCK + 1
2160  GOTO 1880
2170  KCK = KCK + 1
2180  IF ER = EOR THEN 2370
2185  IF ER > EOR THEN 2230
2190  IF W(1) = WATH THEN 2370
2200  BOT = W(1)
2210  W(1) = (W(1) + TOP) * .5
2220  GOTO 2260
2230  IF W(1) = WATL THEN 2370
```

```
2240  TOP = W(1)
2250  W(1) = (W(1) + BOT) * .5
2260  J = INT((W(1) - T(1)) / DELW) + 1
2270  BB = (W(1) - T(J)) / DELW
2280  H(1) = (P(J + 1) - P(J)) * BB + P(J)
2290  TWW = (W(1) + Y(1)) * .5
2300  J = INT((TWW - T(1)) / DELW) + 1
2310  BB = (TWW - T(J)) / DELW
2320  DIFFA = (D(J + 1) - D(J)) * BB + D(J)
2330  HI = (P(J + 1) - P(J)) * BB + P(J)
2340  GOTO 2000
2350  B(I) = (D(J + 1) - D(J)) / (P(J + 1) - P(J))
2360  IF I = 1 THEN 2030
2370  TWW = TW
2380  HI = GI
2390  DIFFA = DIFFB
2400  TW = (W(I + 1) + Y(I + 1)) * .5
2410  J = INT((TW - T(1)) / DELW) + 1
2420  C(I + 1) = DELW / (P(J + 1) - P(J))
2430 NEXT I
2440  KCK = 1: KCP = 0
2450 REM ****COMPUTE TRANSPIRATION BY DEPTHS
2460  ETPL = ET
2470  IF ET >= 0 THEN 2850
2480  IF EOR > 0 THEN 2510
2490  ETPL = ET - EOR
2500  'IF ABS(ETPL) < .0001 THEN 2850
2510   HHOLD = HROOT
2520  SINK = 0
2530  DSINK = SINK
2540  FOR I = 2 TO K
2550  E(I) = G(I) - DD(I) * RRES
2560  NEXT I
2570  LCNT = 0
2580  DSAVE = DSINK
2590  DSINK = 0
2600  SINK = ETPL
2610  FOR I = 2 TO K
2620  IF HROOT - E(I) > 0 THEN 2650
2630  SINK = SINK + B(I) * RDF(I) * E(I)
2640  DSINK = DSINK + B(I) * RDF(I)
2650  NEXT I
2660  IF DSINK <> 0 THEN 2700
2670  IF HROOT = HLOW THEN 2760
2680  HROOT = HLOW
2690  GOTO 2580
2700  IF DSINK = DSAVE THEN 2760
2710  HROOT = SINK / DSINK
```

SOIL EVAPORATION AND TRANSPIRATION

```
2720 IF HROOT < HLOW THEN HROOT = HLOW
2730 LCNT = LCNT + 1
2740 IF LCNT <= 20 THEN 2580
2750 PRINT #2, "TOO MANY TIMES IN TRAN EST LOOP"
2760 SINK = 0
2770 FOR I = 2 TO K
2780 IF HROOT - E(I) > 0 THEN 2820
2790 A(I) = B(I) * 2 * RDF(I) * (HROOT - E(I)) /
*    (DD(I + 1) - DD(I - 1))
2800 SINK = SINK + RDF(I) * B(I) * (HROOT - E(I))
2810 GOTO 2830
2820 A(I) = 0
2830 NEXT I
2840 GOTO 2890
2850 FOR I = 2 TO K
2860 SINK = 0
2870 A(I) = 0
2880 NEXT I
2890 REM*WATER FLOW TRIDIAGONAL MATRIX SOLUTION
2900 FOR I = 2 TO K
2910   POT = (DD(I + 1) - DD(I - 1)) / (2 * DELT)
2920 DLXA = (DD(I) - DD(I - 1))
2930   DLXB = (DD(I + 1) - DD(I))
2940 AA = B(I - 1) / DLXA
2950 CC = B(I) / DLXB
2960   BB = C(I) * POT / TT + CC + AA
2970   DA = (C(I) * POT * G(I) + CC * (TM * (G(I +
*    1) - G(I)) - DLXB) + AA * (TM * (G(I - 1) -
*    G(I)) + DLXA) + A(I) * (DD(I + 1) - DD(I - 1))*
*    .5) / TT
2980   IF I > 2 THEN 3050
2990   IF H(1) >= HWET OR H(1) <= HDRY THEN DA = DA
*    + AA * H(1): GOTO 3020
3000   DA = DA - (AA * (TM * (G(I - 1) - G(I)) +
*    DLXA)) / TT + EOR / TT
3010   BB = BB - AA
3020   F(I) = DA / BB
3030   E(I) = CC / BB
3040   GOTO 3080
3050   IF I >= K THEN 3090
3060   E(I) = CC / (BB - AA * E(I - 1))
3070   F(I) = (DA + AA * F(I - 1)) / (BB - AA * E(I
*    - 1))
3080 NEXT I
3090   IF ITAA = 0 THEN EX = (G(I) - G(I + 1) +
*    DLXB) * B(I) / DLXB ELSE EX = 0
3100   BB = BB - CC
3110   DA = DA + CC * ((G(I) - G(I + 1)) * TM +
*    DLXB) / TT - EX / TT
```

```
3120 H(I) = (DA + AA * F(I - 1)) / (BB - AA * E(I
*  - 1))
3130 I = I - 1
3140 H(I) = E(I) * H(I + 1) + F(I)
3150 IF I > 2 THEN 3130
3160 IF ITAA = 0 THEN 3190
3170 H(KK) = H(K) + DD(KK) - DD(K)
3180 G(KK) = G(K) + DD(KK) - DD(K)
3190 IF TAA = 2 AND H(KK) >= HWET THEN ITAA = 0
*  ELSE ITAA = 1
3200 FOR I = 2 TO K
3210  IF H(I) > HWET THEN 3330
3220 NEXT I
3230 REM Compute water cont. vs matric heads.
3240 IF H(1) <= HDRY OR H(1) >= HWET THEN 3270
3250 WFDD = EOR
3260 GOTO 3310
3270 WFDD = (B(1) * (H(1) * TT - H(2) * TT -G(2)*
*  TM + G(1) * TM + DD(2))) / DD(2)
3280 IF H(1) >= HWET THEN W(1) = WATH
3290 IF H(1) <= HDRY THEN W(1) = WATL
3300 GOTO 3600
3310 HI = (EOR * DD(2) / B(1) + H(2) * TT - G(1) *
*  TM + G(2) * TM - DD(2)) / TT
3320 IF HI > HDRY AND HI < HWET THEN 3410
3330 IF KCP >= KCPMAX THEN 3370
3340 KCP = KCP + 1
3350 DELT = DELT * .5
3360 GOTO 2900
3370 IF HI < HDRY THEN H(1) = HDRY
3380 IF HI > HWET THEN H(1) = HWET
3390 WFDD = (B(1) * (H(1) * TT - H(2) * TT - G(2)*
*  TM + G(1) * TM + DD(2))) / DD(2)
3400 GOTO 3280
3410 H(1) = HI
3420 I = 1
3430 IF ABS(H(I) - G(I)) < .0001 THEN 3590
3440 NHI = ND
3450 NLO = 1
3460 J = INT(ND / 2)
3470 IF H(I) = P(J) THEN 3560 ELSE IF H(I) > P(J)
*  THEN 3500
3480 NHI = J
3490 GOTO 3510
3500 NLO = J
3510 JT = J
3520 J = INT((NHI - NLO) / 2) + NLO
3530 IF J <> JT THEN 3470
```

```
3540 IF H(I) >= P(J) THEN 3560
3550 J = J - 1
3560 WAT = (H(I) - P(J)) * DELW / (P(J + 1) -
   *  P(J)) + T(J)
3570 W(I) = WAT
3580 GOTO 3600
3590 W(I) = Y(I)
3600 FOR I = 2 TO KK
3610 W(I) = C(I) * (H(I) - G(I)) + Y(I)
3620 IF W(I) > WATH THEN W(I) = WATH
3630 IF W(I) < WATL THEN W(T) = WATL
3640 NEXT I
3650 SUM3 = 0
3660 SUM2 = 0
3670 SUM1 = 0
3680 FOR I = 2 TO K
3690 SUM1 = W(I) + SUM1
3700 SUM2 = Y(I) + SUM2
3710 IF ABS(SUM1 - SUM2) <= ABS(SUM3) THEN 3730
3720 SUM3 = SUM1 - SUM2
3730 NEXT I
3740 IF ABS(SUM3) <= ABS(CONQ) THEN 3780
3750 IF DELT <= DETT * .1 THEN 3780
3760 DELT = .5 * DELT
3770 GOTO 2900
3780 SUM1 = 0
3790 SUM2 = 0
3800 WFUU = B(NB) * ((H(NB) - H(NB + 1)) * TT +
   *  (G(NB) - G(NB + 1)) * TM + DD(NB + 1) - DD(NB))
   *  / (DD(NB + 1) - DD(NB))
3810 FOR I = 2 TO K
3820 SUM1 = W(I) * (DD(I + 1) - DD(I - 1)) / 2 +
   *  SUM1
3830 SUM2 = Y(I) * (DD(I + 1) - DD(I - 1)) / 2 +
   *  SUM2: NEXT I
3840 CWF = SUM1 - PIT
3850 WFRDD = (SUM1 - SUM2) / DELT
3860 CUMS = WFDD * DELT + CUMS
3870 IF EOR > 0 THEN SIR = EOR * DELT + SIR
3880 IF EOR < 0 THEN EVAP = WFDD * DELT + EVAP
3890 IF EOR > 0 THEN RPI = RPI + WFDD * DELT
3900 IF EOR > 0 THEN RUNOF = (EOR - WFDD) * DELT +
   *  RUNOF
3910 CUMB = WFUU * DELT + CUMB
3920 CUMET = CUMET + ET * DELT
3930 HRFLUX = WFUU
3940 SUMA = SUMA + SINK * DELT
3950 CTRAN = CTRAN + ETPL * DELT
3960 CWFLX = (SUM1 - SUM2)
```

```
3970 KB = K - 1
3980 TIME = TIME + DELT
3990 WATBAL = SIR + EVAP - RUNOF - CUMB - CWF +
*    SUMA
4000 PRINT USING "####.#### "; TIME; CWF; SIR;
*    SUMA; EOR; EVAP; WATBAL; SINK
4010 PRINT #2, USING "#.###^^^^ "; TIME; CWF; SIR;
*    SUMA; EOR; EVAP; WATBAL; SINK; HROOT
4020 REM ---CHANGE DELT HERE
4030 IF ABS(SUM3 - 0) > .0001 THEN 4060
4040 DELT = 3 * DELT
4050 GOTO 4150
4060 TW = ABS(CONQ * DELT / SUM3)
4070 IF TW >= .1 * DETT THEN 4100
4080 TW = .1 * DETT
4090 GOTO 4120
4100 IF TW <= 1000 * DETT THEN 4120
4110 TW = 1000 * DETT
4120 IF TW > 2 * DELT THEN 4040
4130 DELT = TW
4140 REM  SEE IF EVAP OR RAIN ETC. HAS CHANGED
4150 IF IDELT = 1 THEN DELT = DELT1
4160 IDELT = 0
4170 IF DELT < DETT THEN DELT = DETT
4180 IF DELT > 6 THEN DELT = 6
4190 IF TIME - V(KC + 1) < 0 THEN 4310
4200 PRINT #2, "  WATER CONTENT VS DEPTH"
4210 FOR I = 1 TO KK: PRINT #2, USING " #.### ";
*    W(I); : NEXT I
4220 PRINT #2,
4230 PRINT #2,
4240 PRINT #2, "    TIME       CWF      IRR+RAIN
*   TRAN        EPOT       EVAP      WATBAL
*   TRRAT       HROOT"
4250 EOR = V(KC + 2)
4260 IR = INT((KC + 2) / 2)
4270 KC = KC + 2
4280 MTIME = 0
4290 DELT = DETT
4300 GOTO 4330
4310 IF (TIME + DELT) <= V(KC + 1) THEN 4330
4320 DELT = V(KC + 1) - TIME
4330 IF V(KC) > 0 THEN 4580
4340 'ET = TET(IR): GOTO 4580 ' USE TO BYPASS
4350 TIMEA = TIME / 24 - INT(TIME / 24)
4360 TIMED = (TIME + DELT) / 24 - INT((TIME +
*    DELT) / 24)
```

```
4370 IF TIMED < TIMEA THEN 4510
4380 IF .5 - TIMEA < .0001 THEN 4510
4390 IF TIMED < .5 THEN 4440
4400 TIMED = .5
4410 DELT1 = DELT
4420 IDELT = 1
4430 DELT = (.5 - TIMEA) * 24
4440 IF MTIME = 1 THEN 4470
4450 MTIME = 1
4460 IR = (KC + 1) / 2
4470 ETNEW = (COS(TIMEA * PIEB2) - COS(TIMED *
*  PIEB2))
4480 EOR = ETNEW * 12 * V(KC) / DELT
4490 ET = ETNEW * 12 * TET(IR) / DELT
4500 GOTO 4580
4510 IF TIME >= .5 THEN 4550
4520 DELT1 = DELT
4530 IDELT = 1
4540 DELT = (1 - TIMEA) * 24
4550 ET = -1E-10
4560 EOR = -1E-11
4570 MTIME = 0
4580 IF DELT < DETT THEN DELT = DETT
4590 IF TIME >= CUMT THEN 4790
4600 Y(1) = (W(1) + Y(1)) * .5
4610 J = INT((Y(1) - T(1)) / DELW) + 1
4620 BB = (Y(1) - T(J)) / DELW
4630 IF ABS(EOR - 0) < .0001 THEN 4650
4640 G(1) = (P(J + 1) - P(J)) * BB + P(J)
4650 FOR I = 2 TO KK
4660 J = INT((W(I) - T(1)) / DELW) + 1
4670 BB = (W(I) - T(J)) / DELW
4680 G(I) = (P(J + 1) - P(J)) * BB + P(J)
4690 TW = (W(I) - Y(I)) + W(I)
4700 IF TW > WATH THEN 4740
4710 IF TW >= WATL THEN 4750
4720 TW = WATL
4730 GOTO 4750
4740 TW = WATH
4750 Y(I) = W(I)
4760 W(I) = TW
4770 NEXT I
4780 GOTO 1500
4790 END
```

* denotes a continuation of the previously line if the first character in a line is *.

INPUT FILE

EVAPTRAN.DAT FOR MONOGRAPH- MILLVILLE-
* EVAPOTRANSPIRATION
20,12,20,26,3
-.0416,24,-0.0416,48,-.0416,72,-.0416,96,-.0416,
* 120,-.0416,240
0,5,10,15,20,25,30,35,40,45,50,55,60,65,70,75,80,
* 85,90,95,100
-2.50E+06,-4.00E+05,-6.40E+04,-2.47E+04,-9.50E+03,
* -3.77E+03,-1.50E+03,-7.65E+02
-3.90E+02,-3.06E+02,-2.40E+02,-2.02E+02,-1.70E+02,
* -1.43E+02,-1.20E+02,-1.00E+02
-8.40E+01,-6.86E+01,-5.60E+01,-4.23E+01,-3.20E+01,
* -1.96E+01,-1.20E+01,-3.46E+00
0,10000
1.0E-11,1.0E-10,1.6E-09,1.0E-08,2.8E-08,1.9E-07,
* 9.1E-07,5.9E-06
4.50E-05,1.8E-04,5.2E-04,2.0E-03,3.0E-03,4.5E-03,
* 7.8E-03,1.3E-02
2.1E-02,3.4E-02,5.1E-02,8.3E-02,1.3E-01,2.2E-01,
* 7.0E-01,9.3E-01
1.2E+00,1.2E+00
.22,.22,.21,.21,.21,.2,.2,.2,.18,.1,.1,.1,.1,.1,
* .1,.1,.1,.1,.1,.1,.1
.0024,.038,0,0.,1.,240
-4.0E5,0,.02,.48,-.15E5,0.02
.1,24.,.01,.1,.9,1.05,0
0,.1,.1,.1,.1,.1,.1,.1,.1,.1,.1,0,0,0,0,0,0,0,0,
* 0

CONTENTS OF FILE

Line 1-- Label
Line 2-- K,IER,NB,ND,KCPMAX
Line 3-- V Array   upper boundary condition
Line 4-- DD Array   depth increments
Line 5-8 P Array   Matric potential-water content
* starting at 0.0
Line 9-12 E Array   Hydraulic conductivity-water
* content starting at 0.0
Line 13-14 W Array Initial water content
* corresponding to depth array
Line 15--DETT,CONQ,TAA,TIME,TT,CUMT
Line 16--HDRY,HWET,WATL,WATH,HLOW,DELW
Line 17--RDFDAY,RDFDEL,ESTART,ESTOP,AK1,RRES,HHI
Line 18--RDFSAV Array   Relative root distribution
* profile

SOIL EVAPORATION AND TRANSPIRATION

NOTE THAT DEPTHS ARE IN CM
MATRIC POTENTIAL IS IN CM (HEAD)
HYDRAULIC CONDUCTIVITY IS IN CM/HR
TIME IS IN HOURS
WATER CONTENTS ARE VOLUME FRACTIONS

* Denotes a continuation of the previous line if the first character of the line is *.

```
                        OUTPUT FILE
EVAPTRAN.DAT FOR MONOGRAPH- MILLVILLE-
*  EVAPOTRANSPIRATION  01-23-1990      14:26:56
     K    IER    NB    ND   KCPMAX
     20   12     20    26     3
 WAT CONT   MAT HEAD   CONDUCTI   DIFFUSIV    WAT CONT
*  MAT HEAD   CONDUCTI   DIFFUSIV
 0.00E+00   -.25E+07   0.10E-10   0.21E-04    0.26E+00
*  -.14E+03   0.45E-02   0.37E+00
 0.20E-01   -.40E+06   0.10E-09   0.23E-03    0.28E+00
*  -.12E+03   0.78E-02   0.55E+00
 0.40E-01   -.64E+05   0.16E-08   0.77E-03    0.30E+00
*  -.10E+03   0.13E-01   0.81E+00
 0.60E-01   -.25E+05   0.10E-07   0.12E-02    0.32E+00
*  -.84E+02   0.21E-01   0.11E+01
 0.80E-01   -.95E+04   0.28E-07   0.16E-02    0.34E+00
*  -.69E+02   0.34E-01   0.17E+01
 0.10E+00   -.38E+04   0.19E-06   0.27E-02    0.36E+00
*  -.56E+02   0.51E-01   0.23E+01
 0.12E+00   -.15E+04   0.91E-06   0.47E-02    0.38E+00
*  -.42E+02   0.83E-01   0.34E+01
 0.14E+00   -.77E+03   0.59E-05   0.91E-02    0.40E+00
*  -.32E+02   0.13E+00   0.48E+01
 0.16E+00   -.39E+03   0.45E-04   0.26E-01    0.42E+00
*  -.20E+02   0.22E+00   0.75E+01
 0.18E+00   -.31E+03   0.18E-03   0.41E-01    0.44E+00
*  -.12E+02   0.70E+00   0.13E+02
 0.20E+00   -.24E+03   0.52E-03   0.75E-01    0.46E+00
*  -.35E+01   0.93E+00   0.21E+02
 0.22E+00   -.20E+03   0.20E-02   0.15E+00    0.48E+00
*   0.00E+00   0.12E+01   0.25E+02
 0.24E+00   -.17E+03   0.30E-02   0.25E+00    0.50E+00
*   0.10E+05   0.12E+01   0.12E+05
  DEPTH      C(I)       W(I)       H(I)     RDFSAV(I)
    5.     0.62E-03    0.220    -.202E+03    0.100
   10.     0.53E-03    0.210    -.221E+03    0.100
..............................................
..............................................
```

```
   90.   0.88E-05   0.100   -.377E+04   0.000
   95.   0.88E-05   0.100   -.377E+04   0.000
  100.   0.88E-05   0.100   -.377E+04   0.000
 TIME END      SOIL FLUX    ET FLUX
 0.240E+02   -.416E-02   -.416E-01
 0.480E+02   -.416E-02   -.416E-01
 0.720E+02   -.416E-02   -.416E-01
 0.960E+02   -.416E-02   -.416E-01
 0.120E+03   -.416E-02   -.416E-01
 0.240E+03   -.416E-02   -.416E-01
     DETT       CONQ        TAA        TIME         TT
*    CUMT       DELW
   0.0024     0.0380     0.0000     0.0000     1.0000
*  240.0000    0.0200
     HDRY       HWET        WATL        WATH       HLOW
*    HHI
-.400E+06 0.000E+00 0.200E-01 0.480E+00 -.150E+05
*  0.000E+00
    RDFDAY     RDFDEL     ESTART      ESTOP         AK1
0.100E+00 0.240E+02 0.100E-01 0.100E+00 0.900E+00
     TIME       CWF       IRR+RAIN      TRAN       EPOT
*    EVAP      WATBAL      TRRAT       HROOT
0.240E-02-.105E-040.000E+000.000E+00-.416E-02
*   -.998E-050.504E-060.000E+00-.202E+03
..............................................
..................................
0.240E+02-.102E+010.000E+00-.917E+00-.100E-10
*   -.102E+00-.161E-05-.549E-09-.331E+03
   WATER CONTENT VS DEPTH
  0.169    0.170    0.175    0.178    0.179    0.178    0.178
*   0.176    0.164    0.127    0.102    0.100    0.100    0.100
*   0.100    0.100    0.100    0.100    0.100    0.100    0.100
     TIME       CWF       IRR+RAIN      TRAN       EPOT
*    EVAP      WATBAL      TRRAT       HROOT
0.240E+02-.102E+010.000E+00-.917E+00-.411E-05
*   -.102E+00-.263E-05-.370E-04-.333E+03
..............................................
..................................
0.480E+02-.202E+010.000E+00-.183E+01-.100E-10
*   -.189E+00-.227E-050.000E+00-.600E+03
   WATER CONTENT VS DEPTH
  0.141    0.142    0.147    0.149    0.150    0.150    0.150
*   0.151    0.148    0.135    0.105    0.100    0.100    0.100
*   0.100    0.100    0.100    0.100    0.100    0.100    0.100
     TIME       CWF       IRR+RAIN      TRAN       EPOT
*    EVAP      WATBAL      TRRAT       HROOT
0.480E+02-.202E+010.000E+00-.183E+01-.411E-05
*   -.189E+00-.240E-05-.370E-04-.606E+03
```

```
..................................................
..............................
 0.720E+02-.288E+010.000E+00-.265E+01-.100E-10
*  -.232E+00-.113E-05-.720E-10-.116E+04
    WATER CONTENT VS DEPTH
  0.121    0.121    0.128    0.129    0.130    0.130    0.129
*  0.123    0.128    0.128    0.107    0.101    0.100    0.100
*  0.100    0.100    0.100    0.100    0.100    0.100    0.100
    TIME        CWF      IRR+RAIN     TRAN       EPOT
*   EVAP       WATBAL      TRRAT      HROOT
 0.720E+02-.288E+010.000E+00-.265E+01-.411E-05
*  -.232E+00-.115E-05-.370E-04-.120E+04
..................................................
..............................
..................................................
..............................
 0.240E+03-.406E+010.000E+00-.369E+01-.100E-10
*  -.363E+00-.866E-06-.111E-09-.323E+04
    WATER CONTENT VS DEPTH
  0.088    0.088    0.101    0.104    0.105    0.105    0.105
*  0.104    0.104    0.103    0.099    0.101    0.101    0.100
*  0.100    0.100    0.100    0.100    0.100    0.100    0.100
```

* Denotes a continuation of the previous line if the first character of the line is *.

REFERENCES

Childs, S.W., and R.J. Hanks. 1975. Model of soil salinity effects on crop growth. Soil Sci. Soc. Am. Proc. 39:617-622.

Doorenbos, J., and W.O. Pruitt. 1977. Guidelines for predicting crop water requirements. FAO Irrig. Drain. Paper 24. FAO, Rome, Italy.

Hanks, R.J. 1974. Model for predicting plant yield as influenced by water use. Agron. J. 66:660-665.

Hanks, R.J. 1983. Yield and water use relationships: An overview. p. 393-411. *In* H.M. Taylor et al. (ed.) Limitations to efficient water use in crop production. ASA, CSSA, and SSSA, Madison, WI.

Hanks. R.J. 1984. Prediction of crop yield and water consumption under saline conditions. p. 272-283. *In* Soil salinity under irrigation. Springer-Verlag, New York.

Jensen, M.E. (ed.) 1973. Consumptive use of water and irrigation water requirements. Irrigation and Drainage Div., Am. Soc. Civ. Eng., New York.

Molz, F.J. 1981. Models of water transport in the soil-plant system: A review. Water Res. 17:1245-1260.

Nimah, M.N., and R.J. Hanks. 1973. Model for estimating soil water and atmospheric interrelations. Soil Sci. Soc. Am. Proc. 37:522-532.

Retta, A., and R.J. Hanks. 1980. Manual for using model PLANTGRO. Utah Agric. Exp. Stn. Res. Rep. 46.

Stewart, J.I., R.E. Danielson, R.J. Hanks, E.B. Jackson, R.M. Hagan, W.O. Pruitt, W.T. Franklin, and J.P. Riley. 1977. Optimizing crop production through control of water and salinity levels in the soil. Utah Water Res. Lab. PR 151-1, Utah State Univ., Logan.

12 Simulation of Water Uptake by Plant Roots

G. S. CAMPBELL

Washington State University
Pullman, Washington

Models of water uptake by plant roots generally have one of two purposes. Either they produce estimates of transpirational water loss for water budget models or they provide estimates of plant water status for predicting water stress. The most detailed models are useful for both purposes.

This chapter will develop equations that can be used to describe plant water uptake from soil. Simple analytical models will be derived and a computer algorithm will be developed that correctly simulates both root water extraction and plant water status for a given soil-plant-atmosphere system.

Detailed reviews and analyses of plant water uptake have been published (Hillel, 1980; Molz, 1981). The development presented here, therefore, will not be a detailed review of the subject. Only those equations needed to support the algorithms will be given.

I. WATER UPTAKE FROM A UNIFORMLY ROOTED SOIL LAYER

In the development of a theory to describe plant water uptake and loss, it has been useful to employ electrical analogs of the system (Gradmann, 1928; Van den Honert, 1948; Gardner, 1960; Cowan, 1965). Here, the model of Cowan (1965) will be followed. Figure 12-1 shows an electrical analog of a plant withdrawing water from a uniformly rooted soil layer. Only resistances in the liquid phase are shown. Steady flow is assumed in both the soil and the plant over the period of the calculation. Transpiration rate, E, is regulated mainly by vapor phase resistances and driving forces. Therefore, the plant water potentials in Fig. 12-1 are primarily the result of an imposed value of E, not the determinant of E. Leaf water potential does, however, have an indirect effect, since stomatal (vapor phase) resistance increases when leaf water potentials become low.

The major resistances in the system are the soil resistance (R_s), the soil-root interfacial resistance (R_I), the root endodermis resistance (R_r), the root and stem xylem resistance (R_x), and the leaf resistance (R_L). For most species growing in well-watered, medium-textured soil, the important resistances are R_r and R_L. In round numbers, a representative midday trans-

---

Copyright © 1991 ASA-CSSA-SSSA, 677 S. Segoe Rd., Madison, WI 53711, USA. *Modeling Plant and Soil Systems*—Agronomy Monograph no. 31.

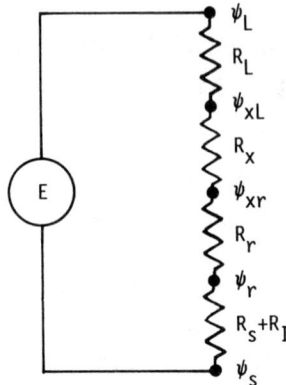

Fig. 12-1. Potentials and resistances in the soil-plant system. Transpiration rate is shown as a current generator, E.

piration rate (E) of 2.3×10^{-4} kg m^{-2} s^{-1} in a mature plant produces leaf water potentials of about -1500 J kg^{-1} in exposed leaves and -1000 J kg^{-1} in covered, darkened leaves. Thus, when $\psi_s \simeq 0$, $\psi_L \simeq -1500$ J kg^{-1} and $\psi_{xL} \simeq -1000$ J kg^{-1}. If xylem and soil resistance are assumed negligible, then, using Ohm's law and the circuit in Fig. 12-1, we can calculate $R_r = (\psi_s - \psi_{xL})/E = 1000/2.3 \times 10^{-4} = 4.3 \times 10^6$ m^4 s^{-1} kg^{-1} and $R_L = (\psi_{xL} - \psi_L)/E = 500/2.3 \times 10^{-4} = 2.2 \times 10^6$ m^4 s^{-1} kg^{-1}. Axial resistance in the root, where important, is included in the R_r value.

The root resistance just calculated is the total resistance of the root system. If the length of root is known, the resistance per unit length of root can be calculated. Using the data from Bristow et al. (1984), values of about 2.5×10^{10} m^3 kg^{-1} s^{-1} are obtained for resistance per unit length.

The soil and interfacial resistances are unimportant in wet soil, but become increasingly important as the soil dries (Herkelrath et al., 1977; Bristow et al., 1984). Interfacial resistance is most important in coarse-textured soils (Bristow et al., 1984). It will not be included in this analysis, but could easily be added.

Following the analysis of Cowan (1965) we can write the differential equation for water uptake by a single root as

$$q/A = -k \, d\psi/dr \qquad [1]$$

where q is the flux of water (kg s^{-1}), k is the soil hydraulic conductivity (kg s m^{-3}), ψ is the matrix potential (J kg^{-1}), and r is the radial distance from the root axis (m). The area for water flow is $A = 2\pi r l$ where l is the length of root. Hydraulic conductivity can be related to water potential using

$$k = k_s \, (\psi_e/\psi)^n \qquad [2]$$

where k_s is saturated conductivity (kg s m^{-3}), ψ_e is air entry potential (J kg^{-1}) and n is a constant that depends on soil texture and ranges typically from 2 to 3.5 (Campbell, 1974). Equations [1] and [2] can be combined and integrated from the bulk soil at r_s, where the potential is ψ_s, to the root surface at r_r, where the potential is ψ_r to obtain (Campbell et al., 1976)

$$(q/2\pi l)\ln(r_r/r_s) = (k_s\psi_s - k_r\psi_r)/(1 - n). \qquad [3]$$

The variables in Eq. [3] can be related to measurable values if we note that

$$q/l = E/L\Delta z \qquad [4]$$

where L is rooting density (m^{-2}) and Δz is rooting depth (m). Following Gardner (1960) we can also write

$$r_s = (\pi L)^{-1/2} \qquad [5]$$

Substituting these into Eq. [3] yields

$$E = (k_r\psi_r - k_s\psi_s)/B \qquad [6]$$

where

$$B = (1 - n) \ln(\pi r_r^2 L)/(4\pi L \Delta z). \qquad [7]$$

Equation [7] can be used to obtain estimates of R_s, since

$$R_s = (\psi_s - \psi_r)/E. \qquad [8]$$

The soil resistance is, therefore,

$$R_s = B(\psi_s - \psi_r)/(k_r\psi_r - k_s\psi_s) \qquad [9]$$

Since B depends on rooting depth and root density, soil resistance will be a function of these variables. It will also vary with soil water potential and, to some extent (since ψ_r depends on E), transpiration rate. Figure 12-2 compares soil with soil plus root resistance for a high and a low rooting density at three transpiration rates. The high rooting density ($L = 10$ cm cm^{-3}) is roughly comparable to that found in a potted plant; the low rooting density ($L = 0.1$ cm cm^{-3}) would be found deep in a soil profile. As pointed out by Newman (1969), soil resistance is probably always negligible when root density is high ($L = 10$ in Fig. 12-2). If, in nature, uniform, low rooting densities were present, Figure 12-2 ($L = 0.1$) indicates that R_s would become important at relatively high water potentials.

II. WATER UPTAKE FROM SOIL WITH SPATIALLY VARYING ROOT DENSITY

The model developed in the last section is not particularly useful because it assumes a constant rooting density with respect to depth in the soil.

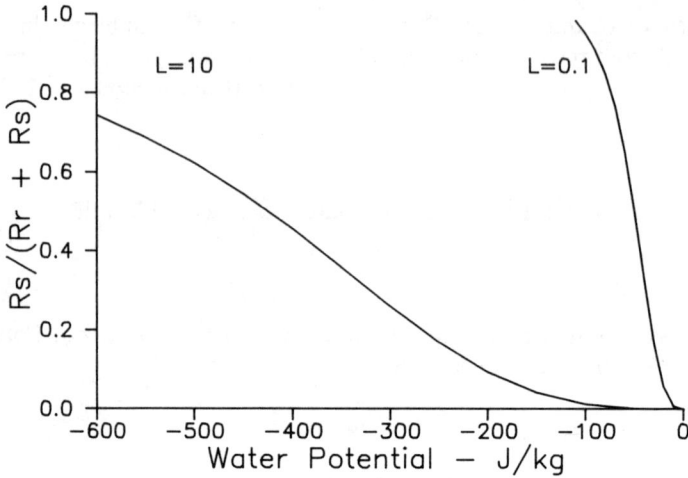

Fig. 12-2. Ratio of soil resistance to total (soil and root) resistance as a function of soil water potential for a 15-cm deep soil with root density of 10 cm^{-2} and 0.1 cm^{-2}. Calculations were made using k_s = 0.001 kg s m^{-3}, ψ_e = 2 J kg^{-1}, E = 0.0001 kg m^{-2} s^{-1}, n = 3, r_r = 0.001 m.

Few rooting systems would fit this assumption. To extend the model to include variable rooting density conditions, the root volume is assumed to be made up of zones with constant root densities. Water uptake from each zone can be described by the equations in the previous section. The total uptake is the sum of the uptake from each zone. Therefore, we write

$$E = \sum_i [(\psi_{si} - gz_i - \psi_x)/(R_{si} + R_{ri})] \quad [10]$$

where ψ_{si} is the soil matric potential of Zone i, g is the gravitational constant (9.8 m s^{-2}), z_i is the depth of Layer i, ψ_x is the xylem water potential, R_{si} is the soil resistance in Zone i (from Eq. [9]), and R_{ri} is the root resistance, which can be calculated from:

$$R_{ri} = R_r(\Sigma L_i)/L_i \quad [11]$$

Here, L_i is the rooting density of soil zone, i. Equation [11] expresses the assumption that the root resistance in any layer, R_{ri}, is directly proportional to the total root resistance, R_r, and inversely proportional to the fraction of the root system that is in that layer. The total root resistance is calculated from water potential and transpiration measurements, as was previously shown. An additional axial resistance can be added to Eq. [11] when appropriate, although uncertainty exists about the magnitude of this parameter.

The ψ_x in Eq. [10] represents the crown water potential of the plant. We can solve Eq. [10] for ψ_x, obtaining

$$\psi_x = \frac{\sum[(\psi_{si} - qz_i)/(R_{si} + R_{ri})] - E}{\sum[1/(R_{si} + R_{ri})]} \quad [12]$$

The leaf water potential is calculated by

$$\psi_L = \psi_x - ER_L \qquad [13]$$

Combining Eq. [12] and [13], we obtain

$$\psi_L = \overline{\psi_s} - E(R_L + R_{sr}) \qquad [14]$$

where

$$\overline{\psi_s} = \frac{\Sigma[(\psi_{si} - gz_i)/(R_{si} + R_{ri})]}{\Sigma[1/(R_{si} + R_{ri})]} \qquad [15]$$

and can be thought of as a weighted mean soil-water potential.

The combined soil-root resistance for the entire root volume, R_{sr}, is calculated as

$$R_{sr} = 1/\Sigma[1/(R_{si} + R_{ri})]. \qquad [16]$$

Equation [14] must be solved simultaneously with another equation that expresses the effect of ψ_L on E (through stomatal conductance). A reasonable approximation to the $E(\psi_L)$ relationship is (Campbell, 1985)

$$E = E_p/[1 + (\psi_L/\psi_c)^a] \qquad [17]$$

where E_p is the potential transpiration rate, ψ_c is the water potential at which $E = E_p/2$, and a is a species-dependent constant with a typical value of about 10. Equation [17] is combined with Eq. [14] and solved for ψ_L. Note that, in Eq. [17], $E \simeq E_p$ until ψ_L approaches ψ_c. Therefore, these equations provide a realistic approximation to the prediction of both atmospheric and soil-limited transpiration rate.

III. NUMERICAL ALGORITHMS FOR ROOT WATER UPTAKE

The program shown in Fig. 12-3 is a short BASIC simulation of water flow, plant water uptake, and plant-water relations. The main program is from Campbell (1985) and simulates water flow in the soil. Similar programs are found in Ch. 9 and 11 of this volume. The subroutine beginning on Line 1000 simulates the upper boundary and arbitrarily divides potential ET into potential transpiration (TP) and potential evaporation (EP). It is called at the beginning of each time step. More detailed algorithms for the upper boundary condition are described by Norman and Campbell (1983) and Campbell (1985). Chapter 11 describes a method, based on crop coefficients, for dividing ET into EP and TP throughout the growing season.

The subroutine beginning on Line 1500 initializes the variables for the root water uptake subroutine. The number of nodes, M, and the node depths, $Z(I)$, must already have been initialized before this subroutine is called (note that variable names used here refer to the variables in Fig. 12-3). The total

```
10 M=12:X=M+1
20 X=M+1
30 DIM A(X),B(X),C(X),F(X),P(X),Z(X),V(X),DP(X),W(X),WN(X),K(X),CP(X),H(X),
      DV(X),JV(X),DJ(X)
40 INPUT"INITIAL WATER POTENTIAL-J/KG";P:P=-ABS(P)
50 INPUT"SATURATED CONDUCTIVITY-KG S/M^3";KS
60 INPUT"AIR ENTRY POTENTIAL-J/KG";PE:PE=-ABS(PE)
70 INPUT"SOIL B VALUE";B
80 BD=1.3:WS=1-BD/2.6:DZ=.15
90 INPUT"POTENTIAL EVAPOTRANSPIRATION RATE - MM/DAY";ET
100 B1=1/B:N=2+3/B:N1=1-N:WD=1000:DT=3600
110 Z(0)=0:Z(1)=0:Z(2)=.01:Z(3)=.075:MW=.018:T=293:R=8.3143:DA=1
120 FOR I=1 TO M
130     P(I)=P
140     W(I)=WS*(PE/P(I))^B1:WN(I)=W(I):H(I)=EXP(MW*P(I)/(R*T))
150     K(I)=KS*(PE/P(I))^N
160     IF I>2 THEN Z(I+1)=Z(I)+DZ
170     V(I)=WD*(Z(I+1)-Z(I-1))/2
180 NEXT
190 P(M+1)=P(M):H(M+1)=H(M):Z(0)=-1E+10:K(M+1)=KS*(PE/P(M+1))^N
200 GOSUB 1500
210 GR=9.8:IM=.000001:DV=.000024:VP=.017
220 P(0)=P(1):K(0)=0:HA=.5:Z(M+1)=1E+20
230 TI=TI+DT/3600:IF TI>24 THEN TI=TI-24:DA=DA+1
240 GOSUB 1000:GOSUB 2000
250 SE=0:FOR I=1 TO M:K(I)=KS*(PE/P(I))^N:NEXT
260 JV(0)=EP*(H(1)-HA)/(1-HA):DJ(0)=EP*MW*H(1)/(R*T*(1-HA))
270 FOR I=1 TO M
280     KV=.66*DV*VP*(WS-(WN(I)+WN(I+1))/2)/(Z(I+1)-Z(I))
290     JV(I)=KV*(H(I+1)-H(I)):DJ(I)=MW*H(I)*KV/(R*T)
300     CP(I)=-V(I)*WN(I)/(B*P(I)*DT)
310     A(I)=-K(I-1)/(Z(I)-Z(I-1))+GR*N*K(I-1)/P(I-1)
320     C(I)=-K(I+1)/(Z(I+1)-Z(I))
330     B(I)=K(I)/(Z(I)-Z(I-1))+K(I)/(Z(I+1)-Z(I))+CP(I)-
              GR*N*K(I)/P(I)+DJ(I-1)+DJ(I)
340     F(I)=((P(I)*K(I)-P(I-1)*K(I-1))/(Z(I)-Z(I-1))-(P(I+1)*K(I+1)-P(I)*K(I))
              /(Z(I+1)-Z(I)))/N1+V(I)*(WN(I)-W(I))/DT-GR*(K(I-1)-K(I))
              +JV(I-1)-JV(I)+E(I)
350     SE=SE+ABS(F(I))
360 NEXT
370 FOR I=1 TO M-1
380     C(I)=C(I)/B(I)
390     F(I)=F(I)/B(I)
400     B(I+1)=B(I+1)-A(I+1)*C(I)
410     F(I+1)=F(I+1)-A(I+1)*F(I)
420 NEXT
430 DP(M)=F(M)/B(M):P(M)=P(M)-DP(M):IF P(M)>PE THEN P(M)=PE
440 FOR I=M-1 TO 1 STEP -1
450     DP(I)=F(I)-C(I)*DP(I+1):P(I)=P(I)-DP(I)
460     IF P(I)>PE THEN P(I)=(P(I)+DP(I)+PE)/2
470 NEXT
480 FOR I=1 TO M:WN(I)=WS*(PE/P(I))^B1:H(I)=EXP(MW*P(I)/(R*T)):NEXT
490 PRINT SE;:IF SE>IM THEN GOTO 250
500 H(M+1)=H(M)
510 PRINT:PRINT "DAY=";DA;"TIME=";TI;"HRS"
520 PRINT "DEPTH (M)","WC (M^3/M^3)","MP (J/KG)","R-SOIL/(R-SOIL+R-ROOT)":SW=0
530 FOR I=1 TO M:SW=SW+V(I)*(WN(I)-W(I)):W(I)=WN(I)
540 PRINT Z(I),WN(I),P(I),RS(I)/(RS(I)+RR(I)):NEXT
550 FL=EP*(H(1)-HA)/(1-HA)
560 PRINT "EVAP RATE =";FL;"TSP RATE = ";TR;" WC CHANGE = ";SW/DT
570 PRINT "DRAIN RATE = ";GR*K(M);"  E/EP = ";FL/EP;" T/TP = ";TR/TP;
          "PSI LEAF =";PL
```

Fig. 12-3. BASIC program to simulate water flow in soil, water uptake by plant roots, and plant water potentials (from Campbell, 1985).

```
580 GOTO 230
590 END
600 '
1000 ' EVAPOTRANSPIRATION SUBROUTINE
1010 E=2.3*ET*(.05+(SIN(.0175*7.5*TI))^4)/86400!
1020 EP=.1*E:TP=E-EP
1030 'RAIN OR IRRIG. CAN BE ADDED BY SETTING E(0) TO - THE DESIRED FLUX DENSITY
1040 RETURN
1050 '
1500 ' SUBROUTINE TO INITIALIZE ROOT WATER UPTAKE VARIABLES
1510 ' M AND Z(I) MUST BE KNOWN BEFORE CALLING THIS SUBROUTINE
1520 DIM RR(M),L(M),E(M),RS(M),PR(M),BZ(M)
1530 RR=4.3E+6:PC=-1500:RL=2.2E+6:PI=3.14159:SP=10:R1=.001:LSUM=0
1540 FOR I=1 TO M:READ L(I):L(I)=!(I)*1E4:LSUM=LSUM+L(I):NEXT
1550 DATA 0,4,4,1.9,0.8,0.8,0.4,0.4,0.2,0.1,0,0
1560 FOR I=1 TO M
1570    IF L(I)>0 THEN RR(I)=RR*LSUM/(L(I)):
                      BZ(I)=N1*LOG(PI*R1*R1*L(I))/(2*PI*L(I)*(Z(I+1)-Z(I-1)))
                 ELSE RR(I)=1E+20:BZ(I)=0
1580 NEXT
1590 RETURN
1610 '
2000 ' PLANT WATER UPTAKE SUBROUTINE
2010 PB=0:RB=0
2020 FOR I=1 TO M
2030    RS(I)=BZ(I)/K(I)
2040    PB=PB+(P(I)-GR*Z(I))/(RR(I)+RS(I)):RB=RB+1/(RS(I)+RR(I))
2050 NEXT
2060 PB=PB/RB:RB=1/RB
2070 IF PL>PB THEN PL=PB-TP*(RL+RB)
2080 XP=(PL/PC)^SP
2090 SL=TP*(RL+RB)*SP*XP/(PL*(1+XP)*(1+XP))-1.05
2100 F=PB-PL-TP*(RL+RB)/(1+XP):PRINT "DPL=";F;
2110 PL=PL-F/SL:IF ABS(F)>10 THEN GOTO 2080
2120 TR=TP/(1+XP)
2130 FOR I=1 TO M
2140 E(I)=(P(I)-GR*Z(I)-PL-RL*TR)/(RR(I)+RS(I))
2150 NEXT
2160 RETURN
```

Fig. 12-3. Continued.

root resistance (RR), the potential for stomatal closure (PC), leaf resistance (RL), the power for the stomatal closure function (SP) (variable a is Eq. [17]), and the root radius (R1) are initialized. Root densities ($L[I]$, data in Line 1550) are read in as cm cm^{-3}, converted to m m^{-3}, and summed in Line 1540. The root resistance (Eq. [11]) and BZ (B in Eq. [7]) are calculated in Line 1570. Where no roots are present, the resistance is set to a large number to avoid overflow problems in later computations. The subroutine beginning on Line 1500 is called once at the start of the program.

The subroutine beginning on Line 2000 calculates the soil resistances (RS[I]), the weighted mean soil-water potential (PB), the mean soil resistance (RB), the leaf water potential (PL), and the water uptake from each node (E[I]). The soil resistance is a nonlinear function of the water potential at the root surface (Eq. [9]), which, in turn, depends on the water uptake rate from the layer and, therefore, the total loss from the plant. An iterative method could be used to find the required potentials and resistances, but such an approach is almost never necessary. The soil resistance is seldom a significant part of the total soil-root resistance, so little error results from calculating the soil resistance from just the soil water potential. This is done in Line 2030. Once RS(I) is known, PB and RB (Eq. [15] and [16]) can be calculated. Summations are done in Line 2040 and final values are calculated in Line 2060.

The Newton-Raphson iteration (Campbell, 1985) is used to find leaf water potential. The iteration is in Lines 2080 to 2110 and continues until ψ_L is within 10 J kg^{-1} of the mass-balance value. Actual transpiration is calculated in Line 2120 and water uptake for each node is calculated in Line 2140.

IV. MODEL INPUT-OUTPUT AND VALIDATION

Figure 12-4 shows input data and the resulting output of the first time step using the program in Fig. 12-3. Campbell (1985) provides methods for finding saturated conductivity, air entry potential, and soil B value (pore size distribution index) from texture and bulk density data. The values given in Fig. 12-4 are for a Palouse silt loam (fine-silty, mixed, mesic, pachic Ultic Haploxeroll). The transpiration rate, leaf water potential, and soil water potential over an 8 d simulation are shown in Fig. 12-5.

Several aspects of the simulation in Fig. 12-5 are worth noting. First, transpiration (upper curve) differs little from day to day during the first few days of the simulation. Here, water loss is controlled by atmospheric demand. Beginning about Day 5, transpiration decreases each succeeding day as midday leaf water potentials approach ψ_c and stomates close. If the simulation had been run longer, midday leaf water potential would have decreased to values below ψ_c, further closing stomates and reducing transpiration.

The dirunal pattern of leaf and soil-water potential variation is also important. The range of leaf water potentials observed are typical of measurements one makes in the field (Turner & Burch, 1983; Campbell et al., 1976). The observed variation in soil water potential is also interesting. The trace shown is the weighted mean soil-water potential for all soil layers, but its value is strongly influenced by the densely rooted layers at the top of the soil profile. The fact that mean water potential increases during the night means that the water potential in these layers increases overnight as well. The model does not artificially restrict movement of water from the roots to the soil. Recent work by Baker and van Bavel (1986) indicates that resistance to water flow is the same for water flowing out of roots as it is for water flowing in, which is consistent with the assumptions in this model. The soil water potential fluctuations predicted by the model over a diurnal cycle are similar to those measured by Richards and Caldwell (1987) in the field. The model, therefore, appears to correctly simulate water uptake and loss as well as the internal water relations of the plant.

SIMULATION OF WATER UPTAKE BY PLANT ROOTS

Input

INITIAL WATER POTENTIAL-J/KG? -30
SATURATED CONDUCTIVITY-KG S/M^3? .001
AIR ENTRY POTENTIAL-J/KG? -2
SOIL B VALUE? 3
POTENTIAL EVAPOTRANSPIRATION RATE - MM/DAY? 7

Output for hr 1

```
DPL=-3.814697E-06  1.227508E-05  1.267466E-07
DAY= 1 TIME= 1 HRS
DEPTH (M)    WC (M^3/M^3)   MP (J/KG)    R-SOIL/(R-SOIL+R-ROOT)
0            .2020523       -30.3074     0
.01          .2022681       -30.21051    5.48327E-05
.075         .2026246       -30.05134    1.912837E-05
.225         .2027082       -30.01414    1.862891E-05
.375         .2027274       -30.00562    2.43458E-05
.525         .2027285       -30.00515    2.43458E-05
.675         .2027344       -30.00253    2.892686E-05
.825         .2027348       -30.00235    2.892686E-05
.9749999     .2027375       -30.00114    3.350789E-05
1.125        .2027389       -30.00054    3.808887E-05
1.275        .20274         -30.00002    0
1.425        .2027401       -30          0
EVAP RATE = 9.367599E-07 TSP RATE = 8.434616E-06  WC CHANGE = -1.227412E-05
DRAIN RATE = 2.903704E-06  E/EP = .9995522  T/TP = 1 PSI LEAF =-86.68895
```

Output for hr 12

```
DPL=-34.52088 DPL=-.2683106  1.893327E-04  4.429883E-06  1.313031E-08
DAY= 1 TIME= 12 HRS
DEPTH (M)    WC (M^3/M^3)   MP (J/KG)    R-SOIL/(R-SOIL+R-ROOT)
0            .1732068       -48.11113    0
.01          .175378        -46.34623    1.60191E-04
.075         .1903968       -36.22107    3.033564E-05
.225         .1988656       -31.78782    2.144571E-05
.375         .2011516       -30.71635    2.57809E-05
.525         .2013749       -30.6143     2.558762E-05
.675         .2019927       -30.33424    2.971822E-05
.825         .2020668       -30.30086    2.964404E-05
.9749999     .2023809       -30.16       3.394556E-05
1.125        .2025648       -30.07797    3.833085E-05
1.275        .2027261       -30.00618    0
1.425        .202739        -30.00046    0
EVAP RATE = 1.955141E-05 TSP RATE = 1.668988E-04  WC CHANGE = -1.89332E-04
DRAIN RATE = 2.903572E-06  E/EP = .9992891  T/TP = .9478156  PSI LEAF =
-1122.638
```

Fig. 12-4. Example input, and output at 1 and 12 h on Day 1 for the program in Fig. 12-3.

This model has been validated in several studies. Stockle and Campbell (1985) used it in a model of seasonal response of corn (*Zea mays* L.) to water stress, obtaining good agreement between predicted and measured water uptake and soil water profiles. Stockle (1985) also used this model to simulate water relations of dryland wheat (*Triticum aestivum* L.). Figure 12-6 shows an example of the agreement between predicted and measured water content profiles.

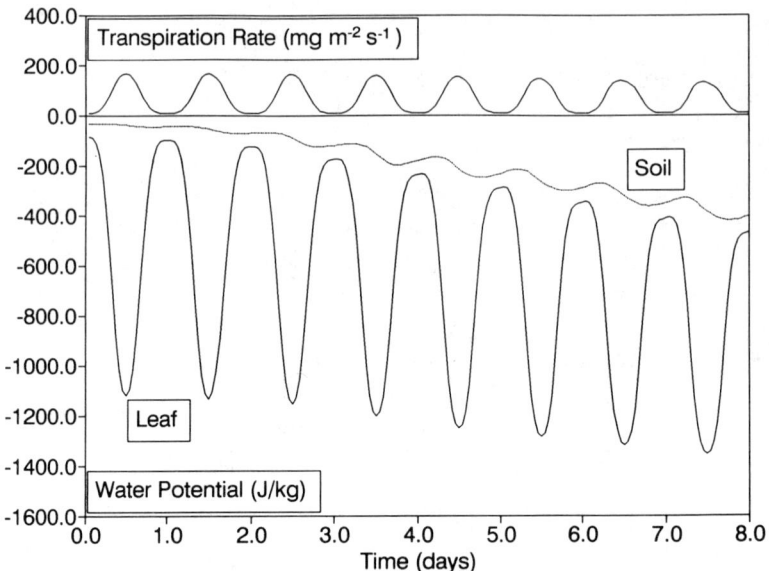

Fig. 12-5. Simulated transpiration, leaf water potential, and soil water potential over 8 d using the program in Fig. 12-3 and the inputs in Fig. 12-4.

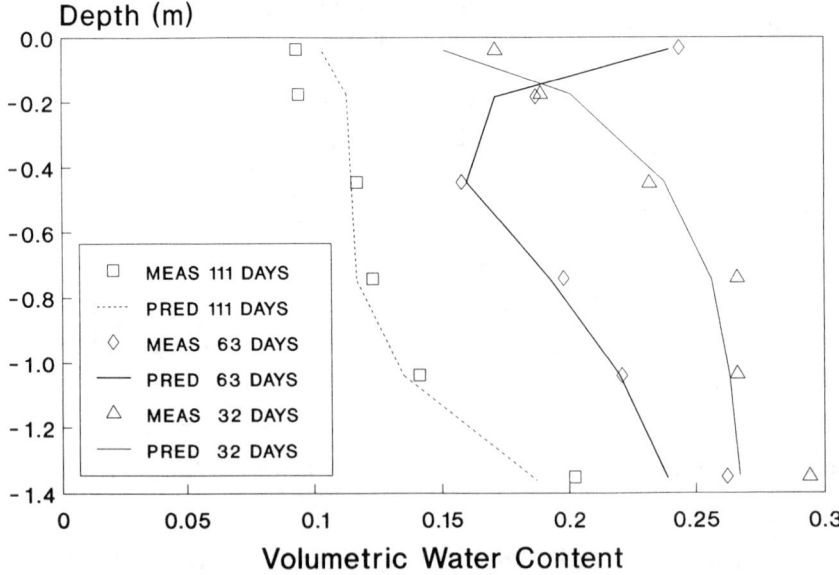

Fig. 12-6. Comparison of modeled (PRED) and measured (MEAS) soil water content profiles for spring wheat (from Stockle, 1985).

V. CRITICAL ASSUMPTIONS

The assumptions used in deriving this model were stated or implied earlier, but are restated here.

1. Water moves from the soil, through the plant, to the evaporating surfaces in the substomatal cavities in response to gradients in water potential. Only matric and gravitational potential were explicitly accounted for in this derivation.
2. Resistances to flow in the liquid phase are mainly in the soil, the root endodermis, and the leaf. These resistances are assumed constant and known during the period of the simulation. Interfacial and xylem resistances could have been included, but were assumed negligible for this model.
3. Low leaf water potential reduces transpiration.
4. The root resistance to water uptake in a soil layer is inversely proportional to the fraction of the total root system present in that layer.
5. Steady flow was assumed during the period of one time step.

Several assumptions have been made in the derivation of the soil resistance. The soil resistance, however, is generally not a large fraction of the total resistance (Campbell & Campbell, 1982), so these assumptions should not be regarded as critical.

VI. RESEARCH NEEDS

The simple electrical analog, presented here to model water flow through plants, has been used for many years and has, in a general way, been validated in a number of studies. Additional validation can and should be performed. We now have the equipment to simultaneously measure transpiration, and leaf water, xylem water, and soil water potentials. A complete study, correctly measuring all of these parameters, has not been performed. Such an experiment would allow assumptions about the constancy of the resistances to be checked.

Additional information is also needed regarding the relationship between resistance to water uptake and root density. We know little about how resistance to water uptake by a root changes over time and with drying of the soil. We also need additional data on magnitudes of axial and xylem resistances, and how these might change when plants become diseased.

Finally, the whole area of stomatal control of transpiration needs to be better understood. Equation [17] has obvious shortcomings. The critical leaf water potential changes with osmotic adjustment, and possibly other factors. An improved water relations model will require a better understanding of this important relationship.

VII. SUMMARY

When correct values are supplied for resistances, rooting densities, and transpiration rates, this model correctly simulates diurnal leaf, xylem, and soil water potential changes and stomatal closure due to drought stress. Water extraction patterns agree with observations that water is extracted from the layers with high root density and high water potential first. It is not necessary to arbitrarily specify unidirectional flow to roots, as is done in many water uptake models. When resistances are correct, little redistribution of water occurs through the root system. The model is, therefore, consistent with observations by Baker and van Bavel (1986) that resistances for water loss from roots are similar to those for water uptake.

This model operates on a short time step. One to two hours is best, since longer periods will miss the midday stress period. The nonlinearity of the transpiration rate-water potential response requires this detail in the simulation.

Modification of the program for either simpler or more complex simulations is straightforward. For example, soil resistance is almost always negligibly small. The speed of the program could be increased, with little loss of accuracy, by eliminating the soil resistance calculation. The model could be extended to include two- and three-dimensional rooting density variations. The summations would then cover all of the rooting volumes, rather than just depth. Axial resistances could be included to delay water uptake in certain parts of the root zone.

The effect of root growth can be simulated by changing the $L(I)$ values with time using measured values or a simple model such as that of Stockle and Campbell (1985). Root resistance should also change with time to reflect increases in root length. Effects of changing soil temperature or aeration on root resistance could also be included.

VIII. ACKNOWLEDGMENT

This manuscript was prepared while the author was Visiting Professor, Univesity of Nottingham, School of Agriculture. Support from the University of Nottingham is gratefully acknowledged.

REFERENCES

Baker, J.M., and C.H.M. van Bavel. 1986. Resistance of plant roots to water loss. Agron. J. 78:641–644.

Bristow, K.L., G.S. Campbell, and C. Calissendorff. 1984. The effects of texture on the resistance to water movement within the rhizosphere. Soil Sci. Soc. Am. J. 48:226–270.

Campbell, G.S. 1974. A simple method for determining unsaturated conductivity from moisture retention data. Soil Sci. 117:311–314.

Campbell, G.S. 1985. Soil physics with BASIC: Transport models for soil-plant systems. Elsevier, Amsterdam.

Campbell, G.S., and M.D. Campbell. 1982. Irrigation scheduling using soil moisture measurements: theory and practice. Adv. Irrig. 1:25–42.

Campbell, M.D., G.S. Campbell, R. Kunkel, and R.I. Papendick. 1976. A model describing soil-plant-water relations for potatoes. Am. Potato J. 53:431–441.

Cowan, I.R. 1965. Transport of water in the soil-plant-atmosphere system. J. Appl. Ecol. 2:221–239.

Gardner, W.R. 1960. Dynamic aspects of water availability to plants. Soil Sci. 89:63–73.

Gradmann, H. 1928. Unterscuchungen uber die wasserverhaltnisse des bodens als grundlage des pflanzenwachstums. Jahrb. Wiss. Bot. 69:1–100.

Herkelrath, W.N., E.E. Miller, and W.R. Gardner. 1977. Water uptake by plants. II. The root contact model. Soil Sci. Soc. Am. J. 41:1039–1043.

Hillel, D. 1980. Applications of soil physics. Academic Press, New York.

Molz, F.J. 1981. Models of water transport in the soil-plant system: A review. Water Resour. Res. 17:1245–1260.

Newman, E.I. 1969. Resistance to water flow in soil and plant. J. Appl. Ecol. 6:1–12.

Norman, J.M., and G.S. Campbell. 1983. Application of a plant-environment model to problems in irrigation. Adv. Irrig. 2:155–188.

Richards, J.H., and M.M. Caldwell. 1987. Hydraulic lift: substantial nocturnal water transport between soil layers by *Artemisia tridentata* roots. Oecologia 73:486–489.

Stockle, C.O. 1985. Simulation of the effect of water and nitrogen stress on the growth and yield of spring wheat. Ph.D. thesis. Washington State Univ., Pullman.

Stockle, C.O., and G.S. Campbell. 1985. A simulation model for predicting effect of water stress on yield: An example using corn. Adv. Irrig. 3:283–311.

Turner, N.C., and G.J. Burch. 1983. The role of water in plants. p. 73–125. *In* I.D. Teare and M.M. Peet (ed.) Crop-water relations. John Wiley and Sons, New York.

Van den Honert, T.J. 1948. Water transport in plants as a catenary process. Discuss. Faraday Soc. 3:146–153.

13 Nitrogen Dynamics in Soil-Plant Systems

D. C. GODWIN

International Fertilizer Development Center
Muscle Shoals, Alabama

C. ALLAN JONES

Texas Agricultural Experiment Station
Temple, Texas

Nitrogen plays a key role in plant nutrition. It is the mineral element required in the greatest quantity by cereal crop plants and it is the nutrient most often deficient. As a result of its critical role and low supply, the management of N resources is an extremely important aspect of crop production (Novoa & Loomis, 1981). Nitrogen is currently the most widely used fertilizer nutrient and the demand for it is likely to grow in the foreseeable future.

Despite this large investment in fertilizer N, the efficiency with which crops use it is poor. Allison (1965) suggested the average recovery of fertilizer N in the aboveground parts of crops is about 50%. Power (1981) has indicated that the general range of recovery in plant parts is between 20 to 90% of the fertilizer applied.

The N that is not recovered by the crop may be lost to the atmosphere through volatilization of ammonia, denitrification, or leaching, or made unavailable to the plant through immobilization in the soil. It may also become inaccessible to the plant through lack of water. The magnitude of each of the various transformations affecting the use of N is influenced by many climatic, edaphic, and agronomic factors. The myriad of transformation pathways and the multitude of factors affecting transformation rates renders N as one of the most complex plant nutrients to study. Quantifying these factors and predicting the response to added N is, thus, a very difficult task. The fraction of N that is lost from the cropping system is a source of some of the environmental pollution associated with fertilization. Thus, for economic and environmental reasons, a major thrust of current fertilizer research seeks to improve the efficiency of its use.

Traditionally, information used to generate a response curve is obtained from field experiments that test several increasing dressings of each nutrient to show how the fertilizer effects crop yields. This response information then forms the basis for deciding an optimum application rate. Regression models of fertilizer response, however, are static in nature and are unable to account for within-season variability in the supply of either water or nutrients.

Copyright © 1991 ASA-CSSA-SSSA, 677 S. Segoe Rd., Madison, WI 53711, USA. *Modeling Plant and Soil Systems*—Agronomy Monograph no. 31.

Computer simulation models that are able to capture the nuances of weather and the effects of various soil properties and agronomic practices on nutrient dynamics and crop growth processes, potentially could make a large contribution to furthering our understanding of fertilizer behavior in cropping systems. Such models, with the capability of readily simulating various crop and fertilizer management strategies, should lead to a great improvement in the efficacy of fertilizer decision making. Optimizing fertilization strategies, given the uncertainties of climate, is generally difficult. The problem is compounded in some less-developed regions of the world where fertilizer data are sparse. Where adequate climatic, soil, and crop data exist, simulation models will allow some extrapolation into these less-developed areas and, thus, provide some insights into fertilizer behavior in various environments.

Many different simulation models exist that describe some or all of the N-cycle processes occurring in cropping systems. Some of these were the subject of the proceedings of a workshop (Frissel & van Veen, 1981), and of reviews by Tanji (1982). Several of the models cited in these reviews are concerned with specific aspects of the N cycle, such as ammonia volatilization (Parton et al., 1981), leaching (Addiscott, 1981; Burns, 1980), and denitrification (Smith, 1981; Leffelaar, 1981). Most of the models are primarily concerned with the major soil processes in the N cycle and few consider the N dynamics of a growing crop. Several very comprehensive models have also been developed that contain details that link together all the major soil transformations. Among them, those of van Veen (1977) and Molina et al. (1983) provide considerable detail about the functioning of the various transformations at the microbiological level.

Other models have been developed with the specific purpose of examining aspects of pollution from organic wastes and from excessive fertilization (Rao et al., 1981; Selim & Iskandar, 1981; Donigian & Crawford, 1976). Some models, such as those of Watts and Hanks (1978), Tillotson et al. (1980), and Tanji et al. (1981), simulate most of the major soil transformations of N as well as the uptake by the crop, but fall short of fully describing the system in that the crop growth and yield response is not incorporated. A simulation model that predicts dry-matter production as limited by water and N, PAPRAN (Seligman & van Keulen, 1981) is used to make predictions for pastures in semiarid environments.

Many of these models were designed for a specific purpose and do not lend themselves to integration with a general crop and water balance model. To simulate N dynamics adequately in a range of diverse cropping environments, a model capable of describing the major soil N transformations, as well as the plant component, is required. The CERES-N model described below is the nitrogen component of the CERES-WHEAT (Jones & Kiniry, 1986) and CERES-MAIZE (Godwin & Vlek, 1985) models. The CERES models simulate growth, phenology, water and N balance, and yield. They were designed to have widespread applicability in diverse environments and to require only a minimum data set of commonly available field data as inputs.

I. THE CERES-N MODEL

The N component of the model is not designed to operate in a stand-alone mode, but as a component of the CERES models. Complete documentation of the CERES models has been published elsewhere (for maize [*Zea mays* L.] see Jone & Kiniry, 1986, and for wheat [*Triticum aestivum* L.] see Ritchie et al., 1988). The CERES models can be run with an option set to simulate the condition in which N is not limiting. This version describes evapotranspiration, soil water balance, and crop development as influenced by temperature, vernalization (wheat only) and photoperiod, vegetative growth, root growth, and grain growth. The N components add to this the description of mineralization and/or immobilization of N associated with the decay of crop residues, nitrification, denitrification, urea hydrolysis, leaching of nitrate, and the uptake and use of N by the crop. The N model uses the layered soil-water balance model described by Ritchie (1985) and the soil temperature component of the EPIC model (Williams et al., 1983). Both these submodels are integrated into the CERES models.

The N model comprises four subroutines describing N movement, soil N transformations, N uptake, and plant N stress factors. A listing of these subroutines can be found in Appendix 13-1 and a glossary of the variables is listed in Appendix 13-2. The effects of N deficiency on plant growth processes are incorporated into the main growth subroutine (GROSUB) of the CERES models. Additional subroutines are used within the CERES models for input and output of data, and for the initialization of the various N pools. Since the model is not designed to operate in a stand-alone manner, it is difficult to estimate the computer time required for execution of the N components. By comparing CERES model runs with the option set to simulate or not simulate the N balance, an approximate estimate can be obtained. On a VAX 11/750 computer, simulation of crop growth and water balance for one complete season requires approximately 7 s of CPU time. The additional simulation of the N balance requires approximately 3 s. On an IBM PC AT equipped with floating point coprocessor, corresponding times are 25 and 33 s, respectively.

A. Nitrate and Urea Movement

Leaching of nitrate (NO_3) is probably the most common and the best understood N loss process. Nitrates leaching from soil often become a source of contamination of ground water. This has generated more recent interest in leaching from an environmental standpoint (Wild & Cameron, 1980).

The many approaches to modeling leaching was the subject of a review by Addiscott and Wagenet (1985). Many of these approaches are based on numerical techniques, which require solution in a manner inappropriate for use in a management-level model such as CERES. In the CERES model, only the movement of nitrate and urea is considered. Ammonium is assumed not to be transported across soil layers. The same procedures for simulating nitrate movement are used for simulating urea movement.

Nitrate movement in the soil profile is dependent upon water movement. In the water balance component of the model the volume of water moving from a layer (L) to the layer below [FLUX(L)] is calculated. The volume of water present in the layer before drainage occurred is also calculated from the volumetric water content [SW(L)] and the depth of the layer [DLAYR(L)]. Nitrate lost from each layer (NOUT) may then be calculated as a function of the water that is retained and the water that moves as

$$\text{NOUT} = \text{SNO3(L)} \times \text{FLUX(L)}/[\text{SW(L)} \times \text{DLAYR(L)} + \text{FLUX(L)}] \qquad [1]$$

where SNO3(L) is the quantity of nitrate present in layer L (kg N/ha). Thus, a fraction of the mass of nitrate [SNO3(L)] present in each layer moves with each drainage event. A simple cascading approach is used in which the nitrate lost from one layer is added to the layer below. When the concentration of nitrate in a layer falls to 1.0 μg NO_3/g of soil, then no further leaching from that layer is allowed to occur. The implicit assumption is that all the nitrate present in a layer is uniformly and instantaneously in solution in all of the water in the layer.

Nitrate is more readily displaced from sands since the volume of water that can move is large in comparison with the retained water. Most of the difference in the simulated leaching rate between soils of different texture is explained by this difference in the proportion of water that is mobile. Some difference is also attributable to the rate at which the profile can drain.

Similar procedures are used to model the rate of upward movement of nitrate and urea with evaporation of water from the surface layers. In this case, the water balance routine calculates the upward flow of water [FLOW(L)] and the amount of upward movement (NUP) is calculated as for NOUT as

$$\text{NUP} = \text{SNO3(L)} \times \text{FLOW(L)}/[\text{SW(L)} \times \text{DLAYR(L)} \times \text{FLOW(L)}] \qquad [2]$$

This movement of N is confined to the upper layers of the soil profile and is generally small, due to the small volume of moving water. No upwards loss from the top layer occurs by this process.

B. Soil Nitrogen Transformations

In subroutine NTRANS, the CERES model simulates the decay of organic matter and the subsequent mineralization and/or immobilization of N, the nitrification of ammonium, and denitrification. Fertilizer addition and urea hydrolysis are also performed in this subroutine.

C. Fertilizer Additions and Urea Hydrolysis

Fertilizer N is partitioned in the model between nitrate, ammonium, and urea pools according to the nature of the fertilizer used. The assumption is made that the fertilizer is uniformly incorporated into the soil layer into which it is placed. Surface fertilizer applications are treated as being uniformly incorporated into the top layer. Up to 10 split applications can be accommodated by the model.

To simulate urea hydrolysis, a maximum hydrolysis rate is estimated from the soil organic carbon and pH. Data to estimate this rate were drawn from several laboratory studies (Tabatabai & Bremner, 1972; McGarity & Myers, 1967; Myers & McGarity, 1968; Zantua et al., 1977; Beri & Brar, 1978). This maximum hydrolysis rate is then scaled downward according to the prevailing moisture and temperature conditions. The functions used are:

$$AK = -1.12 + 1.31 \times OC(L)$$
$$+ 0.203 \times PH(L) - 0.155 \times OC(L) \times PH(L) \quad [3]$$

where AK is the maximum hydrolysis rate (dimensionless rate constant), OC(L) is the organic carbon in layer L (%), PH(L) = soil pH in layer L, and AK has an assumed minimum value of 0.25. The actual hydrolysis rate (UHYDR) is estimated as:

$$UHYDR = AK \times AMIN1(SWF,TF) \times UREA(L) \quad [4]$$

where SWF,TF represents zero to unity indices for soil water and temperature, respectively, UREA(L) is the quantity of urea present in a layer (kg urea/ha), and AMIN1 is a FORTRAN function in which the minimum of the arguments is selected. The temperature and soil water indices are designed to simulate the effects of soil moisture and temperature on urea hydrolysis reported in a laboratory study by Vlek and Carter (1983). Any urea remaining after 21 d is assumed to have hydrolyzed. The N released by the hydrolysis of urea is added to the soil ammonium pool.

D. Mineralization and Immobilization

Mineralization refers to the net release of mineral N with the decay of organic matter, and immobilization refers to the transformation of inorganic compounds to the organic state. Both processes are microbial in origin. Immobilization occurs when soil microorganisms assimilate inorganic N compounds and use them in the synthesis of the organic constituents of their cells. A balance exists between the two processes. When crop residues with a high C:N ratio are added to soil, the balance can shift, resulting in net immobilization for a period of time. After some of the soil carbon has been consumed by respiration, net mineralization may resume. Nitrogen mineralized from the soil organic pool often constitutes a large part of the N available to the crop.

The perceived application for the CERES models in studies examining crop growth and fertilizer management requires that a mineralization model be simple, require few inputs, and able to work on a diversity of soils. Simulation studies examining the effects of crop residues also require that the model be capable of simulating the fate of residues of various compositions. Other studies examining the potential role of nitrification inhibitors require a model wherein the processes of ammonification and nitrification are separated. The approach used in the CERES-WHEAT model is based on a modified version of the mineralization and immobilization component of the PAPRAN model (Seligman & van Keulen, 1981). This model attempts to maintain some of the functionality of the microbiological-level models, but does so at a very simplified level. The modifications to the model have been performed to first simulate nitrification rather than assume it as an instantaneous process. Secondly, the fresh organic matter pools simulated were partitioned further, so that an interface to the denitrification procedures could be constructed. Modifications were also made to temperature and water indices to fit the CERES water balance and soil temperature routines. Unless indicated otherwise, the coefficients used for the mineralization/immobilization functions described below were drawn from the PAPRAN model. The mineralization and immobilization routine simulates the decay of two types of organic matter. These are fresh organic matter (FOM), which includes crop residues or green manure, and a more stable organic or humic pool (HUM). The FOM pool is further subdivided into three pools: carbohydrate, cellulose, and lignin.

In PAPRAN, FOM is simulated as one pool and the decay rate constant is selected according to the proportion of the initial amount of FOM remaining. In CERES, separation of FOM into three pools enabled a better estimate of soluble carbon, which is used in the denitrification routine.

Initially, the FOM(L) contains 20% carbohydrate, 70% cellulose, and 10% lignin. The model requires, as input data, the amount of straw added, its C/N ratio, and its depth of incorporation (if any). An estimate of the amount of root residue from the previous crop is also required. These data are used to initialize FOM and the N contained within (FON) for each layer. The FOM is measured in units of kilograms of organic matter per hectare and FON is measured in kilograms of N per hectare. The mineralization routine also requires the soil organic carbon in each layer [OC(L)] as an input. This is used to calculate HUM(L) and, together with a simplifying assumption of bulk soil C/N ratio of 10, is used to estimate the N associated with this fraction [NHUM(L)].

Each of the three FOM pools has a different decay rate [RDECR(1 to 3)]. Under nonlimiting conditions, the decay constants as reported by Seligman and van Keulen (1981) are 0.80, 0.05, and 0.0095 for carbohydrate, cellulose, and lignin, respectively. The decay constant for carbohydrate implies that, under nonlimiting conditions, 80% of the pool will decay in 1 d. Nonlimiting conditions very seldom occur in soils since one or all of soil temperature, soil moisture, or residue composition will limit the decay process. To quantify these limits, three (zero to unity) factors are calculated. A water factor (MF) is first determined from the volumetric soil water content [SW(L)] relative to the lower limit (LL) and drained upper limit (DUL) (Fig. 13-1).

The function for MF follows the observations reported by Myers et al. (1982) and Linn and Doran (1984) on moisture effects on ammonification.

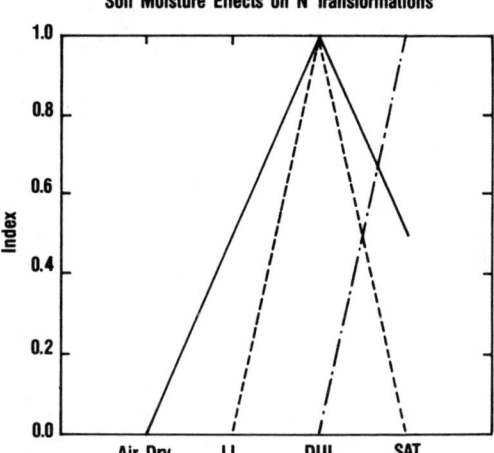

Fig. 13-1. Soil water indices used to modify soil N transformation rates (LL = lower limit moisture content, DUL = drained upper limit moisture content, SAT = saturation moisture content, —— = ammonification, --- = nitrification, —·— = denitrification).

Under very wet conditions (100% of water filled porosity), ammonification proceeds at approximately half of the rate of ammonification at field capacity (Linn & Doran, 1984).

The temperature factor (TF) approximates the soil temperature effects on ammonification reported by others (Stanford et al., 1973; Myers, 1975). The C/N ratio (CNR) imposes the third limit on decay rate. In this case, CNR is calculated as the carbon contained in FOM divided by the N available for the decay process. This N available for decay is the sum of the N contained in the FOM and the extractable mineral N present in the layer. From CNR, an index (CNRF) is calculated, which has a critical CNR of 25 (Fig. 13-2). In low N containing residues (e.g., freshly incorporated wheat straw) with a high CNR, the N available for the decay process will greatly limit the decay rate.

For each of the FOM pools, a decay rate appropriate for that pool (JP) can be calculated by multiplying the rate constant by the three indices as

$$G1 = TF \times MF \times CNRF \times RDECR(JP) \qquad [5]$$

where G1 is then the proportion of the pool that decays in 1 d. The amount of material that has decayed is then the product of G1 and the pool size. The gross mineralization of N associated with this decay (GRNOM) is then calculated according to the proportion of the pool that is decaying as

$$GRNOM = G1 \times FPOOL(L,JP)/FOM(L) \times FON(L) \qquad [6]$$

where FPOOL(L,JP) is the pool of either carbohydrate (JP = 1), cellulose (JP = 2), or lignin (JP = 3) present in layer L (μg/ha). The GRNOM is summed for each of three pools in each layer. Similarly, GRCOM, the amount of decaying organic matter, is determined as the sum of three pool fractions.

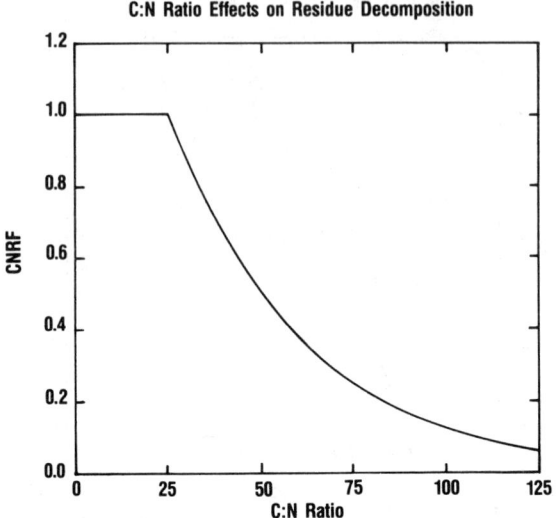

Fig. 13-2. Effect of C/N ratio on the index (CNRF) for residue decay rate.

The procedure used for calculating the N released from the humus (RHMIN) also uses TF and MF. In this case, CNRF is not used and the potential decay rate constant (DMINR) is very small (8.3×10^{-5}). A further index (DMOD) was added to the RHMIN calculations to adjust the mineralization rate for certain atypical soils. On certain volcanic ash soils mineralization proceeds slowly, necessitating a less than unity value of DMOD so that mineralization is not overestimated. On freshly cultivated virgin soils, a slightly greater than unity value has been found necessary to account for the sudden increase in mineralization activity. In all other circumstances a value of 1.0 is used for DMOD. Satisfactory alternatives for estimating DMOD are currently being sought. The procedure for calculating RHMIN, then, is the product of the various indices and the N contained within the humus [NHUM(L)].

$$\text{RHMIN} = \text{NHUM(L)} \times \text{DMINR} \times \text{TF} \times \text{MF} \times \text{DMOD} \qquad [7]$$

These calculations also allow for the transfer of 20% of the gross amount of N released by the mineralization of FON(L) to be incorporated into NHUM(L). This accounts for N incorporated into microbial biomass (Seligman & van Keulen, 1981).

As organic matter decomposes, some N is required by the decay process and may be incorporated into microbial biomass. The N that is immobilized in this way (RNAC) is calculated as the minimum of the soil-extractable mineral N (TOTN) and the demand for N by the decaying FOM(L) as

$$\text{RNAC} = \text{AMIN1(TOTN,GRNOM)} \times [0.02 - \text{FON(L)/FOM(L)}] \qquad [8]$$

where 0.02 is the N requirement for microbial decay of a unit of FOM(L). The value of 0.02 is the product of the fraction of carbon in the FOM(L) (40%), the biological efficiency of carbon turnover by the microbes (40%), and the N/C ratio of the microbes (0.125). The FOM(L) and FON(L) are then updated as

$$FOM(L) = FOM(L) - GRCOM \qquad [9]$$

$$FON(L) = FON(L) + RNAC - GRNOM \qquad [10]$$

The balance between RNAC and GRNOM determines whether net mineralization or immobilization occurs. The net N released from all organic sources (NNOM) is

$$NNOM = 0.8 \times GRNOM + RHMIN - RNAC. \qquad [11]$$

Note that only 80% of GRNOM enters this pool since the remaining 20% was incorporated into NHUM(L). The NNOM can then be used to update the ammonium pool [SNH4(L)].

$$SNH4(L) = SNH4(L) + NNOM. \qquad [12]$$

If net immobilization occurs (NNOM negative) ammonium is first immobilized and, if there is not sufficient ammonium to retain this pool with a concentration of 1 ppm, then withdrawals are made from the nitrate pool.

E. Nitrification

Nitrification refers to the process of oxidation of ammonium to nitrate. It is a biological process and occurs under aerobic conditions. The main factors that limit nitrification are substrate NH_4^+, oxygen, soil pH, and temperature (Focht & Verstraete, 1977).

The approach used in the CERES models has been to calculate a potential nitrification rate and a series of zero to unity environmental indices to reduce this rate. The potential nitrification rate is a Michaelis-Menten kinetic function dependent only on ammonium concentration and thus, is independent of soil type. A further index, termed a nitrification capacity index, is introduced, which is designed to introduce a lag effect on nitrification if conditions in the immediate past (previous 2 d) have been unfavorable for nitrification. Actual nitrification capacity is calculated by reducing the potential rate by the most limiting of the environmental indices and the capacity index. The functions reported below were found to be appropriate across the range of data sets tested.

The key functions are

$$A = AMIN1(RP2, WFD, TF, PHN) \qquad [13]$$

where RP2 is the zero to unity index for nitrification potential based on the immediate past (unitless), WFD is the zero to unity index for the soil moisture effect on nitrification rate (unitless), TF is the zero to unity index for soil temperature effects on nitrification rate, and PHN is the zero to unity function for the soil pH effect on nitrification rate. A Michaelis-Menten function described by McLaren (1970) was modified to incorporate these environmental indices and to determine the proportion of the soil ammonium pool that nitrifies in 1 d.

$$\text{RNTRF} = [A \times 40.0 \times \text{NH}_4(L)]/[\text{NH}_4(L) + 90.0] \times \text{SNH}_4(L) \qquad [14]$$

where RNTRF is the rate of nitrification (kg N ha^{-1} d^{-1}), NH$_4$(L) is the concentration of ammonium N in layer L (μg N/g soil), and SNH$_4$(L) is the amount of ammonium in layer L (kg N/ha).

F. Denitrification

Denitrification is the dissimilatory reduction of nitrate (or nitrite) to gaseous products including NO, N$_2$O, and N$_2$ (Knowles, 1981). Denitrification is a microbial process that occurs under anaerobic conditions and is influenced by organic carbon content, soil aeration, temperature, and soil pH (Focht & Verstraete, 1977).

The approach adopted in the CERES models has been to adapt the functions described by Rolston et al. (1980) to fit within the framework of the model, and to match inputs derived from the water balance and mineralization components of CERES. The basic function used by these authors was also used by Davidson et al. (1978) and was the subject of field testing under a variety of conditions in California. Predicted rates of denitrification compared favorably with direct measures of gaseous losses in the field experiments.

Denitrification calculations are performed only when the soil water content (SW) exceeds the drained upper limit (DUL). A zero to unity index (FW) (see Fig. 13-1) for soil water in the range from DUL to saturation (SAT) is calculated as

$$\text{FW} = 1.0 - [\text{SAT}(L) - \text{SW}(L)]/(\text{SAT}(L) - \text{DUL}(L)) \qquad [15]$$

Linn and Doran (1984) used percentage of water-filled porosity as an index of soil-water availability effects on soil N transformations. In their studies, denitrification commenced with a water-filled porosity of 60% and increased linearly up to 100% water-filled porosity. This approximates the linear increase in FW as SW increases from DUL to SAT.

A factor for soil temperature (FT) is also calculated.

$$\text{FT} = 0.1 \times \exp[0.046 \times \text{ST}(L)] \qquad [16]$$

where ST(L) is the temperature (°C) of layer L. Rolston et al. (1980) estimate the water-extractable carbon in soil organic matter (CW) as

$$CW = 24.5 + 0.0031 \times SOILC \qquad [17]$$

In the CERES model, SOILC is calculated as 58% of the stable humic fraction. To this is added the carbon contained in the carbohydrate fraction of the FOM pool. Denitrification rate (DNRATE) is then calculated from the nitrate concentration and converted to a kg N/ha basis for the mass balance calculations.

$$DNRATE = 6.0 \times 1.0 \times 10^{-5}$$
$$\times CW \times NO_3(L) \times FW \times FT \times DLAYR(R)$$

where FT is the temperature factor effect on denitrification (unitless), FW is the water factor effect on denitrification (unitless), $NO_3(L)$ is the nitrate concentration in layer L (μg N/g soil), CW is the total water extractable carbon in the soil layer (μg carbon/g soil), and DNRATE is the denitrification rate (kg N ha^{-1} d^{-1}).

G. Plant Nitrogen Concentrations

Plant growth is greatly affected by the supply of N. Typically the supply of N to plants at the beginning of the season is relatively high and becomes lower as the plant reaches maturity (Jones, 1983). The concentration of N in plant tissues also changes as the plant ages. During early growth, N concentrations are usually high due to synthesis of large amounts of organic N compounds required by the biochemical processes constituting photosynthesis and growth (Novoa & Loomis, 1981). As the plant ages, less of this new material is required and export from the old tissue to new tissue occurs, lowering the whole-plant N concentration. At any point in time there exists a critical N concentration in the plant tissue below which growth will be reduced. These concentrations are determined as a function of crop ontogenetic age and are used within the model as part of the procedure to simulate the effects of N deficiency. Critical N concentration functions related to plant age have been determined for maize (Jones, 1983) and for wheat (D.C. Godwin, 1987, unpublished data). These authors also defined critical concentrations for plant roots for these two crops, and minimum concentrations for tops and roots in both crops.

The critical and minimum concentrations are used to define a N factor (NFAC), which ranges from zero to slightly above unit. The NFAC is the primary mechanism used within the model to determine the effect of N on plant growth. It is an index of deficiency relating the actual concentration in tops (TANC) to these critical concentrations. The NFAC has a value of zero when TANC is at its minimum value of TMNC and increases to 1.0 as concentration increases toward the critical concentration. It is calculated as

$$NFAC = 1.0 - (TCNP - TANC)/(TCNP - TMNC) \qquad [18]$$

where TCNP is the tops critical N concentration (g N/g DW). Since all plant growth processes are not equally affected by N stress, a series of indices based on NFAC are used. The shape and nature of these differ between the maize and wheat models. For the calculation of these indices, NFAC has a maximum value of 1.0. This implies that when TANC exceeds TCNP no extra growth occurs. The relationship between these deficiency indices and NFAC is depicted in Fig. 13-3.

The first of these indices (NDEF1) is used to describe the effects of N deficiency on photosynthesis per unit leaf area. The second (NDEF2) describes the effect of N deficiency on the rate of leaf area expansion and leaf senescence. In the wheat model a third index (NDEF3) is used to describe the effect of N deficiency on tillering, and a fourth factor (NDEF4) is used to modify the rate of grain N accumulation. In the growth component of the models, these N deficiency indices, together with similarly defined zero to unity indices for soil water stress, are used to reduce potential rates of the growth processes. The more limiting of either soil water or N stress is selected in each case to modify the process.

H. Nitrogen Uptake

The approach used in the CERES models calculates the components of crop demand for N and the soil supply of N separately, and then uses the lesser of these two to determine the actual rate of uptake. Demand can be considered to have two components. First, there is a deficiency demand. This is the amount of N required to restore the actual N concentration in the plant (TANC for tops) to the critical concentration (TCNP for tops). Critical concentrations for shoots and roots were previously defined. This deficiency demand can be quantified as the product of the existing biomass and the concentration difference as

$$\text{TNDEM} = \text{TOPWT} \times (\text{TCNP} - \text{TANC}) \qquad [19]$$

where TOPWT is the weight of plant tops (kg/ha) and TNDEM is the plant tops N demand (kg N/ha). Similary, for roots the discrepancy in concentration (difference between RCNP and RANC) is multiplied by the root biomass (RTWT) to calculate the root N demand

$$\text{RNDEM} = \text{RTWT} \times (\text{RCNP} - \text{RANC}). \qquad [20]$$

Luxury consumption of N can occur when the supply of N to the plant is large and the growth of the plant is slowed due to some environmental washout, such as low light flux density. If luxury consumption of N has occurred such that TANC is greater than TCNP, then these demand components have negative values. If total N demand is negative, then no uptake is performed on that day. As biomass increases with crop growth, plant N tends to become diluted and the concentration falls. Thus, when TANC is greater than TCNP, a period of growth will generally cause TANC to fall toward or below TCNP.

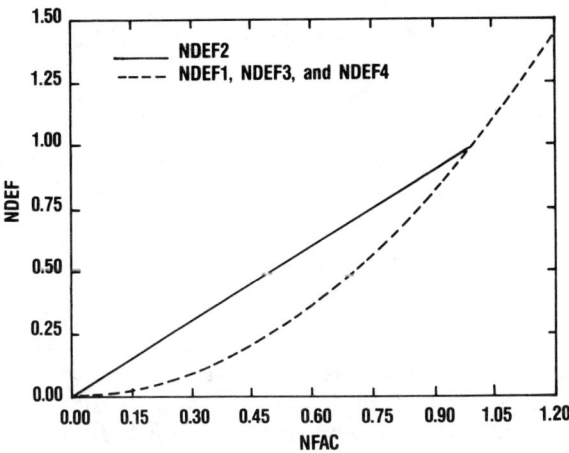

Fig. 13-3. Relationship between N deficiency index (NFAC) and the N deficit factors (NDEF) used to modify plant growth processes.

The second component of N demand is that needed by the new growth. Here the assumption is made that the plant would attempt to maintain a critical N concentration in the newly formed tissues. To calculate the new growth demand a potential amount of new growth is first estimated in the GROSUB subroutine. New growth is estimated from potential photosynthesis (PCARB) and is partitioned into a potential root growth (PGRORT) and a potential tops growth (PDWI). These potential growth increments provide a mechanism for the tops actual N concentration (TANC) to exceed TCNP. This occurs when some stress prevails and the actual growth increment is less than the potential. New growth demand for tops (DNG) is calculated as

$$DNG = PDWI \times TCNP \qquad [21]$$

and the new growth demand for roots is calculated as the product of

$$PGRORT \times RCNP. \qquad [22]$$

During the early stages of plant growth, the new growth component of N demand will be a large proportion of the total demand. As the crop biomass increases, the deficiency demand becomes the larger component. During grain filling, the N required by the grain is removed from the vegetative and root pools to form a grain N pool. The resultant lowering of concentration in these pools may lead to increased demand. The total plant N demand (ANDEM) is the sum of all of these demand components.

To calculate the potential supply of N to the crop, zero to unity availability factors for nitrate (FNO3) and ammonium (FNH4) are calculated from the soil concentrations of the respective ions as

$$\text{FNO3} = 1.0 - \exp[-0.0275 \times \text{NO3(L)}] \qquad [23]$$

$$\text{FNH4} = 1.0 - \exp[-0.025 \times \text{NH4(L)}] \qquad [24]$$

The coefficients used in these two functions, obtained by trial and error, were found to be appropriate for several data sets from the literature.

A zero to unity soil-water factor (SMDFR), which reduces potential uptake, is calculated as a function of the relative availability of soil water

$$\text{SMDFR} = [\text{SW(L)} - \text{lL(L)}/\text{ESW(L)} \qquad [25]$$

where ESW(L) is the extractable soil water in the layer L. To account for increased anaerobiosis and declining root function at moisture contents above the drained upper limit, SMDFR is reduced as saturation is approached. The maximum potential N uptake from a layer may be calculated as a function of the maximum daily uptake per unit length of root and the total amount of root present in the layer. The first of these is a temporary variable (RFAC), which integrates the effects of root length density [RLV(L)] (cm root/cm^3 soil), the soil water factor described above, and the depth of the layer [DLAYR(L) (cm) as

$$\text{RFAC} = \text{RLV(L)} \times \text{SMDFR} \times \text{SMDFR} \times \text{DLAYR(L)} \times 100.0 \qquad [26]$$

The second of these equations incorporates the ion concentration effect (FNO3) and the maximum daily uptake per unit length of root (0.006 kg N/ha cm root) to yield a potential uptake of nitrate from the layer [RNO3U(L)].

$$\text{RNO3U(L)} = \text{RFAC} \times \text{FNO3} \times 0.006. \qquad [27]$$

Thus, [RNO3U(L)] is the potential uptake of nitrate from layer L in kg N/ha constrained by the availability of water, the root length density, and the concentration of nitrate. Initial estimates for the maximum uptake per unit length of root coefficient were obtained from the maize root data of Warncke and Barber (1974). This estimate was the subject of continuing modification during early model development. The value reported here appears to be appropriate across a broad range of data sets. The effect of each of these parameters on determining potential uptake can be seen in Fig. 13-4. A similar function is employed to calculate the potential uptake of ammonium [RNH4U(L)].

$$\text{RNH4U(L)} = \text{RFAC} \times \text{FNH4} \times 0.006 \qquad [28]$$

To account for declining root function with increasing plant age during grain filling, a further term to reduce RFAC is also introduced.

Potential N uptake from the whole profile (TRNU) is the sum of RNO3U(L) and RNH4U(L) from all soil layers where roots occur. Thus, TRNU represents an integrated value that is sensitive to (i) rooting density;

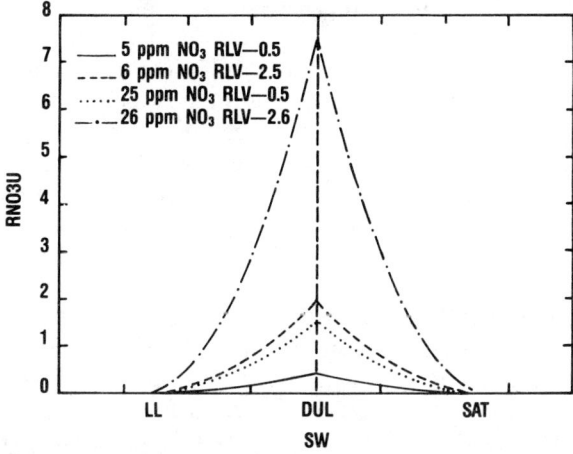

Fig. 13-4. Simulated effect of changing soil water status (SW) on uptake of nitrate from a layer (RNO3U). The LL, DUL, SAT represent the lower limit, drained upper limit, and saturation moisture contents, respectively. The four curves refer to differing concentrations of nitrate (ppm NO3) and differing root length densities (RLV in cm root/cubic cm of soil).

(ii) the concentration of the two ionic species; and (iii) their ease of extraction as a function of the soil-water status of the different layers. This method of determining potential uptake enables the condition of nutritional drought to be simulated. Nutritional drought occurs when nutrients and roots are concentrated in the upper layers of the soil profile, but sufficient water for growth and uptake is present only in the lower layers.

If the potential N supply from the whole profile (TRNU) is greater than the crop N demand (ANDEM), a N uptake factor (NUF) is calculated and used to reduce the N uptake from each layer to the level of demand.

$$\text{NUF} = \text{ANDEM}/\text{TRNU}. \qquad [29]$$

This could occur when plants are young and have a high N supply. If the demand is greater than the supply, then NUF has a value of 1.0. When NUF is <1.0, uptake from each layer is reduced.

Under conditions of luxury N upake (TANC > TCNP) exudation of organic N compounds can occur. Rovira (1969) found that changes in the shoot environment, which cause more rapid growth, can increase exudation and Bowen (1969) reported that N deficiency can cause it to decrease. In the CERES-N model this exuded N is added to the fresh organic N pool [FON(L)], and can be mineralized and subsequently made available to the plant again. The amount of N that can be lost daily from the plant in this manner is calculated as 5% of the N contained in the roots/day.

Following uptake, concentrations of N in each of the shoots and roots are updated. Partitioning of the N taken up between shoot and root parts occurs on the basis of the proportions of the total plant demand arising from shoots and roots, respectively.

I. Grain Nitrogen Concentration

In many crop-growing areas, when the crop reaches the grain-filling stage soil supplies of N are very low. In these cases the N requirement of the developing grains is largely satisfied by remobilization of protein N from vegetative organs. When N supply is increased, the proportion of grain N arising from remobilization declines and the proportion from uptake increases (Vos, 1981). Many studies (Benzian et al., 1983, Terman et al., 1969) have found negative correlations between grain yield and grain protein concentration. Temperature and soil moisture also affect the grain N content.

Grain N determination is simulated differently in the maize and wheat models. In maize, the concentration of N in each day's new grain growth is modeled as a function of a temperature index, a soil water index, and an index based on NFAC. This is accomplished as follows.

First, two factors, TFAC and SFAC, are calculated and used to estimate the effects of mean temperature (TEMPMN) and drought stress (SWDF2) on the concentration of nitrogen (GNP) in GROGRN for the day. The AMAX1 is a FORTRAN function subprogram that finds the maximum of two or more values.

$$\text{TFAC} = 0.69 + 0.125 \times \text{TEMPM} \quad [30]$$

$$\text{SFAC} = 1.125 - 0.125 \times \text{SWDF2} \quad [31]$$

$$\text{GNP} = (0.007 \times 0.010 \times \text{NDEF2}) \times \text{AMAX1}(\text{TFAC,SFAC}) \quad [32]$$

Thus, high mean temperatures and drought stress can cause the value of AMAX1(TFAC,SFAC) to exceed 1.0. Nitrogen deficiency can cause NDEF2 (calculated in subroutine NFAC) to be <1.0. The net effect of these equations is to allow GNP to range from <0.010 when N deficiency is severe to about 0.018 when adequate N is available. However, high temperatures or drought stress limit grain growth.

In the wheat model the N accumulation rate of single grains is simulated as a function dependent on the daily maximum and minimum temperature and is quite similar to the functions for the rate of grain mass accumulation. Grain N accumulation is more sensitive to temperature than grain mass accumulation, thus, at high temperatures grain N concentrations will tend to be higher. Moisture stress affects crop photosynthesis, which supplies assimilate to developing grains. Moisture stress will affect the simulated grain N concentration in wheat by altering the dilution of N in the grain by assimilate.

In both models, a daily grain N sink is determined based on this daily concentration of N in new grain growth and the amount of new grain growth. In the wheat model, the sink is adjusted according to the N status of the plant. This will influence grain N concentration.

Two pools of N within the plant are available for translocation: a shoot pool (NPOOL1) and a root pool (NPOOL2). These pools are determined from the N concentration (VANC or RANC) relative to the critical concentration (VMNC or RMNC) and the biomass of the pool (RTWT or TOPWT).

$$\text{NPOOL1} = \text{TOPWT} \times (\text{VANC} - \text{VMNC}) \quad [33]$$

$$\text{NPOOL2} = \text{RTWT} \times (\text{RANC} - \text{RMNC}) \quad [34]$$

The total N available for translocation (NPOOL) is the sum of these two pools. When NPOOL is not sufficient to supply the grain N demand (NSINK), NSINK is reduced to NPOOL. If NSINK is greater than that which can be supplied by the tops (NPOOL1), then NPOOL1 is set to zero and the tops N concentration is set to its minimum value (VMNC). The remaining NSINK is then satisfied from the root N pool and the N pool is updated accordingly.

$$\text{NPOOL2} = \text{NPOOL2} - (\text{NSINK} - \text{NPOOL1}) \quad [35]$$

The root N concentration is then updated.

$$\text{RPOOLN} = \text{RTWT} \times \text{RMNC} \times \text{NPOOL2} \quad [36]$$

$$\text{RANC} = \text{ROOTN}/\text{RTWT} \quad [37]$$

When NSINk < NPOOL1, it is totally satisfied from tops N

$$\text{NPOOL1} = \text{NPOOL1} - \text{NSINK} \quad [38]$$

and the tops concentrations are updated accordingly.

$$\text{TOPSN} = \text{NPOOL1} + \text{VMNC} \times \text{TOPWT} \quad [39]$$

$$\text{VANC} = \text{TOPSN}/\text{TOPWT} \quad [40]$$

The total amount of N contained in the grain can tehn be accumulated

$$\text{GRAINN} = \text{GRAINN} + \text{NSINK}. \quad [41]$$

This routine, together with the remainder of the growth routine and the N deficiency indices, thus provides several pathways by which N stress during grain filling can affect grain yield and grain protein content. First, as N is removed from the vegetative tissues, NFAC will become lower. This will, in turn, lower NDEF4 and lower the sink size for N, thus providing for the capability of reduced grain N concentration. Lowering NFAC will also lower NDEF1, which will cause the rate of crop photosynthesis to fall, thus lowering the assimilate available for grain filling. A declining NFAC will also speed the rate of senescence, which will reduce the leaf area available for photosynthesis. Soil water stress during grain filling can also increase the grain N concentration since it will reduce photosynthesis. This would lower assimilate availability and, thus, not dilute grain N as much as would occur in an unstressed crop.

II. MODEL INPUTS

To facilitate model usage across a wide range of locations and model applications, the required input data have been kept to a minimum. In addition to the data required by the CERES crop growth models, the N model requires extractable soil nitrate and ammonium concentrations, organic carbon, bulk density, and pH in each layer. The model also requires an estimate of the amount of crop residue present and its C/N ratio, as well as fertilizer information including the date, amount, and depth of all applications made to the crop, and the type of fertilizer used.

III. MODEL EVALUATION

Testing of the N component of the CERES models has occurred simultaneously with the testing of the crop growth components of the model. Detailed documentation of this testing can be found in Otter-Nacke et al. (1985) and Godwin (1987, unpublished data) for the wheat model and in Jones and Kiniry (1986) and Singh (1985, unpublished data) for the maize model. In the case of both models, test data sets for these two crops were obtained from diverse locations spanning the world's cropping regions. These data sets also represented a diversity of soil types, fertilizer practices, and climatic conditions. In each data set, pests, diseases, and nutrients other than N were not considered to be limiting to growth. The testing of each model (Fig. 13-5, 13-6, and 13-7) indicates that the model generally provides reasonable predictions of grain-yield N uptake by the aboveground plant and the partitioning of this N into grain.

IV. MODEL LIMITATIONS AND USES

It should be stressed that the model can only account for variations in the factors defined in the model's description and this assumes that other potentially important factors, such as other nutrients, pests, and diseases, are nonlimiting. The model is unable to account for volatile losses of ammonia from the soil surface, leaching of ammonium N, or direct losses of N from plants into the atmosphere. Where these losses are substantial, the model will be inaccurate. Since the soil water and N components of the model

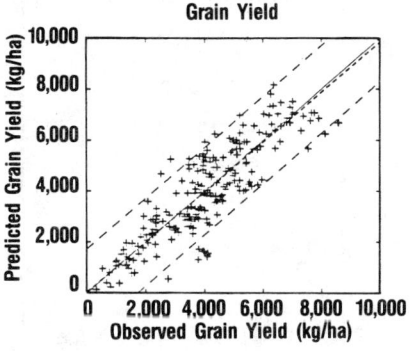
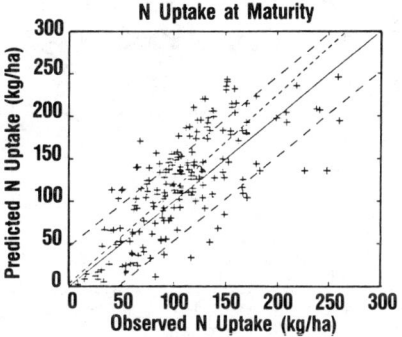
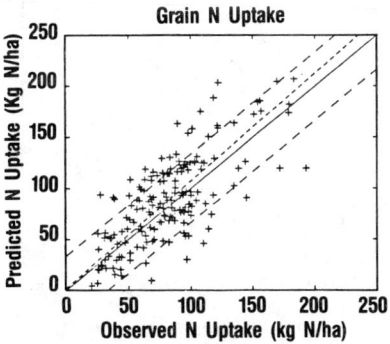

Fig. 13-5. Comparison of predictions of the CERES-WHEAT model with observed data from experiments: (a) Grain yield; (b) Nitrogen uptake at maturity; and (c) Grain N uptake (from Otter-Nacke et al., 1985).

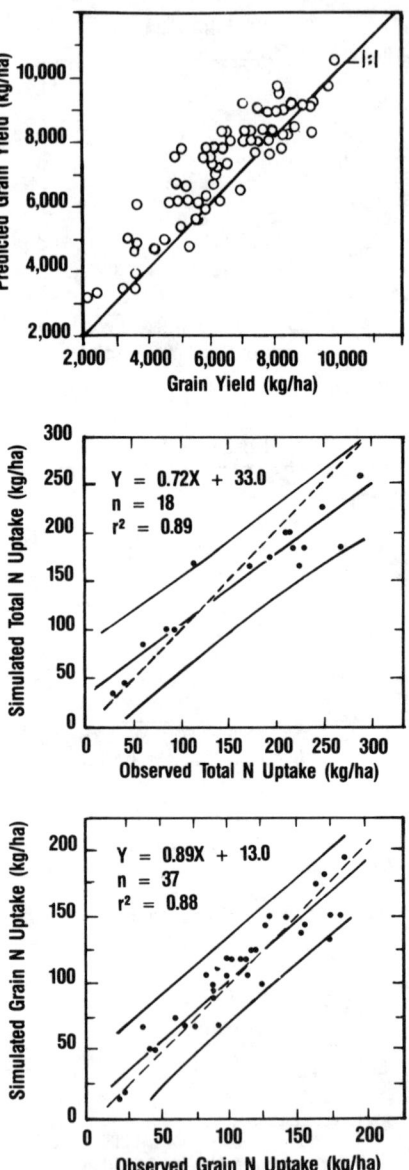

Fig. 13-6. Comparison of predictions of the CERES-MAIZE model with observed data from field experiments: (a) Grain yield (from Singh, 1985, unpublished data); (b) Crop N uptake at maturity for four cultivars at four locations (from Jones and Kiniry, 1986); and (c) Grain N uptake at maturity for nine cultivars at seven sites.

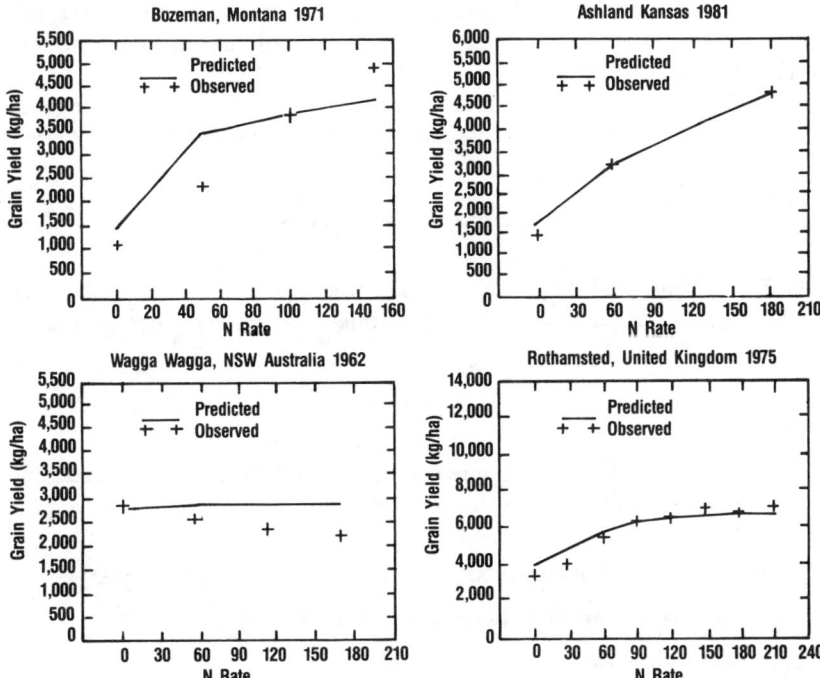

Fig. 13-7. Comparison of predicted (lines) and observed (pluses) grain yield response to applied N in individual data sets using the CERES-Wheat model. (a) Bozeman, MT (observed data from Christensen & Killorn, 1981); (b) Ashland, KS (observed data from Wagger, 1983); (c) Wagga Wagga, NSW, Australia (observed data from Storrier, 1965); and (d) Rothamsted, U.K. (observed data from Pearman et al., 1978).

operate in a one-dimensional manner, placement patterns such as side banding and point placement can only be simulated by assuming uniform incorporation into a layer. Under certain conditions this may lead to erroneous predictions. These various problems will be the subject of future efforts to further enhance the model's versatility. From the validation studies cited above, it appears that the model is able to explain most of the observed variation in yield and other key parameters where these conditions of loss and placement do not apply.

The model, particularly when coupled with long-term weather data or generated climatic data, is a valuable tool for providing insights into the behavior of many aspects of a cropping system. Running the model with long-term weather data enables a quantification of the temporal variability in yield and response to fertilizer. The frequency and nature of N losses from the system leading to poor fertilizer efficiency can be identified and their relative significance evaluated. Using the model, experiments can be performed to predict the effects of varying fertilizer rates, timing, placement depths, fertilizer sources, planting times, etc. In this way, a fertilizer strategy to maximize efficiency under the uncertainties of climatic variability can be readily obtained. Studies such as this have been reported by Godwin and Vlek (1985).

V. ACKNOWLEDGMENTS

The authors wish to gratefully acknowledge the researchers who generously provided unpublished data for model development and testing.

VI. APPENDIX 1

A listing of the four subroutines used in the CERES models for simulating N transformations, uptake, movement, and deficiency factors.

```
C****************DRAINAGE AND LEACHING ROUTINE********************
C
      SUBROUTINE NFLUX(iflag,skgn,sppm,fac)
C
$Include: 'Whea3.Blk'
$Include: 'Nmove.Blk'
C
      Dimension skgn(15),sppm(15),fac(15)
      if(iflag.eq.1) go to 300
      KPFLG=0
90    OUTN=0.0
      DO 200 L=1,NLAYR
          skgn(L)=skgn(L)+OUTN
          sppm(L)=skgn(L)*FAC(L)
          xmin=0.25/fac(l)
100       OUTN=(skgn(L)-xmin)*FLUX(L)/(SW(L)*DLAYR(L)+FLUX(L))
          IF(KPFLG.EQ.1)OUTN=OUTN*0.5
          skgn(L)=skgn(L)-OUTN
          sppm(L)=skgn(L)*FAC(L)
          If(L.EQ.Nlayr) Tlch=Tlch+outn
200   CONTINUE
      RETURN
300   continue
      OUTN=0.0
      KPFLG=0
      DO 500 J=1,MU
          K=MU+1-J
          skgn(K)=skgn(K)+OUTN
          IF(K.EQ.1) GO TO 500
          IF(FLOW(K-1).GE.0.0)GO TO 400
          OUTN=0.0
          KPFLG=1
          GO TO 500
400       xmin=0.25/fac(k)
          OUTN=(skgn(K)-xmin)*(FLOW(K-1)/(SW(K)*DLAYR(K)+FLOW(K)))*0.5
          skgn(K)=skgn(K)-OUTN
500   CONTINUE
      IF(KPFLG.EQ.0)RETURN
      OUTN=0.0
      DO 700 L=1,NLAYR
```

NITROGEN DYNAMICS IN SOIL-PLANT SYSTEMS

```
              FLUX(L)=0.0
              IF(FLOW(L).LT.0.0)FLUX(L)=FLOW(L)*(-1.0)
  700   CONTINUE
        GO TO 90
        END
C
C*******************NITROGEN DEFICIENCY FACTOR ROUTINE************
C
        SUBROUTINE NFACTO
C
$Include: 'Whea1.Blk'
$Include: 'Whea2.Blk'
$Include: 'Ntrc1.Blk'
C
        REAL NFAC
C***** CONVERT XSTAGE TO ZADOKS GROWTH STAGE
        IF(XSTAGE.LE.2.0)ZSTAGE=XSTAGE
        IF(XSTAGE.GT.2..AND.XSTAGE.LE.3.0)ZSTAGE=2.0+2.0*(XSTAGE-2.0)
        IF(XSTAGE.GT.3..AND.XSTAGE.LE.4.)ZSTAGE=4.0+1.7*(XSTAGE-3.0)
        IF(XSTAGE.GT.4..AND.XSTAGE.LE.4.4)ZSTAGE=5.7+0.8*(XSTAGE-4.0)
        IF(XSTAGE.GT.4.4.AND.XSTAGE.LE.6.)ZSTAGE=6.02+1.8625*(XSTAGE-4.4)
        YSTAGE=XSTAGE
        zs2=zstage*zstage
        zs3=zs2*zstage
        zs4=zs3*zstage
        IF(P1V.ge.0.03)THEN
          TCNP=-5.0112400-6.350677*ZSTAGE+14.95784*SQRT(ZSTAGE)
     1 +0.2238197*ZS2
        ELSE
          TCNP=7.4531813-1.7907829*ZSTAGE+0.6092849*SQRT(ZSTAGE)
     1 +0.0933967*ZS2
        ENDIF
        IF(ZSTAGE.GT.6.0)TCNP=TCNP-(ZSTAGE-6.0)*0.140
        TCNP=TCNP/100.0
        TMNC=(2.97-0.455*XSTAGE)/100.0
C       RCNP=(-0.25*XSTAGE+2.5)/100.0
        rcnp=(2.10-0.14*sqrt(zstage))*0.01
        IF (ISTAGE.LT.5) GO TO 5
        TANC=VANC
        VMNC=TMNC
  5     NFAC=1.0-(TCNP-TANC)/(TCNP-TMNC)
        if(nfac.lt.0.02)nfac=0.02
        IF (XSTAGE.LE.1.10) NFAC=1.0
        ndef4=nfac*nfac
        if(ndef4.gt.1.5)ndef4=1.5
        IF (TANC.GE.TCNP) NFAC=1.0
        NDEF3=NFAC*NFAC
        NDEF1=0.10+2.*NFAC
        NDEF2=NFAC
        IF(NDEF2.GT.1.0)NDEF2=1.0
        IF(NDEF1.GT.1.0)NDEF1=1.0
        IF(NDEF3.GT.1.0)NDEF3=1.0
        CNSD1=CNSD1+1.0-Ndef1
        CNSD2=CNSD2+1.0-NDEF2
        RETURN
C
  10    FORMAT (' EXTREME NITROGEN DEFICIENCY INCURRED ON DAY',I4)
        END
C
C
C
C
C
C
        SUBROUTINE SOLT
C
```

```
$Include: 'Whea1.blk'
$Include: 'Whea4.Blk'
$Include: 'Whea3.Blk'
$Include: 'Ntrc2.Blk'
      XI=DOY
      ALX=(XI-HDAY)*0.0174
      ATOT=ATOT-TMA(5)
      K=5
  100 TMA(K)=TMA(K-1)
      K=K-1
      IF (K.GT.1) GO TO 100
      TMA(1)=(1.-ALBEDO)*(Tempm+(TEMPMX-Tempm)*SQRT(SOLRAD*.03))
    1 +ALBEDO*TMA(1)
      ATOT=ATOT+TMA(1)
      AW=PESW
      IF(AW.LE.0.0)AW=0.01
      WC=AW/(WW*depmax*10.)
      F=EXP(B*((1.-WC)/(1.+WC))**2)
      DD=F*DP
      TA=TAV+AMP*COS(ALX)/2.
      DT=ATOT/5.-TA
      DO 200 L=1,NLAYR
           ZD=-Z(L)/DD
           ST(L)=TAV+(AMP/2.*COS(ALX+ZD)+DT)*EXP(ZD)
  200 CONTINUE
      RETURN
      END
C
      SUBROUTINE NUPTAK
C
$Include: 'Whea1.blk'
$Include: 'Whea2.Blk'
$Include: 'Whea3.Blk'
$Include: 'Whea4.Blk'
$Include: 'Ntrc1.Blk'
$Include: 'Ntrc2.Blk'
$Include: 'Nmove.Blk'
C
      REAL NUF,ndem
      DIMENSION RNO3U(15), RNH4U(15)
      TRNLOS=0.0
      ranc=rootn/rtwt
      tanc=topsn/topwt
      trlv=0.
      Ano3=0.0
      Anh4=0.0
      DO 100 L=1,NLAYR
           rno3u(l)=0.
           rnh4u(l)=0.
           trlv=trlv+rlv(l)
           NO3(L)=SNO3(L)*FAC(L)
           NH4(L)=SNH4(L)*FAC(L)
           Ano3=ano3+Sno3(l)
           Anh4=anh4+Snh4(L)
           TOTN=NO3(L)+NH4(L)
  100 CONTINUE
      IF (PDWI.EQ.0.) PDWI=1.
      DNG=PDWI*TCNP
      IF (XSTAGE.LE.1.2) DNG=0.0
      TNDEM=TOPWT *(TCNP-TANC)+DNG
      RNDEM=RTWT*(RCNP-RANC)+PGRORT*RCNP
      NDEM=TNDEM+RNDEM
      ANDEM=NDEM*PLANTS*10.0
      DROOTN=0.0
      DSTOVN=0.0
```

NITROGEN DYNAMICS IN SOIL-PLANT SYSTEMS

```
      TRNU=0.0
      TNUP=0.0
      nuf=0.
      DO 200 L=1,NLAYR
         IF (RLV(L).EQ.0.0) GO TO 300
         L1=L
         FNH4=1.0-EXP(-0.025*(NH4(L)-0.5))
         FNO3=1.0-EXP(-0.0275*NO3(L))
         IF (FNO3.LT.0.03) FNO3=0.0
         IF (FNO3.GT.1.0) FNO3=1.0
         IF (FNH4.LT.0.03) FNH4=0.0
         IF (FNH4.GT.1.0) FNH4=1.0

            SMDFR=(SW(L)-LL(L))/ESW(L)
            if(smdfr.lt.0.0)smdfr=0.
            if(smdfr.gt.1.0)smdfr=(sat(1)-sw(1))/(sat(1)-dul(1))
            RFAC=RLV(L)*SMDFR*SMDFR*DLAYR(L)*100
    c          if(xstage.gt.5.0)rfac=rfac*(6.0-xstage)
            RNO3U(L)=RFAC*FNO3*0.009
            RNH4U(L)=RFAC*FNH4*0.009
            TRNU=TRNU+RNO3U(L)+RNH4U(L)
  200 CONTINUE
  300 If(andem.le.0.)then
          Trnu=0.
          Nuf=0.
          Go to 400
      Endif
      IF (ANDEM.GT.TRNU) ANDEM=TRNU
      IF (TRNU.EQ.0.0) GO TO 600
      NUF=ANDEM/TRNU
        Trnu=0.0
  400 Continue
      DO 500 L=1,L1
          UNO3=RNO3U(L)*NUF
          UNH4=RNH4U(L)*NUF
          Xmin=0.25/FAC(L)
          if(uno3.gt.sno3(1)-xmin)uno3=sno3(1)-xmin
          SNO3(L)=SNO3(L)-UNO3
          Xmin=0.5/FAC(L)
          if(unh4.gt.(snh4(1)-xmin))unh4=snh4(1)-xmin
          SNH4(L)=SNH4(L)-UNH4
          NO3(L)=SNO3(L)*FAC(L)
          NH4(L)=SNH4(L)*FAC(L)
          Rnloss=0.0
          if(tanc.gt.tcnp)rnloss=ranc*rtwt*0.05*plants*rlv(1)/trlv
          TRNLOS=TRNLOS+RNLOSS
          fon(1)=fon(1)+rnloss
          trnu=trnu+uno3+unh4
  500 CONTINUE
      TRNU=TRNU/(PLANTS*10.0)
      if(ndem.gt.trnu)Then
          xndem=trnu
          Factor=xndem/ndem
          ndem=xndem
          tndem=tndem*factor
          rndem=rndem*factor
      Endif
      DTOPSN=TNDEM/NDEM*TRNU-ptf*trnlos/(plants*10)
      DROOTN=RNDEM/NDEM*TRNU-(1.0-ptf)*trnlos/(plants*10)
      TOPSN=TOPSN+DTOPSN
      TANC=TOPSN/TOPWT
      ROOTN=ROOTN+DROOTN
      RANC=ROOTN/(RTWT-0.01*RTWT)
      tno3=0.0
      tnh4=0.0
      do 599 l=1,nlayr
```

```
              tno3=tno3+sno3(1)
              tnh4=tnh4+snh4(1)
599     Continue
  600 RETURN
        END
C
C
C
C
C*****************N Transformations Subroutine***********************
C
        SUBROUTINE NTRANS
C
$Include: 'Whea1.blk'
$Include: 'Whea3.Blk'
$Include: 'Whea4.Blk'
$Include: 'Ntrc1.Blk'
$Include: 'Ntrc2.Blk'
$Include: 'Nmove.Blk'
        Real Nnom,Mf
C
        Dimension rdecr(3),PROF(15)
        data rdecr/0.2,0.05,0.0095/
        tifon=0.0
        thumn=0.0
        IF (nFERT .eq. 0) GO TO 900
        DO 10 L=1,NLAYR
          PROF(L)=0.
10      CONTINUE
        DEPTH=0.0
        KMAX=1
        DO 100 J=1,NFERT
          IF (DOY .eq. Fday(J)) THEN
            M=IFTYPE(J)
            IF(M.EQ.17) M=12
            KMAX=IDLAYR(NLAYR,DLAYR,DFERT(J))
            CD=0.0
            IF(KMAX.EQ.1) THEN
              PROF(1)=1.0
C             *** Surface layer incorporation ***
            ELSE
              CD=DLAYR(1)
              PROF(1)=DLAYR(1)/DFERT(J)
              DO 20 L=2,KMAX
                CD=CD+DLAYR(L)
                IF (DFERT(J).LE.CD) THEN
                  PROF(L)=(DFERT(J)-(CD-DLAYR(L)))/DFERT(J)
                ELSE
                  PROF(L)=DLAYR(L)/DFERT(J)
                ENDIF
20            CONTINUE
            ENDIF
            GO TO (500,400,500,400,300,400,400,600,400,700,400,600), M
C           FERTILIZER TYPES AS GIVEN IN APPENDIX 4, TECHNICAL REPORT 1,
C             IBSNAT (1986).
C             1      =AMMONIUM NITRATE
C             2      =AMMONIUM SULPHATE
C             3      =AMMONIUM NITRATE-SULPHATE
C             4      =ANHYDROUS AMMONIA
C             5      =UREA (HYDROLYSIS TO BE ADDED LATER)
C             6      =DIAMMONIUM PHOSPHATE
C             7      =MONOAMMONIUM PHOSPHATE
C             8      =CALCIUM NITRATE
C             9      =AQUA AMMONIA
C             10     =UREA AMMONIUM NITRATE
```

```
C               11    =CALCIUM AMMONIUM NITRATE
C               17    =POTASSIUM NITRATE
C
  300         DO 350 K=1,KMAX
                UREA(K)=UREA(K) + AFERT(J)*PROF(K)
  350         CONTINUE
              IUON=.TRUE.
              IUOF=DOY + 21
              IF(DOY.GT.344) IUOF=21-(365-DOY)
              GO TO 900
  400         DO 450 K=1,KMAX
                SNH4(K)=SNH4(K)+AFERT(J)*PROF(K)
  450         CONTINUE
              GO TO 900
  500         DO 550 K=1,KMAX
                SNH4(K)=SNH4(K)+0.5*AFERT(J)*PROF(K)
                SNO3(K)=SNO3(K)+0.5*AFERT(J)*PROF(K)
  550         CONTINUE
              GO TO 900
  600         DO 650 K=1,KMAX
                SNO3(K)=SNO3(K)+AFERT(J)*PROF(K)
  650         CONTINUE
              GO TO 900
  700         DO 750 K=1,KMAX
                SNO3(K)=SNO3(K)+AFERT(J)*0.25*PROF(K)
                SNH4(K)=SNH4(K)+AFERT(J)*0.25*PROF(K)
                UREA(K)=UREA(
                                K)+AFERT(J)*0.50*PROF(K)
  750         CONTINUE
              IUON=.TRUE.
              IUOF=DOY + 21
              IF(DOY.GT.344) IUOF=21-(365-DOY)
              GO TO 900
            ENDIF
  100   CONTINUE
  900   CONTINUE
        IF(DOY.EQ.IUOF) THEN
C ***** Assume all urea was hydrolyzed.
          DO 1000 L=1,NLAYR
                SNH4(L)=SNH4(L) + UREA(L)
                UREA(L)=0.0
                UPPM(L)=0.0
 1000     CONTINUE
          IUON=.FALSE.
        ENDIF
        DO 1500 l=1,NLAYR
            ad=ll(l)
            if(l.eq.1)ad=ll(l)*swef
            mf=(sw(l)-ad)/(dul(l)-ad)
            wfd=mf
            if(sw(l) .gt. dul(l))then
                xl=(sw(l)-dul(l))/(sat(l)-dul(l))
                mf=1.0-0.5*xl
                wfd=1.0-xl
            endif
            if(wfd .lt. 0.0)wfd=0.0
            NO3(l)=SNO3(l)*FAC(l)
            NH4(l)=SNH4(l)*FAC(l)
            UPPM(L)=UREA(L)*FAC(L)
            IF(IUON) THEN
C ***** DO SOME UREA HYDROLYSIS.
                SWF=MF+0.20
                IF(SWF.GT.1.0) SWF=1.0
                if(swf.lt.0.) swf=0.
                TF=(ST(L)/40.0)+0.20
```

```
                IF(TF.LT.0) TF=0.0
                AK=-1.12+1.31*OC(L)+0.203*PH(L)-0.155*OC(L)*PH(L)
                IF(AK.LT.0.25) AK=0.25
                UHYDR=AK*AMIN1(SWF,TF)*UREA(L)
                IF(UHYDR.GT.UREA(L)) UHYDR=UREA(L)
                UREA(L)=UREA(L)-UHYDR
                SNH4(L)=SNH4(L)+UHYDR
            ENDIF
            tf=(st(1)-5.0)/30.0
            if(st(1) .lt. 5.0)tf=0.0
            TOTN=SNO3(1)+SNH4(1)-0.5/FAC(1)
            IF (TOTN .lt. 0.0) TOTN=0.0
            CNR=(0.4*FOM(1))/(FON(1)+TOTN)
            CNRF=EXP(-0.693*(CNR-25)/25.0)
            IF (CNRF .gt. 1.0) CNRF=1.0
            fom(1)=0.0
            grcom=0.0
            grnom=0.0
            if(mf.lt.0.)mf=0.
            do 1200 jp=1,3
                fom(1)=fom(1)+fpool(1,jp)
                if(fom(1).lt.5.0)then
                    g1=0.0
                    go to 1100
                endif
                g1=tf*mf*cnrf*rdecr(jp)
                x=fpool(1,jp)/fom(1)
                fpool(1,jp)=fpool(1,jp)-fpool(1,jp)*g1
 1100           grcom=grcom+g1*x*fom(1)
                grnom=grnom+g1*x*fon(1)
 1200       continue
            RHMIN=NHUM(1)*DMINR*TF*MF*dmod
            HUM(1)=HUM(1)-RHMIN*10.0+0.2*GRNOM/0.04
            NHUM(1)=NHUM(1)-RHMIN+0.2*GRNOM
            RNAC=AMIN1(TOTN,grcom*(0.02-FON(1)/FOM(1)))
            IF(RNAC .lt. 0.)RNAC=0.
            FOM(1)=FOM(1)-grcom
            FON(1)=FON(1)+RNAC-GRNOM
            NNOM=0.8*GRNOM+RHMIN-RNAC
            tifon=tifon+fon(1)
            thumn=thumn+nhum(1)
            XMIN=0.5/FAC(L)
c     net mineralization added here
c     immobilization if nnom is less than 0.
c     first got N from ammonium- when too low get remainder from nitrate
            if(nnom.gt.0)then
                snh4(1)=snh4(1)+nnom
            else
                if(abs(nnom).gt.(snh4(1)-xmin))then
                    nnom=nnom+snh4(1)-xmin
                    Snh4(1)=xmin
                else
                    Snh4(1)=snh4(1)+nnom
                    nnom=0.0
                endif
                Sno3(1)=sno3(1)+nnom
            Endif
            tmin=tmin+nnom
c           c************nitrification section
 1300       sanc=1.0-exp(-0.01363*snh4(1))
            ELNC=AMIN1(TF,WFD,SANC)
            RP2=CNI(1)*EXP(2.302*ELNC)
            IF (RP2 .lt. 0.05) RP2=0.05
            IF (RP2 .gt. 1.0) RP2=1.0
            CNI(1)=RP2
```

NITROGEN DYNAMICS IN SOIL-PLANT SYSTEMS

```
            A=AMIN1(RP2,WFD,TF,phn(1))
            NH4(1)=SNH4(1)*FAC(1)
            bb=(a*40.0*NH4(1)/(NH4(1)+90.0))
            if(bb.gt.0.80)bb=0.80
            bb=bb*snh4(1)
            b2=(nh4(1)-0.5)/fac(1)
              rntrf=amin1(bb,b2)
            SNH4(1)=SNH4(1)-RNTRF
            SNO3(1)=SNO3(1)+RNTRF
            SARNC=1.0-EXP(-0.1363*SNH4(1))
            XW=AMAX1(WFD,WFY(1))
            XT=AMAX1(TF,TFY(1))
            CNI(1)=CNI(1)*AMIN1(XW,XT,SARNC)
            IF (CNI(1) .le. 0.05) CNI(1)=0.05
            WFY(1)=WFD
            TFY(1)=TF


c*********denitrification section
            dnrate=0.0
            NO3(1)=SNO3(1)*FAC(1)
            IF(NO3(L).lt.1.0.OR.SW(1).le.dul(1).or.st(1).lt.5) GO TO 1400
            FW=0.0
            SOILC=0.58*HUM(L)
            CW=FAC(L)*(SOILC*0.0031+0.4*fpool(1,1))+24.5
            FW=1.0-(SAT(L)-SW(L))/(SAT(L)-DUL(L))
            FT=0.1*EXP(0.046*ST(L))
            DNRATE=6.0*1.E-05*CW*NO3(L)*BD(L)*FW*FT*DLAYR(L)
c c            If(Dnrate.lt.0.) Dnrate=0.0
            xmin=0.25/fac(1)
            IF(DNRATE.GT.(SNO3(L)-xMIN)) DNRATE=SNO3(L)-xMIN
            SNO3(L)=SNO3(L)-DNRATE
            Tnox=Tnox+dnrate
 1400       Continue
            NO3(L)=SNO3(L)*FAC(L)
            NH4(L)=SNH4(L)*FAC(L)
 1500 continue
      RETURN
      END
C*****
      Function Idlayr(Nlayr,Dlayr,Fdepth)
      Dimension Dlayr(Nlayr)
c****** routine used to identify layer where fertilizer is placed
      Depth=0
      Idlayr=1
      Do 10 l=1,nlayr
      Depth=Depth+Dlayr(l)
      If(Fdepth.gt.Depth)go to 10
      Idlayr=L
      Return
 10   Continue
      Idlayr=Nlayr
      Return
      End
```

VII. APPENDIX 2

Glossary of the variables used in the N subroutines of the CERES models.

A	Zero to unity factor for relative nitrification rate (unitless)
AFERT(J)	Amount of N added as fertilizer on JFDAY(J) (kg N/ha)
AK	Nondimensional constant defining proportion of urea which hydrolyzes in 1 d
ANDEM	Crop N demand (kg N/ha)
BB	Maximum rate of nitrification (kg N/ha per day)
BD	Bulk density of soil (g/cm**3)
CNI(L)	Capacity for nitrification index in layer L. This is a zero to unity number indicating the relative capability for nitrification to proceed
CNR	C/N ratio calculated as (kg C in FOM)/(kg N in FOM + kg mineral N)
CNRF	Zero to unity C/N ratio factor for decomposition rate
CNSD1	Accumulates N deficit factor (NFAC) in each state and is printed at the end of each stage as a daily average
CNSD2	Accumulates N deficit factor (NDEF2) in each stage and is printed at the end of each stage as a daily average
CW	Water soluble carbon content of soil (ppm)
DEF	Interim variable used to ensure soil N pools remain positive
DEPTH	Depth to the bottom of a layer from the surface (cm)
DEFERT(J)	Depth of incorporation of fertilizer application on Julian date (JFD)
DLAYR(L)	Depth increment of soil layer L – cm
DMINR	Humic fraction decay rate (1/days)
DMOD	Zero to unity dimensionless factor used to decrease to rate of mineralization in soils with chemically protected organic matter
DNG	N demand of potential new growth of tops (g N/plant)
DNRATE	Denitrification rate (kg N/ha per day)
DOY	Day of the year
DROOTN	Daily change in plant root N content (g N/plant)
DTOPSN	Daily change in plant tops N content (g N/plant)
DUL(L)	Drained upper limit soil water for soil layer L – volume fraction
ELNC	Environmental limit on nitrification capacity (zero to unity unitless factor)
ESW(L)	Extractable soil water content for soil layer L (the difference between DUL and LL – volume fraction)
FAC(L)	Conversion factor for ppm N to kg N/ha for layer L
FDAY(J)	Day of the year when fertilizer application J was made
FLOW(L)	Volume of water moving from layer L due to unsaturated flow (CM) positive indicates upward movement and negative value indicates downward movement
FLUX(L)	Water moving downward from layer L with drainage (cm)
FNH4	Unitless soil ammonium supply index
FNO3	Unitless soil nitrate supply index
FOM(L)	Fresh organic matter (residue) in layer L (kg/ha)
FON(L)	N in fresh organic matter in layer L (kg N/ha)
FPOOL(L,J)	Fresh organic matter in layer L (kg OM/ha). If J = 1 pool is comprised of carbohydrates, if J = 2 pool is comprised of cellulose, and if J = 3 pool is comprised of lignin
FT	Temperature factor affecting denitrification rate
FW	Unitless soil moisture factor affecting denitrification rate

G1	Genetic specific constant related to rate of vegetative expansion growth during Stage 1
GRCOM	Gross release of carbon from organic matter decomposition (kg C/ha)
GRNOM	Gross release of N from organic matter decomposition (kg N/ha per day)
GRORT	Daily root growth (g/day)
HUM(L)	Stable humic fraction material in layer L (kg/ha)
IFLAG	Switch variable used to direct control to either the leaching component or the upward flux component of subrouting NFLUX
IFTYPE	Code number for fertilizer type
ISTAGE	Phenological stage
	= 1 Emergence to terminal spikelet
	= 2 Terminal spikelet to end of vegetative growth
	= 3 End of vegetative growth to end of pre-anthesis ear growth
	= 4 Pre-anthesis ear growth to beginning of grain fill (anthesis occurs during this phase)
	= 5 Beginning of grain fill to physiological maturity
	= 6 Physiological maturity to fallow (harvest)
	= 7 Fallow to sowing
	= 8 Sowing to germination
	= 9 Germination to emergence
IUOF	Switch variable to indicate urea hydrolysis is not active
IUON	Switch variable to indicate urea hydrolysis is active
L1	The number of soil layers to the bottom of the root zone
LL(L)	Lower limit soil water content for soil layer L − volume fraction
MF	Zero to unity moisture factor for residue decomposition rate
MU	Loop variable to indicate layer below the current layer
NDEF1	Zero to unity N deficiency factor for photosynthetic rate
NDEF2	Zero to unity N deficiency factor for expansion growth
NDEF3	Zero to unity N deficiency factor for tiller number
NDEF4	Zero to unity N deficiency factor for grain N determination
NDEM	Plant N demand (g/plant)
NFAC	Zero to unity factor based on actual and critical N concentrations
NFERT	Number of fertilizer applications made
NH4(L)	Soil ammonium (ppm) in layer L
NHUM(I)	N associated with the stable humic fraction in layer I (kg N/ha)
NLAYR	Number of layers in soil
NNOM	Net N released from all organic sources in a layer (kg N/ha)
NO3(L)	Soil nitrate (ppm) in layer L
NUF	Plant N supply/demand ratio used to modify uptake
OC(L)	Organic carbon in layer L (%)
OUTN	Nitrate N leaching from a layer (kg N/ha)
PIV	Genetic specific coefficient that determines sensitivity to vernalization
PDWI	Potential increment of new shoot growth (g/plant)
PGRORT	Potential increment of new root growth (g/plant)
PH(L)	Soil pH in layer L
PHN(L)	Zero to unity factor describing the effect of soil pH or nitrification rate on layer L
PLANTS	Number of plants per square meter
PTF	Fraction of photosynthesis partitioned to aboveground plant parts
RANC	Root actual N concentration (g N/g root dry weight)

RCNP	Root critical N concentration (g N/g root dry weight)
RDECR(J)	The maximum rate constant for decay of residue components (1/days)
RFAC	Interim variable describing the effects of root length density on potential N uptake from a layer
RHMIN	N mineralized from humus in a layer (kg N/ha)
RLV(L)	Root length per unit soil volume for layer L (cm/cm^3)
RNAC	Immobilization rate of N associated with the decay of residues (kg N/ha per day)
RNDEM	Plant root demand for N (g/plant)
RNH4U(L)	Potential ammonium uptake from layer L (kg N/ha)
RNLOSS	Loss of N from the plant via root exudation in one layer (g N/plant)
RNO3U(L)	Potential nitrate uptake from layer L (kg N/ha)
RNTRF	Amount of ammonium nitrified in a layer (kg N/ha per day)
ROOTN	Plant root N content (g N/plant)
RP2	Temporary variable used in nitrification calculations
RTWT	Root weight (g/m^2)
SANC	Supply of ammonium effect on nitrification capacity
SARNC	Supply of ammonium effect on the reduction of nitrification capacity (zero to unity, unitless)
SAT(L)	Field saturated soil water content in layer L (cm/cm)
SKGN(L)	Dummy variable for either urea or nitrate present in layer L (kg N/ha)
SMDER	Soil moisture deficit factor affecting N uptake
SMIN	Interim variable to prevent soil N pools from becoming < 1 ppm
SNH4(L)	Soil ammonium in layer L (kg N/ha)
SNO3(L)	Soil nitrate in layer L (kg N/ha)
SOILC	Soil carbon content (kg C/ha)
SPPM(L)	Dummy variable for either urea or nitrate concentration present in layer L (ppm N)
ST(L)	Soil temperature in layer L (°C)
SW(L)	Actual soil water content in layer L (cm/cm)
TANC	Tops actual N concentration (g N/g dry wt.)
TCNP	Tops critical N concentration (g N/g dry wt.)
TF	Temperature factor for nitrification on mineralization
TEY(L)	Yesterday's temperature factor for nitrification in layer L
TMNC	Plants tops minimum N concentration (g N/g dry wt.)
TNDEM	Plant tops demand for N (g N/plant)
TNUP	Total N uptake from the profile on 1 d (kg N/ha)
TOPSN	N contained in plant tops excluding grain (g N/plant)
TOPWT	Weight of plant tops excluding grain (g)
TOTN	Total mineral N in a layer (kg N/ha)
TRLV	Total root length density variable
TRNLOS	Total plant N lost by root exudation (g N/plant)
TRNU	Total potential root nitrogen uptake from the soil (kg N/ha)
UHYDR	Urea hydrolysis rate (kg urea N/ha per day)
UNH4	Plant uptake of ammonium from a layer (kg N/ha)
UNO3	Plant uptake of nitrate from a layer (kg N/ha)
UP1	Interim variable used to prevent soil N pools from becoming < 1 ppm
UPPM(L)	Concentration of urea N in layer L (ppm)
UREA(L)	Urea N in layer L (kg N/ha)
VANC	Plant vegetative actual N concentration (g N/plant)
VMNC	Plant vegetative minimum N concentration (g N/g dry wt.)

WFD	Today's water factor for nitrification
WEY(L)	Yesterday's water factor for nitrification in layer L
XL	Temporary variables used to determine soil
XSTAGE	Non-integer growth stage indicator ranging from zero to six
XT	Temperature effect on nitrification capacity
XW	Moisture effect on nitrification capacity
ZS2	Interim variables used in critical concentration calculations
ZS3	Interim variables used in critical concentration calculations
ZSTAGE	Zadoks' growth stage

REFERENCES

Addiscott, T.M. 1981. Leaching of nitrate in structured soils. In M.J. Frissel and J.A. van Veen (ed.) Simulation of nitrogen behavior of soil-plant systems. PUDOC, Wageningen, Netherlands.

Addiscott, T.M., and R.J. Wagenet. 1985. Concepts of solute leaching in soils: A review of modelling approaches. J. Soil Sci. 36:411-424.

Allison, F.E. 1965. Evaluation of incoming and outgoing processes that affect soil nitrogen. In W.V. Bartholomew and F.E. Clark (ed.) Soil nitrogen. Agronomy 10:573-606.

Benzian, B., R.J. Derby, P. Lane, F.V. Widdowson, and L.M.J. Verstraeten. 1983. Relationship between N concentration of grain and grain yield in recent winter wheat experiments in England and Belgium, some with large yields. J. Sci. Food Agric. 34:685-695.

Beri, V., and S.S. Brar. 1978. Urease activity in sub-tropical, alkaline soils of India. Soil Sci. 126(6):330-335.

Bowen, G.D. 1969. Nutrient status effects on loss of amides and amino acids from pine roots. Plant Soil 30:139-142.

Burns, I.G. 1980. A simpler model for predicting the effects of leaching of fertilizer nitrate during the growing season on the nitrogen fertilizer needs of crops. J. Soil Sci. 31:175-202.

Christensen, N.W., and R.J. Killorn. 1981. Wheat and barley growth and N fertilizer utilization under sprinkler irrigation. Agron. J. 73:307-312.

Davidson, J.B., D.A. Graetz, P.S.C. Rao, and H.M. Selim. 1978. Simulation of nitrogen movement, transformation, and uptake in plant root zone. U.S. Environ. Protect. Agency EPA-600/3-78-029. Office Res. Develop., Athens, GA.

Donigian, A.S., and N.H. Crawford. 1976. Modelling pesticides and nutrients on agricultural lands. U.S. Environ. Protect. Agency EPA-600/2-76-043. Office Res. Develop., Athens, GA.

Focht, D.D., and W. Verstraete. 1977. Biochemical ecology of nitrification and denitrification. Adv. Microb. Ecol. 1:135-214.

Frissel, M.J., and J.A. van Veen (ed.). 1981. Simulation of nitrogen behavior of soil-plant systems. PUDOC, Wageningen, Netherlands.

Godwin, D.C., and PL.G. Vlek. 1985. Simulation of nitrogen dynamics in wheat cropping systems. p. 311-332. In W. Day and R.K. Atkin (ed.) Wheat growth and modelling. Plenum Publ. Corp., New York.

Jones, C.A. 1983. A survey of the variability of tissue nitrogen and phosphorus concentrations in maize and grain sorghum. Field Crop Res. 6:133-147.

Jones, C.A., and J.R. Kiniry (ed.). 1986. CERES-Maize. A simulation model of maize growth and development. Texas A&M Univ. Press, College Station, TX.

Knowles, R. 1981. Denitrification. Ecol. Bull. 33:315-329.

Leffelaar, P.A. 1981. A model to simulate partial anaerobiosis. p. 254-258. In M.J. Frissel and S.A. van Veen (ed.) Simulation of nitrogen behaviour of soil-plant systems. PUDOC, Wageningen, Netherlands.

Linn, D.M., and J.W. Doran. 1984. Effect of water-filled pore space on carbon dioxide and nitrous oxide production in tilled and nontilled soils. Soil Sci. Soc. Am. J. 48:1267-1272.

McGarity, J.W., and M.G. Myers. 1967. A survey of urease activity in soils of northern New South Wales. Plant Soil 27:217-238.

McLaren, A.D. 1970. Temporal and vectorial reactions of nitrogen in soil: A review. Can. J. Soil Sci. 50:97-109.

Molina, J.A.E., C.E. Clapp, M.J. Shaffer, F.W. Chichester, and W.E. Larson. 1983. NCSOIL, a model of nitrogen and carbon transformations in soil: description, calibration, and behaviour. Soil Sci. Soc. Am. J. 47:85-91.

Myers, M.G., and J.W. McGarity. 1968. The urease activity in profiles of five great soil groups from northern New South Wales. Plant Soil 28:25-37.

Myers, R.J.K. 1975. Temperature effects on ammonification and nitrification in a tropical soil. Soil Biol. Biochem. 7:83-86.

Myers, R.J.K., C.A. Campbell, and R.L. Weier. 1982. Quantitative relationship between net nitrogen mineralization and moisture content of soils. Can. J. Soil Sci. 62:111-124.

Novoa, R., and R.S. Loomis. 1981. Nitrogen and plant production. Plant Soil 58:177-204.

Otter-Nacke, S., D.C. Godwin, and J.T. Ritchie. 1985. Yield model development: Testing and validating the CERES-WHEAT model in diverse environments. Agristars Publ. YM-15-00407, JSC 20244. Johnson Space Center, Houston, TX.

Parton, W.J., W.D. Gould, F.J. Adamson, S. Torbit, and R.G. Woodmansee. 1981. NH_3 volatilization model. p. 233-244. In M.J. Frissel and J.A. van Veen (ed.) Simulation of nitrogen behaviour of soil-plant systems. PUDOC, Wageningen, Netherlands.

Pearman, I., S.M. Thomas, and G.N. Thorne. 1978. Effect of nitrogen fertilizer on growth and yield of semi-dwarf and tall varieties of winter wheat. J. Agric. Sci. (Cambridge) 91:31-43.

Power, J.F. 1981. Nitrogen in the cultivated ecosystem. Ecol. Bull. 33:529-546.

Rao, P.S., J.M. Davidson, and R.E. Jessup. 1981. Simulation of nitrogen behaviour in the root zone of cropped land areas receiving organic wastes. p. 81-95. In M.J. Frissel and J.A. van Veen (ed.) Simulation of nitrogen behavior of soil-plant systems. PUDOC, Wageningen, Netherlands.

Ritchie, J.T. 1985. A user-oriented model of the soil water balance in wheat. p. 293-306. In W. Day and R.K. Atkin (ed.) Wheat growth and modelling. Plenum Publ. Corp., New York.

Ritchie, J.T., D.C. Godwin, and S. Otter-Nacke. 1988. CERES-Wheat. A simulation model of wheat growth and development. Texas A&M Univ. Press, College Station, TX.

Rolston, D.E., A.N. Sharpley, D.W. Toy, D.L. Hoffman, and F.E. Broadbent. 1980. Denitrification as affected by irrigation frequency of a field soil. EPA-600/2-80-06. U.S. Environmental Protection Agency, Ada, OK.

Rovira, A.D. 1969. Plant root exudates. Bot. Rev. 35:35-57.

Seligman, N.C., and H. van Keulen. 1981. PAPRAN: A simulation model of annual pasture production limited by rainfall and nitrogen. p. 192-221. In M.J. Frissel and J.A. van Veen (ed.) Simulation of nitrogen behaviour of soil-plant systems. PUDOC, Wageningen, Netherlands.

Selim, H.M., and I.K. Iskandar. 1981. Modeling nitrogen transport and transformations in soils: I. Theoretical considerations. Soil Sci. 131:233-241.

Smith, K.A. 1981. A model of denitrification in aggregated soils. p. 259-266. In M.J. Frissel and J.A. van Veen (ed.) Simulation of nitrogen behavior of soil-plant systems. PUDOC, Wageningen, Netherlands.

Stanford, C., M.H. Frere, and D.H. Schwaninger. 1973. Temperature coefficient of soil nitrogen mineralization. Soil Sci. 115:321-323.

Storrier, R.R. 1965. Excess soil nitrogen and the yield and uptake of nitrogen by wheat in southern New South Wales. Aust. J. Exp. Agric. Anim. Husb. 5:317.

Tabatabai, M.A., and J.M. Bremner. 1972. Assay of urease activity in soils. Soil Biol. Biochem. 4:479-487.

Tanji, K.K. 1982. Modelling of the nitrogen cycle. In J.F. Stevenson (ed.) Nitrogen in agricultural soils. Agronomy 22:721-772.

Tanji, K.K., M. Mehran, and S.K. Gupta. 1981. Water and nitrogen fluxes in the root zone of irrigated maize. p. 51-66. In M.J. Frissel and H. van Veen (ed.) Simulations of nitrogen behavior in soil-plant systems. PUDOC, Wageningen, Netherlands.

Terman, C.L., R.E. Ramig, A.F. Dreier, and R.A. Olson. 1969. Yield-protein relationships in wheat grain, as affected by nitrogen and water. Agron. J. 61:755-759.

Tillotson, W.R., C.W. Robbins, R.J. Wagenet, and R.J. Hanks. 1980. Soil water, solute and plant growth simulation. Utah Agric. Exp. Stn. Bull. 502.

van Veen, J.A. 1977. The behavior of nitrogen in soils. A computer simulation model. Ph.D. thesis. Vrije Univ., Amsterdam.

Vleck, R.G.L., and M.F. Carter. 1983. The effect of soil environment and fertilizer modifications on the rate of urea hydrolysis. Soil Sci. 136:56-63.

Vos, J. 1981. Effects of temperature and nitrogen supply on post-floral growth of wheat; measurements and simulations. Agric. Res. Rep. 911. PUDOC, Wageningen, Netherlands.

Wagger, M.G. 1983. Nitrogen cycling in the plant-soil system. Ph.D. thesis. Kansas State Univ., Manhattan (Diss. Abstr. 83-28137).

Warncke, D.D., and S.A. Barber. 1974. Root development and nutrient uptake by corn grown in solution culture. Agron. J. 66:514–516.

Watts, D.C., and R.J. Hanks. 1978. A soil-water-nitrogen model for irrigated corn on sandy soils. Soil Sci. Soc. Am. Proc. 42:492–499.

Wild, A., and K.C. Cameron. 1980. Soil nitrogen and nitrate leaching. *In* P.B. Tinker (ed.) Soils and agriculture. Vol. 2. Blackwell Sci. Publ., Oxford, England.

Williams, J.R., P.T. Dyke, and C.A. Jones. 1983. EPIC—A model for assessing the effects of erosion on soil productivity. p. 553–572. *In* W.K. Lavenroth et al. (ed.) Proc. Third Int. Conf. State-of-the-Art in Ecological Modelling, Fort Collins, CO. 24–28 May 1982. Elsevier Sci. Publ. Co., New York.

Zantua, M.I., L.C. Dumenil, and J.M. Bremner. 1977. Relationships between soil urease activity and other soil properties. Soil Sci. Soc. Am. J. 41:350–352.

14 Modeling Phosphorus Dynamics in the Soil-Plant System

C. ALLAN JONES
Texas Agricultural Experiment Station
Temple, Texas

ANDREW N. SHARPLEY
USDA-ARS
Durant, Oklahoma

J. R. WILLIAMS
USDA—ARS
Temple, Texas

Phosphorus (P) is an essential element for plant growth and, in many cases, low soil P levels limit crop yields. Consequently, P is often added to the soil-plant system as mineral fertilizer or manure. Although considerable information is available on the complex chemical and physiological transformations of P in the soil and its uptake by plants, attempts to formulate comprehensive, management-oriented simulation models of P cycling have been limited. Lack of information about processes such as root growth and activity has slowed the development of comprehensive models, thus descriptions of the complex interaction between environment and P cycling over time and space is difficult. A brief, but not exhaustive discussion of models simulating the soil-plant P cycle follows. The types of available models are outlined, ranging from complex mathematical representations of soil P pools and plant uptake to relatively simple predictions of fertilizer P requirements for farm management decisions.

I. PREVIOUS MODELS

Mathematical models simulating P flux to plant roots have been developed (Claassen & Barber, 1976; Cole et al., 1977; Cushman, 1982; Helyar & Munns, 1974; Olsen & Kemper, 1967; Schenk & Barber, 1979a). In general, these models follow the work of Nye and Marriott (1969) to describe the movement of P to individual roots by mass flow and diffusion. Absorption kinetics of the root are assumed to follow Michaelis-Menten kinetics. For

---

Copyright © 1991 ASA-CSSA-SSSA, 677 S. Segoe Rd., Madison, WI 53711, USA. *Modeling Plant and Soil Systems*—Agronomy Monograph no. 31.

example, Claassen and Barber (1976) proposed a model based on theoretical consideration of the processes of potassium (K) uptake by plant roots growing in soil. This model was subsequently used to predict P uptake by maize (*Zea mays* L.) (Schenk & Barber, 1979a) and soybean [*Glycine max* (L.) Merr.] (Silberbush & Barber, 1983). The following equation from Nye and Marriott (1969) was used to describe P flux to the root:

$$J_r = D_e \frac{\partial C_s}{\partial r} + V_o C_L \qquad [1]$$

where J_r is flux to the root, D_e is the effective diffusion coefficient, r is the radial distance, C_s is the concentration of P in the solid phase that readily equilibrates with the concentration of P in the soil solution (C_L), and V_o is the rate of water flux into the root. Phosphorus uptake by the root follows Michaelis-Menten kinetics, after substracting a term for efflux (E) as described by Claassen and Barber (1974):

$$J_r = \frac{I_{max} C_L}{K_m + C_L} - E \qquad [2]$$

where I_{max} is the rate of influx at infinite concentration and K_m is the Michaelis-Menten constant. Soil parameters required in the model include effective average diffusion coefficient, initial P concentration in solution, and buffer power. Plant parameters include maximum influx rate, the Michaelis-Menten constant, water influx, root radius, initial root length, and rate of root growth. The mathematical model for P uptake has been subsequently developed to account for: (i) nonlinear root boundary conditions (Cushman, 1984); (ii) arbitrary spatial and temporal changes in the buffer power and diffusion coefficients; (iii) nonlinear sources and sinks of nutrients in both the rhizosphere and bulk soil; and (iv) arbitrary, nonlinear initial conditions.

Simulated uptake of P was closely related to observed uptake by plants grown in controlled environments for periods up to 28 d. Although the model does not simulate the cycling of P in inorganic (solution and adsorbed) and organic pools, it does provide a quantitative assessment of the importance of several factors, such as root morphology (Schenk & Barber, 1979a, b), soil temperature (Mackay & Barber, 1984), and fertilizer placement (Anghinoni & Barber, 1980) on P uptake by plant roots.

A mechanistic multistep model was proposed by Mansell et al. (1977) to describe the transformation and transport of applied P under steady water flow conditions in uniform soils. Inorganic soil P was considered in five phases (Ryden et al., 1977): solution, physically adsorbed, chemisorbed, immobilized, and precipitated. First-order kinetics are used to describe flow among the phases. Simulated results provide a detailed description of the retention and transport of applied P in the root zone during the growing season. Fertilizer-P distribution depends on the rate constants that describe P movement among pools, which in this study were experimental variables. For practical applications of the model, a method is needed to assign rate constants based on soil properties, and to simulate organic P transformations and their effects on inorganic pools.

Cole et al. (1977) constructed a simulation model of the P cycle in semi-arid grasslands. Basically, P supply to the plant is assumed to be limited by diffusion. As plant uptake reduces the P concentration at the root surface, a concentration gradient is established in the surrounding water films, and phosphate ions flow toward the root. The soluble P pool is replenished from slightly soluble phosphate minerals, phosphate adsorbing surfaces, and organic P mineralization (Cole et al., 1977). This model was used as a submodel in ELM, a grassland simulation model developed for the U.S. International Biological Program Grassland Biome study (Cole, 1976). The ELM simulates abiotic, producer, decomposer, and consumer processes, which all interact with the P submodel (Anway, 1976).

The P-cycling simulation model of Cole et al. (1977) and other similar models (Chapin et al., 1978; Harrison, 1978; Mishra et al., 1979) have incorporated the effects of moisture, temperature, soil properties, plant phenology, and organic matter decomposition on P flows. These models simulate many soil and plant processes, but require input data that are not readily available. However, they have revealed gaps in our knowledge of processes such as organic P mineralization (Stewart & McKercher, 1981). Consequently, these models provide valuable direction for future research.

A more user-oriented, simple, predictive model of the effects of soil-P buffering capacity on residual fertilizer-P effectiveness was developed by Bennett and Ozanne (1972) and subsequently modified by Helyar and Godden (1976). Bennett and Bowden (1976) described a response-curve prediction, named Decide, that makes fertilizer recommendations, in which the chosen response curve is the exponential or Mitscherlich curve:

$$Y = A[1 - B \exp(-CX)] \qquad [3]$$

where Y is the yield per unit area, A is the maximum yield per unit area, and B is the relative response to the applied P. The rate of P applied (X) is standardized to kilograms P per hectare and the curvature coefficient or buffer capacity (C) has reciprocal dimensions to X. The coefficients are adjusted for soil type, past soil fertility, and yield goals.

The Decide model was developed on the highly weathered, leached soils of Western Australia and is used routinely by the Western Australia Department of Agriculture advisory services. It is presently being modified for use on the younger, slightly weathered soils of eastern Australia. Consequently, the model estimates fertilizer P recommendations only for a limited range of soil types at the present time.

Several soil and plant P models have been developed, each designed to meet certain objectives. However, none of these models provide long-term simulations of soil organic, inorganic, and plant processes affecting P fertility for a wide variety of soils and crop rotations and using readily available data inputs. A soil and plant P model with these characteristics was developed as a component of the Erosion-Productivity Impact Calculator (EPIC), a model designed to simulate soil erosion and its long-term effects on crop productivity for a wide variety of soils, climates, crops, and soil conservation practices (Williams et al., 1984). This chapter summarizes the characteristics of the P components of EPIC, which have been described in more detail elsewhere (Jones et al., 1984a, b; Sharpley et al., 1984), although a few modifications have since been made.

II. MODEL STRUCTURE

The logical structure of the soil and plant P model is illustrated in Fig. 14-1 and its variables are defined in the Appendix. Soil inorganic P consists of labile, active, and stable P pools. Tiessen et al. (1984) performed sequential extractions on 78 soil samples used to derive equations for estimation of initial labile P and a P sorption index (Sharpley et al., 1984, 1985). Tiessen et al. (1984) obtained estimates of resin (Sibbesen, 1978) and bicarbonate-extractable inorganic P (Colwell, 1963), which were thought to consist of inorganic P adsorbed on surfaces of more crystalline P compounds, sesquioxides, and carbonates (Mattingly, 1975). These forms of P are considered more readily available to plants than inorganic P removed by stronger (hydroxide and acid) extractants. Analysis of data in Sharpley et al. (1985) revealed that, for calcareous and slightly weathered soils, bicarbonate-extractable inorganic P was approximately half of resin-extractable P. For more highly weathered soils, bicarbonate-extractable inorganic P was as large or up to 3.5 times larger than resin-extractable P. However, Tiessen et al. (1984) showed that the solubility of bicarbonate-extractable inorganic P is lower in Ultisols than in Mollisols. For operational purposes, therefore, the labile and active P pools are assumed to be in rapid equilibrium, with the labile pool twice the size of the active pool.

More stable crystalline species of inorganic P can act as sinks as well as long-term reservoirs of labile P (Murrman & Peech, 1969; Dalal & Hallsworth, 1976). The model's stable, inorganic P pool is the operational equivalent to stable, crystalline, inorganic P forms and occluded organic P (Tiessen et al., 1984).

Sharpley et al. (1984) described a fertilizer P availability index (F_1) ranging from <0.1 in soils that sorb large amounts of P to >0.7 in soils of low P sorption capacity. The F_1 index can be estimated from soil taxonomic and chemical data (Sharpley et al., 1984). Whenever P is added to the labile pool (by fertilizer or mineralization of organic P), a fraction of the added P (F_1) remains in the labile P pool and the remainder is transferred to the active mineral P pool. If, following the transfer, the ratio of the active to the labile P pool >0.5, enough P is transferred from the active to the stable inorganic P pool to maintain that ratio. Whenever plant uptake removes P from the labile pool, enough P is transferred to the labile pool

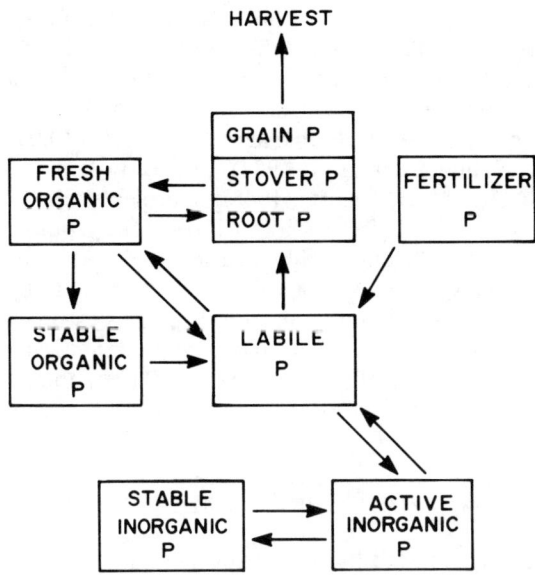

Fig. 14-1. Pools and flows of P in the EPIC model (from Jones et al., 1984a).

to maintain the ratio. If plant uptake reduces the concentration of P in the labile pool to very low values, the stable mineral pool can serve as a long-term source of labile P. Movement from the stable mineral pool is sufficient to maintain labile P concentrations of 2 to 5 mg/kg, concentrations at which crops cannot obtain sufficient P for maximum growth and thus, yield is reduced.

The plant absorbs P from the labile pool and distributes it to root, stover, and grain pools. When the crop is harvested, residue P in the stover is returned to the soil and is available for mineralization. Transformations of organic P in crop residues and soil organic matter are similar to the transformations of crop residues, soil organic matter, and organic N in the PAPRAN, a model developed by Seligman and van Keulen (1981). Organic P is divided into the fresh residue P pool, consisting of P in the microbial biomass and undecomposed residues, and the stable organic P pool, consisting of P in stable organic matter. Stable organic matter is divided into mineralizable and nonmineralizable fractions whose relative sizes depend on the amount of time that the land has been under cultivation.

III. MODEL INPUTS

The EPIC model is designed to use soil chemical, physical, and taxonomic data available in U.S. Soil Conservation Service (SCS)/State Agricultural Experiment Station Soil Survey Investigative Reports and SCS pedon descriptions. Soil P data are not normally included in these sources. Therefore, information describing P fertilizer and P sorption characteristics of the soil must be obtained elsewhere or estimated from other soil characteristics.

The P model requires estimates of initial sizes of the labile, active inorganic, stable inorganic, residue, and stable mineralizable P pools. An estimate of the P availability index of the soil is also required. Initial labile P can be estimated from soil test P extracted by the Olsen bicarbonate (0.5 M NaHCO$_3$; Olsen et al., 1954), double acid or Mehlich I (0.05 M HCl and 0.0125 M H$_2$SO$_4$; Sabbe & Breland, 1974), and Bray I methods (0.03 M NH$_4$F and 0.25 M HCl; Bray & Kurtz, 1945). Sharpley et al. (1984) provide equations to calculate labile P from soil test P for three groups of soils: calcareous, slightly weathered, and highly weathered.

Residue P content is estimated from the type and amount of plant residue present. The size of the stable mineralizable P pool is estimated from the size of the organic N pool and the fraction of that pool that is mineralizable. In virgin grassland soils, up to 50% of the organic P is assumed to be mineralizable. The percentage declines as the number of years under cultivation increases.

IV. MODEL PERFORMANCE

The model has been tested by comparing simulated results with results measured in a number of field experiments throughout the continental USA (Jones et al., 1984b). Results of long-term (30-70 yr) studies of organic carbon, N, and P concentrations in the Great Plains (Haas et al., 1961; Sharpley & Smith, 1983) were used to test model predictions of the effects of climate, soil, and crop rotation on organic P concentrations of the topsoil and the whole profile. In general, the model produced accurate estimates of the decline in topsoil organic P due to cultivation (Table 14-1). After cultivation, organic P decreased more slowly in the subsoil than in the topsoil (Fig. 14-2).

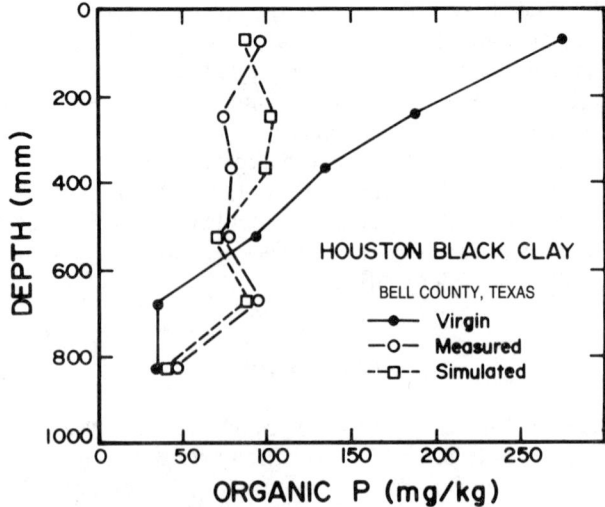

Fig. 14-2. Changes in profile organic-P content after ~60 yr of cultivation on a Houston black clay (fine, montmorillonitic, thermic Udic Pellustert). Measured data from Sharpley and Smith (1983).

Table 14–1. Measured and simulated changes in surface soil organic P, total N, labile P, and organic carbon in the Great Plains. Measured data from Haas et al. (1957, 1961).

Location	Duration of study	Rotation	Organic P			Total N			Labile P		
			Virgin	Cultivated		Virgin	Cultivated		Virgin	Cultivated	
				Measured	Simulated		Measured	Simulated		Measured	Simulated
	yr					mg kg^{-1}					
Havre, MT	31	SWF†	157	102	108	1510	900	1135	11	13	15
Moccasin, MT	39	SWF	308	183	169	3000	2050	1787	14	14	20
Dickinson, ND	41	SWF	292	148	174	2930	1490	1957	10	12	14
Mandan, ND	31	SWF	139	132	97	1600	1160	1172	9	12	7
Sheridan, WY	30	SWF	120	93	86	1590	1210	1149	12	14	9
Laramie, WY	34	SWF	142	91	96	1220	820	900	13	24	9
Akron, CO	39	SWF	115	82	81	1340	800	911	26	45	19
Colby, KS	31	WWF†	158	61	92	1650	1050	952	34	30	27
Hays, KS	30	Winter wheat	174	97	108	2200	1220	1360	11	40	8
Lawton, OK	28	Winter wheat	128	71	73	1540	740	904	8	9	8
Dalhart, TX	29	Maize	84	39	53	670	420	444	17	13	9
Big Spring, TX	41	Winter wheat	55	30	29	600	410	328	12	12	6
Mean	34		156	94	97	1654	1023	1083	15	20	13

† SWF and WWF represent spring wheat-fallow and winter wheat-fallow, respectively.

The model should accurately simulate the effects of P fertilization on soil-test P levels, the amount of fertilizer P needed to maintain adequate levels of soil-test P, and the rate of increase or decrease in soil-test P when rates of fertilizer P are excessive or inadequate. Data from several fertilizer P experiments (0–10 yr) were used to test model estimates of maintenance fertilizer-P requirements and changes in soil-test P. Where available, original soil information was input as initial conditions and coefficients determined by the equations of Sharpley et al. (1984). The model produced estimates of maintenance fertilizer-P requirements that were similar to those recommended by Ibach and Adams (1968) for maize and wheat grown on several major land resource areas in the USA (Fig. 14-3). The model also produced accurate estimates of changes in soil-test P over time due to variation in fertilizer P application rate (Fig. 14-4 and 14-5). Figure 14-4 illustrates a problem that can occur during medium- to long-term simulation of soil-test P. Over a period of several years, soil characteristics such as labile P content, pH, base saturation, and organic carbon content can change significantly due to management. These changes can affect soil P sorption and the forms of soil P. This contributes to the rather abrupt changes in the rate of increase or decrease in soil test P when a particular rate of fertilizer-P

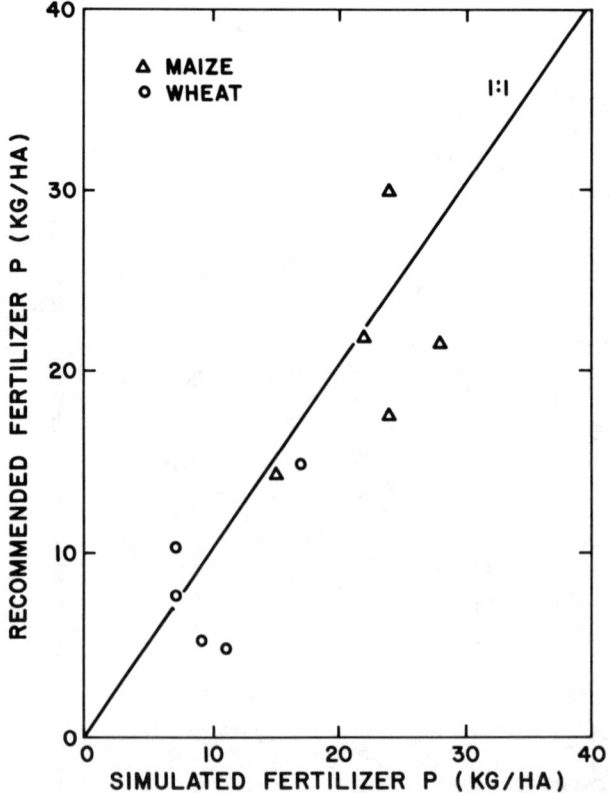

Fig. 14-3. Relationship between simulated mean annual fertilizer-P application and recommended annual fertilizer-P application (Ibach & Adams, 1968) in 10 major land resource areas in the USA.

application is maintained for a number of years (Cope, 1981; Hooker et al., 1983). Future model improvement will require accurate simulation of changes in soil characteristics and their effects on P sorption and forms of P in the soil.

Another important aspect of model performance is the ability to simulate the effects of P deficiency on crop growth and yield. Although the model uses a simple method of estimating P availability and uptake by the plant, the effects of P deficiency on maize and wheat yields (Fig. 14-6) can be accurately simulated.

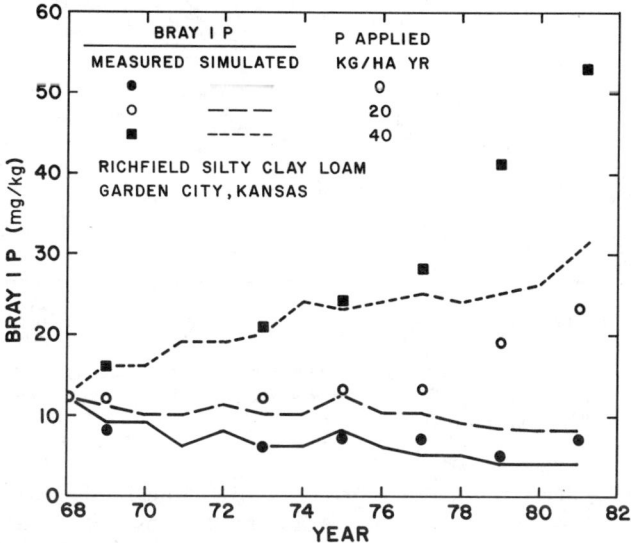

Fig. 14-4. Effect of fertilizer-P application on Bray I P content of Richfield silty clay loam (fine, montmorillonitic, mesic Aridic Argiustoll) under corn at Garden City, KS. Measured data from Hooker et al. (1984) (from Jones et al., 1984b).

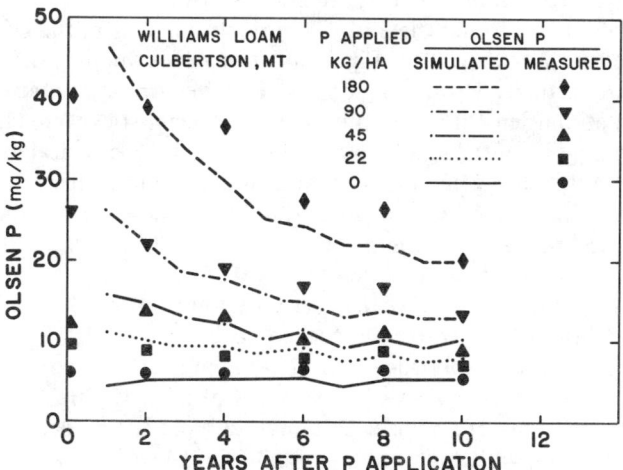

Fig. 14-5. Effect of a single application of fertilizer P (five rates) on Olsen P content of a Williams loam (fine-loamy, mixed Typic Argiboroll) under spring wheat at Culbertson, MT. Measured data from Black (1982) (from Jones et al., 1984b).

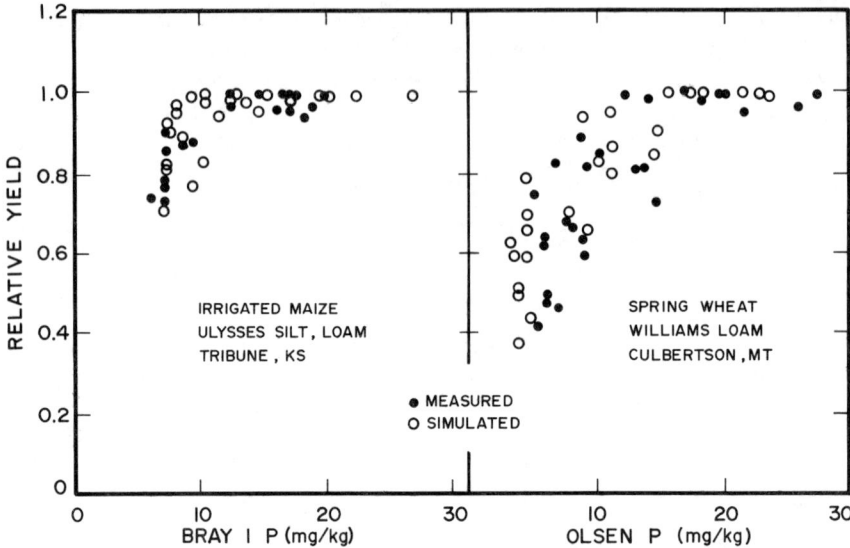

Fig. 14-6. Comparison of measured and simulated wheat yield response to Olsen P at Culbertson, MT on a Ulysses silt loam (fine, silty, mixed, mesic Aridic Haplustoll). Measured data from Black (1982) (from Jones et al., 1984b).

Sensitivity analysis is another method of evaluating model performance. The remainder of this section demonstrates model sensitivity to several important soil and management variables. The model uses a fertilizer P availability index (F_1) to quantify the amount of fertilizer P that remains in the labile P pool after rapid adsorption is complete. Soils with a high P sorption capacity, have lower values of F_1 than soils of lower P sorption capacity. Sharpley et al. (1984) found that, for 78 agricultural soils in the USA, topsoil F_1 ranged from 0.06 to 0.74. Figure 14-7 provides the simulated effects of F_1 and the fertilizer P rate on soil-test P under irrigated maize over a 14-yr period at Garden City, KS. Initial soil characteristics were chosen to produce values of F_1 of 0.25 and 0.50. The value of 0.25 is typical of a highly weathered soil, which generally sorb large amounts of fertilizer P. A value of 0.50 is typical of a moderately weathered soil with low P fixation. As would be expected, the rate of increase in soil test P due to fertilizer addition was greater when F_1 was 0.50 than when it was 0.25.

The EPIC model also allows the user to maintain P fertility with variable yearly additions of fertilizer P needed to bring soil-test P back to a predetermined level. Phosphorus fixation has a large effect on maintenance P fertilizer requirements. Soils with low values of F_1 (0.12, high P fixation) require larger amounts of fertilizer P to maintain predetermined soil-test P levels than do soils with high values of F_1 (0.50, low P fixation) (Table 14-2).

Crops grown on soils with high net mineralization rates may require little or no fertilizer P. For example, no fertilizer P response may be found for

Table 14-2. Sensitivity of simulated, maintenance fertilizer P (FP) for irrigated maize in Kansas to initial levels of topsoil organic P (ORGP), fertilizer-P availability index (F_1), and years of cultivation).

F_1	Topsoil organic P	Cultivation	Fertilizer P
	mg/kg	yr	kg/ha yr
Effect of F_1			
0.50	100	75	18.5
0.25	100	75	25.8
0.12	100	75	44.1
Effect of organic P			
0.5	400	75	14.1
0.5	100	75	18.5
0.5	25	75	19.3
Effect of years of cultivation			
0.50	200	3	8.8
0.50	200	15	14.1
0.50	200	75	18.1

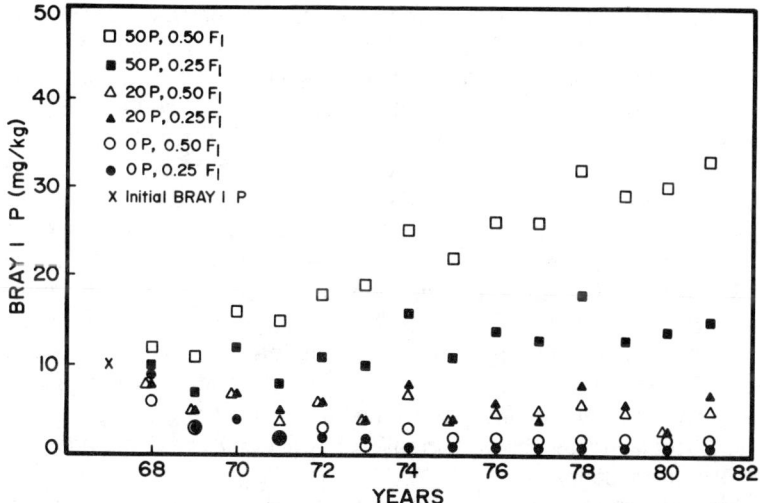

Fig. 14-7. Effect of fertilizer-P rate (0, 20, and 50 kg P/ha yr) and fertilizer-P availability index (F_1; 0.25 and 0.50) on long-term changes in topsoil Bray P for irrigated maize in Kansas.

many years after virgin soils are cultivated, due in part to mineralization of organic P, which supplies adequate levels of labile P (Haas et al., 1961). Hypothetical cases, using soils with three initial organic P contents (25, 100, and 400 mg P/kg soil), were used to simulate 14 yr of irrigated maize production in Kansas. Annual amounts of fertilizer P required to maintain initial levels of topsoil labile P (15 mg P/kg soil) were greater for the soils with low rather than high initial organic P contents (Table 14-2).

Finally, the model predicts two pools of soil organic matter, one of which is more active than the other. Soils that have been cultivated for long periods of time normally have a larger fraction of their total organic matter in the more stable pool. Thus, at the same concentration of organic P, soils that have recently been broken from sod have higher rates of fertilizer-P mineralization and, therefore, lower maintenance fertilizer-P requirements (Table 14-2).

V. CONCLUSIONS

Several soil-plant P simulation models are available. They were developed for differing reasons and range from detailed simulation models of basic soil chemical reactions, and kinetics of P diffusion and uptake, to more general models predicting crop response to fertilizer P. These models have been successfully used to predict specific soil P reactions, crop P uptake, and crop yield. In doing this, the models of Barber (1984) and Cole et al. (1977) have elucidated and pinpointed gaps in certain pathways and reactions of the soil-plant P cycle, which are not fully researched. In contrast, the more user-orientated Decide model (Bennett & Ozanne, 1972) has been used as a basis for examining fertilizer use in terms of both physical and economic efficiency. The model is now used by the Western Australian Department of Agriculture's advisory services in making fertilizer P recommendations.

No one model, however, describes the complete soil-plant cycle including P transformations, reactions, uptake, and effects on crop growth and yield. As this was required as a submodel of EPIC (Williams et al., 1984), we designed the soil-plant P model described in this chapter to simulate P dynamics and crop fertilizer P requirements for a wide range of soils, climates, and management alternatives. Input data availability as well as computational speed also greatly affected the content and complexity of the model. We attempted to simulate organic, inorganic, and plant P dynamics accurately and in a manner consistent with accepted theory. However, where necessary, empiricisms were used to obtain reaction coefficients and initial pool sizes. The model produced accurate estimates of long-term decreases in topsoil organic P due to cultivation, medium-term changes in soil test P due to fertilizer addition, crop response to soil test P, and maintenance fertilizer-P requirements.

Additional testing is needed to verify model accuracy for a wider range of soil, crop, and weather conditions. In addition, further model development is needed in several areas. By including soil solution P as an additional nutrient pool, model sensitivity to root length density could increase and might allow more realistic estimates of the effects of banding P fertilizer. The addition of a P dissolution component would allow realistic simulation of the use of rock phosphate and partially acidulated products. Partial coupling of N and P uptake would allow simulation of the synergistic effects of placing N and P in the same band. More accurate simulation of the effects of lime and acidifying fertilizers on soil pH and P sorption is also needed. In many soils, mineralization of organic P is an important source of labile P (Tiessen et al., 1984). Therefore, more accurate simulation of organic P cycling would probably improve estimation of fertilizer P requirements.

MODELING P DYNAMICS

Model development should also include interactions between P and other nutrients that may affect P availability or uptake; for example, enhanced P uptake occurs when soil ammonium-N levels are high.

Work is currently in progress to improve our estimation of P sorption and pool sizes in soils with extreme properties such as high (>10%) calcium carbonate and low pH (>4.5), and to improve our estimation of the effects of lime and acid-forming fertilizers on soil pH and P sorption.

VI. APPENDIX

The following FORTRAN code details the major components of the P submodel in EPIC. This represents 162 lines of code from a model consisting of over 5000 lines. The P submodel interacts with many other components of EPIC, including crop growth and yield, P loss in runoff, automatic fertilizer, and residue fall and decomposition. These interactions are not described, thus the following FORTRAN code will not simulate P cycling in isolation from the other components of EPIC.

```
      SUBROUTINE NMNIM
C     THIS SUBROUTINE ESTIMATES DAILY N AND P MINERALIZATION AND
C     IMMOBILIZATION CONSIDERING FRESH ORGANIC MATERIAL (CROP RESIDUE)
C     AND ACTIVE AND STABLE HUMUS MATERIAL.
      COMMON /BK2/ DIR(12,16),HRLT(12),RMX(12),TAMX(12),TAV(12),WVL(12),
     1 AL1,ANG,BAS,CF,CO2,ELEV,HR1,PI2,R,RA,RM,SML,SNO,SNOF,TH,TMN,TMX,
     2 TX,WAL,WB,WV
      COMMON /BK7/ QIN(12),LID(11),FC(10),PO(10),S15(10),SC(10),SEV(10),
     1 ST(10),ADEO,ADRF,AWC,QINT,SAT,SW,SWW,WTBL,WTMN,WTMX,IR,LD1,NE,NS
      COMMON /BK8/ HK(10),O(10),RT(10),SSF(10),CRSP,SEP,SLO,SST,SU,K
      COMMON /BK9/ RTN(10),WMN(10),WN(10),WP(10),WT(10),ER,THK,Y,YAP,YN,
     1 YP
      COMMON /BK10/ AP(10),BK(10),OP(10),PMN(10),PSP(10),RWT(10),UN(10),
     1 UP(10),SUN,SUP,UST,L
      COMMON /BK11/ BN(4,11),BP(4,11),REG,UN1,UN2,UNO3,UP1,UP2,UPP,WFX,
     1 IRO,JE
      COMMON /BK13/ SSO3(10),WNO3(10),APB,SSFN,TFON,TNO3,TNOR,TWN,YNO3
      COMMON /BK14/SOL(6,10),SID(11),ALS(10),BD(10),BDD(10),BDM(10),
     1 BDP(10),CAC(10),CBN(10),CEC(10),CLA(10),PH(10),ROK(10),SAN(10),
     2 SIL(10),SMB(10),Z(10),LORG(10),RZ,TLA,ZF,ZQT,LC
      COMMON /BK15/ NC(13),IDA,IDS,IFA,IRI,IY,IYT,KDA,KR1,KR2,KR3,KR4,
     1KR5,KR6,KR7,KW,KW1,KW2,KW3,LYR,MAT,MO,NBG,NCL,ND,NII,NOP,NRO,NT
      COMMON /BK17/ CNY(11),CPY(11),FON(10),FOP(10),HUM(10),RSD(10),
     1AGP,BTN,BTP,CDG,CMN,HMN,RMNR,STD,STDN,STDP,SUT,
     2 TRSD,WDN,WIM,WIP,WMP,JD
      COMMON /BK22/ SMM(58,12),EST(17,10),LY(30,4),SMY(58),STDA(4,12),
     1 TIL(20,2),NCP(30),NTL(30),SCRP(14,2),COTL(20),SF(5,4),IHC(20),
     2 PARM(14),TRA(12),COSD(11),PST(11),SDW(11),TB(11),WCY(11),
     3 SFMO(5),CAW(4),CSTF(4),DMF(4),ETG(4),HUF(4),RDF(4),
     4VIR(4),YLDF(4),YLNF(4),BFT,COIR,COL,CON,COP,EFI,FFC,PSTF,RZSW,SDN,
     5SHM,SHRL,SIM,SIP,SMP,SMR,SN,SP,TSFN,WS,IPD,IRR,NPST
      TKG=RSD(K)*1000.
      CS=SQRT(CDG*SUT)
      RWN=.1E-4*(WMN(K)*(1./RTN(K)-1.)-WN(K))
      WN(K)=WN(K)+RWN
      WIM=0.
      WIP=0.
      HMN=CMN*CS*WMN(K)/(BDP(K)*BDP(K))
      XX=WN(K)+WMN(K)
      HMP=1.4*HMN*WP(K)/XX
      IF(TKG.GT.1.) GO TO 1
      HUM(K)=HUM(K)*(1.-HMN/XX)
      WMN(K)=WMN(K)-HMN-RWN
      WP(K)=WP(K)-HMP
      RMNR=HMN
```

```
              WNO3(K)=WNO3(K)+RMNR
              AP(K)=AP(K)+HMP
              WMP=HMP
              GO TO 6
            1 R4=.58*TKG
              CNR=R4/(FON(K)+WNO3(K))
              CPR=R4/(FOP(K)+AP(K))
              CNRF=EXP(-.693*(CNR-25.)/25.)
              CPRF=EXP(-.693*(CPR-200.)/200.)
              CA=AMIN1(CNRF,CPRF,1.)
              DECR=.05*CA*CS
              RMN=DECR*FON(K)
              RMP=DECR*FOP(K)
              RM2=.2*RMN
              HUM(K)=HUM(K)*(1.+(RM2-HMN)/XX)
              WMN(K)=WMN(K)+RM2-HMN-RWN
              WP(K)=WP(K)-HMP+.2*RMP
              RDC=DECR*TKG
              RSD(K)=.001*(TKG-RDC)
              RMNR=.8*RMN+HMN
              WMP=.8*RMP+HMP
              WIM=.0232*RDC-RMN
              XX=RMNR-WIM+WNO3(K)
              IF(XX.LT.0.)WIM=RMNR+WNO3(K)
              WIP=.0029*RDC-RMP
              XX=WMP-WIP+AP(K)
              IF(XX.LT.0.)WIP=WMP+AP(K)
            5 FOP(K)=FOP(K)+WIP-RMP
              FON(K)=FON(K)+WIM-RMN
              WNO3(K)=WNO3(K)-WIM+RMNR
              AP(K)=AP(K)-WIP+WMP
            6 IF(K.NE.LD1) GO TO 7
              ZZ=.01*(1.+R/10.)
              IF(STD.GT..001) CALL NFALL (ZZ)
       C      WRITE(KW,2)KDA,MO,STD,TKG,RSD(K),WNO3(K),AP(K),CS,CDG,SUT,FON(K),
       C     1HMN,CNR,CPR,CNRF,CPRF,CA,DECR,RMN,RMP,RDP,RDC,WIM,WIP,RMNR
            2 FORMAT(2I4,9E13.5/(8X,9E13.5))
            7 RETURN
              END

              SUBROUTINE NPMIN
       C      THIS SUBROUTINE COMPUTES P FLUX BETWEEN THE LABILE, ACTIVE MINERAL
       C      AND STABLE MINERAL P POOLS.
              COMMON /BK9/ RTN(10),WMN(10),WN(10),WP(10),WT(10),ER,THK,Y,YAP,YN,
             1 YP
              COMMON /BK10/ AP(10),BK(10),OP(10),PMN(10),PSP(10),RWT(10),UN(10),
             1 UP(10),SUN,SUP,UST,L
              COMMON /BK17/ CNY(11),CPY(11),FON(10),FOP(10),HUM(10),RSD(10),
             1AGP,BTN,BTP,CDG,CMN,HMN,RMNR,STD,STDN,STDP,SUT,
             2 TRSD,WDN,WIM,WIP,WMP,JD
              COMMON /BK18/ WFT(12),T(10),T0(5),ABD,AVT,BCV,DD,DST0,ST0
              COMMON /BK22/ SMM(58,12),EST(17,10),LY(30,4),SMY(58),STDA(4,12),
             1 TIL(20,2),NCP(30),NTL(30),SCRP(14,2),COTL(20),SF(5,4),IHC(20),
             2 PARM(14),TRA(12),COSD(11),PST(11),SDW(11),TB(11),WCY(11),
             3 SFMO(5),CAW(4),CSTF(4),DMF(4),ETG(4),HUF(4),RDF(4),
             4VIR(4),YLDF(4),YLNF(4),BFT,COIR,COL,CON,COP,EFI,FFC,PSTF,RZSW,SDN,
             5SHM,SHRL,SIM,SIP,SMP,SMR,SN,SP,TSFN,WS,IPD,IRR,NPST
              S5=.1*SUT*EXP(.115*T(L)-2.88
              RTO=PSP(L)/(1.-PSP(L))
              RMN=S5*(AP(L)-PMN(L)*RTO)
              IF(RMN.LT.0.) RMN=RMN*.1
              ROC=BK(L)*(4.*PMN(L)-OP(L))
              IF(ROC.LT.0.) ROC=ROC*.1
              OP(L)=OP(L)+ROC
              PMN(L)=PMN(L)-ROC+RMN
              AP(L)=AP(L)-RMN
              RETURN
              END

              SUBROUTINE NPUP
       C      THIS SUBROUTINE CALCULATES THE DAILY P DEMAND FOR OPTIMAL PLANT
       C      GROWTH.
              COMMON /BK11/ BN(4,11),BP(4,11),REG,UN1,UN2,UNO3,UP1,UP2,UPP,WFX,
             1 IRO,JE
              COMMON /BK19/PHU(12,30),DLAP(12,2),FRST(12,2),WAC2(12,2),RIN(20),
             1ALT(12),DLAI(12),DMLA(12),GSI(12),HI(12),HMX(12),RBMD(12),
             2RDMX(12),RLAD(12),SWH(12),SWP(12),TG(12),VPD2(12),VPTH(12),WA(12),
             3WAVP(12),WSYF(12),IDC(12),IHU(12),NCR(12),NHU(12),CAF(11),AJHI,
             4AJWA,CHT,DDM,DM,DM1,HU,RW,STL,SYP,WLV,XDLAI,YLD,YLN,YLP,IG,IHV,
```

```
     5IPL,JDHU,JPL,KC
      CPT=BP(2,JE)+BP(1,JE)*EXP(-BP(4,JE)*SYP)
      UP2=CPT*DM*1000.
      IF(UP2.LT.UP1) UP2=UP1
      UPP=UP2-UP1
      RETURN
      END

      SUBROUTINE NUSE
C     THIS SUBROUTINE CALCULATES THE DAILY POTENTIAL SOIL SUPPLY OF P
C     FOR EACH LAYER.
      COMMON /BK7/ QIN(12),LID(11),FC(10),PO(10),S15(10),SC(10),SEV(10),
     1 ST(10),ADEO,ADRF,AWC,QINT,SAT,SW,SWW,WTBL,WTMN,WTMX,IR,LD1,NE,NS
      COMMON /BK9/ RTN(10),WMN(10),WN(10),WP(10),WT(10),ER,THK,Y,YAP,YN,
     1 YP
      COMMON /BK10/ AP(10),BK(10),OP(10),PMN(10),PSP(10),RWT(10),UN(10),
     1 UP(10),SUN,SUP,UST,J
      COMMON /BK11/ BN(4,11),BP(4,11),REG,UN1,UN2,UN03,UP1,UP2,UPP,WFX,
     1 IRO,JE
      COMMON /BK12/ U(10),COST,GX,RD,S,SX,TYN,TYP,UB1,UOB,UX
      COMMON /BK13/ SSO3(10),WNO3(10),APB,SSFN,TFON,TNO3,TNOR,TWN,YNO3
      COMMON /BK19/PHU(12,30),DLAP(12,2),FRST(12,2),WAC2(12,2),RIN(20),
     1ALT(12),DLAI(12),DMLA(12),GSI(12),HI(12),HMX(12),RBMD(12),
     2RDMX(12),RLAD(12),SWH(12),SWP(12),TG(12),VPD2(12),VPTH(12),WA(12),
     3WAVP(12),WSYF(12),IDC(12),IHU(12),NCR(12),NHU(12),CAF(11),AJHI,
     4AJWA,CHT,DDM,DM,DM1,HU,RW,STL,SYP,WLV,XDLAI,YLD,YLN,YLP,IG,IHV,
     5IPL,JDHU,JPL,KC
      COMMON /BK22/ SMM(58,12),EST(17,10),LY(30,4),SMY(58),STDA(4,12),
     1 TIL(20,2),NCP(30),NTL(30),SCRP(14,2),COTL(20),SF(5,4),IHC(20),
     2 PARM(14),TRA(12),COSD(11),PST(11),SDW(11),TB(11),WCY(11),
     3 SFMO(5),CAW(4),CSTF(4),DMF(4),ETG(4),HUF(4),RDF(4),
     4VIR(4),YLDF(4),YLNF(4),BFT,COIR,COL,CON,COP,EFI,FFC,PSTF,RZSW,SDN,
     5SHM,SHRL,SIM,SIP,SMP,SMR,SN,SP,TSFN,WS,IPD,IRR,NPST
      XX=1.5*UPP/RW
      DO 1 L=1,IR
      J=LID(L)
      UN(J)=WNO3(J)*U(J)/(ST(J)+.001)
      SUN=SUN+UN(J)
      F=1000.*AP(J)/WT(J)
      F=F/(F+EXP(SCRP(11,1)-SCRP(11,2)*F))
      UP(J)=XX*F*RWT(J)
      IF(UP(J).GE.AP(J)) UP(J)=.9*AP(J)
    1 SUP=SUP+UP(J)
      RETURN
      END
```

REFERENCES

Anghinoni, I., and S.A. Barber. 1980. Predicting the most efficient phosphate placement for corn. Soil Sci. Soc. Am. J. 44:1016-1020.

Anway, J.C. 1976. Introduction. p. 1-5. In G.W. Cole (ed.) ELM: Version 2.0. Colorado State University Range Sci. Dep. Sci. Ser. 20.

Barber, S.A. 1984. Soil nutrient bioavailability: A mechanistic approach. Wiley-Interscience, New York.

Bennett, D., and J.W. Bowden. 1976. "Decide"—An aid to efficient use of phosphorus. p. 77-81. In G.J. Blair (ed.) Prospects for improving efficiency of phosphorus utilization. Proc. of Symp. at Univ. of New England, Armidale, N.S.W. Australia. Reviews in Rural Sci. III.

Bennett, D., and P.G. Ozanne. 1972. Australia, CSIRO division of plant industry Annual Report. Commonwealth Scientific and Industrial Res.

Black, A.L. 1982. Long-term N-P fertilizer and climate influences on morphology and yield components of spring wheat. Agron. J. 74:651-657.

Bray, R.H., and L.T. Kurtz. 1945. Determination of total, organic, and available forms of phosphorus in soils. Soil Sci. 59:39-45.

Chapin, F.S., R.J. Barsdate, and D. Barel. 1978. Phosphorus cycling in Alaska coastal tundra: A hypothesis for the regulation of nutrient-cycling. Oikos 31:181-199.

Claassen, N., and S.A. Barber. 1974. A method for characterizing the relation between nutrient concentration and flux into roots of intact plants. Plant Physiol. 54:564-568.

Claassen, N., and S.A. Barber. 1976. Simulation model for nutrient uptake from soil by a growing plant root system. Agron. J. 68:961-964.

Cole, C.V., G.S. Innis, and J.W.B. Stewart. 1977. Simulation of phosphorus cycling in semi-arid grasslands. Ecology 58:1-15.

Cole, G.W. 1976. Nutrient submodels. p. 304-397. In G.W. Cole (ed.) ELM: Version 2.0. Colorado State Univ. Range Sci. Dep. Sci. Ser. 20.

Colwell, J.D. 1963. The estimation of the phosphorus fertilizer requirements of wheat in southern New South Wales by soil analysis. Aust. J. Exp. Agric. Anim. Husb. 3:190-197.

Cope, J.T., Jr. 1981. Effects of 50 years of fertilization with phosphorus and potassium on soil test levels and yields at six locations. Soil Sci. Soc. Am. J. 45:342-347.

Cushman, J.H. 1982. Nutrient transport inside and outside the root rhizosphere: Theory. Soil Sci. Soc. Am. J. 46:704-709.

Cushman, J.H. 1984. Nutrient transport inside and outside the root rhizosphere: Generalized model. Soil Sci. 138:164-171.

Dalal, R.C., and E.G. Hallsworth. 1976. Evaluation of the parameters of soil phosphorus availability factors in predicting yield response and P uptake. Soil Sci. Soc. Am. J. 40:541-546.

Haas, H.J., C.E. Evans, and E.F. Miller. 1957. Nitrogen and carbon changes in Great Plains soils as influenced by cropping and soil treatments. USDA Tech. Bull. 1164. U.S. Gov. Print. Office, Washington, DC.

Haas, H.J., D.L. Grunes, and G.A. Reichman. 1961. Phosphorus changes in Great Plains soils as influenced by cropping and manure applications. Soil Sci. Soc. Am. Proc. 24:214-218.

Harrison, A.F. 1978. Phosphorus cycles of forest and upland grassland systems and some effects of land management practices. p. 175-195. In Phosphorus in the environment: Its chemistry and biochemistry. CIBA Foundation Symp. 57, Amsterdam. Elsevier, North Holland.

Helyar, K.R., and D.P. Godden. 1976. The phosphorus cycle—What are the sensitive areas? p. 23-30. In G.J. Blair (ed.) Prospects for improving efficiency of phosphorus utilization. Proc. of Symp. at Univ. of New England, Armidale, N.S.W. Australia Reviews in Rural Sci. III.

Helyar, K.R., and D.N. Munns. 1975. Phosphate fluxes in the soil-plant system: A computer simulation. Hilgardia 43:103-130.

Hooker, M.L., R.E. Gwin, G.M. Herron, and P. Gallagher. 1983. Effects of long-term, annual applications of N and P on corn grain yields and soil chemical characteristics. Agron. J. 75:94-99.

Ibach, D.R., and J.R. Adams. 1968. Crop yield response to fertilizer in the United States. USDA, Econ. Res. Serv. and Stat. Rep. Serv., U.S. Gov. Print. Office, Washington, DC.

Jones, C.A., C.V. Cole, A.N. Sharpley, and J.R. Williams. 1984a. A simplified soil and plant phosphorus model: I. Documentation. Soil Sci. Soc. Am. J. 48:800-805.

Jones, C.A., A.N. Sharpley, and J.R. Williams. 1984b. A simplified soil and plant phosphorus model: III. Testing. Soil Sci. Soc. Am. J. 48:810-813.

Mackay, A.D., and S.A. Barber. 1984. Soil temperature effects on root growth and phosphorus uptake by corn. Soil Sci. Soc. Am. J. 48:818-823.

Mansell, R.S., H.M. Selim, and J.G.A. Fiskell. 1977. Simulated transformations and transport of phosphorus in soil. Soil Sci. 124:102-109.

Mattingly, G.E.G. 1975. Labile phosphate in soils. Soil Sci. 119:369-375.

Mishra, B., P.K. Khanna, and B. Ulrich. 1979. A simulation model for organic phosphorus transformation in a forest soil ecosystem. Ecol. Modell. 6:31-46.

Murrman, R.P., and M. Peech. 1969. Relative significance of labile and crystalline phosphates in soil. Soil Sci. 107:149-155.

Nye, P.H., and F.H.C. Marriott. 1969. A theoretical study of the distribution of substances around roots resulting from simultaneous diffusion and mass-flow. Plant Soil 30:459-472.

Olsen, S.R., C.V. Cole, F.S. Watanabe, and L.A. Dean. 1954. Estimation of available phosphorus in soils by extraction with sodium bicarbonate. USDA Circ. 939. U.S. Gov. Print. Office, Washington, DC.

Olsen, S.R., and W.D. Kemper. 1967. Movement of nutrients to plant roots. Adv. Agron. 20:91-151.

Ryden, J.C., J.R. McLaughlin, and J.K. Syers. 1977. Mechanisms of phosphate sorption by soils and hydrous ferric oxide gel. J. Soil Sci. 28:72-92.

Sabbe, W.E., and H.L. Breland. 1974. Procedures used by state soil testing laboratories in the southern region of the United States. South. Coop. Ext. Ser. Bull. 190.

Schenk, M.K., and S.A. Barber. 1979a. Phosphate uptake by corn as affected by soil characteristics and root morphology. Soil Sci. Soc. Am. J. 43:880-883.

Schenk, M.K., and S.A. Barber. 1979b. Root characteristics of corn gneotypes as related to P uptake. Agron. J. 71:921-924.

Seligman, N.G., and H. van Keulen. 1981. PAPRAN: A simulation model of annual pasture production limited by rainfall and nitrogen. p. 192-221. *In* M.J. Frissel and J.A. van Veen (ed.) Simulation of nitrogen behavior in soil-plant systems. PUDOC, Wageningen, The Netherlands.

Sharpley, A.N., C.A. Jones, C. Gray, and C.V. Cole. 1984. A simplified soil and plant phosphorus model: II. Prediction of labile, organic, and sorbed phosphorus. Soil Sci. Soc. Am. J. 48:805-809.

Sharpley, A.N., C.A. Jones, C. Gray, C.V. Cole, H. Tiessen, and C.S. Holzhey. 1985. A detailed characterization of seventy-eight soils. USDA-ARS, Ser. Publ. ARS-31.

Sharpley, A.N., and S.J. Smith. 1983. Distribution of phosphorus forms in virgin and cultivated soil and potential erosion losses. Soil Sci. Soc. Am. J. 47:581-586.

Sibbesen, E. 1978. An investigation of the anion-exchange resin method for soil phosphate extraction. Plant Soil 50:305-321.

Silberbush, M., and S.A. Barber. 1983. Prediction of phosphorus and potassium uptake by soybean with a mechanistic mathematical model. Soil Sci. Soc. Am. j. 47:262-265.

Stewart, J.W.B., and R.B. McKercher. 1981. Phosphorus cycle. p. 221-238. *In* R.G. Burns and J.H. Slater (ed.) Experimental microbial ecology. Blackwell Sci. Publ., Oxford, England.

Tiessen, H., J.W.B. Stewart, and C.V. Cole. 1984. Pathways of phosphorus transformations in soils of differing pedogenesis. Soil Sci. Soc. Am. J. 48:853-858.

Williams, J.R., C.A. Jones, and P.T. Dyke. 1984. A modeling approach to determining the relationship b etween erosion and soil productivity. Trans. ASAE 27:129-144.

15 Nitrogen Solute Transport

K. K. TANJI AND M. NOUR EL DIN
University of California
Davis, California

The flow of solutes in a soil-plant-water system is dynamic and complex. Chapter 13 presented N transformations and movement in wheat (*Triticum aestivum* L.) and maize (*Zea mays* L.) cropping systems. Although N transformations and movement are considered, the emphasis in Ch. 13 is on grain yields, N plant uptake, and grain N contents. Chapter 16 presented solute transport in the soil profile for nonreactive and reactive solute species. The flow of chloride ion is the simplest case since it is assumed to be nonreactive. Other solute species (Na, Ca, Mg, SO_4, HCO_3, etc.) are more difficult to simulate since they participate in cation exchange or mineral solubility. This chapter complements Ch. 13 by presenting more details on N transformations and transport in the soil zone. It also complements Ch. 16 by considering transformations and transport of a primary nutrient solute species.

I. MODEL DESCRIPTION

To simulate the transport of solutes in the soil zone of a soil-plant-water system, it is necessary to first simulate soil-water fluxes. Figure 15–1 describes a dynamic simulation model used to predict water and N flows in an irrigated, fertilized maize field. The water-flow submodel considers the seasonal soil-water cycle, including irrigation and precipitation, soil hydraulic properties, and evapotranspiration. The outputs from this submodel serve as inputs to the N transport submodel, which considers N transformations, N plant uptake, and convective-diffusion transport of N. This chapter focuses on reactivity and mobility of N solute species (Tanji et al., 1981), with a brief description of water flow (Gupta et al., 1978).

A. Modeling Soil Water Flow

The one-dimensional, nonsteady flow of water in the root zone (Chapter 11; Nimah & Hanks, 1973) can be expressed as:

$$C(\theta) \frac{\partial h}{\partial t} = \frac{\partial}{\partial z} K(h) \frac{\partial H}{\partial z} - S \quad [1]$$

Copyright © 1991 ASA-CSSA-SSSA, 677 S. Segoe Rd., Madison, WI 53711, USA. *Modeling Plant and Soil Systems*—Agronomy Monograph no. 31.

DYNAMIC SIMULATION MODEL

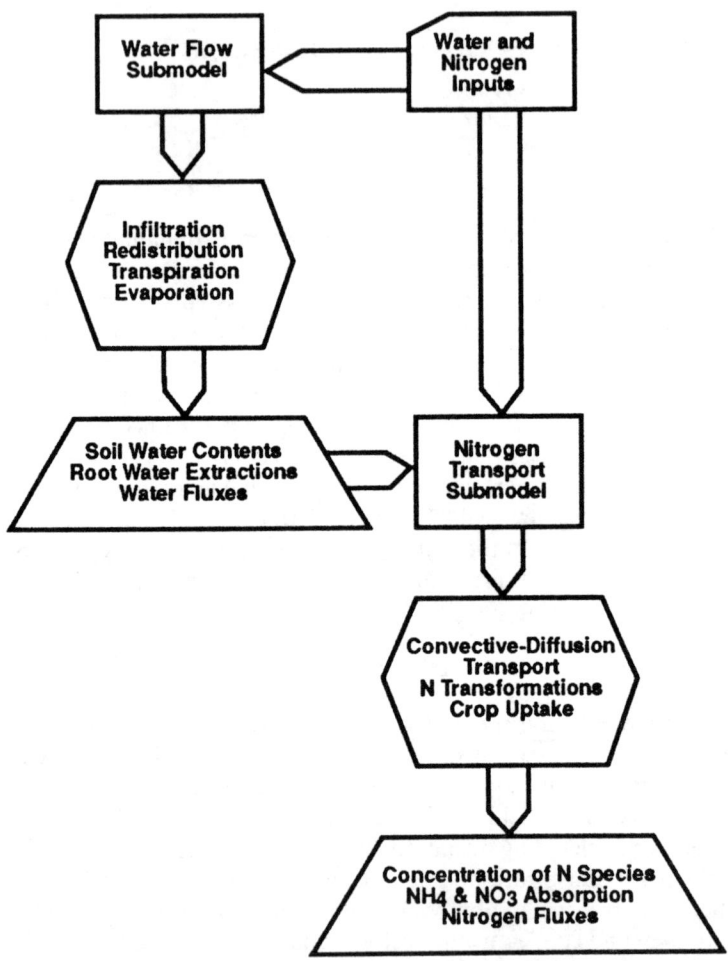

Fig. 15-1. A simplified block diagram of the waterflow and N transport submodels (Tanji, 1982).

where $C(\theta)$ is the differential soil-water capacity ($\partial\theta/\partial h$), θ is the volumetric soil water content, t is time, z is the vertical distance from the soil surface, K is the hydraulic conductivity, H is the hydraulic head (sum of pressure head h and gravity head z), and S is the sink term for root water extraction.

The solution of this nonlinear partial differential equation requires certain initial and boundary conditions. Gupta et al. (1978) solved Eq. [1] for a semi-infinite soil depth and used an implicit finite-difference approximation method. The nonlinearity of Eq. [1] was treated by an iterative scheme that considers daily mass balance of water among the convergence criteria.

The water flow submodel has been validated with experimental field data on changes in θ over an entire growing season (Gupta et al., 1978). It has also been used to obtain further insight into the complexities of the field soil-water cycle, including the evaluation of spatial variability of K (Biggar et al., 1977).

As mentioned earlier, the N transport submodel requires input data from the water flow submodel. Other appropriate water flux models may be used (e.g., those described in Ch. 9, 10, 11, and 12, as well as in Tanji [1982]).

B. Modeling Soil Nitrogen Transformations and Transport

The flow of N in the soil profile (Tanji et al., 1981) may be described by

$$\frac{\partial(\theta C)}{\partial t} = D\frac{\partial^2 C}{\partial z^2} - v\frac{\partial C}{\partial z} + \frac{C\lambda(z,t)}{\theta} - \frac{\rho}{\theta}\frac{\partial E}{\partial t} + S_c \quad [2]$$

where C is the concentration of the mobile N species, D is the apparent diffusion coefficient, v is the pore water velocity, ρ is the soil bulk density, E is the concentration of N on the exchanger phase, λ is the dimensionless root absorption coefficient, and S_c is the sources and sinks of the mobile N species. The term S_c includes a number of N transformation processes as sources and sinks affecting the N species. As a first approximation, Mehran and Tanji (1974) assumed that NH_4^+ ion exchange obeys reversible first-order kinetics and that nitrification, denitrification, mineralization, and immobilization obey irreversible first-order kinetics. It is also assumed that NH_4^+ and NO_3^- are the only mobile species and that they are differentially taken up by the maize plant.

Differential equations describing the rate of change of NH_4^+, NO_3^-, exchangeable NH_4^+, organic N, and gaseous N are, respectively,

$$(1 + R_f)\frac{\partial A}{\partial t} = D_A\frac{\partial^2 A}{\partial z^2} - v\frac{\partial A}{\partial z}$$

$$+ \frac{A\lambda(z,t)}{\theta} - [K_1(z) + K_4(z)]A + K_5(z)\frac{\rho}{\theta} \quad [3]$$

$$\frac{\partial B}{\partial t} = D_B\frac{\partial^2 B}{\partial z^2} - v\frac{\partial B}{\partial z}$$

$$+ \frac{B\lambda(z,t)}{\theta} - [K_2(z) + K_3(z)]B + K_1(z)A \quad [4]$$

$$\frac{\partial E}{\partial T} = K_{ex} \frac{\partial A}{\partial t} \qquad [5]$$

$$\frac{\partial F}{\partial t} = K_3(z) \frac{\theta}{\rho} B + K_4(z) \frac{\theta}{\rho} A - K_5(z) F \qquad [6]$$

$$\frac{\partial G}{\partial t} = K_2(z) \frac{\partial}{\rho} B \qquad [7]$$

where A is the solution NH_4^+ concentration, B is NO_3^- concentration, E is exchangeable NH_4 concentration, F is organic N concentration, G is gaseous N ($N_2 + N_2O$) concentration, K_{ex} is the coefficient for exchange reaction, R_f is $\rho K_{ex}/\theta$, D_A and D_B are the respective apparent diffusion coefficients for NH_4^+ and NO_3^-, and K_1, K_2, K_3, K_4, and K_5 are the respective transformation rate constants for nitrification of NH_4^+, denitrification of NO_3^-, immobilization of NO_3^-, immobilization of NH_4^+, and mineralization of organic N.

The computations for N-leaching losses at any soil depth for all times require the simultaneous solution of the above differential equations. Equations [3] to [7] are subject to a number of initial and boundary conditions (Tanji & Mehran, 1979) and can be solved by an implicit finite-difference scheme. To ensure the accuracy of this numeric scheme, convergence is based in part on attaining a daily mass balance of NH_4^+ and NO_3^-. This mass balance is checked for each soil depth increment.

Equation [2] is simplified by dropping the last three terms on the right-hand side for transport of nonreactive solutes like Cl^-. Other sink/source terms may be added to Eq. [2] for those solute species which participate in cation exchange, clay fixation, specific adsorption, and mineral solubility (Iskandar, 1981).

C. Input, Output, and Verifiable Parameters

1. Water Flow Submodel Input Parameters

This submodel requires the following:

1. Number of nodes, depth segments, and material properties (layers with distinct hydraulic properties).
2. Crop data (dates of planting and harvest, and number of days roots take to grow to various depths).
3. Factors to convert daily potential evapotranspiration to hourly values.
4. Irrigation and rainfall (amount, time, and date).
5. Parameters for hydraulic conductivity (initial values of θ and h, θ at steady-state infiltration, and saturated and dry K).

6. Matric potential (suction) and water content limits (total initial θ, maximum depth of ponding, maximum and lowest limits of suction, and saturation and dry θ).
7. Controlling factors for time steps and iteration (maximum time steps during irrigation and nonirrigation periods, minimum limit of time step if mass balance or iteration does not converge, and maximum difference in mass balance).
8. Coefficients for spatial variability of K.

2. Nitrogen Flow Submodel Input Parameters

This submodel requires the following: (i) output data from water flow submodel (θ, q [flux], and S); (ii) initial profile distribution of solution and exchangeable NH_4^+, NO_3^-, organic N, and gaseous N; (iii) number of nodes and depth increments; (iv) specification of initial time and space increments; (v) root absorption coefficient for NH_4^+ and NO_3^-; (vi) apparent diffusion coefficients for NH_4^+ and NO_3^-; (vii) soil bulk density; (viii) first-order transformation rate constants and NH_4^- ion exchange constant; (iv) applied concentrations of NH_4^+ and NO_3^- in irrigation water and rainfall; (x) and applied fertilizer N.

3. Output and Verifiable Variables

The output and verifiable parameters include: water content, water-flux, and water uptake; concentration of all N species in the soil profile; mass of NH_4^+ and NO_3^- taken up by corn, and mass of NH_4^+ and NO_3^- leached past the root zone.

4. File Listing

A program listing of the N flow submodel that considers N transformations and transport is found in the Appendix 1. The output file listing presented in Appendix 2 is a simulation run for the experimental 180 kg N ha^{-1}, 5/3 ET (evapotranspiration) treatment. The output also contains all of the required input data, initial conditions, and other control variables. Similar listings for the water flow submodel are available from the authors (Gupta et al., 1978).

5. Running Time and Cost

Table 15-1 lists the typical running time and costs for operating this model on a medium-sized computer for a 131-day growing season.

D. Critical Assumptions for the Nitrogen Submodel

The N transport model uses outputs (water flux, volumetric soil-water content, and root water extraction) from the water flow submodel throughout the growing season. The solute transport of mobile N species (NH_4^+ and NO_3^-) is obeyed by a convective-diffusion equation. Nitrogen transforma-

Table 15-1. Running time and costs for the water flow and N flow submodels on a Burroughs B6700 system for 131 daily outputs for water flow and nine output dates (0, 13, 26, 42, 56, 70, 84, 98, and 112 d after planting) for N flow simulation.

Description	Water flow submodel		Nitrogen flow submodel	
	Quantity	Charge	Quantity	Charge
Processor time (min.)	5.06	$10.12	3.45	$ 7.11
Virtual memory (kilowords/min.)	8.43	$ 0.67	21.12	$ 1.74
Save and buffer memory (kilowords/min.)	18.59	$ 2.97	12.48	$ 2.06
Other costs		$ 7.77		$ 5.46
Total costs		$21.53		$16.37

tion processes obey first-order kinetics and plant uptake of NH_4^+ and NO_3^- is constrained by variable absorption coefficients.

Other approaches to modeling N transformations, uptake, and transport have been reviewed by Tanji (1982). Dutt et al. (1972) used regression equations for nitrification, mineralization-immobilization and urea hydrolysis, an equilibrium cation exchange equation for NH_4^+ exchange, and assumed N plant uptake proportional to water uptake. Later, Shaffer et al. (1977) revised the denitrification portion of the Dutt et al. (1972) model with zero-order kinetics. Davidson et al. (1978) described nitrification, mineralization, and immobilization by first-order kinetics modified by soil-water pressure head, denitrification by first-order kinetics modified by pressure head, water content and organic matter content, NH_4^+ sorption by linear partition model, and NH_4^+ and NO_3^- uptake by Michaelis-Menten kinetics. Watts and Hanks (1978) used Stanford and Smith's (1972) first-order rate equation for mineralization adjusted for soil water content and fraction of fillable pore space containing water, a temperature-dependent first-order equation for nitrification, and the Dutt et al. (1972) urea hydrolysis model. Tillotson et al. (1980) also used first-order kinetics for urea hydrolysis, nitrification and NH_3 volatilization, NH_4^+ sorption by linear partition model, and NH_4^+ and NO_3^- plant uptake involving diffusion to roots.

Beek and Frissel (1973) and Frissel and van Veen (1978; 1981) chose a more biologically based N transformation model. They started initially with modeling the growth of nitrifiers and ammonifiers by Michaelis-Menten and Monod kinetics. The C/N ratio of the organic material was assumed to control mineralization and immobilization. Later soil organic carbon was partitioned into proteins, sugars, cellulose, lignin, humus, and living biomass to describe differences in the rate of decomposition. At this stage, the C/N ratio was no longer assumed to control organic matter decomposition, but by carbon uptake by the growing microbial biomass. In the final stage, the mineralization-immobilization portion of the model was revised to six pools of carbon, three of which contained N. These pools more closely corresponded to typical N and carbon soil analysis (e.g., decomposable crop residue, slowly decomposable crop residue, recalcitrant organic matter, etc.).

A similar modeling approach was taken by Juma and Paul (1981), who used N and carbon isotopes to obtain rate constants.

A variety of modeling approaches have been taken for N transformations and plant uptake. In most instances the models developed by these authors were for site-specific conditions and experiments.

II. MODEL VALIDATION

The experimental data base for the simulation runs used here was derived from a corn (*Zea mays* L.) monoculture field experiment conducted for 5 yr at the University of California-Davis campus (Broadbent & Carlton, 1979; Biggar et al., 1979; Nielsen et al., 1980).

The water flow submodel was used to simulate three sprinkler irrigation treatments of 20, 60, and 100 cm (irrigated seven times) during the 1975 growing season (May through September). These three treatments correspond, respectively, to 1/3, 3/3, and 5/3 of the average corn ET at Davis. The N transport submodel was used to simulate 0, 90, 180, and 360 Kg N ha^{-1} fertilizer treatments of the 5/3 ET irrigation regime of the 1975 growing season.

The water flow model was validated by Tanji et al. (1981), who found that the simulated values of the moisture content were well within the standard deviations of the measured data obtained from 12 plots replicated four times for irrigation and fertilizer treatments.

Figures 15-2A, 15-2B, and 15-2C show the simulated solution NH_4^+ concentrations as a function of time across the 131-d growing season for the 30, 60 and 120 cm depths, respectively, in the 180 Kg N ha^{-1} and 5/3 ET treatment. The $(NH_4)_2SO_4$ fertilizer was applied as a band at a 5 cm soil depth. Note that solution NH_4^+ tends to be confined in the upper 30 cm of the profile (Fig. 15-2A) because of adsorption to the soil exchange complex and rapid nitrification. The peaks in NH_4^+ concentration in Fig. 15-2B, plotted at a different scale than Fig. 15-2A, represent solute transport following the seven irrigation-water applications. Solution NH_4^+ was virtually at zero concentration at the 120 cm depth (Fig. 15-2C).

Figures 15-3A, 15-3B, 15-3C, and 15-3D show simulated NO_3^- concentrations as a function of time for the 30, 60, 180, and 240 cm depths, respectively, in the 180 Kg N ha^{-1} and 5/3 ET treatment. A note should be made that pulses of solute transport, especially near the soil surface, (e.g. Fig. 15-3B) are related to the irrigation scheduling. Because of the significant mineralization of organic N in this Yolo fine sandy loam (fine-silty, mixed, nonacid, thermic Typic Xerorthents), a sensitivity analysis was carried out to ascertain the most likely reaction rate constants for mineralization as well as denitrification using the 0 Kg N ha^{-1} treatment (Tanji et al., 1981).

Measured data are also plotted on the concentration-time curves. Samples of soil solutions were taken by suction probes after each irrigation. At some profile depths and at certain times, measured data are not available, due to problems in obtaining sufficient volume for chemical analyses.

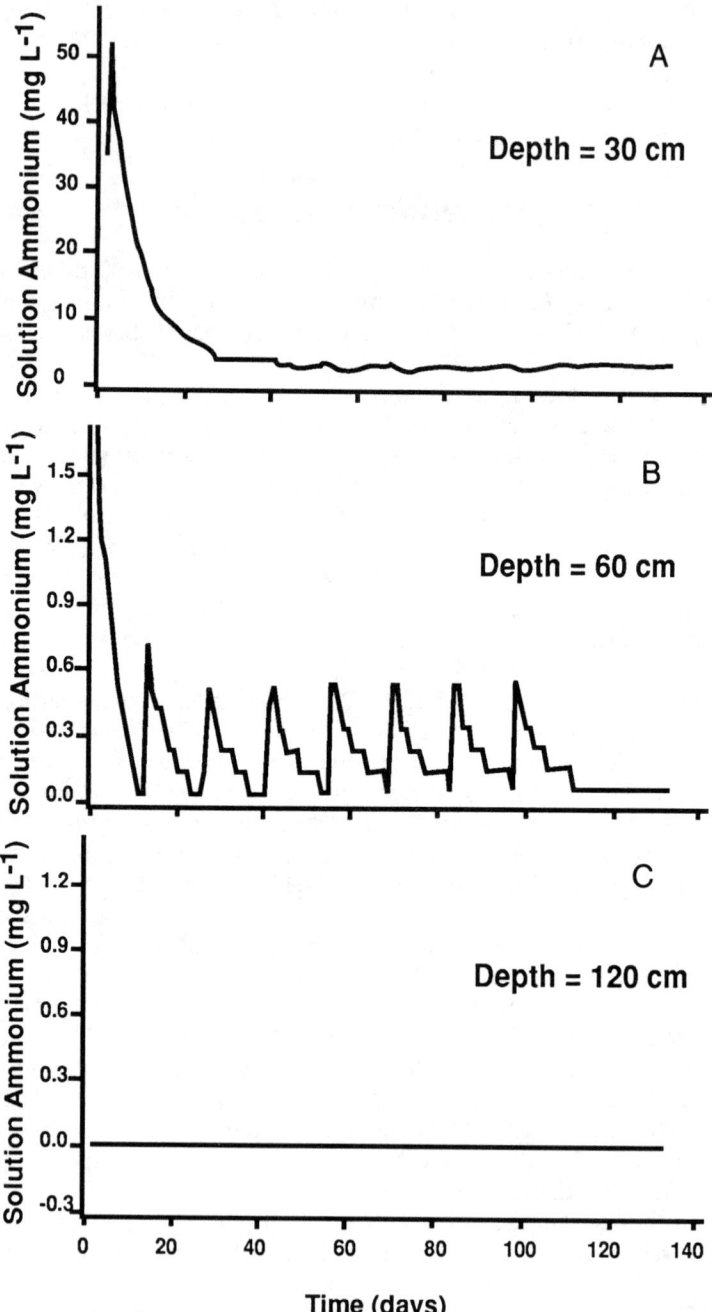

Fig. 15-2. Simulated solution NH_4^+ concentrations (mg L^{-1}) over the 131-d growing season for the (A) 30-cm, (B) 60-cm, and (C) 120-cm soil depths in the 5/3 ET, 180 kg N ha^{-1} fertilizer treatment of a maize crop.

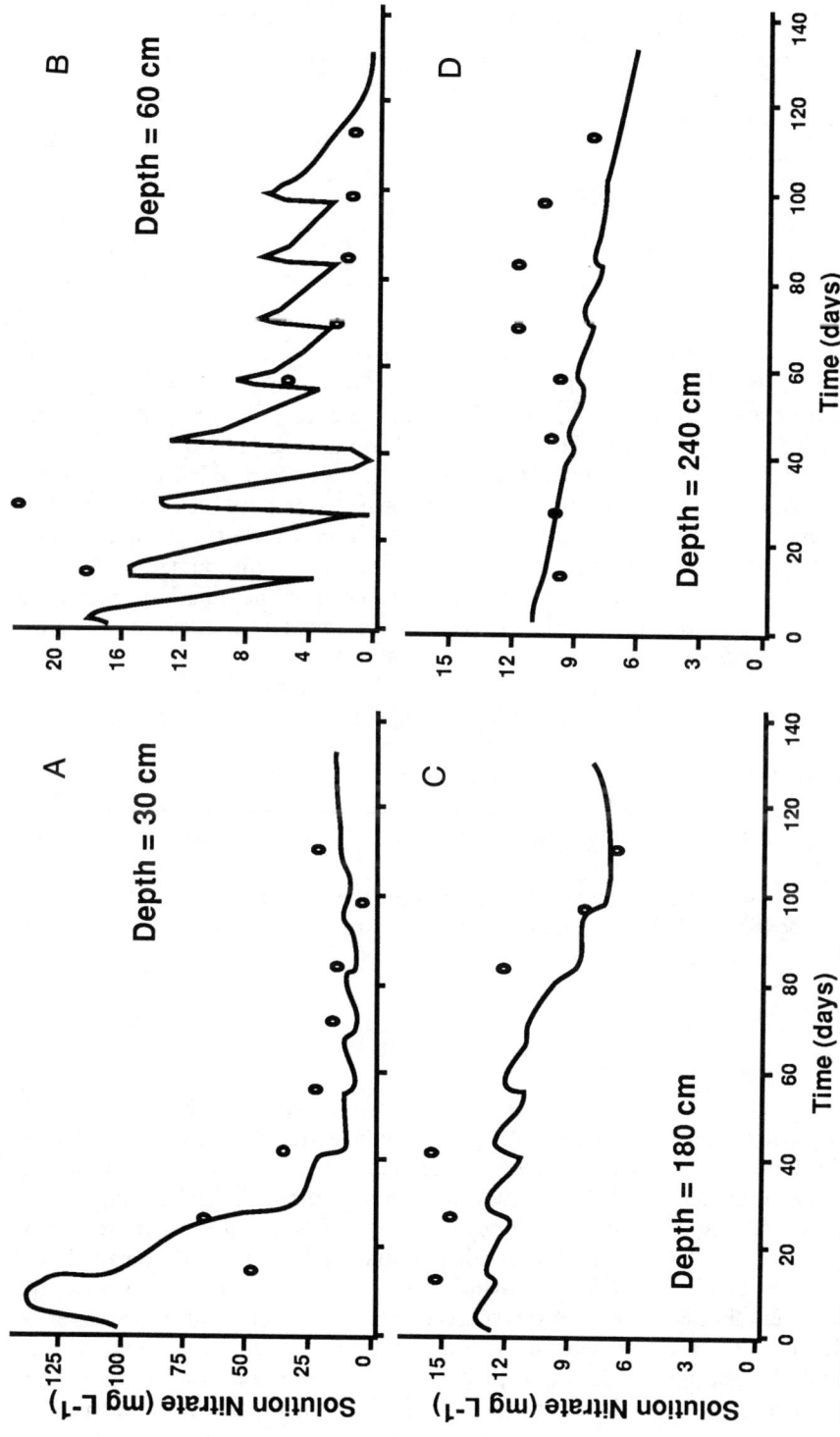

Fig. 15-3. Simulated NO_3^- concentrations (mg L^{-1}) over the 131-d growing season for the (A) 30-cm (B) 60-cm, (C) 180-cm, and (D) 240-cm soil depths in the 5/3 ET, 180 kg N ha^{-1} fertilizer treatment of a maize crop. Measured data on NO_3^- are also plotted as circles.

Table 15-2. Comparison between cumulative measured and computed results for the 5/3 ET irrigation regime in kg ha^{-1} N.

N application rate	Plant uptake		Residual inorganic nitrogen		Calculated leaching losses
	Measured	Simulated	Measured	Simulated	
			kg N/ha		
0	77	57	117	69	70
90	155	121	137	91	77
180	214	202	134	89	123
360	283	260	295	260	300

Taking into account the usual spatial variability encountered in the field, a reasonable fit between the simulated and measured data for NO_3 was obtained. Large discrepancies such as calculated and observed NO_3 concentration at Day 13 in the 30 cm depth may be attributed to problems of sampling the soil solution near the soil surface and beneath the fertilizer band. Another reason could be the spatial variability of the rate constants used in computing the denitrification process. The simulated results undoubtedly could be improved if more information becomes available on the functional relations of transformation rates with environmental parameters, and conditions such as temperature, pH, aeration, and microbial biomass.

Table 15-2 summarizes the cumulative N plant uptake and leaching losses as well as the residual inorganic N at harvest from the simulation model. These are compared to measured values. From this table it may be concluded that the model can closely simulate uptake of N, but not the residual organic N. A probable reason for deviation of simulated from measured values originates from the water flow submodel, which apparently overestimates the seepage beyond the root zone.

A number of researchers have obtained the program listing of this model for possible application to their field plot studies, but the authors are not aware of its utility and validation results.

III. SUMMARY

A computationally efficient N solute transport model that considers N transformations, N plant uptake, and convective-diffusion transport is described herein. The modeling approach assumes that NH_4^+ exchange obeys reversible first-order kinetics, while nitrification, denitrification, mineralization, and immobilization obey irreversible first-order kinetics. It is also assumed that NH_4^+ and NO_3^- are differentially taken up by the maize plant.

The results of model validation indicate reasonable, but not precise, fit between simulated and observed soil profile NO_3 concentrations over a 131-d maize growing season. The discrepancies noted were attributed to spatial variability in the field, problems in sampling the soil solution near the soil surface and directly beneath the fertilizer band, and lack of consideration for environmental parameters and conditions such as temperature, pH, aeration status, and microbial biomass. However, cumulative (seasonal) computed results on N plant uptake for four fertilizer application rates compared favorably with measured data.

IV. APPENDIX 1

```
C----------------------------------------------------------------
C                               *
C                              ***
C                             *****
C                              ***
C                               *
C            N I T R O G E N  T R A N S F O R M A T I O N  A N D
C                      T R A N S P O R T  M O D E L
C                               *
C                              ***
C                             *****
C                              ***
C                               *
C----------------------------------------------------------------
C     THIS PROGRAM COMPUTES CONCENTRATION AND MASS OF SOLUTION AMMONIA
C     AND NITRATE SIMULTANEOUSLY IN THE SOIL PROFILE AS A FUNCTION Of
C     TIME. EXCHANGEABLE AMMONIA,ORGANIC MATTER,AND GASEOUS N ARE ALSO
C     COMPUTED. SOLUTION AMMONIA AND NITRATE ARE SUBJECT TO TRANSPORT AND
C     TRANSFORMATION. WRITTEN BY M.MEHRAN AND MODIFIED BY M.NOUR EL-DIN
C----------------------------------------------------------------
C     KSE=RATE CONSTANT FOR CONVERSION OF [NH4]S TO [NH4]E
C     KES=RATE CONSTANT FOR CONVERSION OF [NH4]E TO [NH4]S
C     K1 =RATE CONSTANT FOR NITRIFICATION OF [NH4]S TO [NO3]S
C     K2 =RATE CONSTANT FOR DENITRIFICATION OF [NO3]S TO [N2O+N2]G
C     K3 =RATE CONSTANT FOR IMMOBILIZATION OF [NO3]S TO [ORGN]
C     K4 =RATE CONSTANT FOR IMMOBILIZATION OF [NH4]S TO [ORGN]
C     KK4=RATE CONSTANT FOR MINERALIZATION OF [ORGN] TO [NH4]S
C     C  =CONCENTRATION OF SOLUBLE AMMONIA=[NH4]S
C     CN =CONCENTRATION OF SOLUBLE NITRATE=[NO3]S
C     CEXCH=CONCENTRATION OF EXCHANGEABLE AMMONIA=[NH4]E
C     ORGN=CONCENTRATION OF SOLUBLE ORGANIC MATTER=[ORGN]
C     GN =CONCENTRATION OF GASEOUS NITROGEN=[N2O+N2]G
C     NPT=NUMBER OF SPACE NODES
C     NT =NUMBER OF TIME NODES
C     NODE=NUMBER OF NODE FOR PRINTED OUTPUT
C     DELZ=SPACE INCREMENT
C     DELT=TIME INCREMENT
C     S=RATE OF WATER UPTAKE BY PLANTS
C     CINPT=INPUT CONCENTRATION OF [NH4]S
C     CINPTN=INPUT CONCENTRATION OF [NO3]S
C     LAMDA=ABSOPTION COEFFICIENT FOR UPTAKE OF [NH4]S
C     LAMDAN=ABSORPTION COEFFICIENT FOR UPTAKE OF [NO3]S
C----------------------------------------------------------------
C     UNITS:
C     CONCENTRATION - MILLIGRAM N / LITER OF SOIL SOLUTION
C----------------------------------------------------------------
      DOUBLE PRECISION SUP(25)
      REAL KSE(25),KES(25),K1(25),K2(25),K3(25),K4(25),KK4(25),MORGN(25)
```

```
      REAL MCI,MCNI,MEXCI,MORGI,MGNI,MIN,MOUT,MST,MBAL,MTRF
      REAL MSTI,MINN,MOUTN,MSTN,MSTIN,MBALN,MGN
      REAL LAMDA,LAMDAN,LLM,LLD
      DIMENSION CINPT(750),CINPTN(750),NODE(25),FLUXB(750),TO(25)
      DIMENSION FLUXO(25),FLUXN(25),THETAO(25),THETAN(25),V(25),SO(25),
     #DO(25),DN(25),DNO(25),DNN(25),DNO(25),AVED(25),SORCC(25),SORCN
     #(25),ZD(25),R1(25),R2(25),R3(25),R4(25),R5(25),R6(25),
     #SUB(25),DIAG(25),W(25),SN(25)
      DIMENSION CO(25),CN(25),CNO(25),CNN(25),CEXCHO(25),CEXCHN(25),
     #ORGNO(25),ORGNN(25),BD(25),GNO(25),GNN(25)
      INTEGER DAY, HR
      CHARACTER*6 A6
      CHARACTER*14 A14
      CHARACTER*18 A18
      CHARACTER*28 A28
C=====================================================================
      OPEN (UNIT=1 , FILE='THETA.DAT',STATUS='OLD')
      OPEN (UNIT=2 , FILE='FLUX.DAT' ,STATUS='OLD')
      OPEN (UNIT=3 , FILE='SINK.DAT' ,STATUS='OLD')
      OPEN (UNIT=5 , FILE='NG.DAT'   ,STATUS='OLD')
      OPEN (UNIT=6 , FILE='SOL.DAT'  ,STATUS='NEW')
      OPEN (UNIT=7 , FILE='NCNTNT.DAT',STATUS='NEW')
      OPEN (UNIT=8 , FILE='NBALAN.DAT',STATUS='NEW')
C---------------------------------------------------------
      READ(5,115)NPT,NT,NPTPRT,NPTI,LPRINT,INTPOL, IGRFS
      READ(5,116)DELT,DELZ,TOTZ,DZ
      READ(5,116)LAMDA, LAMDAN
      READ(5,117) MFLUXB
      READ(5,118) CZERO,CNZERO
      READ(5,119)DM,DMN,ALFA,ALFAN,EN,ENN
      READ(5,120) (NODE(I),I=1,NPTPRT)
C---------------------------------------------------------
C   INITIAL CONDITIONS FOR DIFFERENT NITROGEN SPECIES
C---------------------------------------------------------
      READ(5,121) (CO(IZ),IZ=2,NPT)
      READ(5,121) (CNO(IZ),IZ=2,NPT)
      READ(5,121) (CEXCHO(IZ),IZ=2,NPT)
      READ(5,121) (ORGNO(IZ),IZ=2,NPT)
      READ(5,121) (GNO(IZ),IZ=2,NPT)
      READ(5,121) (BD(IZ),IZ=2,NPT)
      READ(5,122) ULM,LLM,LLD,ULD
      READ(5,123) APLDAM
      write(6,*)'This run has fertilizer ',APLDAM ,' Kg '
C--------------------------------------
C   INPUT FROM WATER FLOW PROGRAM
C--------------------------------------
      COEFV=0.25
      LL=0
      DO 1 J=1,3
 1    READ(J,630) A28 ,(ZD(IZ),IZ=1,NPT)
      READ(1,646) A6, DAY,HR,A14, (THETAO(IZ),IZ=1,NPT)
      READ(2,661) A6, DAY,HR,FLUXO(1),A6, (FLUXO(IZ),IZ=2,NPT-1)
      READ(3,701) A18, DAY,HR,A6, (SO(IZ),IZ=1,NPT)
C = = = = = = = = = = = = = = = = = = = = = = = = = = =
      DO 55 IZ=1,NPT
      THETAN(IZ)=THETAO(IZ)
      FLUXN(IZ)=FLUXO(IZ)
      SN(IZ)=SO(IZ)
      SO(IZ)=SO(IZ)/DELT
 55   FLUXO(IZ)=FLUXO(IZ)/DELT
      FLUXB(1)=FLUXO(1)
      NPT1=NPT-1
```

NITROGEN SOLUTE TRANSPORT

```
            TOTZ=DELZ * NPT1
            DO 61 IZ=1,NPT
            TO(IZ)=THETAO(IZ)
    61    CONTINUE
C---------------------------------------
C     RATE CONSTANTS FOR TRANSFORMATIONS
C---------------------------------------
            DO 11 IZ=1,NPT
            KSE(IZ)=1.0
            KES(IZ)=1.0
            K1(IZ)=0.4
            K2(IZ)=LLD+(ZD(IZ)*(ULD-LLD))/TOTZ
cc          k2(iz) = 0.005                              !for test runs
            KK4(IZ)=ULM-(ZD(IZ)*(ULM-LLM))/TOTZ
            K3(IZ)=0.0
            K4(IZ)=0.0
    11    CONTINUE
            DO 58 IZ=2,NPT
            CI=CI+CO(IZ)*THETAO(IZ)*DELZ*0.1
            CNI=CNI+CNO(IZ)*THETAO(IZ)*DELZ*0.1
            ORGNI=ORGNI+ORGNO(IZ)
    58    CONTINUE
            DO 400 IZ=1,3
   400    K2(IZ)=0.08
            DO 401 IZ=1,NPT
   401    KK4(IZ)=0.0
            KK4(2)=0.0003
            SUMI=CI+CNI+ORGNI+APLDAM
            CO(2)=CO(2)+(APLDAM*10.0)/(THETAO(2)*DELZ)
            DO 56 IZ=2,NPT
            ORGNO(IZ) = (ORGNO(IZ)*10.0)/(THETAO(IZ)*DELZ)
            CEXCHO(IZ)=(140.0*CEXCHO(IZ)*BD(IZ))/THETAO(IZ)
    56    CONTINUE
C---------------------------
C     UPPER BOUNDARY CONDITION
C---------------------------
            IF(MFLUXB.EQ.1) GO TO 8
            CO(1)=CZERO
            CNO(1)=CNZERO
C-------------------------------------------
C     INPUT INORGANIC N TO THE SOIL SURFACE
C-------------------------------------------
C *** READ INPUT CONCENTRATIONS WHEN GIVEN FOR KNOWN TIMES
     8    DO 17 JT=1,NT
            CINPT(JT)=0.0
            CINPTN(JT)=0.0
    17    CONTINUE
            WRITE(6,241)
            WRITE(6,242) (IZ,ZD(IZ),IZ=1,NPT)
            WRITE(6,300)DELZ,TOTZ
            WRITE(6,301)DELT
            WRITE(6,302)LAMDA
            WRITE(6,306)LAMDAN
            WRITE(6,303)
            WRITE(6,304)
            DO 10 I=1,NPTPRT
            IZ=NODE(I)
            WRITE(6,305)ZD(IZ),KSE(IZ),KES(IZ),K1(IZ),K2(IZ),K3(IZ),K4(IZ),
           1KK4(IZ)
    10    CONTINUE
C***
```

```
  801 WRITE(6,239)
      WRITE(6,228) (ZD(NODE(I)),I=1,NPTPRT)
      WRITE(6,240)
C ***
C *** TIME UNIT IS DAYS.
      TIMEO=0.0
      AVED(1)=DM
      AVEDN(1)=DMN
      JT=1
   33 WRITE(6,227) TIMEO, (CO(NODE(I)),I=1,NPTPRT)
      WRITE(6,226) (CNO(NODE(I)),I=1,NPTPRT)
      WRITE(6,225) (CEXCHO(NODE(I)),I=1,NPTPRT)
      WRITE(6,229) (ORGNO(NODE(I)),I=1,NPTPRT)
      WRITE(6,230) (GNO(NODE(I)),I=1,NPTPRT)
      IF(FILES.EQ.'YES')WRITE(7) TIMEO,NPT, (CO(I),CNO(I),CEXCHO(I),
     # ORGNO(I),GNO(I),I=1,NPT)
C-----------------------------------------------------------------
C     HEADING FOR PLOTTING FILES . . . . .
      IF (IGRFS .EQ. 0) GO TO 709
      IF (IGRFS .GE. 1) THEN
      IJK = 11                      ! WRITE FOR [NO3]s and [NH4]s only
      OPEN (UNIT=10,FILE='NO3s.plt' ,STATUS='NEW')
      OPEN (UNIT=11,FILE='NH4s.plt' ,STATUS='NEW')
      END IF
      IF (IGRFS .EQ. 2) THEN
      IJK = 14                      ! write also for [NH4]e, N , ORGN
      OPEN (UNIT=12,FILE='NH4e.plt' ,STATUS='NEW')
      OPEN (UNIT=13, FILE='NTRG.plt' ,STATUS='NEW')
      OPEN (UNIT=14, FILE='ORGN.plt' ,STATUS='NEW')
      END IF
      DO 707  IJ = 10,IJK
  707 WRITE (IJ,222) (ZD(NODE(I)) , I =1,NPTPRT)
C-----------------------------------------------------------------
  709 TMIN=0.0
      TMOUT=0.0
      TMINN=0.0
      TMOUTN=0.0
      TSCP=0.0
      TSCPN=0.0
      TMGN=0.0
      TCNN=0.0
      TORGN=0.0
      GO TO 32
C-----------------------------
C    START OF [NH4]S COMPUTATION
C-----------------------------
   3  READ(1, 646, END=50) A6,DAY,HR,A14, (THETAN(IZ),IZ=1,NPT)
      READ(2, 661) A6,DAY,HR, FLUXn(1),A6, (FLUXN(IZ),IZ=2,NPT-1)
      READ(3, 701) A18,DAY,HR, A6, (SN(IZ),IZ=1,NPT)
C...............................................................
   32 TIME=DAY-1. +HR/24.
      DELT=TIME-TIMEO
      DO 23 IZ=1,NPT
      FLUXN(IZ)=FLUXN(IZ)/DELT
      SN(IZ)=SN(IZ)/DELT
      IF(SN(IZ).EQ.0.0 .AND.HR.EQ.1.0)SN(IZ)=SO(IZ)/24.0
      IF(SN(IZ).EQ.0.0 .AND.HR.GT.1.0)SN(IZ)=SO(IZ)
      R1(IZ)=(2.0-DELT*KES(IZ))/(2.0+DELT*KES(IZ))
      R2(IZ)=(DELT*KSE(IZ))/(2.0+DELT*KES(IZ))
      R3(IZ)=(1.0-0.5*DELT*KK4(IZ))/(1.0+0.5*DELT*KK4(IZ))
```

```
              R4(IZ)=DELT/(2.0+DELT*KK4(IZ))
              R5(IZ)=KSE(IZ)+K1(IZ)+K4(IZ)
              R6(IZ)=K2(IZ)+K3(IZ)
              V(IZ)=0.5*((FLUXN(IZ)/THETAN(IZ))+(FLUXO(IZ)/THETAO(IZ)))
       23     CONTINUE
              MFLUX=1
              R=DELZ*V(1)+AVED(1)
              RN=DELZ*V(1)+AVEDN(1)
              JT=JT+1
              JTT=JT-1
              FLUXB(JT)=FLUXN(1)
              CINPUT=CINPT(JTT)
              IF(V(1).GT.0.)GO TO 22
              CINPUT=0.
              MFLUX=0.
       22  DO 4 IZ=2,NPT1
              IZZ=IZ+1
              IZY=IZ-1
              IF(JT.EQ.2) DO (IZ)=DM +ALFA *((FLUXO(IZ)/THETAO(IZ)) )
              DN (IZ)=DM +ALFA *(V(IZ) )
              AVED(IZ)= 0.5*(DN(IZ)+DO(IZ))
              A=(.25*DELT*AVED(IZ))/DELZ**2
              B=(0.25 *DELT*V(IZ))/DELZ
              SINKO=LAMDA*DELT*SO(IZ)/(2.*THETAO(IZ))
              SINK=LAMDA*DELT*SN(IZ)/(2.*THETAN(IZ))
              SUB(IZ)=-A-B
              DIAG(IZ)=1.+2.*A+B-SINK+0.5*DELT*R5(IZ)
              SUP(IZ)=-A
              W1=(A+B)*CO(IZY)
              W2=(1.-2.*A-B+SINKO)*CO(IZ)-0.5*DELT*R5(IZ)*CO(IZ)
              W3=A*CO(IZZ)
              W4=(KES(IZ)*CEXCHO(IZ)    +KK4(IZ)*ORGNO(IZ))*DELT
              IF(IZ.NE.2) GO TO 4
              IF(MFLUX.EQ.1) GO TO 99
              SUB(2)=0.
C ***   CONSTANT INPUT CONCENTRATION CONDITION.
              CN(1)=CO(1)*THETAO(1)/THETAN(1)
              W1=W1-SUB(2)*CN(1)
              GO TO 4
       99     SUB(2)=0.
              DIAG(2)=DIAG(2)-A
              W1=W1+B*CINPUT
       4   W(IZ)=W1+W2+W3+W4+SORCC(IZ)
C------------------------------------
C     LOWER BOUNDARY CONDITION FOR [NH4]S
C------------------------------------
              DIAG(NPT1)=DIAG(NPT1)+SUP(NPT1)
C------------------------------------
C     SOLUTION OF TRIDIAGONAL SYSTEM
C------------------------------------
              CALL TRIDAG(2,NPT1,SUB,DIAG,SUP,W,CN)
              IF(MFLUX.EQ.0) GO TO 16
              CN(1)=(AVED(1)*CN(2)+DELZ*V(1)*CINPUT)/R
              IF(CN(1).GE.0.) GO TO 16
              MFLUX=0
              GO TO 22
       16     CN(NPT)=CN(NPT1)
C=============================
C     START OF [NO3]S COMPUTATION
C=============================
```

```
          MFLUX=1
          CINPUT=CINPTN(JTT)
          IF(V(1).GT.0.)GO TO 44
          CINPUT=0.0
          MFLUX=0
       44 DO 6 IZ=2,NPT1
          IZZ=IZ+1
          IZY=IZ-1
          IF(JT.EQ.2) DNO(IZ)=DMN+ALFAN*((FLUXO(IZ)/THETAO(IZ)) )
          DNN(IZ)=DMN+ALFAN*(V(IZ) )
          AVED(IZ)= 0.5*(DNN(IZ)+DNO(IZ))
          A=(.25*DELT*AVED(IZ))/DELZ**2.0
          B=(0.25 *DELT*V(IZ))/DELZ
          SINKO=LAMDAN*DELT*SO(IZ)/(2.*THETAO(IZ))
          SINK=LAMDAN*DELT*SN(IZ)/(2.*THETAN(IZ))
          SUB(IZ)=-A-B
          DIAG(IZ)=1.+2.*A+B-SINK+0.5*DELT*R6(IZ)
          SUP(IZ)=-A
          W1=(A+B)*CNO(IZY)
          W2=(1.-2.*A-B+SINKO)*CNO(IZ)-0.5*DELT*R6(IZ)*CNO(IZ)
          W3=A*CNO(IZZ)
          W4=K1(IZ)*CO(IZ)*DELT
          IF(IZ.NE.2) GO TO 6
          IF(MFLUX.EQ.1) GO TO 98
          SUB(2)=0.
C *** CONSTANT INPUT CONCENTRATION CONDITION.
          CNN(1)=CNO(1)*THETAO(1)/THETAN(1)
          W1=W1-SUB(2)*CNN(1)
          GO TO   6
       98 SUB(2)=0.
          DIAG(2)=DIAG(2)-A
          W1=W1+B*CINPUT
        6 W(IZ)=W1+W2+W3+W4+SORCN(IZ)
C------------------------------------
C     LOWER BOUNDARY CONDITION FOR [NO3]S
C------------------------------------
          DIAG(NPT1)=DIAG(NPT1)+SUP(NPT1)
C------------------------------------
          CALL TRIDAG(2,NPT1,SUB,DIAG,SUP,W,CNN)
          IF(MFLUX.EQ.0) GO TO 21
          CNN(1)=(AVED(1)*CNN(2)+DELZ*V(1)*CINPUT)/RN
          IF(CNN(1).GE.0.) GO TO 21
          MFLUX=0
          GO TO 44
       21 CNN(NPT)=CNN(NPT1)
C------------------------------------
C     START OF [NH4]E, [N2O+N2]G, [ORGN] COMPUTATION
C------------------------------------
          DO 13 IZ=2,NPT
          CEXCHN(IZ) = R1(IZ)*CEXCHO(IZ)+R2(IZ)*(CO(IZ)+CN(IZ))
          ORGNN(IZ)  = R3(IZ)*ORGNO(IZ)+R4(IZ)*(K4(IZ)*(CO(IZ)+CN(IZ))
        # +K3(IZ)*(CNO(IZ)+CNN(IZ)))
          ORGNN(IZ)=ORGNN(IZ)*(THETAO(IZ)/THETAN(IZ))
          GNN(IZ)  = GNO(IZ)+0.5*DELT*K2(IZ)*(CNO(IZ)+CNN(IZ))
          GNN(IZ)=GNN(IZ)*(THETAO(IZ)/THETAN(IZ))
       13 CONTINUE
          MIN=0.0
          MOUT =0.0
          MINN=0.0
          MOUTN=0.0
```

```
      MIN =0.5*DELT*(FLUXB(JT)*CINPT (JT)+FLUXB(JTT)*CINPT (JTT))*0.1
      MINN=0.5*DELT*(FLUXB(JT)*CINPTN(JT)+FLUXB(JTT)*CINPTN(JTT))*0.1
      MOUT =0.5*DELT*(THETAN(NPT)*CN  (NPT)+THETAO(NPT)*CO (NPT))*0.1
     #*V(NPT)
      MOUTN=0.5*DELT*(THETAN(NPT)*CNN(NPT)+THETAO(NPT)*CNO(NPT))*0.1
     #*V(NPT)
      TMIN=TMIN+MIN
      TMOUT=TMOUT+MOUT
      TMINN=TMINN+MINN
      TMOUTN=TMOUTN+MOUTN
      MSTI=0.0
      MST = 0.0
      MSTIN=0.0
      MSTN=0.0
      SIT=0.0
      SOT=0.0
      SITN=0.0
      SOTN=0.0
      SCP = 0.0
      SCPN=0.0
      MGN=0.0
   43 DO 28 IZ=2,NPT
      IZY=IZ-1
      MSTI=MSTI+DELZ*(THETAO(IZ)*CO(IZ))*0.1
      MST=MST+DELZ*(CN(IZ)*THETAN(IZ))*0.1
      MSTIN=MSTIN+DELZ*(THETAO(IZ)*CNO(IZ))*0.1
      MSTN=MSTN+DELZ*(THETAN(IZ)*CNN(IZ))*0.1
      SIT=SIT+.05*R5(IZ)*DELZ*DELT*(CN(IZ)*THETAN(IZ)+CO(IZ)*THETAO(IZ))
      SOT=SOT+.05*DELZ*DELT*(KES(IZ)*(CEXCHN(IZ)*THETAN(IZ)+CEXCHO(IZ)*
     #THETAO(IZ))+KK4(IZ)*(ORGNN(IZ)*THETAN(IZ)+ORGNO(IZ)*THETAO(IZ)))
      SITN=SITN+0.5*R6(IZ)*DELZ*DELT*(CNN(IZ)*THETAN(IZ)+CNO(IZ)
     #*THETAO(IZ))*0.1
      SOTN=SOTN+0.5*DELZ*DELT*K1(IZ)*(CN(IZ)*THETAN(IZ)+CO(IZ)
     #*THETAO(IZ))*0.1
      SCP=SCP+0.5*DELT*DELZ*(CO(IZ)*SO(IZ)+CN(IZ)*SN(IZ))*0.1
      SCPN=SCPN+0.5*DELZ*DELT*(CNO(IZ)*SO(IZ)+CNN(IZ)*SN(IZ))*0.1
      MGN=MGN+0.5*DELT*DELZ*K2(IZ)*(GNN(IZ)*THETAN(IZ)+
     #GNO(IZ)*THETAO(IZ))*0.1
   28 CONTINUE
      MBAL=MSTI-MOUT+MIN-SCP+SOT-SIT
      MBALN=MSTIN-MOUTN+MINN-SCPN+SOTN-SITN
      TSCP=TSCP+SCP
      TSCPN=TSCPN+SCPN
      TMGN=TMGN+MGN
C=================================================================
C     WRITTING FILES FOR PLOTTING CONCENTRATIONS EVERY DAY
      IF (IGRFS .EQ. 0)  GO TO 194
      IF (HR .NE. 24) GO TO 194
      IF (IGRFS .GE. 1)  THEN
      WRITE (10,1227) DAY, (CNN(NODE(I)), I = 1,NPTPRT), MBALN,MSTN   !
[NO3]s
      WRITE (11,1227) DAY, (CN(NODE(I)),  I = 1,NPTPRT), MBAL, MST    !
[NH4]s
      END IF
      IF (IGRFS .EQ. 2)  THEN
      WRITE (12,1227) DAY, (CEXCHN(NODE(I)), I = 1, NPTPRT)    ! [NH4]e
      WRITE (13,1227) DAY, (GNN(NODE(I)),    I = 1, NPTPRT)    ! [N]
      WRITE (14,1227) DAY, (ORGNN(NODE(I)),  I = 1, NPTPRT)    !
[ORGANIC M.]
      END IF
```

```
1227  FORMAT (I5, 9X, 8F7.1, 2(3X,F9.2))
C===============================================================
 194     IF(LPRINT.EQ.0)GO TO 94
         IF(LPRINT.EQ.1)GO TO 96
         IF(LPRINT.EQ.2)GO TO 97
  96     IF(.NOT.(DAY.EQ.13.0 .OR.DAY.EQ.27.0 .OR.DAY.EQ.42.0 .OR.
        #DAY.EQ.56.0 .OR.DAY.EQ.70.0 .OR.DAY.EQ.84.0 .OR.DAY.EQ.98.0
        #.OR.DAY.EQ.112.0))GO  TO 94
  97     IF(HR.NE.24.)GO TO 94
         WRITE(6,228)  (ZD(NODE(I)),I=1,NPTPRT)
         WRITE(6,227)  TIME, (CN(NODE(I)),I=1,NPTPRT)
         WRITE(6,226)  (CNN(NODE(I)),I=1,NPTPRT)
         WRITE(6,225)  (CEXCHN(NODE(I)),I=1,NPTPRT)
         WRITE(6,229)  (ORGNN(NODE(I)),I=1,NPTPRT)
         WRITE(6,230)  (GNN(NODE(I)),I=1,NPTPRT)
         IF(FILES.EQ.'YES')WRITE(7)  TIME,NPT,(CN(I),CNN(I),CEXCHN(I),
        # ORGNN(I),GNN(I),I=1,NPT)
         IF(INTPOL.EQ.1) GO TO 94
         IF(JT.NE.2) GO TO 95
         WRITE(6,236)
         WRITE(6,238)
  95     WRITE(6,233)
         LL=LL+1
         MBALN=MBALN+(LL*5.0)
         MSTN=MSTN+(LL*5.0)
         MSTIN=MSTIN+(LL*5.0)
         WRITE(6,243)MSTI,MOUT,MIN,SCP,SOT,SIT,MBAL,MST
         WRITE(6,244)MSTIN,MOUTN,MINN,SCPN,SOTN,SITN,MBALN,MSTN
         IF(FILES.EQ.'YES')WRITE(8)  TIME,MSTI,MOUT,MIN,SCP,SOT,SIT,MBAL,MST
        # ,MSTIN,MOUTN,MINN,SCPN,SOTN,SITN,MBALN,MSTN
  94     DO 25 IZ=1,NPT
         CO(IZ)=CN(IZ)
         CNO(IZ)=CNN(IZ)
         CEXCHO(IZ)=CEXCHN(IZ)
         ORGNO(IZ)=ORGNN(IZ)
         GNO(IZ)=GNN(IZ)
         DO(IZ)=DN(IZ)
         DNO(IZ)=DNN(IZ)
         FLUXO(IZ)=FLUXN(IZ)
         THETAO(IZ)=THETAN(IZ)
         SO(IZ)=SN(IZ)
  25     CONTINUE
         TIMEO=TIME
         GO TO 3
  50     DO 51 IZ=2,NPT
         TCN=TCN+CN(IZ)*THETAN(IZ)*DELZ*0.1
         TCNN=TCNN+CNN(IZ)*THETAN(IZ)*DELZ*0.1
         TORGN=TORGN+ORGNN(IZ)*THETAN(IZ)*DELZ*0.1
         TEX=TEX+CEXCHN(IZ)*THETAN(IZ)*DELZ*0.1
  51     CONTINUE
         TCNN=TCNN+LL*5.0
         SUMF=TCN+TCNN+TORGN+TSCPN+TSCP+TMGN+TMOUT+TMOUTN+TEX
         WRITE(6,249)TCNN
         WRITE(6,256)TCN,TEX
         WRITE(6,250)TORGN
         WRITE(6,251)TSCPN
         WRITE(6,252)TSCP
         WRITE(6,253)TMGN
         WRITE(6,255)SUMI,SUMF
         WRITE(6,245)
```

NITROGEN SOLUTE TRANSPORT

```
          WRITE(6,246)TMIN,TMOUT
          WRITE(6,247)TMINN,TMOUTN
          IF(FILES.EQ.'YES')WRITE(8,88) TMIN,TMOUT,TMINN,TMOUTN
   88     FORMAT (4F11.4)
          IF(FILES.NE.'YES') GO TO 500
          CLOSE (UNIT =1)
          CLOSE (UNIT =2)
          CLOSE (UNIT =3)
          CLOSE (UNIT =4)
          CLOSE (UNIT =5)
          CLOSE (UNIT =6)
          CLOSE (UNIT =7)
          CLOSE (UNIT =8)
          CLOSE (UNIT =10)
          CLOSE (UNIT =11)
          CLOSE (UNIT =12)
          CLOSE (UNIT =13)
          CLOSE (UNIT =14)
C---------------
C    READ FORMATS
C---------------
  115     FORMAT(7I5)
  116     FORMAT(4F10.3)
  117     FORMAT(I5)
  118     FORMAT(2F10.0)
  119     FORMAT(6F10.3)
  120     FORMAT(16I5)
  121     FORMAT(10F7.0)
  122     FORMAT(4F10.0)
  123     FORMAT(F10.0)
  630     FORMAT (A28,11(X,F8.2))
  646     FORMAT (A6,2I4, A14 ,12(X,F7.3))
  661     FORMAT (A6,2I4, F8.3, A6, 11(X,F8.3))
  701     FORMAT (A18,2I4, A6, 12(X,F7.6))
C---------------
C    WRITE FORMATS
C---------------
  222     FORMAT (1H0,4X,'DEPTH (CM)',8F5.0,'CALCULATED    ESTIMATED')
  225     FORMAT(1H ,12X,'[NH4\U',11F10.3)
  226     FORMAT(1H ,12X,'[NO3]S',11F10.3)
  227     FORMAT(/,F10.3,3X,'[NH4]S',11F10.3)
  228     FORMAT(1H0,9X,'DEPTH(CM)',11F10.1)
  229     FORMAT(1H ,12X,'[ORGN]',11F10.3)
  230     FORMAT(1H ,14X,'[N]G',11F10.3////)
  231     FORMAT(1H0,//,40X,'INITIAL MASS IN THE PROFILE'//)
  232     FORMAT(1H0,25X,5F10.3)
  233     FORMAT(10X,'MASS'/,10X,'BALANCE :')
  234     FORMAT(30X,'[NH4]S',4X,'[NO3]S',4X,'[NH4]E',4X,'[ORGN]',4X,'[N]G'/
         #/)
   236    FORMAT(1H ,23X,80('-')//
  ,23X,'INITIAL',4X,'OUTFLUX',8X,'INFLUX',9
         #X,'PLANT',9X,'TRANSFORMATIONS')
  238     FORMAT(1H ,22X,'STORAGE',35X,'SINK',9X,'SOURCES',4X,'SINKS'//,23X,
         #80('-'))
  239     FORMAT(///,50X,'CONCENTRATION'//)
  240     FORMAT(1H0,'TIME (DAYS)'/)
  241     FORMAT(1H1,45X,'DEPTH DETAILS OF EACH NODE'//,35X,'NODE NUMBER',
         #10X,'DEPTH BELOW SOIL SURFACE'/)
  242     FORMAT(1H ,37X,I5,18X,F5.1)
  243     FORMAT(15X,'[NH4]S=',  F9.2,2X,'-',   F9.2,2X,'+'   ,F9.2,2X,
```

```
        #'-',2X,F9.3,2X,'+',2X,F9.2,2X,'-',F9.2,'=',F9.2,2X,' ESTIMATED='
        #,F9.2)
  244   FORMAT(15X,'[NO3]S=',   F9.2,2X,'-',   F9.2,2X,'+'   ,F9.2,2X,
        #'-',2X,F9.3,2X,'+',2X,F9.2,2X,'-',F9.2,'=',F9.2,2X,' ESTIMATED='
        #,F9.2//)
  245   FORMAT(57X,'CUMULATIVE MASS FLUX'//57X,'IN',14X,'OUT'/)
  246   FORMAT(40X,'[NH4]S',4X,F10.3,8X,F10.3)
  247   FORMAT(40X,'[NO3]S',4X,F10.3,8X,F10.3//)
  249   FORMAT(//,11X,'TOTAL NITRATE IN THE PROFILE(KG N/HA)=',F7.3)
  256   FORMAT(1H0,10X,'TOTAL SOLUTION AMMONIUM IN THE PROFILE(KG N/HA)='
        #,F7.3//,11X,'TOTAL EXCHANGEABLE AMMONIUM (KG N/HA)=',F7.3)
  250   FORMAT(1H0,10X,'TOTAL ORGANIC N IN THE PROFILE (KG N/HA)=',
        #F8.1)
  251   FORMAT(1H0,10X,'TOTAL NITRATE TAKEN UP BY PLANTS (KG N/HA)=',F7.3)
  252   FORMAT(1H0,10X,'TOTAL AMMONIUM TAKEN UP BY PLANTS (KG N/HA)=',F7.3
        #)
  253   FORMAT(1H0,10X,'TOTAL AMOUNT OF GASEOUS N PRODUCED IN THE PROFILE
        #(KG N/HA)=',F9.3//)
  255   FORMAT(1H0,'SUM OF TOTAL N AT TIME ZERO=',F12.3//,1X,'SUM OF TOTAL
        # N AT THE END OF RUN=',F12.3//)
  300   FORMAT(1H0,'SPACE INCREMENT =',F5.2//,1X,'THICKNESS OF THE SOIL LA
        #YER MODELED =',F7.1)
  301   FORMAT(1H0,'INITIAL TIME INCREMENT =',F5.2, ' DAYS')
  302   FORMAT(1H0,'PLANT ABSORPTION COEFFICIENT FOR [NH4]S=',F6.3)
  303   FORMAT(50X,'TRANSFORMATION RATES'//)
  304   FORMAT(1H0,2X,'DEPTH(CM)',8X,'KSE',8X,'KES',8X,'K1',8X,'K2'
        1,8X,'K3',8X,'K4',8X,'KK4'//)
  305   FORMAT(1H ,7F11.4,F12.7)
  306   FORMAT(1H0,'PLANT ABSORPTION COEFFICIENT FOR [NO3]S=',F6.3//)
  500   STOP
        END
        SUBROUTINE TRIDAG(IF,L,A,B,C,D,U)
        DOUBLE PRECISION BETA(50),C(1)
        DIMENSION A(1),B(1),D(1),U(1),GAMMA(50)
        BETA(IF)=B(IF)
        GAMMA(IF)=D(IF)/BETA(IF)
        IFP1=IF+1
        DO 1 I=IFP1,L
        II=I-1
        BETA(I)=B(I)-A(I)*C(II)/BETA(II)
  1     GAMMA(I)=(D(I)-A(I)*GAMMA(II))/BETA(I)
        U(L)=GAMMA(L)
        LAST=L-IF
        DO 2 K=1,LAST
        I=L-K
  2     U(I)=GAMMA(I)-C(I)*U(I+1)/BETA(I)
        RETURN
        END
```

V. APPENDIX 2

DEPTH DETAILS OF EACH NODE

NODE NUMBER	DEPTH BELOW SOIL SURFACE
1	0.0
2	30.0
3	60.0
4	90.0
5	120.0
6	150.0
7	180.0
8	210.0
9	240.0
10	270.0
11	300.0

SPACE INCREMENT = 30.00
THICKNESS OF THE SOIL LAYER MODELED = 300.0
INITIAL TIME INCREMENT = 1.00 DAYS
PLANT ABSORPTION COEFFICIENT FOR [NH4]S=-1.250
PLANT ABSORPTION COEFFICIENT FOR [NO3]S=-1.250

TRANSFORMATION RATES

DEPTH (CM)	KSE	KES	K1	K2	K3	K4	KK4
0.0000	1.0000	1.0000	0.4000	0.0800	0.0000	0.0000	0.0000000
30.0000	1.0000	1.0000	0.4000	0.0800	0.0000	0.0000	0.0003000
60.0000	1.0000	1.0000	0.4000	0.0800	0.0000	0.0000	0.0000000
90.0000	1.0000	1.0000	0.4000	0.0050	0.0000	0.0000	0.0000000
120.0000	1.0000	1.0000	0.4000	0.0050	0.0000	0.0000	0.0000000
180.0000	1.0000	1.0000	0.4000	0.0050	0.0000	0.0000	0.0000000
240.0000	1.0000	1.0000	0.4000	0.0050	0.0000	0.0000	0.0000000
300.0000	1.0000	1.0000	0.4000	0.0050	0.0000	0.0000	0.0000000

CONCENTRATION

TIME (DAYS)		DEPTH(CM)	0.0	30.0	60.0	90.0	120.0	180.0	240.0	300.0
0.000	[NH4]S		0.0	193.6	0.0	0.0	0.0	0.0	0.0	0.0
	[NO3]S		0.0	29.6	16.5	19.4	22.3	11.7	10.9	23.4
	[NH4]E		0.0	0.0	0.0	0.0	0.0	0.0	0.0	0.0
	[ORGN]		0.0	4633.3	2880.0	2417.7	2189.1	1953.2	1073.4	2205.8
	[N]G		0.0	0.0	0.0	0.0	0.0	0.0	0.0	0.0
		DEPTH(CM)	0.0	30.0	60.0	90.0	120.0	180.0	240.0	300.0
13.0	[NH4]S		0.0	11.6	0.7	0.0	0.0	0.0	0.0	0.0
	[NO3]S		0.2	108.9	14.8	18.0	20.3	12.5	10.4	13.4
	[NH4]E		0.0	13.7	0.4	0.0	0.0	0.0	0.0	0.0
	[ORGN]		0.0	4600.5	2880.0	2425.4	2195.4	1959.0	1073.4	2205.8
	[N]G		0.0	122.0	12.3	1.1	1.3	0.8	0.696	0.9

	DEPTH(CM)	0.0	30.0	60.0	90.0	120.0	180.0	240.0	300.0
27.0	[NH4]S	0.1	3.6	0.5	0.1	0.0	0.0	0.0	0.0
	[NO3]S	1.1	39.8	10.4	16.3	18.7	11.9	9.8	12.3
	[NH4]E	0.0	4.4	0.2	0.0	0.0	0.0	0.0	0.0
	[ORGN]	0.0	4117.8	2598.7	2260.4	2110.5	1935.9	1073.4	2205.8
	[N]G	0.0	189.4	19.3	2.2	2.6	1.7	1.4	1.8
	DEPTH(CM)	0.0	30.0	60.0	90.0	120.0	180.0	240.0	300.0
42.0	[NH4]S	0.2	2.7	0.5	0.1	0.0	0.0	0.0	0.0
	[NO3]S	0.9	12.3	6.3	14.1	16.7	12.1	9.2	10.7
	[NH4]E	0.0	2.9	0.4	0.1	0.0	0.0	0.0	0.0
	[ORGN]	0.0	4391.2	2748.7	2322.2	2128.1	1930.2	1073.4	2205.8
	[N]G	0.0	231.9	27.1	3.4	3.9	2.5	2.1	2.7
	DEPTH(CM)	0.0	30.0	60.0	90.0	120.0	180.0	240.0	300.0
56.0	[NH4]S	0.1	2.6	0.5	0.1	0.0	0.0	0.0	0.0
	[NO3]S	0.2	7.7	4.5	6.4	14.1	11.7	8.8	9.7
	[NH4]E	0.0	2.8	0.4	0.1	0.0	0.0	0.0	0.0
	[ORGN]	0.0	4293.1	2691.5	2287.4	2104.7	1918.9	1073.4	2205.8
	[N]G	0.0	239.2	30.1	4.0	4.9	3.3	2.7	3.4
	DEPTH(CM)	0.0	30.0	60.0	90.0	120.0	180.0	240.0	300.0
70.0	[NH4]S	0.1	2.6	0.5	0.1	0.0	0.0	0.0	0.0
	[NO3]S	0.2	7.0	4.0	4.8	8.3	10.9	8.4	8.8
	[NH4]E	0.0	2.8	0.4	0.1	0.0	0.0	0.0	0.0
	[ORGN]	0.0	4341.1	2748.7	2322.2	2122.2	1930.2	1073.4	2205.8
	[N]G	0.0	252.2	33.7	4.5	5.7	4.1	3.3	4.0
	DEPTH(CM)	0.0	30.0	60.0	90.0	120.0	180.0	240.0	300.0
84.0	[NH4]S	0.1	2.6	0.5	0.1	0.0	0.0	0.0	0.0
	[NO3]S	0.2	6.9	3.9	3.9	6.1	8.8	8.0	8.2
	[NH4]E	0.0	2.7	0.4	0.1	0.0	0.0	0.0	0.0
	[ORGN]	0.0	4322.9	2740.4	2322.2	2122.2	1930.2	1073.4	2205.8
	[N]G	0.0	261.1	36.3	4.8	6.2	4.8	3.9	4.6
	DEPTH(CM)	0.0	30.0	60.0	90.0	120.0	180.0	240.0	300.0
98.0	[NH4]S	0.1	2.7	0.4	0.1	0.0	0.0	0.0	0.0
	[NO3]S	0.2	7.4	3.7	3.5	4.7	7.4	7.6	7.6
	[NH4]E	0.0	2.8	0.4	0.1	0.0	0.0	0.0	0.0
	[ORGN]	0.0	4386.0	2808.5	2365.3	2152.1	1941.6	1073.4	2205.8
	[N]G	0.0	275.3	40.0	5.1	6.6	5.4	4.5	5.2
	DEPTH(CM)	0.0	30.0	60.0	90.0	120.0	180.0	240.0	300.0
112.0	[NH4]S	0.1	3.4	0.0	0.0	0.0	0.0	0.0	0.0
	[NO3]S	0.2	12.0	1.3	3.2	4.0	7.1	7.1	7.0
	[NH4]E	0.0	3.3	0.0	0.0	0.0	0.0	0.0	0.0
	[ORGN]	0.0	4724.1	3055.2	2538.2	2260.4	1982.8	1076.1	2205.8
	[N]G	0.0	308.6	46.2	5.7	7.3	6.0	5.0	5.7

```
TOTAL NITRATE IN THE PROFILE (KG N/HA) = 52.544
TOTAL SOLUTION AMMONIUM IN THE PROFILE (KG N/HA) = 3.011
TOTAL EXCHANGEABLE AMMONIUM (KG N/HA) = 3.012
TOTAL ORGANIC N IN THE PROFILE (KG N/HA) = 21655.8
TOTAL NITRATE TAKEN UP BY PLANTS (KG N/HA) = 46.359
TOTAL AMMONIUM TAKEN UP BY PLANTS (KG N/HA) = 4.929
TOTAL AMOUNT OF GASEOUS N PRODUCED IN THE PROFILE (KG N/HA) = 2669.938

SUM OF TOTAL N AT TIME ZERO= 22172.793
SUM OF TOTAL N AT THE END OF RUN= 24435.586
```

```
             CUMULATIVE MASS FLUX

                    IN       OUT
         [NH4]S    0.000    0.000
         [NO3]S    0.000    0.000
```

REFERENCES

Beek, J., and M.J. Frissel. 1973. Simulation of nitrogen behavior in soils. PUDOC, Wageningen, The Netherlands.

Biggar, J.W., J. MacIntyre, K.K. Tanji, and D.R. Nielsen. 1979. Field investigation of water and nitrate movement in Yolo soil. p. 467–485. In P.F. Pratt (ed.) Nitrate in effluents from irrigated agriculture. Univ. of California, Riverside, CA.

Biggar, J.W., K.K. Tanji, C.S. Simmons, S.K. Gupta, J.L. MacIntyre, and D.R. Nielsen. 1977. Theoretical and experimental observation on water and nitrate movement below a crop root zone. p. 71–77. In J.P. Law and G.V. Skogerboe (ed.) Proc. Natl. Conf. Irrigation Return Flow Quality Management. Fort Collins, CO. 16–19 May 1977. Colorado State Univ., Fort Collins, CO.

Broadbent, F.E., and A.B. Carlton. 1979. Field trials with isotopes—plant and soil data for Davis and Kearney sites. p. 433–465. In P.F. Pratt (ed.) Nitrate in effluents from irrigated agriculture. Univ. of California, Riverside, CA.

Davidson, J.B., D.A. Graetz, P.S.C. Rao, and H.M. Selim. 1978. Simulation of nitrogen movement, transformation, and uptake in plant root zone. EPA-600/3-78-029, U.S. Gov. Print. Office, Washington, DC.

Dutt, G.R., M.J. Shaffer, and W.J. Moore. 1972. Computer simulation model of dynamic biophysicochemical processes in soils. Univ. of Arizona Agric. Exp. Stn. Tech. Bull. 196.

Frissel, M.J., and J.A. van Veen. 1978. A critique of "computer simulation modeling for nitrogen in irrigated croplands." p. 145–162. In D.R. Nielsen and J.G. MacDonald (ed.) Nitrogen in the environment. Vol. I. Academic Press, Inc., New York.

Frissel, M.J., and J.A. van Veen. 1981. Simulation model for nitrogen immobilization and mineralization. p. 359 381. In I.K. Iskandar (ed.) Modeling wastewater renovation by land disposal. John Wiley & Sons, Inc., New York.

Gupta, S.K., K.K. Tanji, D.R. Nielsen, J.W. Biggar, C.S. Simmons, and J.L. MacIntyre. 1978. Field simulation of soil-water movement with crop extraction. Univ. of California-Davis, Water Sci. and Eng. Paper 4013.

Iskandar, I.K. (ed.). 1981. Modeling wastewater renovation. Land treatment. A. Wiley-Interscience Publ., New York.

Juma, N.G., and E.A. Paul. 1981. Use of tracers and computer simulation techniques to assess mineralization and immobilization of soil nitrogen. p. 145–154. In M.J. Frissel and J.A. van Veen (ed.) Simulation of nitrogen behavior of soil-plant systems. PUDOC, Wageningen, The Netherlands.

Mehran, M., and K.K. Tanji. 1974. Computer modeling of nitrogen transformations in soils. J. Environ. Qual. 3:391–396.

Nielsen, D.R., J.W. Biggar, J. MacIntyre, and K.K. Tanji. 1980. Field investigation of water and nitrate-nitrogen movement in Yolo soil. p. 145–168. In Soil nitrogen as fertilization or pollutant. Int. Atomic Energy Agency, Vienna, Austria.

Nimah, M.N., and R.J. Hanks. 1973. Model for estimating soil water, plant and atmospheric interrelations: I. Description and sensitivity. Soil Sci. Soc. Am. Proc. 37:522–527.

Shaffer, M.J., R.W. Ribbens, and C.W. Huntly. 1977. Prediction of mineral quality of irrigation return flow. V. Detailed return flow salinity and nutrient simulation model. EPA-600/2-77-179E, EPA, Ada, OK.

Stanford, G., and S.J. Smith. 1972. Nitrogen mineralization potential of soils. Soil Sci. Soc. Am. proc. 36:465–472.

Tanji, K.K. 1982. Modeling of the soil nitrogen cycle. In Nitrogen in agricultural soils. Agronomy 22:721–771.

Tanji, K.K., and M. Mehran. 1979. Conceptual and dynamic models for nitrogen in irrigated croplands. p. 555–646. In P.F. Pratt (ed.) Nitrate in effluents from irrigated agriculture. Univ. of California, Riverside, CA.

Tanji, K.K., M. Mehran, and S.K. Gupta. 1981. Water and nitrogen fluxes in the root zone of irrigated maize. p. 51–67. *In* M.J. Frissel and J.A. van Veen (ed.) Simulation of nitrogen behavior of soil-plant systems. PUDOC, Wageningen, The Netherlands.

Tillotson, W.R., C.W. Robbins, R.J. Wagenet, and R.J. Hanks. 1980. Soil water, solute and plant growth simulation. Utah State Univ. Agric. Exp. Stn. Bull. 502.

Watts, D.G., and R.J. Hanks. 1978. A soil-water-nitrogen model for irrigated corn on sandy soils. Soil Sci. Soc. Am. Proc. 42:492–499.

16 Solute Transport and Reactions in Salt-Affected Soils

CHARLES W. ROBBINS

Soil and Water Management Research Unit
Kimberly, Idaho

Modeling solute transport and reactions in salt- and sodium-affected soils can be considered as three simultaneous processes: (i) solute transport; (ii) precipitation-dissolution reactions; and (iii) cation exchange. Solute transport is the physical movement of ions by convective transport (water transport) and ion dispersion within the solvent system (due to concentration gradients). Precipitation-dissolution reactions are dominated by carbonate or lime and gypsum reactions. Mineral weathering reactions are important in special cases, but are not considered here. Cation exchange models usually consider only calcium (Ca), magnesium (Mg), and sodium (Na) exchange on the negatively charged soil surfaces. However, in some cases it may be necessary to consider potassium (K) exchange if K constitutes a substantial portion of the solute or exchangeable ions. These three processes will be discussed separately and will be presented as separate subroutines that can be called by water flow and plant growth models similar to that described in Ch. 11.

A short program that calculates the cation-exchange selectivity coefficients needed by the cation exchange subroutine is also explained and listed. Other reactions and interactions of importance to salt-affected soil management, which may need to be considered in special cases, are also discussed.

I. MODELING SOLUTE TRANSPORT

The solute transport subroutine was an expansion of Childs and Hanks' (1975) solute model. Two program listings for solute transport, one in FORTRAN and one in BASIC, are shown in Appendices 1 and 3. Soil water flow and root extraction is discussed in Ch. 11. To include solute flow, an additional equation has to be solved after water flow and root extraction is computed. Root extraction rates are dependent on the combined effects of matrix and osmotic potential. This is done by modifying Eq. [4] of Ch. 11 to include osmotic effects as:

$$A(z,t) = \frac{[H_{root} + R_z - h_{(z,t)} - s_{(z,t)}] \, RDF_{(z,t)} \, K}{\Delta z \, \Delta x} \quad [1]$$

Copyright © 1991 ASA-CSSA-SSSA, 677 S. Segoe Rd., Madison, WI 53711, USA. *Modeling Plant and Soil Systems*—Agronomy Monograph no. 31.

where $A(z,t)$ is the root extraction rate, H_{root} is the water potential in the root at the soil surface, R_z is the root resistance term, $h_{(z,t)}$ is the soil and water matric potential, $s_{(z,t)}$ is the osmotic potential of the soil solution, RDF is the fraction of roots in the z depth increment, and K is the hydraulic conductivity of the depth increment. This equation includes only osmotic effects. Specific ion effects are not considered. The osmotic potential is assumed to be related to the soil solution concentration by a constant conversion factor. If the soil solution concentration is measured in moles of charge per cubic meter, the conversion factor is 0.36 to convert to osmotic potential expressed as meters of water.

Another equation needed to solve solute flow is:

$$\delta\,(\theta C)/t = \frac{\delta}{\delta z}\,[D(\theta,q)\,C/z + qC] \qquad [2]$$

where θ is soil water content, C is solute concentration, t is time, z is depth, $D(\theta,q)$ is a combined diffusion and dispersion coefficient, and q is volumetric water flux computed from the solution of the water flow equation. Note that qC is the solute mass flow term and θC is the total salt content. The solution of the above equation is interrelated with water flow because θ and q are dependent on water content. The solution of Eq. [2] assumes that no salt is removed from the soil by plants. The initial solute concentration must be known, as well as solute characteristics at the upper and lower boundaries. These boundary conditions must be consistent with the water flow boundary conditions as described in the water flow subroutine. Equation [2] is solved by numerical approximation using a tri-diagonal matrix solution in the same manner as the water flow equation described in Ch. 11. The solution of Eq. [2] has a problem with numerical dispersion, which causes the solution to be dependent on the size of the depth and time steps. To minimize this problem, a more complete numerical approximation is used here than is used for water flow in Ch. 11 (Bresler, 1973).

Individual ion transport requires an array to include complete, solution ion concentrations at the soil surface (SF1–SF8 arrays) corresponding to the water flux information.

Childs and Hanks' (1975) water flow-salt transport method was expanded from moving bulk dissolved salt to independently moving Ca, Mg, Na, K, Cl, and SO_4 as nonreactive ions, and to calculating HCO_3 and CO_3. During water application to the soil surface, these ions are contained in the SF1 to SF8 arrays (SF1 = Ca, SF2 = Mg, etc.) and represent the soil surface flux of ions in the irrigation or rain water. There are twice as many elements in each SF1 array as in the V or surface water flux array. The elements of the V array are in pairs; the first element is the water flux direction and the second element is the flux duration given in hours. If, for example, the V array contained 1.0, 10.0, -0.04, and 240, the SF1 array would include two elements, such as 20.0 and 0.0. When the water flux reached 1.0 cm/h for 10 h, the Ca concentration in the water surface element would be 20 mmol/L.

SOLUTE TRANSPORT AND REACTIONS IN SOILS

This would be followed by 240 h of evapotranspiration at a rate of -0.04 cm/h and, since Ca will not evaporate from the soil surface, the Ca flux would be zero.

Arrays SS1 to SS8 were added to contain the initial solute ion concentration, at the beginning of each time step, as a function of depth. The SE1 to SE8 arrays were added, and contain the final solute concentration for the end of each time step, as a function of depth. The surface boundary conditions are determined by the water flux boundary conditions. It is assumed that there is no diffusion or salt flow across the soil surface boundary when water is evaporating. Thus, salt can accumulate during evaporation in the depth increment nearest the top, but not at the top since that is the boundary.

Solute flux for the bottom boundary conditions is also necessary and three conditions are provided for. The first is a constant water content (and matrix potential) for the bottom boundary, such as a water table. Solutes flow up or down depending on soil water flux. The net solute flow would depend on the water flow direction and solute concentration at the lower boundary. The second condition provided for in this model, but not in the water flow model of Ch. 11, is a unit hydraulic gradient. This would occur if the soil is quite wet and if steady downward water flow was established. Solute would then flow downward only, and the amount of solute flow would be governed by conditions above the bottom layer and not by the bottom. It would also not matter what concentration was assumed for the lower layer. The third condition is that of zero water flux. This situation would occur for dry subsoils when there was no leaching. All solutes would, thus, be contained in the layers above the bottom boundary.

II. MODELING LIME AND GYPSUM SOLUBILITY REACTIONS

Calcium carbonate (lime) and gypsum precipitation and dissolution reaction models for describing salt- and sodium-affected soil reactions have received considerable attention. Interaction between CO_2, CO_3, HCO_3, Ca, and pH are a major component of these reactions and are reviewed elsewhere (Robbins, 1985). Sulfate and gypsum reactions are also important in many salt-affected soils (Dutt et al., 1972; Tanji, 1969). The importance of these two kinds of reactions often requires simultaneous modeling of lime and gypsum in the same system (Nakayama, 1969; Robbins et al., 1980a). The chemical precipitation-dissolution model described here considers both lime and gypsum reactions.

The CHEM subroutine calls several functions and subroutines during its execution. Soil solution electrical conductivity (EC) is calculated from individual ion concentrations by the ECM3 subroutine (McNeal et al., 1970). A function called ACT calculates monovalent and divalent, mean ion activity coefficients by first calculating ionic strength as 0.0127 multiplied by the EC (Griffin & Jurinak, 1973) and then uses the Davies' equation to calculate the activity coefficients from the ionic strengths (Stumm & Morgan, 1970). The PRECIP subroutine determines the equilibrium status between lime, gypsum, and the soil solution. It then calls the SINK subroutine to determine the amount of lime and/or gypsum that must be precipitated or dissolved to bring the solution phase into equilibrium with the solid phase. These are all short subroutines and are explained by comments at the beginning and throughout the computer listings. The cation exchange subroutine, XCHANG, is also called by the CHEM subroutine under certain conditions and is discussed in the next section.

The CHEM subroutine starts by converting the input ion concentration from millimoles/L to moles/L and estimates a value for HCO_3 ion concentration. Carbon dioxide partial pressure (PCO_2) is converted from percent CO_2 or kilopascals to atmospheres. Lime and gypsum are converted from a weight basis to a solution-concentration basis for ease of mass balance calculation. The ECM3 subroutine and the ACT function are then called to calculate activity coefficients for estimating ion activities. First approximations of ion activities, including H, are made prior to entering the chemical equilibrium loop. Within each loop cycle, new activity coefficients and new ion activities are calculated for Ca, Mg, Na, SO_4, H, HCO_3, and CO_3. The calculated activity values include activity-coefficient and ion-pairing corrections for soils above a pH of 6.5 that contain lime and possibly gypsum (Robbins et al., 1980a). At this point, the PRECIP subroutine is called and the solution phase is equilibrated with solid phase lime and, if present, gypsum. The PRECIP subroutine in turn calls SINK to complete this calculation. New Ca activities, and HCO_3 and CO_3 concentrations are calculated next. Then, a new EC value is calculated and compared with the previous EC value. If the EC has changed $< 1.0\%$ then the program proceeds. Otherwise the equilibrium loop is run again to fine-tune the calculation. On leaving the chemistry equilibrium loop, pH is calculated and the option to call the XCHANG subroutine is exercised. If called, new values are returned for solution and exchangeable cation values. All ions, CO_2, and lime and gypsum values are then converted to their original units, and sodium adsorption ratio (SAR) and exchangeable sodium percentage (ESP) values are calculated. The computer then returns to the main program that called CHEM.

III. MODELING CATION EXCHANGE

The primary reason for modeling cation exchange processes in salt-affected soils concerns the necessity to predict changes in the ESP. The exchangeable cations in a given volume of medium- to fine-textured soils are usually about two orders of magnitude greater than in solution for sodic and slightly to moderately salty soils. This gives the exchangeable ions a tremendous buffering effect on the ion composition in salt-affected soils. The tendency for high exchangeable-sodium concentrations to induce poor physical conditions in soils is a function of ESP and EC (see Bresler et al., 1982 for a review). Models predicting exchangeable cations and ESP in salt-affected soils range in complexity from ESP equivalent to SAR (Jury et al., 1979) through a series of expressions discussed in detail by Oster and Sposito (1980).

An exact relationship between solution Na and other cations, and relationships between SAR and ESP, does not exist for all soils or for different solution compositions in equilibrium with a particular soil (Babcock & Schulz, 1963; Sposito & Mattigod, 1977; Robbins & Carter, 1983).

A three cation exchange model for Ca-Mg-Na exchange in salt-affected soils was used by Dutt et al. (1972). The model described here was expanded to predict Ca-Mg-Na-K exchange in order to include soils high in soluble and exchangeable K (Robbins et al., 1980a). When the irrigation water or the soil solution contains less than four times as much Na as K, on a molar basis, the four-cation-exchange calculation method should be used (Robbins, 1984). The XCHANG subroutine uses the following four equations as the basis of the model:

$$X_{Ca} = CEC\left[\frac{(Mg)^{1/2}}{(Ca)^{1/2}K1} + \frac{(Na)}{(Ca)^{1/2}K2} + \frac{(K)K3}{(Ca)^{1/2}} + 1\right]^{-1} \quad [3]$$

$$X_{Mg} = CEC\left[\frac{(Ca)^{1/2}K1}{(Mg)^{1/2}} + \frac{(Na)}{(Mg)^{1/2}K5} + \frac{(K)K4}{(Mg)^{1/2}} + 1\right]^{-1} \quad [4]$$

$$X_{Na} = CEC\left[\frac{(Ca)^{1/2}K2}{(Na)} + \frac{(Mg)^{1/2}K5}{(Na)} + \frac{(K)K6}{(Na)} + 1\right]^{-1} \quad [5]$$

$$X_K = CEC\left[\frac{(Ca)^{1/2}}{(K)K3} + \frac{(Mg)^{1/2}}{(K)K4} + \frac{(Na)}{(K)K6} + 1\right]^{-1} \quad [6]$$

where X_{Ca}, X_{Mg}, X_{Na}, and X_K are exchangeable cations (meq/100 g or mol_c/Kg), CEC is the cation exchange capacity, (Na), (Ca), (Mg), and (K) molar activities in solution, and the Ki terms are the selectivity coefficient. The Vanselow convention for cation exchange is used here (Robbins et al., 1980a; Sposito, 1977). This model assumes that the exchangeable cations' sum is equal to the CEC. Exchange reactions are assumed to be sufficiently rapid that reaction rates are ignored. This is probably satisfactory, since the soil is continually experiencing wetting and drying cycles. Cation molar activities are needed as input and are calculated by the CHEM subroutine. Exchange selectivity coefficients $K1, K2 \ldots K6$ are calculated by the EXCOEF model described later. These coefficients vary from one soil to another, and are due to differences in clay mineralogy and possibly other factors.

The XCOEF program is listed with the subroutines. It is used to calculate cation-exchange selectivity coefficients needed by the XCHANG subroutine for those soils in which the coefficients are not available. Necessary input data are Ca, Mg, Na, K, Ca, SO_4, HCO_3, and CO_3 concentrations in saturation paste extracts, saturation paste pH, and exchangeable Ca, Mg, Na, and K, and CEC for the soils. Extract-solution ion-concentration data units are entered as milliequivalent's per liter. When millimoles per liter units are used as input data, Z2, the conversion factor to convert to moles per liter, should be changed to equal 1000 rather than 2000. Exchangeable ion and CEC can be entered as milliequivalents per 100 grams, milliequivalents per kilogram, or millimoles of charge per kilogram as long as the units are consistent between exchangeable ions and CEC (Robbins and Carter, 1983). The same assumptions are used for this program as for the CHEM subroutine. The XCOEF program also uses the FUNCTION ACT and the ECM3 subroutines to calculate activity coefficients and solution EC. Sample input and output files are listed with the XCOEF program.

When using cation-exchange selectivity coefficients from the literature, care must be used to determine if cation concentrations or activities were used, as well as the equation form used to calculate the coefficients. Some values may be the reciprocal of the values produced by this subroutine. Other available coefficients will be less reliable if they were calculated from cation concentration data rather than cation activity data (Robbins & Carter, 1983).

Values for $K3$, $K4$, and $K6$ are not required when K is not being modeled. Other expressions containing K, such as $XK, ACK, AK, TK,$ or SK in the XCHANG subroutine, can also be simplified to exclude these variables and any equations in which these variables are calculated can be removed when K exchange is not of interest.

In a steady state system, the ESP can be calculated as

$$ESP = 100(Na) \, [(Na) + K2(Ca)^{1/2} + K5(Mg)^{1/2} + K6(K)]^{-1} \quad [7]$$

when cation activities and selectivity coefficients are available (Robbins, 1984).

IV. CRITICAL ASSUMPTIONS

The solute transport subroutine calculates vertical ion movement in the soil profile and assumes that: (i) the ions are not taken up by plant roots; (ii) ions are concentrated in the zone of water uptake; (iii) water uptake rates are a function of matrix plus osmotic potential; (iv) each ion moves independently of the other ions; (v) salts move up from a water table with upward moving water when present; and (vi) salts accumulate in the surface depth increment during periods of water evaporation from the soil surface.

The chemical precipitation-dissolution model considers both lime and gypsum reactions and assumes that: (i) the soil contains lime; (ii) the soil solution pH is controlled by soil-atmosphere, CO_2 partial pressure and Ca ion activity; (iii) the soil solution is an open system with respect to CO_2, meaning that CO_2 can enter (from roots or other biological activity) or leave (with moving water or air) the system, and rather than that the system is in equilibrium with the atmosphere; (iv) these reactions are thermodynamically rather than rate controlled, because the soil moisture content is continually changing from wetting to drying or drying to wetting cycles, thus, the system is seldom at equilibrium; and (v) Henry's Law constant (KH) for

CO_2 is assumed to be independent of temperature and salt concentration. The Davies' equation is used to calculate single ion activity coefficients and is not valid for solutions more concentrated than 0.5 M (Stumm & Morgan, 1970). Solutions more concentrated than 0.5 M should be modeled by other methods (Van Luik & Jurinak, 1979).

The cation exchange subroutine is constructed on the assumption that: (i) the CEC is equal to the sum of the exchangeable cations; (ii) the CEC is independent of pH; (iii) independent of total solution ion concentration; and (iv) independent of the ratio of each soluble or exchangeable cation to the other cation species; (v) the exchange reaction rates are sufficiently fast that equilibrium can be assumed; and (vi) the selectivity coefficients are constant over the range of the conditions simulated. The XCOEF program used to calculate the cation exchange coefficients is based on the same assumptions as in XCHANG.

V. MODEL VALIDATION

The initial validation data for these subroutines were obtained from a lysimeter study using two calcareous soils from Emery County, Utah. Hunting silty clay loam (fine-silty, mixed, calcareous, mesic Aquic Ustifluvent) did not contain gypsum, while Penoyer loam (coarse-silty, mixed, calcareous, mesic Typic Torrifluvent) did. Low, medium, and high $CaSO_4$ irrigation-water treatments were applied to the soils at 0.10 and 0.25 leaching fractions. The 12 treatments were randomly replicated three times. Soil solution samples were taken through 100-kPa porous, ceramic cups inserted into the lysimeter sides at 0.25, 0.50, and 0.75 m below the soil surface. A sand covered drain was placed in the bottom of each lysimeter (Robbins & Willardson, 1980). The lysimeters were cropped with alfalfa (*Medicago sativa* L.) for water consumption and for concentrating ions in the soil solution (Robbins et al., 1980a, b). Only two treatments will be discussed here. Treatment A consisted of low Ca and SO_4 irrigation water applied at a 0.25 leaching fraction to the Penoyer soil (with 0.07 gypsum by weight). This treatment produced the greatest amount of gypsum dissolution of those used. Treatment B consisted of irrigating the Hunting soil (no gypsum) with the high Ca and SO_4 irrigation water at a 0.10 leaching fraction. This treatment produced the greatest amount of gypsum precipitation in the nongypsiferous soil.

For the lysimeter validation studies, the main program was designed so that one of three calculation-method options could be selected. Option 1 periodically printed the various ions without calculating any chemical or exchange reactions. Only ion transport and dispersion were calculated. Option 2 called the CHEM subroutine, and calculated lime and gypsum precipitation as affected by ion concentration and CO_2 partial pressure. Option 3 called the XCHANG subroutine in addition to the CHEM subroutine and calculated changes in solution and exchangeable ion as a result of changes in cation concentration and ratios in the solution flowing through the soil.

A leaching study conducted in field plots with and without a corn (*Zea mays* L.) crop was used to evaluate the complete model for sensitivity and accuracy by Dudley et al. (1981). Instrumented plots were established on a Millville silt loam (coarse silty, carbonatic, mesic Typic Haploxeroll) with fairly uniform physical properties to below 1.2 m. The five irrigation waters that were used had a variety of ion ratios and concentrations, and EC and SAR values. High and low Ca and SO_4 concentrations were provided to include conditions with and without gypsum precipitation. The cropped plots were irrigated as needed. The remaining plots were either irrigated and then covered with plastic between irrigations, or continuously ponded with low-salt water between irrigations of the high-salt waters to provide a steady state water regime. Solution samples were obtained through suction extractions at 0.15-, 0.30-, 0.60-, and 1.20-m depths. Tensiometers were placed at these depths to follow the matric potential changes. Soil water content was measured using access tubes and a neutron probe.

Individual Ca, Mg, Na, K, Cl, HCO_3, and SO_4 ion and EC concentrations were measured, and SAR was calculated and compared with the predicted values for the lysimeter and field plot studies. On completion of the leaching treatments, soil samples were taken from the lysimeters and several plots at the same depths that the extraction tubes were placed. Water and ammonium acetate extracts were made to determine soluble and exchangeable Ca, Mg, Na, and K concentrations. The six selectivity coefficients were calculated from these data for the three soils.

VI. SUMMARY

In the lysimeter study, Option 1 did not adequately predict EC (Fig. 16-1) or SAR (Fig. 16-2) of the soil solution for any of the treatments. Option 2 was usually better at predicting EC, but was only occasionally better than Option 1 at predicting SAR. Option 3 predicted these two parameters very well when the pH profile was properly adjusted (see Robbins et al., 1980a for more discussion).

Differences in the ability of the three options to predict EC and SAR under different soil and water conditions arise from the differences in types of reactions involving each ion, thus requiring consideration of each ion separately. The chloride ion was considered to be chemically nonreactive, thus the same results were obtained regardless of the calculation method used (Fig. 16-3a). The agreement between predicted and measured Cl concentrations for all treatments after 278 d of irrigation, indicated that the solute transport prediction subroutine was working correctly.

Chemical precipitation and dissolution reactions were required to predict SO_4 concentration when gypsum solubility became a factor. Sulfate was overestimated by Option 1 when gypsum was being precipitated, but Option 2 (data not shown) and Option 3 produced essentially the same results (Fig. 16-3b). Sulfate was underestimated for treatment A (data not shown) when

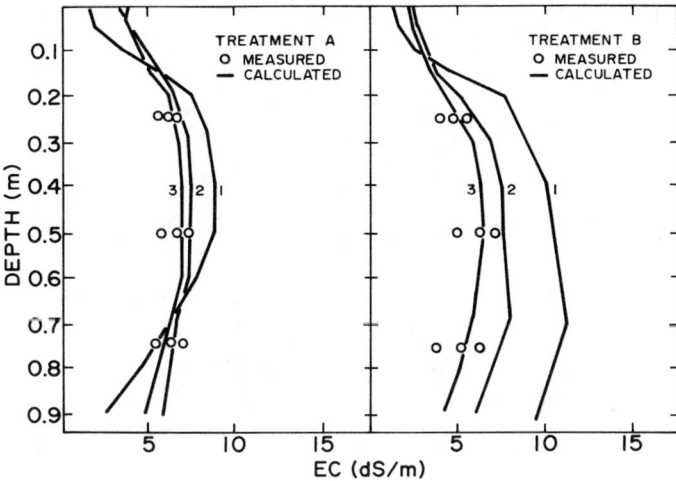

Fig. 16-1. Measured and calculated electrical conductivity (EC) values for Treatments A and B by solute transport only (1); solute transport and chemical precipitation (2); and solute transport, chemical precipitation, and cation exchange (3).

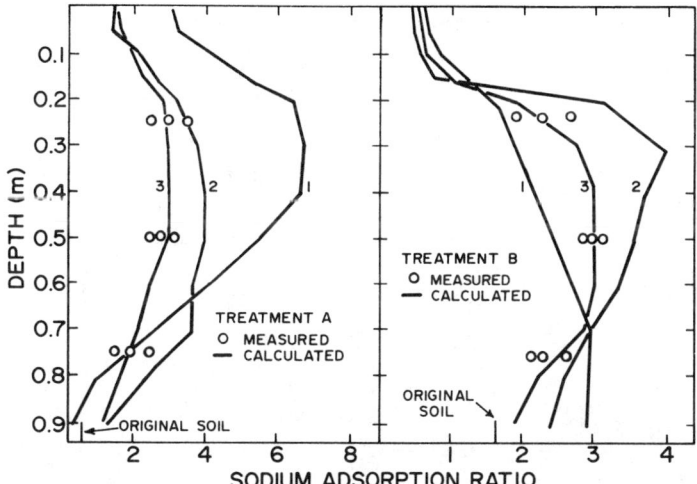

Fig. 16-2. Measured and calculated sodium adsorption ratio (SAR) values for Treatments A and B by solute transport only (1); solute transport and chemical precipitation (2); and solute transport, chemical precipitation, and cation exchange (3).

CHEM was not used, since gypsum was dissolving and releasing SO_4 into solution during irrigation with low SO_4 water.

Predicted Mg, Na, and K values are not affected by the CHEM subroutine alone. These are moved as inert ions by Options 1 and 2. Consequently, both options give the same results for these ions for all treatments. Neither Option 1 nor 2 adequately predicted any of the cations when their ratios or concentration in solution was not in equilibrium with the cation

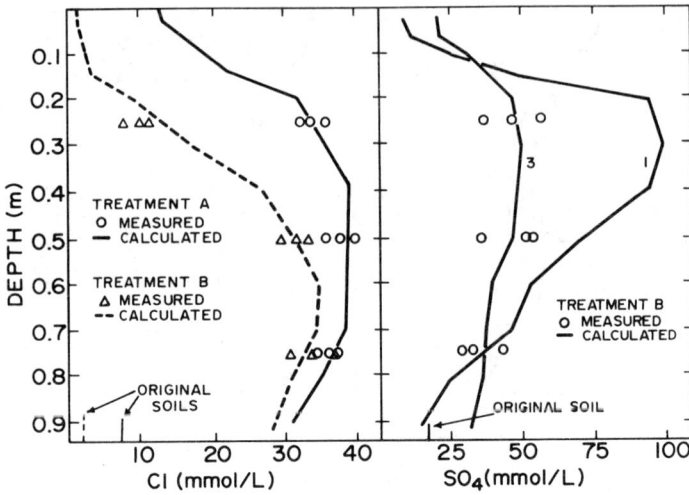

Fig. 16-3. Measured and calculated Cl concentrations for Treatments A and B by all three methods. Measured and calculated SO_4 concentrations by solute transport only (1); and solute transport, chemical precipitation, and cation exchange (3) for Treatment B.

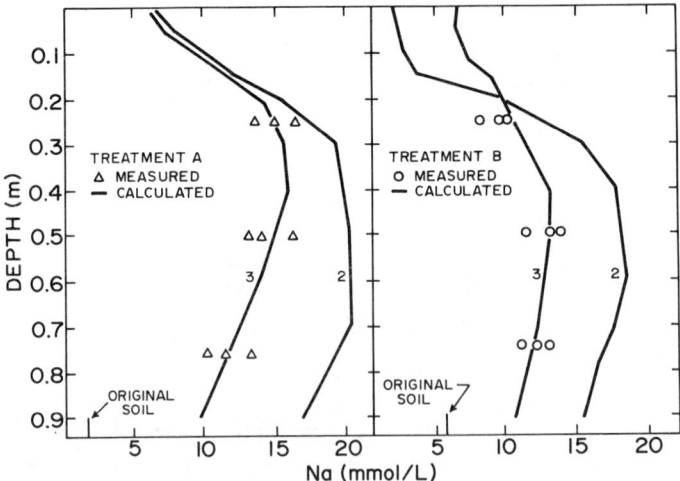

Fig. 16-4. Measured and calculated Na concentrations for Treatments A and B by solute transport and chemical precipitation (2); and solute transport, chemical precipitation, and cation exchange (3).

mix on the exchange sites. Sodium in solution, as measured and calculated, is shown in Fig. 15-4a and 15-4b for Treatments A and B, as an example. In both cases, Na was overestimated by Options 1 (data not shown) and 2, while Option 3 satisfactorily predicted Na solution concentration. Additional Na, Mg and K data are shown elsewhere (Robbins et al., 1980b).

Calcium ion concentration calculation is the most complex of the cations since it is affected by pH, CO_2 partial pressure, CO_3, HCO_3, and SO_4

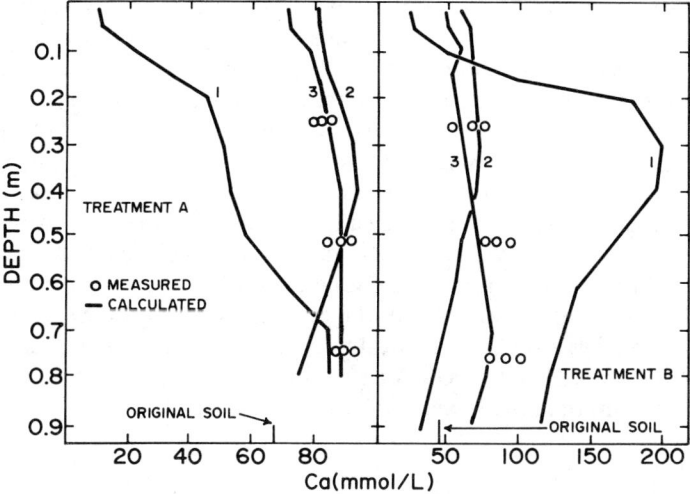

Fig. 16-5. Measured and calculated Ca concentrations for Treatments A and B by solute transport only (1); solute transport and chemical precipitation (2); and solute transport, chemical precipitation, and cation exchange (3).

concentration through dissolution and precipitation reactions and by the other cations through exchange reactions. Consequently, Option 3 was the only calculation method that consistently predicted Ca concentration. Options 1 and 2 would either underestimate or overestimate Ca concentration depending on the concentrations of the other ions (Fig. 16-5). Since Ca concentration in solution and on the exchange sites plays such an active part in soil chemical and physical interaction, Option 3 is necessary to model these changes with time if equilibrium between the irrigation water, soil solution, and exchange phase is modeled.

In the field plot evaluation, Dudley et al. (1981) discussed Cl and Ca concentration and EC in detail. The chloride ion was selected for evaluation because it is considered to be chemically nonreactive and provides an evaluation of ion transport modeling. Calcium evaluation includes the lime and gypsum chemical reactions as well as cation exchange reactions. Overall model performance can be partially evaluated by EC prediction evaluation. Under field plot conditions, they found that Ca and Cl ion concentrations and EC values were not predicted at a given point for the short study time and conditions evaluated in the noncropped treatments. The model did not provide a method of representing the field variability found in the plots. Under cropped conditions, Ca and Cl ion concentrations and measured EC values in the field were more accurately predicted by the model than the noncropped treatments. Growing plants appeared to have an averaging effect on the measured results. This can be explained by the fact that roots of a given plant will extract water from the areas of lowest total potential (i.e., the wettest and least salty locations). Also, the wetting and drying cycles produced by the roots also will cause water to move through areas not otherwise affected by flowing water in a noncropped soil.

The original CHEM subroutine assumed a constant pH for each depth increment, and calculated CO_2 partial pressure, pH, HCO_3, and CO_3 from pH and Ca ion activity (Robbins et al., 1980a). This calculation method was used because of a lack of soil atmosphere CO_2 data for calcareous soils. Now that data are becoming available (Robbins, 1986a), it is possible to use CO_2 data as an input, and pH, HCO_3 and CO_3 can be modeled much simpler and in a more realistic manner, as is done in this version of CHEM. Because of this recent improvement, validation data are presently being obtained, but are not yet available for comparison. The calculation method and the relationships are basically the same, but CO_2 rather than pH is considered the independent variable. In the past, the Ca-CO_3-HCO_3-CO_2 portion of these models have been the weak section of the chemistry calculations, but with the new CO_2 data becoming available, these processes are easier to model in a more realistic manner (Robbins, 1985a, b).

VII. ADDITIONAL RESEARCH NEEDS

In addition to the reactions considered in these models, there are several additional chemical and physical reactions that are of interest under special conditions. This model supplies ion concentrations and activities necessary for additional reactions, thus leading to the systematic addition of other reactions as desired.

Many geothermal springs and wells being developed for irrigation in arid areas contain high fluoride (F) concentrations. There is concern that the F from these water sources might eventually leach into shallow water supplies used for domestic and livestock drinking water. Fluorite precipitation reactions (Tracy et al., 1984) and F adsorption in calcareous soils (Robbins, 1986b) have been sufficiently quantified so that these processes can easily be added to the CHEM subroutine for modeling high F systems.

The subroutines described here also provide the necessary soil solution electrolyte concentrations and exchangeable sodium data needed as input data for a hydraulic conductivity-infiltration rate model that could calculate changes in water flow rates caused by EC and ESP changes. Once the relationships between ESP, EC, pH, and water flow are established for a particular soil, the changes in infiltration and hydraulic conductivity rates as a function of EC and ESP changes could be added to the water flow models (Shainberg et al., 1981a). Soils with high ESPs and low to moderate ECs have been shown to differ in their responses to irrigation with very low electrolyte water or by wetting with rain water. Differences among soil minerals' tendency to release salts when exposed to low electrolyte water has been suggested to be part of the cause for differences in soil dispersion and infiltration, and hydraulic conductivity rates (Shainberg et al., 1981b).

In the past, soil-atmosphere, CO_2 partial-pressure values needed for lime equilibrium calculation have either been calculated from input pH data or, as in the case of this model, CO_2 is read in for each depth increment. In either case, the pH or CO_2 values are held constant throughout the simulation. Recently, more data has become available for CO_2 concentrations and changes in calcareous soil systems (Robbins, 1986b). These data present the opportunity to develop CO_2 concentrations or CO_2 production models that would calculate CO_2 as a function of crop variety, root depth, growth stages and rates, and soil water content. Modeling CO_2 changes in the soil atmosphere could improve salinity prediction models and help increase the present knowledge of this segment of soil chemical reactions.

Under special conditions, it would also be advantageous to be able to predict the movement of such ions as boron, selenium, arsenic, and heavy metal ions. With more basic information about many of these ions, the CHEM subroutine could also be amended to predict movement of many of these ions toward the groundwater.

VIII. APPENDIX 1

PROGRAM LISTINGS

SOLUTE SUBROUTINE

```
*****************************************************************
      SUBROUTINE SOLUTE(K,DIFO,DIFA,DIFB,AMBDA,DD,DELT,KK,
     #SE1,SE2,SE3,SE4,SE5,SE6,SE7,SE8,SS1,SS2,SS3,SS4,SS5,SS6,SS7,SS8,
     #SSE1,SSE2,SSE3,SSE4,SSE5,SSE6,SSE7,SSE8,HEAD,IRRNO,W,Y,WF2,R9,
     #SS1,SS2,SS3,SS4,SS5,SS6,SS7,SS8)
*****************************************************************
*     THIS SUBROUTINE SOLVES THE SOLUTE TRANSPORT EQUATION FOR EIGHT
*     IONS (CA,MG,NA,K,CL,SO4,HCO3,CO3 AND ARE PASSED IN AS SS1 THROUGH SS8
*     AND NEW ION CONCENTRATIONS PASSED BACK OUT AS SE1 THROUGH SE8.
*
*     DIFFUS=BRESLER'S (1973) APPARENT DIFFUSION COEFFICIENT (MM**2/DAY)
*     RHO=BULK DENSITY OF SOIL (KG/DM*3)
*     SE1 TO SE8: IONS,CURRENT TIME STEP (MMOL/L)
*
*     SSE1 TO SSE8: IONS LEACHED OUT OF ROOT ZONE (MMOL)
*     SS1 TO SS8: ION CONCENTRATIONS PREVIOUS TIME STEP (MMOL/L)
*     WF2=FLUX DENSITY OF WATER DURING CURRENT TIME STEP (MM/DAY)
*------------------------
      DIMENSION TW(30),Y(30),W(30),WF2(30),DIFFUS(30)
      REAL DD(30),A1(30),B1(30),C1(30),D1(30),AB(30),BB(30)
      REAL CB(30),DB(30),H(30),G(30),G1(30),F1(30)
      REAL SE1(30),SE2(30),SE3(30),SE4(30),SS1(30),T(30,8),S(30,8)
      REAL SS2(30),SS3(30),SS4(30),WW(30),WW1(30)
      REAL SS5(30),SS6(30),SS7(30),SS8(30),SE5(30),SE6(30),SE7(30)
      REAL SE8(30),SF1(100),SF2(100),SF3(100)
      REAL SF4(100),SF5(100),SF6(100),SF7(100),SF8(100)
      INTEGER R9
*------------------------
      WF21=1.
*     SET SOLUTE CONCENTRATIONS (S VALUES) TO PASSED CONCENTRATIONS
      DO 3 I=1,30
      DO 2 J=1,8
      T(I,J)=0.
```

```
      2 CONTINUE
      3 CONTINUE
        DO 5 I=2,KK
        S(I,1)=SS1(I)
        S(I,2)=SS2(I)
        S(I,3)=SS3(I)
        S(I,4)=SS4(I)
        S(I,5)=SS5(I)
        S(I,6)=SS6(I)
        S(I,7)=SS7(I)
        S(I,8)=SS8(I)
      5 CONTINUE
*----------------------
*   SET SURFACE BOUNDARY TO ZERO OR TO SOLUTE CONC. IN APPLIED
*   WATER (SF VALUES) AFTER PARTITION
        IF(HEAD.LE.0.)THEN
        DO 6 M=1,8
      6 S(1,M)=0.
        ELSE
        S(1,1)=SF1(IRRNO)
        S(1,2)=SF2(IRRNO)
        S(1,3)=SF3(IRRNO)
        S(1,4)=SF4(IRRNO)
        S(1,5)=SF5(IRRNO)
        S(1,6)=SF6(IRRNO)
        S(1,7)=SF7(IRRNO)
        S(1,8)=SF8(IRRNO)
        ENDIF
*----------------------
*   CALC DEPTH INCREMENT (FOR EQUAL DEPTHS ONLY
        DLZ=DD(2)-DD(1)
*----------------------
*   CALCULATE TOP AND BOTTOM WATER CONTENTS
        IF(WF2(1).LT.0.)THEN
        WF21=WF2(1)
        WF2(1)=0.
        ELSE
        W1=W(1)
        WF21=WF2(1)
        W(1)=Y(1)-WF2(1)*DELT/DLZ
        ENDIF
        IF(R9.EQ.3)WF2(KK-1)=0.
*----------------------
*   MEAN THETA; DIFFUSION COEFFICIENT; PORE WATER VELOCITY BETWEEN
*   NODES; PORE WATER VELOCITY ACROSS NODES
        DO 10 I=1,KK-1
        TW(I)=(Y(I+1)+W(I+1)+Y(I)+W(I))/4.
        DIFFUS(I)=DIFO*DIFA*EXP(DIFB*TW(I))+AMBDA*ABS(WF2(I)/TW(I))
        IF(I.EQ.1.AND.WF2(1).EQ.0.)DIFFUS(I)=0.
        WW1(I)=WF2(I)/TW(I)
        IF(I.GT.1)WW(I)=((WF2(I-1)+WF2(I))/2.)/((W(I)+Y(I))/2.)
     10 CONTINUE
*----------------------
*   CONSTANTS FOR THE DIFFUSION/CONVECTION EQUATION
        DO 20 I=2,KK-1
        AB(I)=(DIFFUS(I-1)-DLZ*WF2(I-1)/2.-WW(I)*WW1(I-1)*DELT*
       #(W(I-1)-Y(I-1))/8.)/(2.*DLZ*DLZ)
        BB(I)=(DIFFUS(I)-DLZ*WF2(I)/2.-WW(I)*WW1(I)*DELT*
       #(W(I)-Y(I))/8.)/(2.*DLZ*DLZ)
        DB(I)=WF2(I)/(2.*DLZ)
        CB(I)=WF2(I-1)/(2.*DLZ)
```

SOLUTE TRANSPORT AND REACTIONS IN SOILS

```
      20 CONTINUE
*--------------------------
*  DIFFUSION TERMS IGNORED IN BOUNDARY SEGMENTS
      AB(2)=0.
      BB(KK-1)=0.
*--------------------------
      DO 25 M=1,8
      DO 30 I=2,K
*  SET BETA VALUES TO 0 OR 1 ACCORDING TO DIRECTION OF WATER FLOW
      IF(WF2(I).GT.0.)THEN
      BETA2=1.
      BETA3=0.
      ELSE
      BETA2=0.
      BETA3=1.
      ENDIF
      IF(WF2(I-1).GT.0.)THEN
      BETA1=1.
      BETA4=0.
      ELSE
      BETA1=0.
      BETA4=1.
      ENDIF
*--------------------------
*  COEFFICIENTS FOR TRIDIAGONAL MATRIX
      A1(I)=-AB(I)-BETA1*CB(I)
      B1(I)=W(I)/DELT+AB(I)+BB(I)+BETA2*DB(I)-BETA4*CB(I)
      C1(I)=-BB(I)+BETA3*DB(I)
      D1(I)=S(I-1,M)*(AB(I)+BETA1*CB(I))
     #+S(I,M)*(Y(I)/DELT-AB(I)-BETA2*DB(I)
     #+BETA4*CB(I)-BB(I))
     #+S(I+1,M)*(BB(I)-BETA3*DB(I))
   30 CONTINUE
*--------------------------
*  SOLVING TRIDIAGONAL MATRIX
*  I) CALC. F & G COEFFICIENTS FROM NODES 2 TO K
      F1(2)=C1(2)/B1(2)
      G1(2)=(D1(2)-A1(2)*S(1,M))/B1(2)
      DO 40 J=3,K
      F1(J)=C1(J)/(B1(J)-F1(J-1)*A1(J))
   40 G1(J)=(D1(J)-A1(J)*G1(J-1))/(B1(J)-A1(J)*F1(J-1))
*----
*  SET BOTTOM BOUNDARY CONDITION
*  I)WATER TABLE OF CONSTANT CONCENTRATION
      IF(R9.EQ.1)T(KK,M)=S(KK,M)
*  II) UNIT HYDRAULIC GRADIENT
      IF(R9.EQ.2)T(KK,M)=S(KK,M)
*  III) ZERO FLUX (DEPENDS ON WHETHER SALT IS ADDED OR LOST)
      IF(R9.EQ.3)T(KK,M)=0.
*----
*   NEW SOLUTE CONCENTRATIONS : T(I,M) VALUES
      J=KK
   50 J=J-1
      T(J,M)=G1(J)-F1(J)*T(J+1,M)
      IF(J.GT.2)GO TO 50
   25 CONTINUE
*--------------------------
*  ADJUST BOUNDARY VALUES BACK TO ORIGINAL WATER CONTENTS AND FLUXES
      IF(WF21.LE.0.)THEN
      WF2(1)=WF21
      ENDIF
```

```
*------------------------
*----------------------
*   CALCULATE DRAINAGE AS FLUX * CONC. FROM SEGMENT K
      SSE1=SSE1+WF2(K)*DELT*(S(K,1)+T(K,1))/2.
      SSE2=SSE2+WF2(K)*DELT*(S(K,2)+T(K,2))/2.
      SSE3=SSE3+WF2(K)*DELT*(S(K,3)+T(K,3))/2.
      SSE4=SSE4+WF2(K)*DELT*(S(K,4)+T(K,4))/2.
      SSE5=SSE5+WF2(K)*DELT*(S(K,5)+T(K,5))/2.
      SSE6=SSE6+WF2(K)*DELT*(S(K,6)+T(K,6))/2.
      SSE7=SSE7+WF2(K)*DELT*(S(K,7)+T(K,7))/2.
      SSE8=SSE8+WF2(K)*DELT*(S(K,8)+T(K,8))/2.
*------------------------
*   SET PASSED SE VALUES TO NEW SOLUTE CONCENTRATIONS
      DO 60 I=2,KK
      SE1(I)=T(I,1)
      SE2(I)=T(I,2)
      SE3(I)=T(I,3)
      SE4(I)=T(I,4)
      SE5(I)=T(I,5)
      SE6(I)=T(I,6)
      SE7(I)=T(I,7)
      SE8(I)=T(I,8)
      IF(SE1(I).LT.1.0E-15)SE1(I)=0.0
      IF(SE2(I).LT.1.0E-15)SE2(I)=0.0
      IF(SE3(I).LT.1.0E-15)SE3(I)=0.0
      IF(SE4(I).LT.1.0E-15)SE4(I)=0.0
      IF(SE5(I).LT.1.0E-15)SE5(I)=0.0
      IF(SE6(I).LT.1.0E-15)SE6(I)=0.0
      IF(SE7(I).LT.1.0E-15)SE7(I)=0.0
      IF(SE8(I).LT.1.0E-15)SE8(I)=0.0
   60 CONTINUE
*------------------------
      RETURN
      END
**********************************************************


CHEM SUBROUTINE

**************************************************************

      SUBROUTINE CHEM(CACO,CASO,PCO2,BD,TCA,TMG,TNA,TK,TCL,TSO4,
     #THCO3,TCO3,VH2O,EC,SAR,XCA,XMG,XNA,XK,CEC,ESP,NN,K1,K2,K3,K4,K5,
     #K6,PH,DELGYP,DELIME)
**************************************************************
*
*   THE CHEM SUBROUTINE AND THE FIVE DEPENDENT SUBROUTINES ARE INTENDED
*   TO BE USED TOGETHER AND INTERFACED WITH EXISTING STEADY STATE
*   OR TRANSIENT WATER FLOW, SALT TRANSPORT MODELS.  FOR USE WITH STEADY
*   STATE MODELS THE FIRST SECTION OF XCHANG IS USED (NN=1)
*   THESE SUBROUTINES ARE INTENDED TO PROVIDE A MORE THERMODYNAMICALLY
*   RIGOROUS DESCRIPTION OF LIME AND GYPSUM PRECIPITATION AND DISSOL-
*   UTION AND CATION EXCHANGE EQUILIBRIUM IN MINERAL SOILS CONTAINING
*   LIME WITH MEDIUM TO HIGH SALT CONCENTRATIONS.  THIS SUBROUTINE ALSO
*   ASSUMES THAT pH OF EACH INCREMENT CONTROLLED BY CO2 PARTIAL PRESSURE
*
*   REQUIRED DATA IN ORDER OF LISTING IN THE CALLING STATEMENT ARE:
*   CASO=GYPSUM  CACO=LIME   (DECIMAL FRACTION ON WT. BASIS)
*   PCO2=CARBON DIOXIDE PARTIAL PRESSURE (PERCENT CO2 OR kPa)
*   BD=SOIL BULK DENSITY (G/CUBIC CM.)
```

SOLUTE TRANSPORT AND REACTIONS IN SOILS

```
*     TCA=CALCIUM      TMG=MAGNESIUM       TNA=SODIUM
*     TK=POTASSIUM     TCL=CHLORIDE        TSO4=SULFATE
*                 (MMOL/L IN SOLUTION)
*     VH2O=VOLUMETRIC WATER CONTENT FOR DEPTH INCREMENT.
*     CEC=CATION EXCHANGE CAPACITY (MEQ/100G OR MEQ/KG OR MOLES OF CHARGE/
*     KG). IF CEC=0 THEN XCHANG IS NOT CALLED.
*     IF XCHANG IS CALLED AND NN IS NOT EQUAL TO 1, VALUES ARE NEEDED FOR
*     THE CATION EXCHANGE SELECTIVITY COEFFICIENTS K1, K2,....K6 THESE ARE
*     USED TO CALCULATE STARTING VALUES FOR THE EXCHANGEABLE CATIONS, XCA
*     XMG XNA AND XK (UNITS ARE SAME AS CEC). IF NN NOT=1 INPUT VALUES FOR
*     XCA, XMG, XNA AND XK ARE NEEDED. FOR ALL CASES WHERE CEC NOT EQUAL
*     TO ZERO, NEW VALUES ARE CALCULATED FOR XCA XMG XNA AND XK. AND
*     PASSED BACK TO THE CALLING PROGRAM.  AFTER EXECUTION THE CHEM
*     SUBROUTINE ALSO RETURNS NEW VALUES FOR ALL LISTED VARIABLES EXCEPT
*     PCO2, BD, VH2O, CEC, K1, K2, K3, K4, K5, AND K6,
*     THE OTHER OUT PUT VARIABLES ARE:
*     THCO3=BICARBONATE       TCO3=CARBONATE      (MMOLE/L)
*     EC=ELECTRICAL CONDUCTIVITY   (MMHOS/CM OR dS/M)
*     SAR=SODIUM ABSORBTION RATIO
*     ESP=EXCHANGEABLE SODIUM PERCENTAGE
*     DELGYP=CHANGE IN GYPSUM DURING EXECUTION STEP
*     DELIME=CHANGE IN LIME DURING EXECUTION STEP
*
*     OTHER FORTRAN SYMBOLS USED IN THE SUBROUTINE.
*     CHEMICAL CONSTANTS USED IN THE DATA STATEMENT;
*     KH  HENRYS LAW CONSTANT FOR CO2
*     KW  STABILITY CONSTANT FOR WATER
*     KA1 FIRST DISSOCIATION CONSTANT FOR H2CO3
*     KA2 SECOND DISSOCIATION CONSTANT FOR H2CO3
*     KD1 STABILITY CONSTANT OF CACO3
*     KD2 STABILITY CONSTANT OF CAHCO3+
*     KD3 STABILITY CONSTANT OF CAOH+
*     KD4 STABILITY CONSTANT OF CASO4
*     KD5 STABILITY CONSTANT OF MGCO3
*     KD6 STABILITY CONSTANT OF MGHCO3+
*     KD7 STABILITY CONSTANT OF MGOH+
*     KD8 STABILITY CONSTANT OF MGSO4
*     KD9 STABILITY CONSTANT OF NASO4-
*     KD10 STABILITY CONSTANT OF NACO3-
*     SP1 SOLUBILITY PRODUCT OF GYPSUM
*     SP2 SOLUBILITY PRODUCT OF LIME
*     H=HYDROGEN ION ACTIVITY   (MOLES/L)
*     ADJGYP AND ADJLIME CONVERT GYPSUM AND LIME BETWEEN
*     DECIMAL FRACTIONS AND MOLES/L IN SOLUTION.
*     ACT1 AND ACT2 ARE THE ACTIVITY COEFFICIENTS FOR MONO- AND
*     DIVALENT IONS
*     CHEMICAL SYMBOLS PRECEDED BY A, REPRESENT ION ACTIVITIES-
*     (ACA=CALCIUM ACTIVITY ETC.).

*     ACCA IS THE "APPARENT" ACTIVITY COEFFICIENT OF CALCIUM.
*     CAT AND AN ARE THE SUM OF CATIONS AND ANIONS (EQUIV./L).
*
      REAL KH,KW,KA1,KA2,KD1,KD2,KD3,KD4,KD5,KD6,KD7,KD8,KD9,
     #KD10
      REAL K1,K2,K3,K4,K5,K6
      DATA KH/.0339/,KW/.1E-13/,KA1/.5E-6/,KA2/.5E-10/,KD1/.63E-3/,
     #KD2/.055/,KD3/.0425/,KD4/.49E-2/,KD5/.4E-3/,KD6/.069/,
     #KD7/.263E-2/,KD8/.0063/,KD9/.24/,KD10/.0535/,SP1/.24E-4/,
     #SP2/.113E-7/
*
```

```
*   CONCENTRATIONS ARE CONVERTED FROM MMOL/L TO MOLES/L AND
*   APPROXIMATE VALUES ARE GIVEN TO THCO3 AND TCO3.
*
      TCA=TCA/1000.
      TMG=TMG/1000.
      TNA=TNA/1000.
      TK=TK/1000.
      TCL=TCL/1000.
      TSO4=TSO4/1000.
      THCO3=2.0*(TCA+TMG-TSO4)+TNA+TK-TCL
      IF (THCO3.LT.0.0) THCO3=0.0
      TCO3=0.0
*
*   CO2 PARTIAL PRESSURE CONVERTED TO ATMOSPHERES.
*
      PCO2=PCO2/100.
*
*   LIME AND GYPSUM ARE CONVERTED FROM DECIMAL FRACTIONS ON A WEIGHT
*   BASIS TO MOLES/L SOIL SOLUTION.
*
      ADJGYP=BD*5.81/VH2O
      CASO=CASO*ADJGYP
      CASOIN=CASO
      ADJLIM=BD*10./VH2O
      CACO=CACO*ADJLIM
*   STARTING VALUES OF LIME AND GYPSUM RECORDED SO THAT PRECIPITATION
*   AND DISSOLUTION CAN BE CALCULATED
      ALIME=CACO
      AGYP=CASO
*
*   EC IS CALCULATED FROM IONIC CONCENTRATIONS AND USED TO CALCULATE
*   THE MONO- AND DIVALENT ION ACTIVITY COEFFICIENTS.
*
      CALL ECM3(TCA,TMG,TNA,TK,TCL,TSO4,THCO3,TCO3,EC)
      ACT1=ACT(1.,EC)
      ACT2=ACT(2.,EC)
*
*   FIRST APPROXIMATIONS OF ION ACTIVITY ARE MADE FROM ACTIVITY
*   COEFFICIENTS AND THE ION PAIRS THAT CAN BE CONSIDERED AT THIS
*   POINT.
*
      ASO4=TSO4*ACT2
      AK=TK*ACT1
      ANA=TNA*ACT1/(1.0+ASO4/KD9)
      ACA=TCA/(1./ACT2+KW/(KD3*ACT1*H)+ASO4/KD4)
      AMG=TMG/(1./ACT2+KW/(KD7*ACT1*H)+ASO4/KD8)
      ASO4=TSO4/(1./ACT2+ACA/KD4+AMG/KD8+ANA/(ACT1*KD9))
      H=SQRT(PCO2*KH*KA1*KA2*ACA/SP2)
*
*
*   CHEMICAL EQUILIBRIUM LOOP
*
      DO 20 I=1,5
*
*   NEW ACTIVITY COEFFICIENTS ARE CALCULATED FROM THE EC VALUE
*   FROM THE PREVIOUS CYCLE AND A NEW PCO2 VALUE IS CALCULATED.
*
      ACT1=ACT(1.,EC)
      ACT2=ACT(2.,EC)
```

```
*
*     ACTIVITIES FOR CA, MG, NA, SO4, ARE CORRECTED FOR IONIC
*     STRENGTH AND ION PAIRING, AND ACO3 IS CALCULATED.
*
      ACA=TCA/(1./ACT2+KA1*KH*PCO2/(KD2*ACT1*H)+KW/(KD3*ACT1*H)+
     #KA1*KA2*KH*PCO2/(KD1*H*H)+ASO4/KD4)
      AMG=TMG/(1./ACT2+KA1*KH*PCO2/(KD6*ACT1*H)+KW/(KD7*ACT1*H)+
     *KA1*KA2*KH*PCO2/(KD5*H*H)+ASO4/KD8)
      ANA=TNA/(1./ACT1+ASO4/(ACT1*KD9)+KA1*KA2*KH*PCO2/(KD10*ACT2*H*H))
      ASO4=TSO4/(1./ACT2+ACA/KD4+AMG/KD8+ANA/(ACT1*KD9))
      H=SQRT(PCO2*KH*KA1*KA2*ACA/SP2)
      ACO3=KA1*KA2*KH*PCO2/(H*H)
*
*     THE SOIL SOLUTION IS EQUILIBRATED WITH LIME AND GYPSUM.
*
      CALL PRECIP(ACA,ACO3,ASO4,TCA,TSO4,CACO,CASO)
*
      ACA=TCA/(1./ACT2+KA1*KH*PCO2/(KD2*ACT1*H)+KW/(KD3*ACT1*H)+
     *KA1*KA2*KH*PCO2/(KD1*H*H)+ASO4/KD4)
      EQUIV=2.*(TCA+TMG-TSO4)+TNA+TK-TCL
      THCO3=H*EQUIV/(KA2*2.+H)
      TCO3=(EQUIV-THCO3)/2.0
      ECOLD=EC
      CALL ECM3(TCA,TMG,TNA,TK,TCL,TSO4,THCO3,TCO3,EC)
      IF(ABS(EC-ECOLD).LT.EC*0.01)GO TO 22
 20   CONTINUE
 22   PH=-ALOG10(H)
*
 40   IF(CEC.EQ.0.)GOTO 45
      CALL XCHANG(TCA,TMG,TNA,TK,ACA,AMG,ANA,AK,XCA,XMG,XNA,XK,
     *BD,VH2O,CEC,NN,K1,K2,K3,K4,K5,K6)
*
*     MOLES/L IN SOLUTION ARE CONVERTED TO MMOL/L AND GYPSUM AND LIME
*     ARE CONVERTED BACK TO DECIMAL FRACTIONS.  SAR IS ALSO CALCULATED.
*
 45   TCA=TCA*1000.
      TMG=TMG*1000.
      TNA=TNA*1000.
      TK=TK*1000.
      TSO4=TSO4*1000.
      THCO3=THCO3*1000.
      TCO3=TCO3*1000.
      TCL=TCL*1000.
C PRECIPITATION OR DISSOLUTION OF LIME AND GYPSUM DURING TIME STEP
      DELGYP=(CASO-AGYP)*1000.
      DELIME=(CACO-ALIME)*1000.
      CASO=CASO/ADJGYP
      CACO=CACO/ADJLIM
      SAR=TNA/SQRT(TCA+TMG)
      ESP=XNA*100./CEC
*
*     CO2 PARTIAL PRESSURE CONVERTED BACK TO PERCENT OR kPa.
*
      PCO2=PCO2*100.
 60   CONTINUE
      RETURN
      END
**************************************************************
```

FUNCTION ACT

```
************************************************************

*   THE FUNCTION ACT CALCULATES IONIC STRENGTH (IS) USING THE
*   APPROXIMATION OF GRIFFIN AND JURINAK (1973). THE SQUARE ROOT
*   (I) OF (IS) IS THEN USED IN THE DAVIES EQUATION TO CALCULATE
*   THE MONO- (Z=1) AND DIVALENT (Z=2) ION MEAN ACTIVITY COEFFICIENTS.
*
      FUNCTION ACT(Z,EC)
      REAL IS,I
      IS=0.0127*EC
      I=SQRT(IS)
      ACT=10.0**(-0.509*Z*Z*(I/(1.0+I)-0.3*IS))
      RETURN
      END
************************************************************

PRECIP SUBROUTINE

************************************************************

*   THE SUBROUTINE PRECIP USES CATION (CAT) AND ANION (AN) ACTIVITIES
*   AND THE SOLUBILITY PRODUCT (SP) TO DETERMINE IF SOLID PHASE
*   MATERIAL (PPT) MUST DISSOLVE OR PRECIPITATE TO BRING THE SYSTEM
*   INTO CHEMICAL EQUILIBRIUM. IT THEN CALLS THE SINK SUBROUTINE TO
*   DETERMINE THE QUANTITY OF PPT TO BE DISSOLVED OR PRECIPITATED.
*   XX AND YY ARE THE SINK SUBROUTINE STARTING VALUE ON INPUT, AND COME
*   BACK FROM SINK AS THE VALUE THAT CAT, AN AND PPT IS TO BE CHANGED.
*
      SUBROUTINE PRECIP(ACA,ACO3,ASO4,TCA,TSO4,CACO,CASO)
      DATA SP1/.24E-04/,SP2/.113E-07/
      ACCA=ACA/TCA
      ACSO=ASO4/TSO4
      YY=SP1
      XX=SP2
*
*   IF THE SOIL INCREMENT CONTAINS GYPSUM AND IS UNDERSATURATED WITH
*   RESPECT TO GYPSUM, GOTO 30.
*
      IF(ACA*ASO4.LT.SP1.AND.CASO.GT.0.0)GO TO 30
*
*   IF THE INCREMENT IS SUPERSATURATED WITH GYPSUM, GOTO 40
*
      IF(ACA*ASO4.GT.SP1)GO TO 40
*
*   IF THE INCREMENT IS UNDERSATURATED WITH LIME, GOTO 10, OF IF IN
*   EQUILIBRIUM WITH LIME GOTO 50, OR IF SUPERSATURATED WITH LIME GOTO
*   20
*
    5 IF(ACA*ACO3-SP2)10,50,20
   10 CALL SINK(ACA,ACO3,SP2,XX)
      CACO=CACO-XX/ACCA
      TCA=TCA+XX/ACCA
      ACA=ACA+XX
      GO TO 50
   20 CALL SINK(ACA,ACO3,SP2,XX)
      CACO=CACO+XX/ACCA
      TCA=TCA-XX/ACCA
```

```
        ACA=ACA+XX
        GO TO 50
  30    CALL SINK(ACA,ASO4,SP1,YY)
        IF((YY/ACSO).GT.CASO)YY=CASO*ACSO
        CASO=CASO-YY/ACSO
        TCA=TCA+YY/ACSO
        ACA=ACA+YY
        TSO4=TSO4+YY/ACSO
        ASO4=ASO4+YY
*
*   GOTO 5 TO CHECK LIME EQUILIBRIUM.
*
        GO TO 5
  40    CALL SINK(ACA,ASO4,SP1,YY)
        CASO=CASO+YY/ACSO
        TCA=TCA-YY/ACSO
        ACA=ACA-ACSO
        TSO4=TSO4-YY/ACSO
        ASO4=ASO4-YY
*
*   GOTO 5 TO CHECK LIME EQUILIBRIUM.
*
        GO TO 5
  50    CONTINUE
        RETURN
        END
*****************************************************************

SINK SUBROUTINE

*****************************************************************
*   THE SUBROUTINE SINK USES CATION (CAT) AND ANION (AN) ACTIVITIES,
*   SOLUBILITY PRODUCT (SP), AND A STARTING VALUE (X) TO CALCULATE THE
*   CATION AND ANION ACTIVITY CHANGE DUE TO SOLUTION OR PRECIPITATION
*   OF SOLID PHASE TO BRING THE SYSTEM INTO CHEMICAL EQUILIBRIUM FOR
*   A GIVEN SPECIES. THE NEWTON METHOD IS USED TO FIND X.
*
        SUBROUTINE SINK(CAT,AN,SP,X)
        DO 5 N=1,10
*
*   THIS STATEMENT KEEPS THE NEXT FROM DIVIDING BY ZERO
*
        IF(CAT+AN.EQ.2.0*X)X=X*1.1
        XI=X-(X*X-X*CAT-X*AN+CAT*AN-SP)/(2.0*X-CAT-AN)
        IF(ABS(XI-X).LT.ABS(XI*.01)) GOTO 10
        X=XI
  5     CONTINUE
  10    X=ABS(X)
        RETURN
        END
*****************************************************************

XCHANG SUBROUTINE

*****************************************************************
*   THE SUBROUTINE XCHANG IS DIVIDED INTO TWO SEGMENTS. IF NN EQUALS
*   1, INITIAL EXCHANGEABLE CATION CONCENTRATIONS ARE CALCULATED FROM
*   THE CATION EXCHANGE CAPACITY (CEC), AND CATION ACTIVITIES SUPPLIED
```

```
*     BY THE CALLING PROGRAM. IF NN IS NOT EQUAL TO 1, NEW EQUILIBRIUM
*     IS CALCULATED FOR SOLUTION AND EXCHANGEABLE CATIONS USING
*     EXCHANGEABLE CATION AND SOLUTION CATION CONCENTRATIONS AND CATION
*     ACTIVITIES, BULK DENSITY (BD), VOLUMETRIC WATER CONTENT (VH2O),
*     AND CEC VALUES SUPPLIED FROM THE CALLING PROGRAM.
*     TCA, TMG,...ETC ARE MOLES/L OF SOLUTION CATIONS
*     ACA, AMG,...ETC ARE CATION ACTIVITIES
*     XCA,XMG,...ETC ARE EXCHANGEABLE CATIONS, INITIALLY AND FINALLY IN
*     MEQ/100G OF SOIL AND WITHIN THE SUBROUTINE THEY ARE CONVERTED TO
*     AND FROM MOLES/L.
*     OTHER FORTRAN SYMBOLS ARE SELF EXPLANATORY.
*
      SUBROUTINE XCHANG(TCA,TMG,TNA,TK,ACA,AMG,ANA,AK,XCA,XMG,XNA,XK,
     *BD,VH2O,CEC,NN,K1,K2,K3,K4,K5,K6)
*
*     SOME OF THE SELECTIVITY COEFFICIENTS, K1,K2,K3, ETC MAY VARY FROM
*     SOIL TO SOIL.
*
      REAL K1,K2,K3,K4,K5,K6
      IF(NN.EQ.1)GOTO 10
      GOTO 20
*
*     STARTING POINT EXCHANGEABLE CATION VALUES ARE CALCULATED FROM
*     INITIAL INPUT DATA OR EXCHANGEABLE CATIONS ARE CALCULATED FOR STEADY
*     STATE CALCULATIONS.
*
*     STARTING POINT EXCHANGEABLE CATION VALUES ARE CALCULATED FROM
*     INITIAL INPUT DATA OR EXCHANGEABLE CATIONS ARE CALCULATED FOR STEADY
*     STATE CALCULATIONS.
*
   10 ZCA=SQRT(ACA)
      ZMG=SQRT(AMG)
      XCA=CEC/(ZMG/(K1*ZCA)+ANA/(ZCA*K2)+AK*K3/ZCA+1.)
      XMG=CEC/(ZCA*K1/ZMG+ANA/(ZMG*K5)+AK*K4/ZMG+1.)
      XNA=CEC/(ZCA*K2/ANA+ZMG*K5/ANA+AK*K6/ANA+1.)
      XK=CEC/(ZCA/(K3*AK)+ZMG/(K4*AK)+ANA/(AK*K6)+1.)
*
*     THE EXCHANGEABLE CATIONS ARE CORRECTED BY A COMMON FACTOR TO FORCE
*     THE SUM OF EXCHANGEABLE CATIONS TO EQUAL THE CEC.  IN A FEW CASES
*     MACHINE ROUND-OFF ERROR MAKES THIS NECESSARY.
*
      C=CEC/(XCA+XMG+XNA+XK)
      XCA=XCA*C
      XMG=XMG*C
      XNA=XNA*C
      XK=XK*C
      NN=1
      GOTO 50
*
*     ADJUSTMENT FACTORS ARE CALCULATED TO CONVERT EXCHANGEABLE CATION
*     UNITS BETWEEN MEQ/100G OF SOIL AND MOLES/L IN SOLUTION.
*              !!!! WARNING !!!!
*     IF EXCHANGEABLE CATION UNITS ARE MEQ/KG OF SOIL OR MMOLES OF CHARGE
*     PER KG THEN:
*  20 ADJ2=0.0005
*     ADJ1=0.0010
*     OTHERWISE:
*
   20 ADJ2=0.005*BD/VH2O
```

SOLUTE TRANSPORT AND REACTIONS IN SOILS

```
         ADJ1=0.010*BD/VH2O
         XCA=XCA*ADJ2
         XMG=XMG*ADJ2
         XNA=XNA*ADJ1
         XK=XK*ADJ1
         EQU=2.*(TCA+TMG)+TNA+TK
*
*   "APPARENT ACTIVITY COEFFICIENTS" ARE CALCULATED FOR EACH CATION.
*
         ACCA=ACA/TCA
         ACMG=AMG/TMG
         ACNA=ANA/TNA
         ACK=AK/TK
*
*   THE SUM OF EACH SOLUTION PLUS EXCHANGEABLE CATION IS CALCULATED.
*
         SCA=TCA+XCA
         SMG=TMG+XMG
         SNA=TNA+XNA
         SK=TK+XK
*
*   THIS LOOP BRINGS THE NEW EXCHANGEABLE AND SOLUTION CATIONS INTO EQUI-
*   LIBRIUM WITH EACH OTHER, ASSUMING, (1)THAT THE APPARENT ACTIVITY
*   COEFFICIENTS ARE CONSTANT, (2) THAT THE CEC IS CONSTANT AND EQUAL
*   TO THE SUM OF THE EXCHANGEABLE CATIONS, AND (3) THAT EACH EXCHANGE-
*   ABLE PLUS SOLUTION CATION CONCENTRATION REMAINS CONSTANT.
*
         DO 30 I=1,4
         ZCA=SQRT(ACA)
         ZMG=SQRT(AMG)
         XCANU=CEC/(ZMG/(K1*ZCA)+ANA/(ZCA*K2)+AK*K3/ZCA+1.)
         XMGNU=CEC/(ZCA*K1/ZMG+ANA/(ZMG*K5)+AK*K4/ZMG+1.)
         XNANU=CEC/(ZCA*K2/ANA+ZMG*K5/ANA+AK*K6/ANA+1.)
         XKNU=CEC/(ZCA/(K3*AK)+ZMG/(K4*AK)+ANA/(AK*K6)+1.)
         XCANU=XCANU*ADJ2
         XMGNU=XMGNU*ADJ2
         XNANU=XNANU*ADJ1
         XKNU=XKNU*ADJ1
         TCA=TCA*XCA*2./(XCANU+XCA)
         TMG=TMG*XMG*2./(XMGNU+XMG)
         TNA=TNA*XNA*2./(XNANU+XNA)
         TK=TK*XK*2./(XKNU+XK)
         EQUNU=2.*(TCA+TMG)+TNA+TK
         CC=EQU/EQUNU
         TCA=TCA*CC
         TMG=TMG*CC
         TNA=TNA*CC
         TK=TK*CC
         XCA=SCA-TCA
         XMG=SMG-TMG
         XNA=SNA-TNA
         XK=SK-TK
         C=CEC/((XCA+XMG)/ADJ2+(XNA+XK)/ADJ1)
         XCA=XCA*C
         XMG=XMG*C
         XNA=XNA*C
         XK=XK*C
         ACA=TCA*ACCA
         AMG=TMG*ACMG
         ANA=TNA*ACNA
         AK=TK*ACK
```

```
   30 CONTINUE
      XCA=XCA/ADJ2
      XMG=XMG/ADJ2
      XNA=XNA/ADJ1
      XK=XK/ADJ1
   50 RETURN
      END
```
****
ECM3 SUBROUTINE

****
****ELECTRICAL CONDUCTIVITY SUBROUTINE USING THE METHOD OF MCNEAL et al.
*   1970. SOIL SCI. 110:405-414. CONCENTRATION UNITS FOR INPUT IONS MUST
*   BE MOLES/L. IF MEQ/L ARE USED THEN X AND Y EQUAL 1.0.
*
```
      SUBROUTINE ECM3(TCA,TMG,TNA,TK,TCL,TSO4,THCO3,TCO3,EC)
      REAL MG
      X=1000.
      Y=2000.
      CA=TCA
      MG=TMG
      SO4=TSO4
      IF(SO4.GT.CA)GOTO 10
      CASO=SO4
      CA=CA-SO4
      SO4=0.
      GOTO 30
   10 CASO=CA
      SO4=SO4-CA
      CA=0.
      IF(SO4.GT.MG)GOTO 20
      CASO=CASO+SO4
      MG=MG-SO4
      SO4=0.
      GOTO 30
   20 CASO=CASO+MG
      SO4=SO4-MG
      MG=0.
   30 EC=.05641*((CA*Y)**.9202)+.05099*((MG*Y)**.9102)+.04748*((TNA*X)
     #**.9495)+.07263*((TK*X)**.9706)+.069*((SO4*Y)**.8973)+.0733*((
     #TCO3*Y)**.8719)+.04143*((THCO3*X)**.9501)+.07206*((TCL*X)**.9671
     #)+.1133*((CASO*Y)**.8463)
      RETURN
      END
      END$
```
****
EXCHANGE COEFFICIENT PROGRAM

FTN7X
$FILES 0,2
```
      PROGRAM XCOEF
```
*
*   THIS PROGRAM USES SOIL SOLUTION ION CONCENTRATIONS(UNITS=meq/L)
*   EXCHANGEABLE ION CONCENTRATIONS(UNITS=meq/100g, meq/Kg or mmoles
*   OF CHARGE/Kg OF SOIL), PH AND CEC(UNITS SAME AS EXCH. IONS) TO

```
*     CALCULATE CATION EXCHANGE COEFFICIENTS NEEDED FOR OTHER CATION
*     EXCHANGE MODELS. THE METHODS USED FOR THESE CALCULATIONS ARE
*     DESCRIBED BY ROBBINS AND CARTER (1983 IRRIGATION SCIENCE 4:95-102.)
*     THESE VALUES SHOULD NOT BE USED IN MODELS OF ROBBINS et al. WRITTEN
*     PRIOR TO OCT 1985 WITHOUT CHANGING THE EQUATIONS USED TO CALCULATE
*     EXCHANGEABLE CATIONS UNLESS THE RECIPROCAL VALUES FOR K1, K3 AND
*     K4 ARE USED.
*
      CHARACTER*32 INPUT,OUTPUT
      CHARACTER*10 SAMPLE
      REAL KH,KW,KA1,KA2,KD1,KD2,KD3,KD4,KD5,KD6,KD7,KD8,KD9,KD10
      REAL K1,K2,K3,K4,K5,K6
      DATA KH/.0339/,KW/.1E-13/,KA1/.5E-6/,KA2/.5E-10/,KD1/.63E-3/,KD2/.
     #055/,KD3/.0425/,KD4/.0049/,KD5/.0004/,KD6/.069/,KD7/.0026/,KD8/.00
     #63/,KD9/.24/,KD10/.054/,SP1/.24E-4/,SP2/.113E-7/
      WRITE(1,'("WHAT IS THE INPUT FILE NAME.")')
      READ(1,'(A32)')INPUT
      WRITE(1,'("WHAT IS THE OUTPUT FILE NAME")')
      READ(1,'(A32)')OUTPUT
      OPEN(10,FILE=INPUT,STATUS='OLD',IOSTAT=IER)
      if(ier.ne.0) write(1,'("error on open1",i5)')ier
      OPEN(16,FILE=OUTPUT,STATUS='NEW',IOSTAT=IER)
      if(ier.ne.0) write(1,'("error on open2",i5)')ier
      READ(10,100)
100   FORMAT(//)
      READ(10,102)NN
102   FORMAT(30X,I5)
      READ(10,100)
      WRITE(16,60)
60    FORMAT(" CATION EXCHANGE SELECTIVITY COEFFICIENTS CALCULATED ")
      WRITE(16,61)
61    FORMAT(" AS DESCRIBED BY ROBBINS & CARTER, 1983 IRRIGATION SCIENCE
     #")
      WRITE(16,100)
      WRITE(16,62)
62    FORMAT("              K1      K2      K3      K4      K5      K6"
     # )
      WRITE(16,63)
63    FORMAT(" SAMPLE   (CA/MG) (CA/NA) (K/CA) (K/MG) (MG/NA) (K/NA
     #) ")
      DO 50 I=1,NN
      READ(10,104)SAMPLE,TCA,TMG,TNA,TK,TCL,TSO4,THCO3,TCO3,XCA,XMG,XNA
     #,XK,PH,CEC
104   FORMAT(A10,14F5.2)
      Z1=1000.
      Z2=2000.
      TCA=TCA/Z2
      TMG=TMG/Z2
      TNA=TNA/Z1
      TK=TK/Z1
      TCL=TCL/Z1
      TSO4=TSO4/Z2
      THCO3=THCO3/Z1
      TCO3=TCO3/Z2
      IF(PH.LT.5.)GOTO 1
      GOTO 2
1     WRITE(16,51)
      GOTO 50
2     IF(TCA.LE.0.OR.TMG.LE.0.OR.TNA.LE.0.OR.TK.LE.0.OR.TCL.LE.0.OR.TSO4
     #.LE.0.)GOTO 3
```

```
         GOTO 4
    3    WRITE(16,52)
         GOTO 50
    4    IF(CEC.NE.(XCA+XMG+XNA+XK))CEC=XCA+XMG+XNA+XK
         CALL ECM3(TCA,TMG,TNA,TK,TCL,TSO4,THCO3,TCO3,EC)
         H=10.**(-PH)
         ACT1=ACT(1.,EC)
         ACT2=ACT(2.,EC)
*   FIRST ION ACTIVITY APPROXIMATION.
         ASO4=TSO4*ACT2
         AK=TK*ACT1
         ANA=TNA*ACT1/(1.+ASO4/KD9)
         ACA=TCA/(1./ACT2+KW/(KD3*ACT1*H)+ASO4/KD4)
         AMG=TMG/(1./ACT2+KW/(KD7*ACT1*H)+ASO4/KD8)
         ASO4=TSO4/(1./ACT2+ACA/KD4+AMG/KD8+ANA/(ACT1*KD9))
         DO 20 II=1,5
         PCO2=H*H*SP2/(ACA*KH*KA1*KA2)
         ACA=TCA/(1./ACT2+KA1*KH*PCO2/(KD2*ACT1*H)+KW/(KD3*ACT1*H)+
        #KA1*KA2*KW*PCO2/(KD1*H*H)+ASO4/KD4)
         AMG=TMG/(1./ACT2+KA1*KH*PCO2/(KD6*ACT1*H)+KW/(KD7*ACT2*H)+
        #KA1*KA2*KH*PCO2/(KD5*H*H)+ASO4/KD8)
         ANA=TNA/(1./ACT1+ASO4/(ACT1*KD9)+KA1*KA2*KH*PCO2/(KD10*ACT2*H*H))
   20    ASO4=TSO4/(1./ACT2+ACA/KD4+AMG/KD8+ANA/(ACT1*KD9))
         ZCA=SQRT(ACA)
         ZMG=SQRT(AMG)
         K1=ZMG*XCA/(ZCA*XMG)
         K2=ANA*XCA/(ZCA*XNA)
         K3=ZCA*XK/(AK*XCA)
         K4=ZMG*XK/(AK*XMG)
         K5=ANA*XMG/(ZMG*XNA)
         K6=ANA*XK/(AK*XNA)
         WRITE(16,64)SAMPLE,K1,K2,K3,K4,K5,K6
   64    FORMAT(A10,6F8.2)
   50    CONTINUE
         WRITE(16,100)
         WRITE(16,65)
   65    FORMAT(" WARNING!!! K1, K3 AND K4 MAY BE RECIPROCALS OF SELECT-")
         WRITE(16,66)
   66    FORMAT(" IVITY COEFFICIENTS REPORTED EARLIER BY ROBBINS et.al.")
         CLOSE(10)
         CLOSE(16)
   51    FORMAT(" PH VALUE IS TO LOW")
   52    FORMAT(" CATION, CL OR SO4 VALUES ARE TO LOW")
         END
         END$
****************************************************************
```

IX. APPENDIX 2

SAMPLE DATA INPUT AND OUTPUT FOR XCOEF

```
DATA FILE FOR CALCULATING SELECTIVITY COEFFICIENTS
              FOR CATION EXCHANGE
---------------------------------------------------------
NUMBER OF SOIL SAMPLES         10
---------------------------------------------------------
SAMPLE     CA    MG    NA     K    CL   SO4  HCO3 CO3   XCA   XMG  XNA    XK   PH  CEC
---------------------------------------------------------
SAMPLE 1   72    48   862     8   201   130   600   8   689   507  286    18  750 1500
SAMPLE 2  461   384  8344    28  6412  1828  1084  19   680   501  302    17  750 1500
SAMPLE 3 1553 152516396       38 13458  5711   519  16   633   532  313    13  750 1491
SAMPLE 4 2292 203917255       51 944811748    535  17   660   513  300    16  750 1489
SAMPLE 5 2500  1000 8000     100  2500  1000  8000 100 1000   260  200    40  860 1500
CROCK. 1  310    66  1130   580    15  2800  1581  12  1789   146  229   949  835 3113
CROCK.20   80   221 13.3    1080    17  2556  1610 6442  557   107 1400   992  955 2956
PORT. 46 2200   458  1220    45  1600  3700   439   06 2294   298   97    70  825 2759
PORT. 55   90     7  7780  1300   760   950  3962 2420  476    32 1189  1965  980 3662
FREE. 22  840  6483 5500 110032290  4800   937   898  617  139  1409   591  920 2756

CATION EXCHANGE SELECTIVITY COEFFICIENTS CALCULATED
AS DESCRIBED BY ROBBINS & CARTER, 1983 IRRIGATION SCIENCE


             K1      K2      K3      K4      K5      K6
SAMPLE     (CA/MG) (CA/NA) (K/CA)  (K/MG)  (MG/NA) (K/NA)
SAMPLE 1    1.12    1.61    4.20    4.68    1.44    6.76
SAMPLE 2    1.26    5.68    2.92    3.68    4.50   16.57
SAMPLE 3    1.22    5.88    2.97    3.61    4.83   17.46
SAMPLE 4    1.28    6.10    2.81    3.59    4.78   17.15
SAMPLE 5    2.45    4.81    3.31    8.11    1.96   15.92
CROCK. 1    5.86    3.36    2.34   13.70     .57    7.86
CROCK.20    2.69    3.25    2.24    6.03    1.21    7.29
PORT. 46    3.66    4.10    4.66   17.03    1.12   19.08
PORT. 55    4.03    1.97    4.98   20.08     .49    9.80
FREE. 22    3.97    3.62    3.67   14.56     .91   13.28
```

WARNING!!! K1, K3 AND K4 MAY BE RECIPROCALS OF SELECTIVITY COEFFICIENTS REPORTED EARLIER BY ROBBINS et al.

X. APPENDIX 3

SOLUTE TRANSPORT
(in basic)

****

```
4047 REM----------------------SALT LOOP STARTS HERE--------
4050 WFRU=WFDD
4055 IF WFRU<=0 OR EOR<=0 THEN WFRU=0
4060 WATU=(Y(1)*TM+W(1)*TT+Y(2)*TM+W(2)*TT)*.5
4065 ALFA=0
4070 FOR I=2 TO K
4075 DLXA=(DD(I)-DD(I-1))
4080 DLXB=(DD(I+1)-DD(I))
4085 DLXC=(DD(I+1)-DD(I-1))*.5
4090 WFRD=B(I)*((H(I)-H(I+1))*TT+(G(I)-G(I+1))*TM+DLXB)/DLXB
4095 WATD=(Y(I)*TM+W(I)*TT+Y(I+1)*TM+W(I+1)*TT)*.5
4100 BETA=DIFO*DIFA*EXP(DIFB*WATD)+ALAMBA*ABS(WFRD/WATD)
4110 TW=DELT*(W(I)-Y(I))*(WFRD+WFRU)/(8*(W(I)+Y(I)))
4115 AX=TW*WFRU/(DLXA*WATU)+ALFA/DLXA+WFRU*.5
4120 CX=TW*WFRD/(DLXB*WATD)+BETA/DLXB-WFRD*.5
4125 IF I=2 THEN AX=WFRU
4130 BB=W(I)*DLXC/(TT*DELT)+AX-WFRU+CX+WFRD
4135 DA=(Y(I)*SS(I)*DLXC/DELT+TM*(AX*(SS(I-1)-SS(I))+
WFRU*SS(I)-CX*(SS(I)-SS(I+1))+WFRD*SS(I)))/TT
4140 IF I>2 THEN 4160
4145 DA=DA+AX*SS(I-1)
4150 F(I)=DA/BB
4155 E(I)=CX/BB
4160 XI=I
4165 IF I=2 THEN 4185
4170 IF I>=K THEN 4200
4175 E(I)=CX/(BB-AX*E(I-1))
4180 F(I)=(DA+AX*F(I-1))/(BB-AX*E(I-1))
4185 ALFA=BETA
4190 WATU=WATD
4195 WFRU=WFRD
4197 NEXT I
4200 IF ITAA=1 THEN BB=BB-CX
4202 IF ITAA=0 THEN DA=DA+CX*SS(I+1)
4205 SE(I)=(DA+AX*F(I-1))/(BB-AX*E(I-1))
4210 I=I-1
4215 SE(I)=E(I)*SE(I+1)+F(I)
4220 IF I>2 THEN 4210
4222 SE(KK)=SS(KK)
4225 FOR I=2 TO K
4230 IF SE(I)>=SE(I-1) OR SE(I)>=SE(I+1) THEN 4290
4235 IF I=2 THEN 4275
```

```
4240 IF I>=K THEN 4255
4245 K6=K6+1
4250 IF SE(I-1)<=SE(I+1) THEN 4275
4255 TW=(SE(I+1)-SE(I))*W(I)*(DD(I+1)-DD(I-1))*.5
4260 SE(I-1)=SE(I-1)-TW/(W(I-1)*(DD(I)-DD(I-2))*.5)
4265 SE(I)=SE(I+1)
4270 GOTO 4290
4275 TW=(SE(I-1)-SE(I))*W(I)*(DD(I+1)-DD(I-1))*.5
4280 SE(I+1)=SE(I+1)-TW/(W(I+1)*(DD(I+2)-DD(I))*.5)
4285 SE(I)=SE(I-1)
4290 NEXT I
4300 IF ITAA=1 THEN SE(KK)=SE(K)
4305 SCMX=WFUU*SE(KK)*DELT+SCMX
4310 SD(1)=SE(1)*W(1)*.5*DD(2)
4312 SCM=WFRD*SS(K)*DELT*SALTA+SCM
4315 SALT=0
4320 FOR I=2 TO K
4325 SD(I)=SE(I)*W(I)*(DD(I+1)-DD(I-1))*SALTA*.5
4330 SALT=SD(I)+SALT
4332 NEXT I
4335 SD(KK)=SE(KK)*W(KK)*(DD(KK)-DD(K))*.5*SALTA
4345 IF WFDD>0 THEN SCMS=SE(1)*WFDD*DELT+SCMS
4390 REM PRINT#2,"    CUMS    CWF    CUMB    SCMS    SALT    SCM "
4395 REM PRINT#2,CUMS,CWF,CUMB,SCMS,SALT,SCMX
4410 TIME=TIME+DELT
5010 IF ABS(SUM3-0)>0.0001 THEN 5040
5020 DELT=3*DELT
5030 GOTO 5130
5040 TW=ABS(CONQ*DELT/SUM3)
5050 IF TW>=.1*DETT THEN 5080
5060 TW=.1*DETT
5070 GOTO 5100
5080 IF TW<=1000*DETT THEN  5100
5090 TW=1000*DETT
5100 IF TW>2*DELT THEN  5020
5110 DELT=TW
5120 REM TEST TO SEE IF EVAP OR RAIN ETC. HAS CHANGED
5130 IF IDELT=1 THEN DELT=DELT1
5140 IDELT=0
5150 IF DELT<DETT THEN DELT=DETT
5160 IF DELT>6 THEN DELT=6
5170 IF TIME-V(KC+1)<0 THEN 5300
5180 WATBAL=SIR+EVAP-RUNOF-CUMB-CWF+SUMA
5190 PRINT#2,"    TIME    CWF    IR+RA    TRAN    CUMB    EVAP    WATBAL    HROOT    SCMS    SALT    SCMX"
5200 PRINT#2, USING "####.### ##.### ##.### ##.### ##.### ##.### ##.### ######. ##.## ####. ##.##";TIME,CWF,SIR,SUMA,CUMB,EVAP,WATBAL,HROOT,SCMS,SALT,SCMX
5210 PRINT#2, "   WATER CONTENT VS DEPTH"
```

```
5220 FOR I=1 TO KK:PRINT#2, USING " #.### ";W(I);:NEXT I
5230 PRINT#2,
5232 PRINT#2, "  SALT CONCENTRATION VS DEPTH"
5234 FOR I=1 TO KK:PRINT#2, USING " ###.## ";SE(I);:NEXT I
5236 PRINT#2,
5240 EOR=V(KC+2)
5250 IR=INT((KC+2)/2)
5255 SE(1)=SF(IR+1)
5257 ER=EOR
5258 ET=TET(IR+1)
5260 KC=KC+2
5270 MTIME=0
5280 DELT=DETT
5290 GOTO 5320
5300 IF (TIME+DELT)<=V(KC+1) THEN   5320
5310 DELT=V(KC+1)-TIME
5320 LL=LL+1
5330 IF V(KC)>0 THEN 5690
5340 LTIME=INT(TIME/24)
5350 TIMEL=LTIME
5360 TIMEA=TIME/24-TIMEL
5370 LTIME=INT((TIME+DELT)/24)
5380 TIMEL=LTIME
5390 TIMED=(TIME+DELT)/24-TIMEL
5400 IF TIMED<TIMEA THEN 5620
5410 IF .5-TIMEA<.0001 THEN 5620
5420 IF TIMED<.5 THEN 5470
5430 TIMED=.5
5440 DELT1=DELT
5450 IDELT=1
5460 DELT=(.5-TIMEA)*24
```

REFERENCES

Babcock, K.L., and R.K. Schultz. 1963. Effect of anions on the sodium-calcium exchange in soils. Soil Sci. Soc. Am. Proc. 27:630–632.

Bresler, E. 1973. Simultaneous transport of solutes and water under transient unsaturated flow conditions. Water resour. Res. 9:975–986.

Bresler, E., B.L. McNeal, and D.L. Carter. 1982. Saline and sodic soils (Principles-Dynamics-Modeling). Springer-Verlag, New York.

Childs, S.W., and R.J. Hanks. 1975. Model of soil salinity effects on crop growth. Soil Sci. Soc. Am. Proc. 39:617–622.

Dudley, L.M., R.J. Wagenet, and J.J. Jurinak. 1981. Description of soil chemistry during transient solute transport. Water Resour. Res. 17:1498–1504.

Dutt, G.R., M.J. Shaffer, and W.J. Moore. 1972. Computer simulation model of dynamic biophysicochemical processes in soils. Univ. Arizona Agric. Exp. Stn. Tech. Bull. 196.

Griffin, R.A., and J.J. Jurinak. 1973. Estimation of activity coefficients from the electrical conductivity of natural aquatic systems and soil extracts. Soil Sci. 116:26–30.

Jury, W.A., W.M. Jarrell, and D. Devitt. 1979. Reclamation of saline-sodic soils by leaching. Soil Sci. Soc. Am. J. 43:1100–1106.

McNeal, B.L., J.D. Oster, and J.T. Hatcher. 1970. Calculation of electrical conductivity from solution composition data as an aid to in-situ estimation of soil salinity. Soil Sci. 110:405–414.

Nakayama, F.S. 1969. Theoretical consideration of the calcium sulfate-bicarbonate-carbonate interrelation in soil solution. Soil Sci. Soc. Am. Proc. 33:668–672.

Oster, J.D., and G. Sposito. 1980. The Gapon coefficient and the exchangeable sodium percentage—sodium adsorption ratio relation. Soil Sci. Soc. Am. J. 44:258–260.

Robbins, C.W. 1984. Sodium adsorption ratio-exchangeable sodium percentage relationships in a high potassium saline-sodic soil. Irrig. Sci. 5:173–179.

Robbins, C.W. 1985. The $CaCO_3$-CO_2-H_2O system in soils. J. Agron. Educ. 14:3–7.

Robbins, C.W. 1986a. Carbon dioxide partial pressure in lysimeter soils. Agron. J. 78:151–158.

Robbins, C.W. 1986b. Fluoride adsorption by a saline sodic soil irrigated with a high F water. Irrig. Sci. 7:107–112.

Robbins, C.W., and D.L. Carter. 1983. Selectivity coefficients for calcium-magnesium-sodium-potassium exchange in eight soils. Irrig. Sci. 4:95–102.

Robbins, C.W., J.J. Jurinak, and R.J. Wagenet. 1980a. Calculating cation exchange in a salt transport model. Soil Sci. Soc. Am. J. 44:1195–1200.

Robbins, C.W., R.J. Wagenet, and J.J. Jurinak. 1980b. A combined salt transport-chemical equilibrium model for calcareous and gypsiferous soils. Soil Sci. Soc. Am. J. 44:1191–1194.

Robbins, C.W., and L.S. Willardson. 1980. An instrumented lysimeter system for monitoring salt and water movement. Trans. ASAE 23:109–111.

Shainberg, I., J.D. Rhoades, and R.J. Prather. 1981a. Effect of low electrolyte concentration on clay dispersion and hydraulic conductivity of a sodic soil. Soil Sci. Soc. Am. J. 45:273–277.

Shainberg, I., J.D. Rhoades, D.L. Suarez, and R.J. Prather. 1981b. Effect of mineral weathering on clay dispersion and hydraulic conductivity of sodic soils. Soil Sci. Soc. Am. J. 45:287–291.

Sposito, G. 1977. The Gapon and the Vanselow selectivity coefficients. Soil Sci. Soc. Am. J. 41:1205–1206.

Sposito, G., and S.V. Mattigod. 1977. On the chemical foundation of the sodium adsorption ratio. Soil Sci. Soc. Am. J. 41:323–329.

Stumm, W., and J.J. Morgan. 1970. Aquatic chemistry. John Wiley and Sons, Inc., New York.

Tanji, K.K. 1969. Solubility of gypsum in aqueous electrolytes as affected by ion association and ionic strengths up to 0.15M and at 25°C. Environ. Sci. Technol. 3:656–661.

Tracy, P.W., C.W. Robbins, and G.C. Lewis. 1984. Fluorite precipitation in a calcareous soil irrigated with high fluoride water. Soil Sci. Soc. Am. J. 48:1013–1016

Van Luik, A., and J.J. Jurinak. 1979. Equilibrium chemistry of heavy metals in concentrated electrolyte solution. p. 683–710. *In* E.A. Jenne (ed.) Chemical modeling in aqueous systems. Am. Chem. Soc. Symp. Ser. 93. Miami Beach, FL. 11–13 Sept. 1978. Am. Chem. Soc., Washington, DC.

17 Soil Heat Flow

ROBERT HORTON AND SANG-OK CHUNG
Iowa State University
Ames, Iowa

Heat flow modeling offers two major contributions to soil and crop sciences as it: (i) describes energy partitioning at the soil surface; and (ii) describes soil temperature distribution. The surface-energy partitioning and resulting soil temperature distribution directly and indirectly affect plant growth and development, and many of the biological, chemical, and physical soil processes.

Three mechanisms, radiation, convection, and conduction, are responsible, simultaneously, for the transfer of heat in soil. Radiative energy transfer includes incoming direct and diffuse shortwave solar radiation and longwave sky radiation to the soil surface, and longwave radiation emitted outward from the soil surface. It is generally more important at the soil surface than it is below the soil surface.

Convective energy transfer in porous media is associated with a net flux of fluids. Jackson (1960), Wierenga et al. (1970), and O'Neill (1979) studied convective heat transfer associated with liquid water flow in soil, and Westcot and Wierenga (1974) studied convective heat transfer associated with vapor flow in soils. Although convection may be responsible for a major portion of the soil heat transfer during periods of large moisture flux (e.g., during rainfall or irrigation) or when vapor movement occurs when the soil heat flux density is small, conduction is generally the mechanism most responsible for subsurface soil heat transfer.

Philip and De Vries (1957) presented a theory to describe coupled water and heat flow in soil. Van Bavel and Hillel (1975, 1976), Sophocleous (1979), Milly (1982), Bristow et al. (1986), and Nassar & Horton (1989a, b) expanded and/or used the theory to calculate heat and water flow in soil. In many cases, heat flow by radiation, convection, and conduction is modeled by the conduction equation alone (Hanks et al., 1971; Wierenga & De Wit, 1970; Gupta et al., 1981; Horton et al., 1984a, b; Parton, 1974; Persaud & Chang, 1984). Apparent thermal properties rather than real thermal properties (Jackson & Kirkham, 1958) are assumed to account for both conductive and nonconductive heat flow. Many investigators have modeled soil heat transfer in two dimensions (Takakura et al., 1971; Jury & Bellantuoni, 1976a, b; Davis, 1977; Mahrer & Katan, 1981; Mahrer, 1982; Horton et al., 1984a, b; Horton, 1989; Benjamin et al., 1990; Kluitenberg & Horton, 1990).

Copyright © 1991 ASA-CSSA-SSSA, 677 S. Segoe Rd., Madison, WI 53711, USA. *Modeling Plant and Soil Systems*—Agronomy Monograph no. 31.

The purpose of this chapter is to present a relatively simple model that calculates temperatures in the soil profile.

I. HEAT FLOW MODEL

Soil heat transfer is frequently considered as a one-dimensional (vertical) process; however, certain circumstances cause two- or three-dimensional soil heat flow to occur (e.g., incomplete surface cover by plants or mulches, buildings, pipelines or other buried heat sources, and ridges or other nonuniformities in surface configuration). In this chapter, a one-dimensional vertical heat flow model is developed.

An equation describing conductive heat transfer in a vertical one-dimensional system is:

$$C \frac{\partial T}{\partial t} = \frac{\partial}{\partial z} \left(\lambda \frac{\partial T}{\partial z} \right) \qquad [1]$$

where C is the volumetric heat capacity (J/°C/m^3), T is the temperature (°C), t is the time(s), z is the depth (positive downward, m), and λ is the thermal conductivity (W/m/°C). In the presence of a temperature gradient in a moist soil, heat transfer takes place by convection in addition to conduction. For the present case, Eq. [1] is used to describe heat transfer by making λ the apparent thermal conductivity rather than the real thermal conductivity; thus, convective heat transfer is not explicitly modeled. In a moist soil, λ depends on the soil water content, which can vary with both depth and time. An equation describing water movement in a one-dimensional medium is:

$$F \frac{\partial h}{\partial t} = \frac{\partial}{\partial z} \left[K(h) \left(\frac{\partial h}{\partial z} - 1 \right) \right] + S(z,t) \qquad [2]$$

where F is the specific water capacity, $d\theta/dh$, where θ is the volumetric soil water content, h is the soil water pressure head, K is the hydraulic conductivity, S is a source or sink term, and z is the vertical distance positive downward (as discussed in Ch. 11).

Initial and boundary conditions must be specified before Eq. [1] and [2] can be solved. Boundary conditions can be either a value-specified or flux condition. A value-specified condition can be a constant, explicitly or implicitly determined. A flux condition is more difficult to handle than a value-specified condition in numerical modeling, but a flux condition is often required when input data are limited.

In this model, an energy partitioning method can be used to determine the thermal and hydraulic upper boundary conditions. The surface-evaporative flux depends, in part, on the soil surface temperature, and the soil surface temperature is likewise influenced in part by the evaporative flux. Thus, coupled-heat and water-flow surface boundary conditions are determined implicitly by solving a set of temperature-dependent equations. The equations used here are physically based and describe the partitioning of energy at the soil surface.

II. ENERGY PARTITIONING AT THE SOIL SURFACE

The energy balance at the soil surface is described by:

$$R_n - H_s - LE - G = 0 \qquad [3]$$

where R_n is net radiation (positive downard), H_s is sensible heat flux (positive upward), LE is latent heat flux (positive upward), and G is soil heat flux (positive downward). The value of R_n can be estimated as:

$$R_n = (1 - a) R_g + R_1 - \epsilon \sigma (T_s + 273.16)^4 \qquad [4]$$

where R_g is the measured global radiation (W/m^2), R_1 is the longwave sky irradiance (W/m^2), T_s is the soil surface temperature (°C), a is the soil surface albedo, ϵ is the emissivity of the soil, and σ is the Stefan-Boltzmann constant (W/°K^{4}m^2).

The longwave sky irradiance, R_1, can be estimated following Van Bavel and Hillel (1976) from the following form of Brunt's formula:

$$R_1 = \sigma (T_a + 273.16)^4 [0.605 + 0.048 (1370 \, HA)^{0.5}] \qquad [5]$$

where T_a the air temperature (°C), and HA the air humidity (kg/m^3), are measured inputs.

Measured values of global radiation are used for direct computation of net radiation on surfaces in direct sunlight. An empirical equation from Horton et al. (1984b) can be used to estimate R_{gs}, which is the R_g received by the portion of the soil surface shaded by a crop canopy.

The latent and sensible heat fluxes at the surface are calculated by using the equations:

$$E = (HS - HA)/(1000 \, r_a) \qquad [6]$$

$$L = 2.49463 \times 10^9 - 2.247 \times 10^6 \, T_s \qquad [7]$$

$$LE = L \times E \qquad [8]$$

$$H_s = (T_s - T_a) SA/r_a \qquad [9]$$

where E is the evaporative flux (m/s), L is the latent heat of vaporization (J/m^3) (Forsythe, 1964), r_a is the aerodynamic boundary layer resistance (s/m), SA is the volumetric heat capacity of air (J/°C m^3), and HS is the humidity of air at the soil surface (kg/m^3).

The aerodynamic boundary layer resistance, r_a, is calculated according to Van Bavel and Hillel (1976) as

$$r_a = [\ln(2.0/z_0)]^2/0.16 \, WS \qquad [10]$$

where z_0 is the roughness length (m) and WS the wind speed (m/s).

The humidity of the soil surface, HS, is calculated from the equation

$$HS = HO \exp[h/46.97 \, (T_s + 273.16)] \qquad [11]$$

where HO is the saturated humidity at the soil surface (kg/m^3), and h is the soil water pressure head (m).

The saturated humidity, HO, of the air at the soil surface is calculated by using the following equation suggested by Murray (1967):

$$HO = 1.323 \exp[17.27 \, T_s/(237.3 + T_s)]/(T_s + 273.16) \qquad [12]$$

The soil heat flux density, G, at the soil surface can be calculated by

$$G = -\lambda \frac{\partial T}{\partial z} \qquad [13]$$

where λ is the thermal conductivity for the soil surface (W/m °C), and $\partial T/\partial z$ is the vertical soil temperature gradient at the soil surface.

The variables R_n, LE, H_s, and G are all functions of the unknown soil surface temperature. Before the soil surface temperature can be calculated, the right-hand side of Eq. [13] must be estimated numerically. The following approximation (Chung & Horton, 1987) is used in the present model,

$$G = -\lambda \left[\frac{T_2 - T_s}{\Delta z}\right] + (T_s - T_1) \, C \, \Delta z/(2\Delta t) \qquad [14]$$

where T_s is the surface temperature for the present time step, T_1 is the surface temperature from the previous time step, T_2 is the soil temperature from the previous time step for the node at vertical position 2, Δz is the vertical spatial increment, C is the volumetric heat capacity for the soil surface layer, and Δt is the time step increment. Equation [14] approximates the soil heat flux density by adding a term that estimates soil heat flux at a depth of $\Delta z/2$ and a term that estimates the change in heat stored in the soil above $\Delta z/2$. The value of T_s is solved at each numerical time step by Eq. [3] (by using Eq. [4], [8], [9], and [14] to describe R_n, LE, H_s, and G, respectively), using a bisector root-finding algorithm (James et al., 1977).

The predicted soil surface temperature, T_s, from the energy balance partitioning is used as the upper-boundary condition. The soil-water evaporative flux is also predicted at each time step as a result of the surface energy partitioning.

III. FINITE DIFFERENCE EQUATIONS

Either explicit or implicit finite difference methods can be used to solve Eq. [1] and [2] numerically. The explicit, finite difference approximations to the partial differential equations are easily programmed for computer solution. However, the stability criteria sometimes require small incremental time steps and, thus, large amounts of computation time. The implicit finite-difference method does not have stability criteria and generally requires less computer time when modeling heat and water flow in soils. This is the case even though programming the implicit method is more difficult, conceptually, than the straightforward explicit method. The computer program used in this chapter is based on the implicit finite-difference method. The implicit, finite difference equation for Eq. [1] is

$$\frac{T_i^{n+1} - T_i^n}{\Delta t} = \frac{\alpha_{i+1/2}^{n+1/2}(T_{i+1}^{n+1} - T_i^{n+1}) - \alpha_{i-1/2}^{n+1/2}(T_i^{n+1} - T_{i-1}^{n+1})}{(\Delta z)^2} \quad [15]$$

where i is the depth index, n is the time step index, α is the thermal diffusivity, which is thermal conductivity divided by volumetric heat capacity, and z is the depth that increases downward. The internodal thermal diffusivities can be estimated by either the arithmetic mean or geometric mean of the diffusivities at the two adjacent nodes. In this study, the arithmetic mean is used. In Eq. [15], the unknown temperature at node i at time $(n+1)$ is expressed in terms of unknown temperatures at time $(n+1)$. Therefore, we cannot solve for T_i^{n+1} explicitly. One implicit equation can be generated at each node by Eq. [15]. Then, we have a system of equations that must be solved simultaneously. The system of equations provide a tridiagonal matrix, which can be solved efficiently by using a particular form of the Gaussian elimination method known as the Thomas algorithm (Remson et al., 1971). The finite difference equation for water flow (Eq. [2]) can be formulated similarly. The solution process is the same as for the heat flow. A finite difference scheme for water flow allowing either implicit or explicit forms of the equation is shown in Ch. 9.

IV. COMPUTER PROGRAM DESCRIPTION

The principal structure and logical components of the computer program for calculating heat and water flow are described by a flow chart, as shown in the Appendix. The computer program is written in FORTRAN 77. A short user manual and complete program listing is also provided in the Appendix.

The model uses the implicit, finite difference method to solve the soil heat-flow equations as described in the previous section. Users can choose appropriate indicators from several options for flow type (either heat flow only, or heat and water flow), soil type, and boundary condition type. When heat flow is considered alone, the model skips over the water flow portion of the program. Measured values, or explicit or implicit expressions, can be used to describe the boundary conditions. A user must provide input information for the upper boundary conditions. For explicit, boundary-condition input data, a user must provide the temperature or water content values and the total number of inputs per day. With this input, the program uses a step function approach to adjust the boundary values to constant time steps throughout a day. If a user has daily maximum and minimum surface-temperature values available, then a sine function option can be used to describe the soil surface boundary condition. The surface energy partitioning method can be used to determine the soil surface temperature and evaporation rate if weather input data are available. The required weather inputs for this program are daily global radiation, maximum and minimum daily air temperatures, maximum and minimum daily dewpoint temperatures, average daily wind speed, and rainfall. The weather inputs are assumed to be measured at a height of 2 m above the ground. When a user chooses the surface-energy partitioning method to describe the upper-boundary condition for heat flow, the choice also provides the upper-boundary water flow condition because of the energy conservation requirement.

When the energy partitioning method is used for the boundary condition, the soil surface temperature is assumed to be equal to the air temperature during a rainfall period. When a flux boundary condition, either input explicitly or from the energy partitioning method, is used for water flow at the surface, the flux can be assumed to be a source or sink term, as described in Campbell (1985). The model is set up for bare soil with no plants. The model can be adapted to include plants similar to the method of Lascano et al. (1987) or by explicitly specifying soil surface evaporative flux and net radiation within the present code.

Soil parameters required as inputs include soil surface emissivity, soil surface albedo, soil thermal conductivity, soil volumetric heat capacity, soil water capacity, and hydraulic conductivity, all as functions of water content, roughness length, and the soil-water characteristic curve.

A few general inputs are also required to run the computer program. These inputs include solar noon, daylength, time length of simulation, Δz, and Δt.

The weather inputs are used in conjunction with empirical expressions to describe weather conditions as a function of time. Global radiation as a function of time for soil in direct sunlight is described as

$$\text{GR} = (\text{DGR}/\text{DL}) \sin[(t - \text{SN} + \text{DL}/2) \pi/\text{DL}] \qquad [16]$$

where GR is global radiation (W/m^2), DGR is the input daily global radiation (J/m^2), DL is the input daylength (s), SN is the input solar noon time (s), and t is the time of day (s) (Van Bavel & Lascano, 1979, unpublished data). Air temperature, T_a, as a function of time is described as

$$T_a = [(T_{max} + T_{min})/2] + [(T_{max} - T_{min})/2] \sin(\omega t + \pi) \qquad [17]$$

where T_{max} and T_{min} are the input daily maximum and minimum air temperature values (°C), and ω is the angular frequency (7.272 × 10^{-5} rad/s). The dewpoint temperature, DPT, as a function of time is described as

$$DPT = [(DP_{max} + DP_{min})/2] + [(DP_{max} - DP_{min})/2] \sin(\omega t + \pi) \qquad [18]$$

where DP_{max} and DP_{min} are the input daily maximum and minimum dewpoint temperature values (°C). The wind speed is assumed constant throughout the day.

The computer program is presently set up to describe all required soil parameters with equations. Soil surface emissivity, ϵ, follows that used by Van Bavel and Hillel (1976),

$$\epsilon = 0.9 + 0.18\theta \qquad [19]$$

Soil surface albedo, a, follows that used by Van Bavel and Hillel (1976),

$$a = \begin{cases} 0.35 - \theta & ; \quad 0.10 < \theta < 0.25 \\ 0.10 & ; \quad 0.25 < \theta \\ 0.25 & ; \quad \theta < 0.10 \end{cases} \qquad [20]$$

Soil thermal conductivity is assumed to be related to water content by the empirical expression,

$$\lambda = \beta_1 + \beta_2\theta + \beta_3\theta^{0.5} \qquad [21]$$

where β_1, β_2, and β_3 are empirical parameters that are specified for a particular soil. The soil volumetric heat capacity (J/°C m^3) is determined following De Vries (1963) as

$$C = (1 - \theta_s)\, 1.92 \times 10^6 + 4.18 \times 10^6\, \theta \qquad [22]$$

where θ_s is the input saturated water content for a specific soil. Soil water characteristics, hydraulic conductivity, and specific water capacity are described by empirical equations presented by Van Genuchten (1980) as

$$\theta = \theta_r + (\theta_s - \theta_r) \left[\frac{1}{1 + (\alpha_1 h)^n}\right]^{1 - \frac{1}{n}} \qquad [23]$$

$$K(h) = K_s \frac{\{1 - (\alpha_1 h)^{n-1} [1 + (\alpha_1 h)^{\frac{1-n}{n}}]\}^2}{[1 + (\alpha_1 h)^n]^{\frac{n-1}{2n}}} \qquad [24]$$

$$F(\theta,h) = (n - 1)(\theta - \theta_r)\left[1 - \left(\frac{\theta - \theta_r}{\theta_s - \theta_r}\right)^{\frac{n}{n-1}}\right]/h \quad [25]$$

where θ_s and θ_r are saturated and residual water content, respectively, K_s is saturated hydraulic conductivity, h is the absolute value of the pressure head, and α_1 and n are nonlinear regression parameters describing the shape of the soil water characteristic curve.

Users can specify one of three representative soil types (clay, loam, or sand) for which thermal and hydraulic parameters are preprogrammed. For the three representative soil types, previously reported data are used to describe the thermal and hydraulic properties. The parameters in the thermal conductivity equation (Eq. [21]) are determined from published data of De Vries (1963, Table 7.6), Wierenga et al. (1969), and Horton and Wierenga (1984) for clay, loam and sand soils, respectively. The thermal properties of the three representative soils are shown in Fig. 17-1. Soil water retention data for clay, loam, and sand soils are taken from Hillel and Van Bavel (1976). Users also can choose to input thermal and hydraulic parameters for the specific soil of their interest.

V. TEST OF MODEL

The performance of a numerical model should be evaluated to examine its validity because any numerical scheme may introduce instability, truncation, and round-off errors. A model is valid only if the approximate solution is satisfactorily accurate or close to the exact solution, if one exists. The accuracy of a model is also dependent on its convergence and stability.

Convergence is satisfied when the approximate solution approaches the exact solution while step sizes of the spatial and temporal discretization approach zero. A model is said to be stable if the amplification of the error is restricted or has a finite limit as computation marches forward in time. The validity of a model can be tested by comparing the numerical solution with either an analytical solution, if it is available, or observed data. We will test the validity of this model in both ways.

The analytical solution used here is for heat transfer in semi-infinite, homogeneous soil with a periodic surface heat flux and surface temperature. The governing equation and initial boundary conditions for this problem are:

$$\frac{\partial T}{\partial t} = \alpha \frac{\partial^2 T}{\partial z^2} \quad [26]$$

$$T(z,0) = \overline{T} + A \exp[-z(\omega/2\alpha)^{1/2}] \sin[\phi - z(\omega/2\alpha)^{1/2}] \quad [27]$$

$$T(0,t) = \overline{T} + A \sin(\omega t + \phi) \quad [28]$$

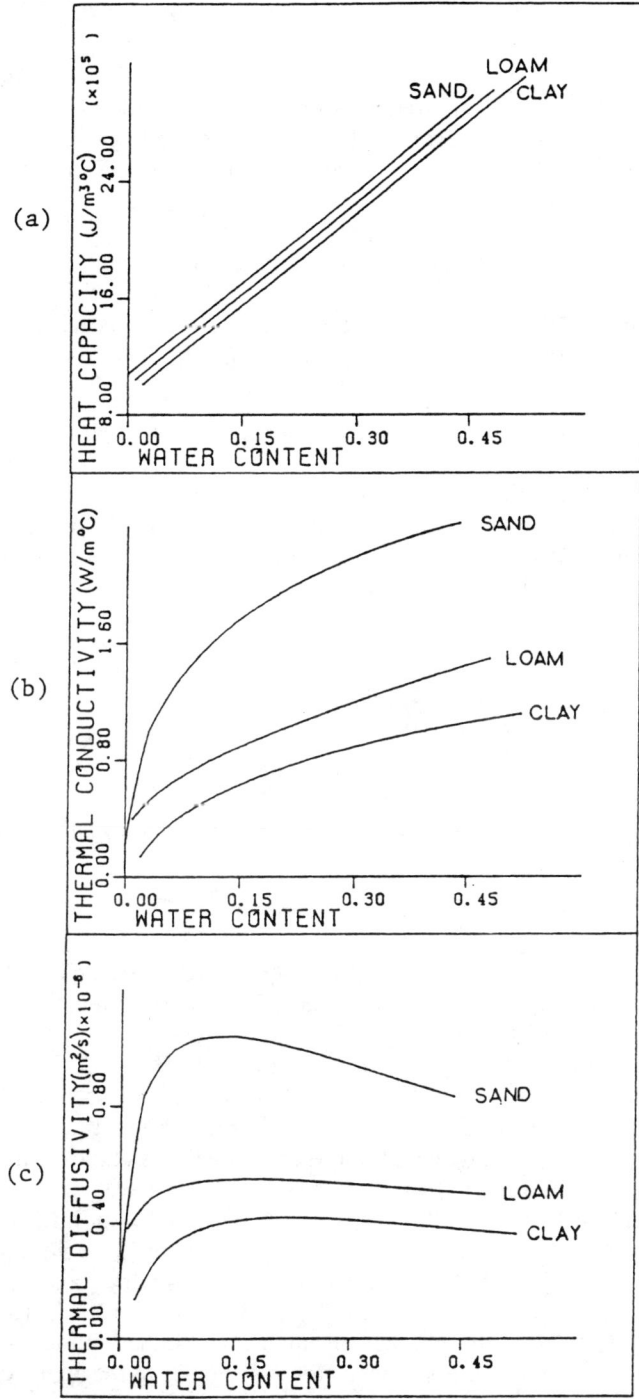

Fig. 17-1. The soil thermal properties as functions of water content for representative soil types: volumetric heat capacity (a), thermal conductivity (b), and thermal diffusivity (c).

$$T(z,t) = \overline{T} \quad [29]$$
$$\lim_{z \to \infty}$$

where $\overline{T}$ is daily average soil temperature, A is amplitude of daily temperature variation, z is depth from soil surface, ω is the angular frequency, α is the thermal diffusivity, and ϕ is time shift. The solution for this problem is given by Van Wijk and De Vries (1963) as

$$T(z,t) = \overline{T} + A \exp[-z(\omega/2\alpha)^{1/2}] \sin[\omega t + \phi - z(\omega/2\alpha)^{1/2}] \quad [30]$$

The soil surface heat flux can be determined by differentiating Eq. [30] with respect to z and multiplying by $-\lambda$, the thermal conductivity, to obtain, at $z = 0$,

$$G(0,t) = A C (\omega\alpha)^{1/2} \sin[\omega t + \phi + \pi/4] \quad [31]$$

where C is volumetric heat capacity.

The test of the energy balance equation to predict soil surface temperature was performed by using simple sine functions for the net radiation, R_n, sensible heat flux, H_s, and latent heat flux, LE, as

$$R_n = 500 \sin(\omega t) \quad [32]$$

$$H_s = 200 \sin(\omega t) \quad (33)$$

$$LE = 200 \sin(\omega t) \quad [34]$$

Then, from the energy balance equation, Eq. [3], the exact soil heat flux is described by

$$G = 100 \sin(\omega t) \quad [35]$$

The soil heat flux and the soil surface temperature were calculated numerically with the energy balance equation, and the subsurface temperatures were calculated numerically using Eq. [15]. Input values selected were $\alpha = 4.0 \times 10^{-7}$ m^2/s, $C = 2.0 \times 10^6$ J/m^3 °C, $\lambda = 0.8$ W/m °C, $\overline{T} = 20$°C, $A = 9.27$°C, and $\omega = 7.27 \times 10^{-5}$ s^{-1}. These values were selected to make the coefficient in Eq. [31] equal to 100, as in Eq. [35], so that a valid comparison between the analytical and numerical methods could be made. The initial condition and the bottom boundary condition based on Eq. [27] and [29], together with $\Delta z = 0.05$ m and $\Delta t = 300$ s, are used in the computation. Results of the numerical method were compared with the exact solutions. Figure 17-2 shows the comparison of soil surface and subsurface temperatures by the numerical method and by the analytical approach. The soil surface heat flux and temperature values, determined numerically using the surface energy balance approach, were very similar to the analytical values obtained using Eq. [28] and [31]. The numerical subsurface temperature values predicted by using Eq. [15] agreed well with the analytical values of

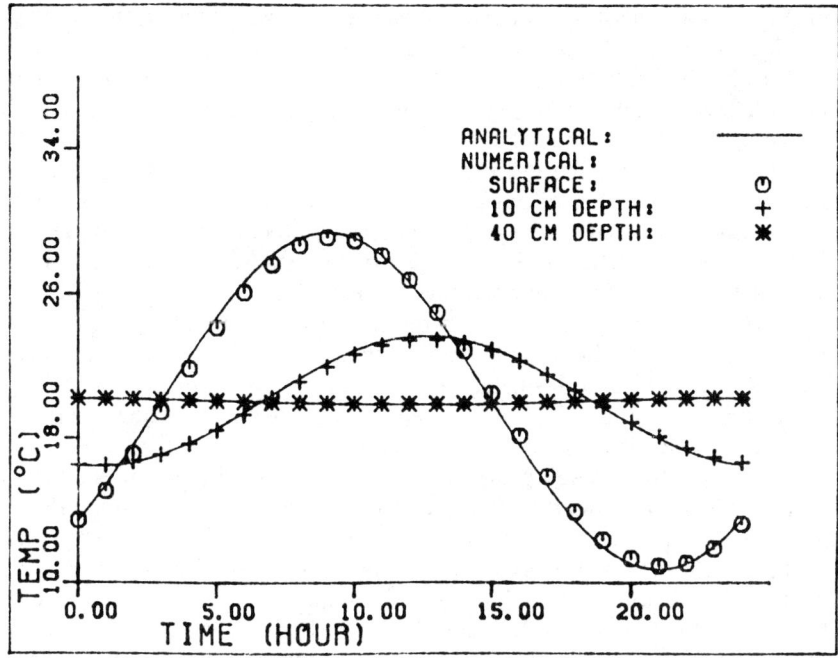

Fig. 17-2. Comparison between analytical (solid curve) and numerical (discrete points) soil temperature values. The space and time increments used for numerical calculations are 5 cm and 300 s, respectively.

Eq. [30]. This test indicates that the numerical heat flow model with the energy balance equations provides a very good approximation of the exact solution.

VI. COMPARISON BETWEEN MEASURED AND PREDICTED TEMPERATURES

A field site was selected and instrumented to obtain soil and weather observations in order to test the ability of this computer program to predict soil temperature. Field measurements of soil temperature were obtained from a soil profile with a bare soil surface located at the Iowa State University Agronomy and Agricultural Engineering Research Center near Ames, IA for the period 17 to 31 Aug. 1985. Appropriate weather data were also collected. The soil in the study area was classified as Nicollet clay loam (fine-silty, mixed, Mesic Aquic Hapludolls).

Soil bulk density, saturated hydraulic conductivity, and the soil-water retention curve were measured in the laboratory using undisturbed cores collected from the field site. The soil thermal-conductivity equation parameters were determined from field measurements using the line heat source method (De Vries, 1952). The measured thermal conductivity values for this site were very close to those for the representative loam soil. Therefore, the representative loam soil, thermal-conductivity parameters were used in the simulation. Weather data collected in the field included air temperature, wind speed, global radiation, and precipitation. Dewpoint temperatures were used that were collected at the Des Moines, IA airport, by the National Oceanic and Atmospheric Administration.

Thermocouples constructed of copper constantan wire were placed at various positions in the soil; at the soil surface, and at 2.5, 5.0, 7.5, 10.0, 15.0, and 30.0 cm below the surface. These thermocouples were connected to a Campbell Scientific CR-5 data logger (Campbell Scientific, Logan, UT) and hourly values of soil temperature were recorded.

In order to test the computer program, a 5-d simulation run was performed. Soil temperature predictions were compared with measured temperature values. Simulation started at 2400 h. The soil-surface boundary condition was determined by the soil-surface energy balance equation using observed weather inputs. A constant temperature was used as the lower boundary condition. A soil profile thickness of 0.6 m was used in the simulation. Table 17-1 shows the input parameter values used in the simulation. All computations were performed using a Zenith 158 personal computer.

Figure 17-3 shows the predicted cumulative net radiation, sensible heat, latent heat, and soil heat during the 5-d simulation of soil heat flow. The weather condition prior to 27 August consisted of two sunny days. On 29 August, a 1.9-cm rainfall occurred from 0230 to 0830 h. Responses to these weather conditions are represented in Fig. 17-3. For the first 2 d soil heat flow and sensible heat flow were quite large. The latent heat flow for these days was relatively small because of a dry soil surface layer. Rainfall on 29 August increased the surface water content, which resulted subsequently in larger latent heat flow (Fig. 17-3c), and reduced soil heat flow and sensible heat flow.

Figure 17-4 presents comparisons of measured and simulated soil temperature variations at various depths during the 5-d simulation. Figure 17-5 shows measured and predicted soil temperature profiles at selected times.

Table 17-1. Parameter values used in a 5-d simulation of soil heat flow.

Description	Computer variable	Input value
Time increment (s)	DELT	300.0
Soil layer thickness (m)	ZLENG	0.6
Solar noon (s)	SNOON	47 770
Daylength (s)	DAYL	48 170
Number of vertical nodes	N	13
Water retention parameter, α_1 (1/m)	AALPA	4.11
Water retention parameter, n	EN	1.28
Saturated hydraulic conductivity (m/s)	CONDS	1.9×10^{-6}
Roughness length (m)	ZO	0.01
Saturated water content (m^3/m^3)	THETAS	0.547
Residual water content (m^3/m^3)	THETAR	0.100
Coefficients for thermal conductivity	b_1	0.243
	b_2	0.393
	b_3	1.536

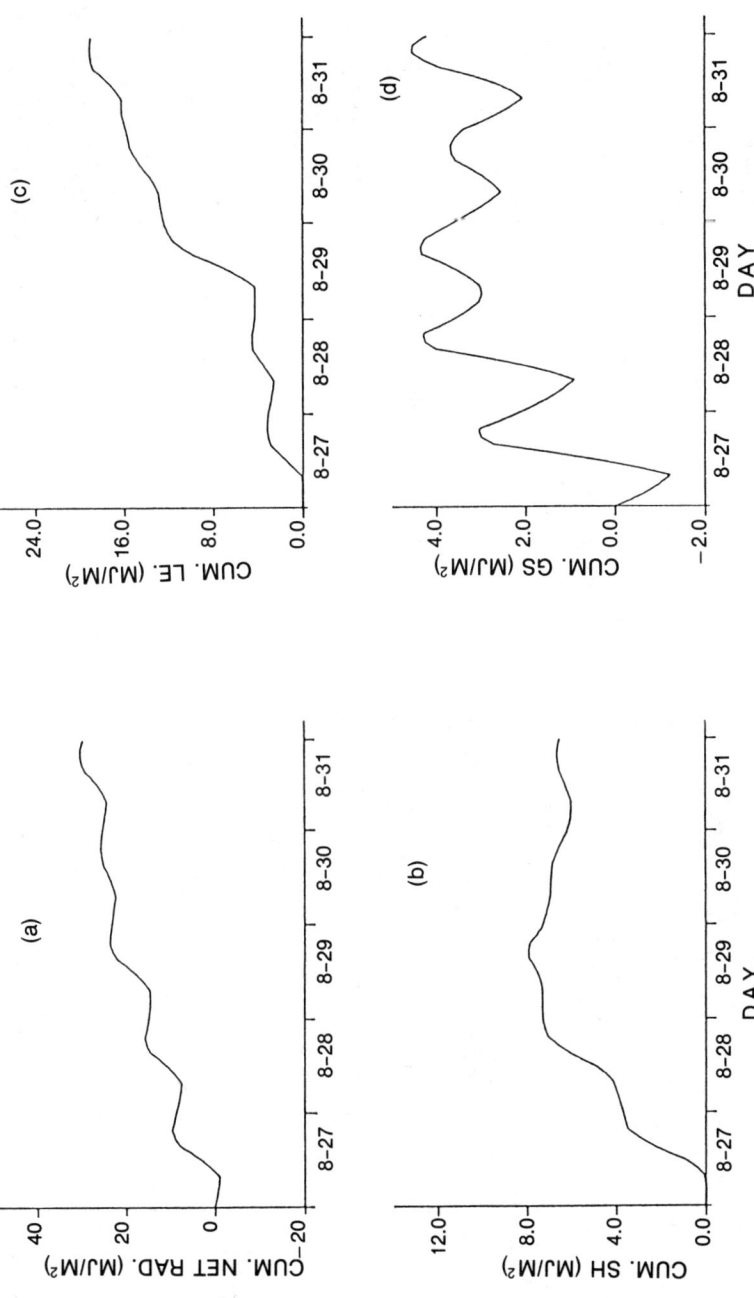

Fig. 17-3. Predicted cumulative net radiation (a), cumulative sensible heat (b), cumulative latent heat (c), and cumulative soil heat (d) during a 5-d simulation of soil heat flow.

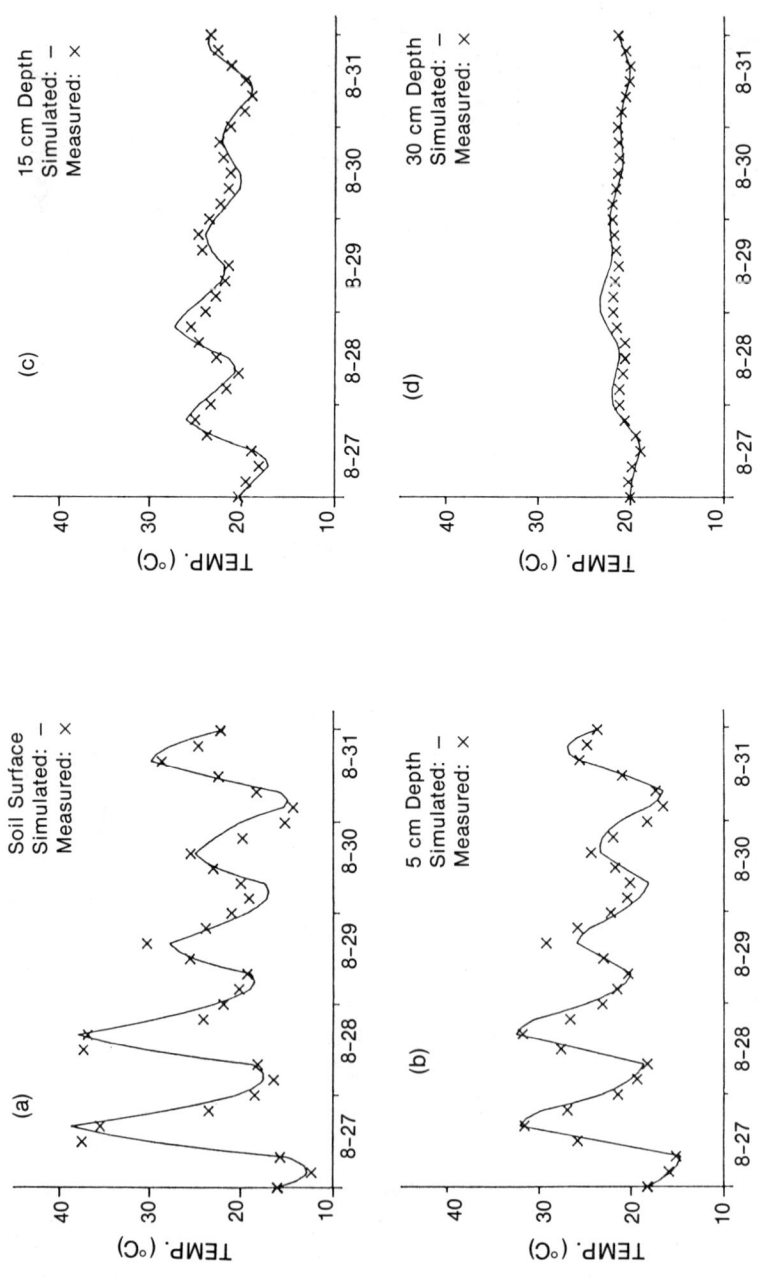

Fig. 17-4. Comparison between field-measured and predicted soil temperature at the surface (a), 5 cm (b), 15 cm (c), and 30 cm (d) during a 5-d simulation of soil heat and water flow. Surface energy partitioning was used.

SOIL HEAT FLOW

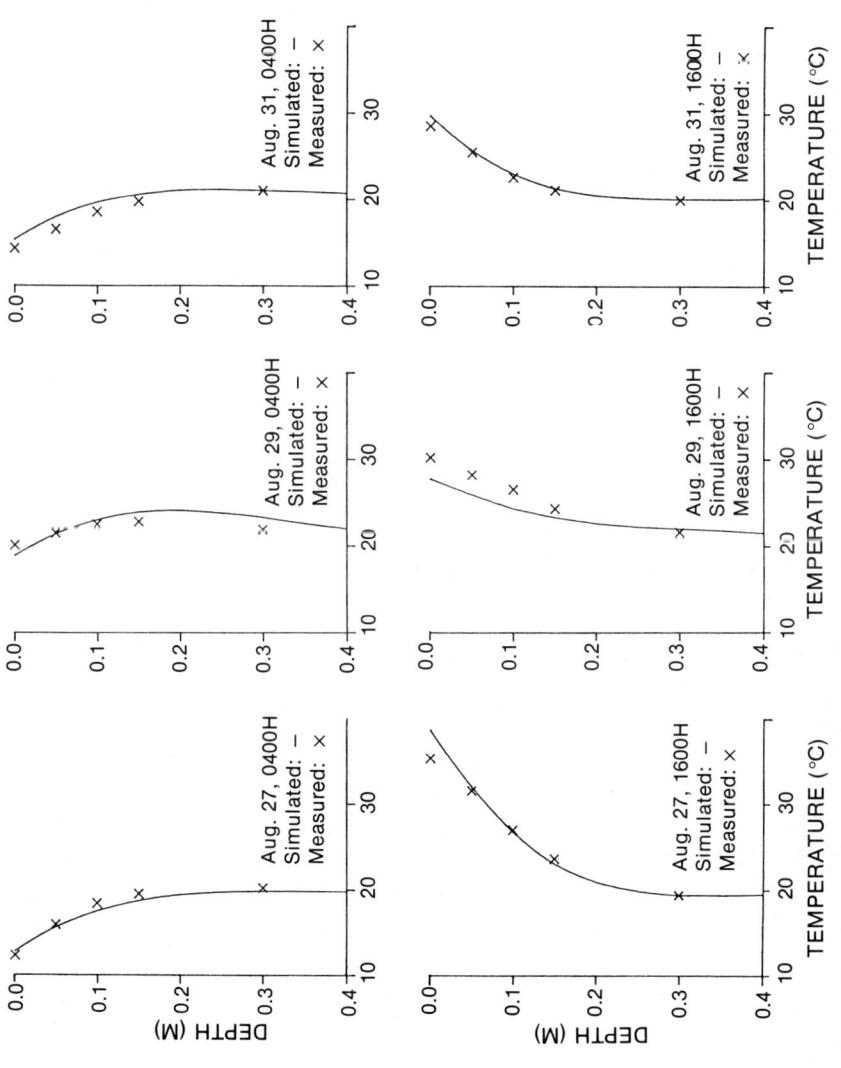

Fig. 17-5. Comparison between field-measured and predicted soil temperature profiles at selected times in July 1985.

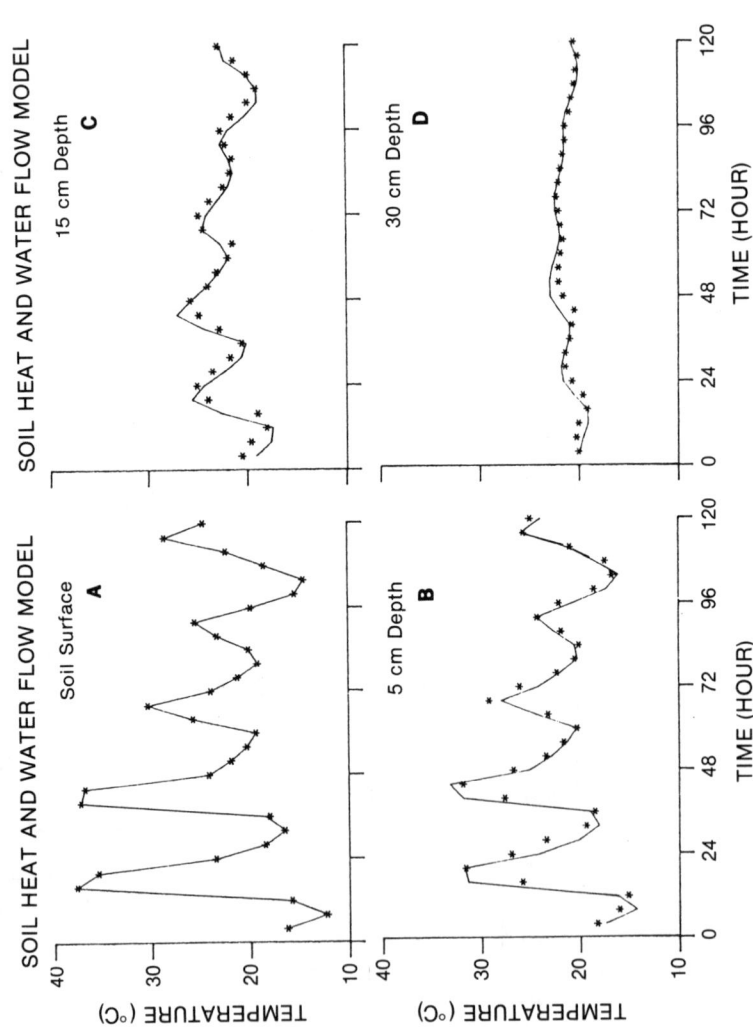

Fig. 17-6. Comparison between field-measured and predicted soil temperatures at the surface (a), 5 cm (b), 15 cm (c), and 30 cm (d) during a 5-d simulation of soil heat flow. Observed surface temperature values were used in the simulation.

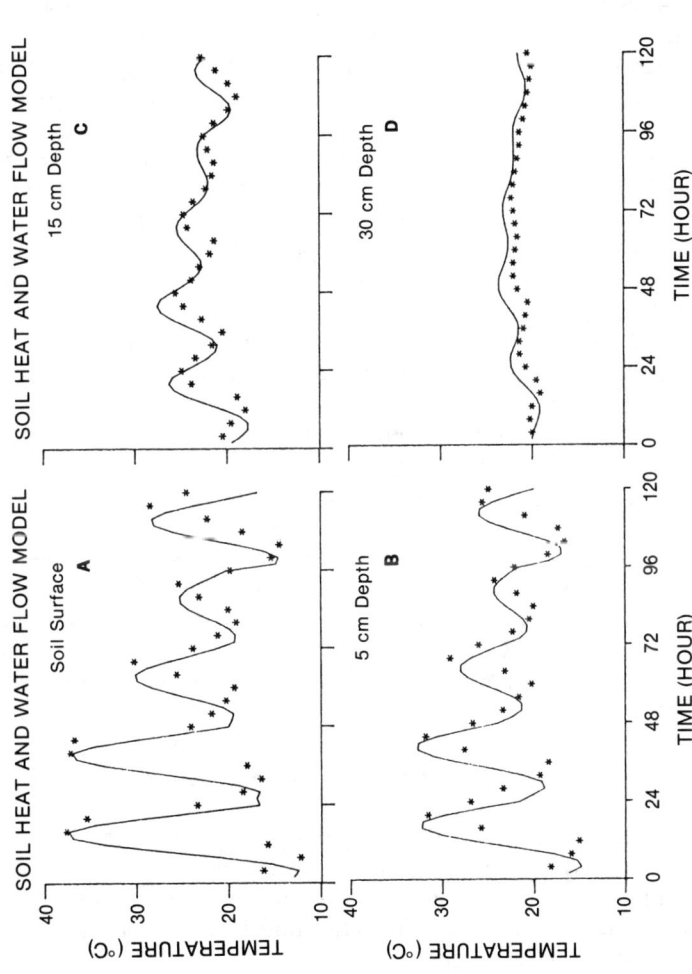

Fig. 17-7. Comparison between field measured and predicted soil temperatures at the surface (a), 5 cm (b), 15 cm (c), and 30 cm (d) during a 5-d simulation of soil heat flow. Daily maximum and minimum surface temperature values were used in the simulation.

Minimum daily surface temperature occurred near 0400 h and maximum temperature occurred near 1600 h. In general, the simulated temperatures are close to the measured temperatures. As expected, the daily temperature variation decreases as soil depth increases, as does the discrepancy between measured and simulated temperatures. In part, the differences between measured and simulated temperatures near the soil surface are caused by the sine function distributions of the weather parameters, which does not exactly represent, temporally, the upper boundary conditions. However the predicted soil temperature values were able to reflect the initial dry period, the rainfall period, and the ensuing drying period.

Although results are not presented, the computer program can provide estimates of soil water contents and potentials, infiltration, evaporation, drainage, and water balance.

Figure 17–6 presents comparisons of measured and simulated soil temperature variations at various depths during a 5-d simulation of soil heat flow without accounting for transient soil water flow. For this simulation, the surface boundary temperature values were supplied to the computer program. The results indicate that soil temperature can be predicted accurately without considering soil water flow if the upper boundary soil temperature is known.

Figure 17–7 presents comparisons of measured and simulated soil temperature variations at various depths during a 5-d simulation of soil heat flow without accounting for transient water flow. For this simulation the daily maximum and minimum surface temperature values were supplied to the computer program. The computer program used these maximum and minimum temperature values to estimate soil surface temperature at required times for the simulation. The predicted surface temperature tended to be out of phase with the observations. Likewise, the subsurface prediction of temperatures were out of phase with observations. Amplitudes of diurnal temperature variations were similar for observations and predictions at all depths. Thus, the heat flow model can provide useful estimates of soil temperature variation when daily maximum and minimum surface temperature values are available.

VII. CONCLUSION

A model for the transport of heat and water in soil has been presented. The most critical assumption to consider in using this model is that only one-dimensional flow is included. The model assumes that conductive heat transfer is dominate below the soil surface. Thus the model is most applicable to moist profiles with flat, bare surfaces.

The model was shown to be able to predict soil temperature quite well with a variety of input boundary information. Soil water transport should be considered when surface energy partitioning is desired in addition to prediction of soil temperature. Soil water transport need not be included when predicting soil temperatures, as long as some surface-boundary temperature values are available as input.

VIII. ACKNOWLEDGMENT

James Hamlett, graduate assistant, Iowa State Univ. (currently assistant professor of agricultural engineering, Penn State) made the soil temperature observations available.

IX. APPENDIX

A. User Manual

Title:	One-dimensional soil heat and water flow.
Programmer:	Sang-Ok Chung, Tai Song Tan, and Robert Horton, Department of Agronomy, Iowa State University
Date:	August 1986
Language:	FORTRAN77
Computer:	Zenith-158 personal computer or other IBM PC compatible with a mathematical processor.
Potential users:	Soil scientists and soil-water management specialists.
Description:	Determine soil temperature and water content in the soil profile. User can choose from several different options for flow type (heat flow only, or heat and water flow), soil type, and boundary condition input data type. Energy balance equation may be used to determine the soil surface boundary conditions. Implicit finite-difference method is used.
Algorithm:	The conductive soil heat flow equation and the pressure-head based, soil-water flow equation are used in this program. The vapor flow is not included, except at the soil surface. Computational procedure is described in detail in the text.
Modification:	This program can be easily modified for the main-frame computer. For different flow domain depths, the DIMENSION declaration and input/output FORMAT must be modified accordingly.
Input:	An input data file is needed in this program. However, one can change that to keyboard input format easily. The SI units are used for the input data. Input data are as follows (for description of I's see program listing):
Line 1	I1: Flow indicator
	I2: Soil type indicator
	I3: Input data type indicator
	I4: Temp. top B.C. indicator
	I5: Temp. bottom B.C. indicator
	I6: Water top B.C. indicator
	I7: Water bottom B.C. indicator

Line 2	Depth of flow region Step size in z-direction Roughness length, z_0.
Line 3	Simulation length in days Time step size in seconds (presently set for >40)
Line 4	Time interval for output in hours
Line 5	Length of daytime in hours Solar noon in hours
Line 6	Only if (I2 = 4) Coefficients in the soil thermal conductivity Eq. Saturated water content Residual water content Fraction of solid in soil Fraction of organic matter in soil
Line 7	Only if (I2 = 4) α_1, and n in Van Genuchten's (1980) retention equation Saturated hydraulic conductivity
Line 8	Temperature initial condition (can be more than one line)
Line 9	Pressure-head or water-content initial condition (for I1 = 2 only)
Line 10	Rainfall indicators for the simulation days (events per day)

Line 11	Top boundary condition for heat flow:
	IF (I4 = 1) number of inputs per day and next line has sequential temperature values
	IF (I4 = 2) daily maximum and minimum temperatures for a day
	IF (I4 = 3) weather data for each day

Line 12	Bottom boundary condition for heat flow:
	IF (I5 = 1) a constant temperature value
	IF (I5 = 2) no. of inputs per day and next line has sequential temperature values
	IF (I5 = 3) computer determines the flux condition

Line 13	Top boundary condition for water flow (I1 = 2 only):
	IF (I6 = 1) number of inputs per day and next line has sequential pressure head or water content values
	IF (I6 = 2) number of inputs per day and next line has sequential flux values in a day
	IF (I6 = 3) no data needed

Line 14	Bottom boundary condition for water flow (I1 = 2 only):
	IF (I7 = 1) number of inputs per day and next line has sequential pressure head or water content values
	IF (I7 = 2 or I7 = 3) no input is needed

Line 15	Rainfall starting time, ending time, and amount if rainfall indicator is not zero (a set for each rain event)
Repeat	Line 11 to Line 15 for each day

SOIL HEAT FLOW

Output: Output data files are generated in this program
They are (user defines all file names):
1. Soil surface energy partitioning
2. Soil temperature
3. Water content
4. Pressure head
(3 and 4 are for I1 = 2 only)

List of files: HEAT. FOR = source file
HEAT. OBJ = object file
HEAT. EXE = executive file
INPUT. DATA = input data file (user specified name)
OUT1. DAT = output data file for soil surface energy partitioning (user specified name)
  Line 1. heat flux density and cumulative heat flux of net radiation, sensible heat, latent heat, and soil heat.
OUT2. DAT = temperature at each node in the soil (user specified name).
OUT3.DAT = water content
OUT4.DAT = pressure head

Variable description: see program listing

Subprogram description:
1. PARAM: determines the values of soil parameters for a given representative soil type.
2. TRIDIA: solves tridiagonal matrix problem without pivoting. Thomas algorithm, which is a special case of Gauss elimination method, is used in this subroutine and the diagonally dominant matrices are safe.
3. STORE: determines volume of soil water stored in the one-dimensional flow region.
4. COEFNT: computes the coefficient values in the flow equations.
5. BARE: determines soil surface temperature and evaporation rate on the bare soil surface by using the energy balance equation. The bisect root-finding method is used in this subroutine and computation is done by iteratively calling subroutine BISEC.
6. BISEC: computes residual in the energy balance equation at the bare soil and atmosphere interface.
7. INFILT: determines soil surface infiltration rate using a Darcy's law-based flow equation.
8. RAINFL: determines rainfall amount during each time step period.

Extension: This program may be extended in many different ways. For example, plant root system, plant shade, source, or sink can be included. Also, this can be extended for two-dimensional flow domain.

Ties with other program = none

Flow chart = See Fig. 17-8.

How to run: (For Zenith-158 personal computer)
1. Boot on
2. Insert or copy HEAT.EXE file in the default disk drive.
3. Type HEAT on the keyboard and return, then supply file names as indicated.
4. Outputs will be displayed on the screen and stored on the diskette in the user specified drive.

Example input and output files: Examples of input and output files for one of the problems discussed in the text (Fig. 17-6) are used for illustration. A 5-d long simulation of heat flow only was made. A 2-h time interval was used for output printing. Sequential temperature values for each day were used for the soil-surface boundary condition. Constant lower boundary temperature and sequential surface water content for each day were used. The input and output FORMAT statements may need to be modified for different flow domains.

SOIL HEAT FLOW

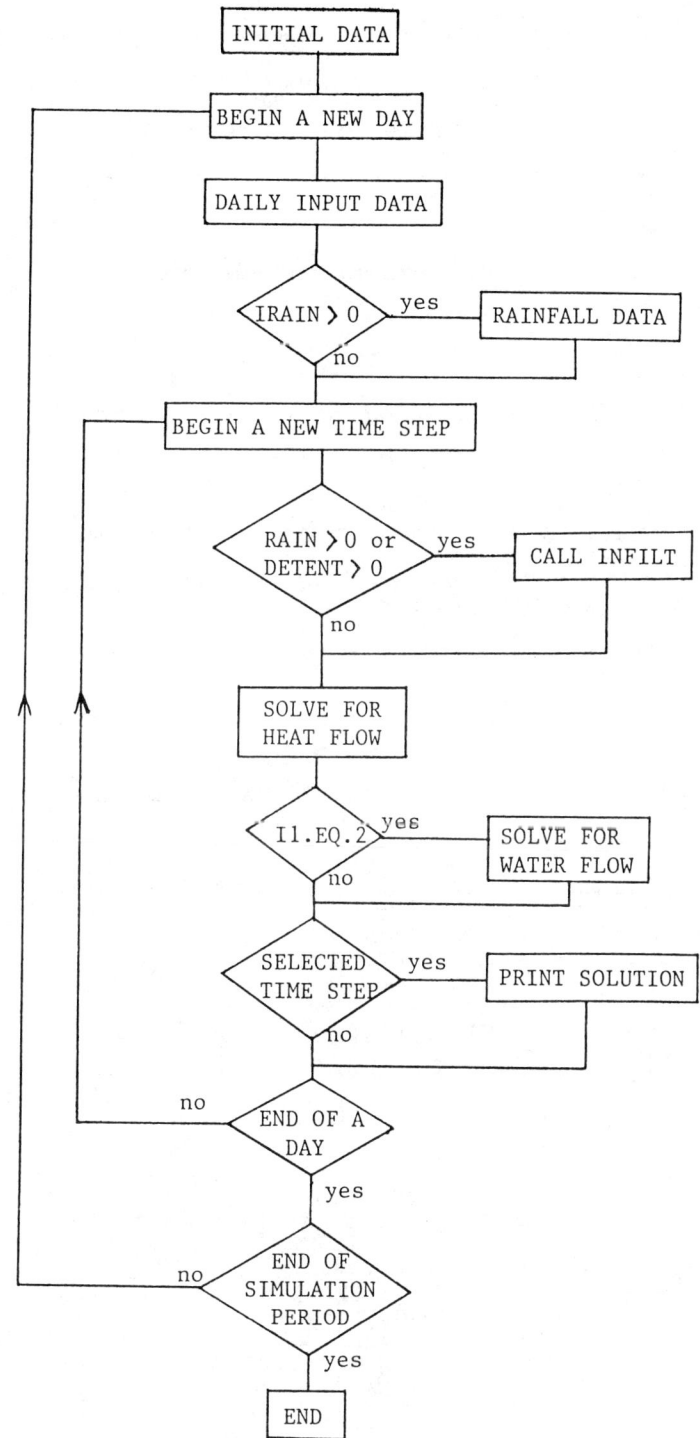

Fig. 17-8. Flowchart of the main program.

```
****************************************************************
*                                                              *
*                         APPENDIX                             *
*                                                              *
*                  COMPUTER PROGRAM LISTING                    *
*                                                              *
****************************************************************
*                                                              *
*                           HEAT                               *
*                                                              *
*                SOIL HEAT AND WATER FLOW MODEL                *
*                                                              *
*                       SEPTEMBER 1986                         *
*                                                              *
*                             BY                               *
*                                                              *
*           SANG-OK CHUNG, TAI SONG TAN AND ROBERT HORTON      *
*                                                              *
*                     IOWA STATE UNIVERSITY                    *
*                                                              *
*   THIS PROGRAM SOLVES ONE-DIMENSIONAL TRANSIENT HEAT AND WATER*
*   FLOW PROBLEMS BY USING IMPLICIT FINITE DIFFERENCE METHOD.  *
*   ONE CAN USE THIS PROGRAM FOR HEAT FLOW ONLY.               *
*   USERS MAY NEED TO CHANGE DIMENSION DECLARATION AND INPUT   *
*   OUTPUT FORMAT FOR DIFFERENT FLOW REGION.                   *
*                                                              *
****************************************************************
*                                                              *
*      VARIABLE DESCRIPTIONS                                   *
*                                                              *
*      VARIABLE              DESCRIPTION                       *
*                                                              *
*      AALPA       COEFFICIENT ALPHA IN THE VAN GENUCHTEN'S RETENTION EQ. *
*      ALES        LATENT HEAT FLUX DENSITY                    *
*      ALECUM      CUMULATIVE LATENT HEAT FLUX                 *
*      ALPA        SOIL THERMAL DIFFUSIVITY                    *
*      B1,B2,B3    COEFFICIENTS IN THE SOIL THERMAL CONDUCTIVITY EQ. *
*      CAPA        VOLUMETRIC SOIL HEAT CAPACITY               *
*      COEF        ELEMENTS OF TRIDIAGONAL MATRIX              *
*      COND        HYDRAULIC CONDUCTIVITY                      *
*      CONDS       SATURATED HYDRAULIC CONDUCTIVITY            *
*      D           RIGHT HAND SIDE OF THE MATRIX               *
*      DAIR        DIFFUSION COEFFICIENT OF WATER VAPOR IN AIR *
*      DAYL        DAY LENGTH                                  *
*      DELZ        SPACE STEP SIZE IN Z-DIRECTION              *
*      DELT        TIME STEP SIZE ADJUSTED FOR A RAINSTORM     *
*      DETENT      WATER DETENTION DEPTH ON THE SOIL           *
*      DETMAX      MAXIMUM VALUE OF WATER DETENTION            *
*      DWDP        SPECIFIC WATER CAPACITY                     *
*      EMIS        EMISSIVITY OF SURFACE                       *
*      EN          COEFFICIENT 'N' IN VAN GENUCHTEN'S RETENTION EQ. *
*      FLUX1       WATER FLUX BOUNDARY CONDITION ACROSS SOIL SURFACE *
*      GS          SOIL HEAT FLUX DENSITY                      *
*      GSCUM       CUMULATIVE SOIL HEAT FLUX                   *
*      HA          ABSOLUTE HUMIDITY OF AIR                    *
*      I1          FLOW INDICATOR:  1 FOR HEAT FLOW ONLY       *
*                                   2 FOR HEAT AND WATER FLOW  *
*      I2          SOIL TYPE INDICATOR:   1   SAND             *
*                                         2   LOAM             *
*                                         3   CLAY             *
*                                         4   USER SUPPLIED SOIL TYPE *
```

```
*     I3           INPUT DATA TYPE:  1  PRESSURE HEAD                   *
*                                    2  WATER CONTENT                   *
*     I4           TEMP TOP B.C.:                                        *
*                       1  READ SEQUENTIAL TEMP VALUES FOR A DAY         *
*                       2  READ DAILY MAX. AND MIN. TEMP.                *
*                       3  READ DAILY WEATHER DATA AND USE               *
*                          ENERGY BALANCE EQ.                            *
*     I5           TEMP. BOTTOM B.C.: 1  READ CONSTANT VALUE             *
*                                     2  READ SEQUENTIAL TEMP VALUES     *
*                                                                        *
**************************************************************************
**************************************************************************
*                                                                        *
*                             FOR A DAY                                  *
*                       3  FLUX BOUNDARY - CALCULATED BY                 *
*                          COMPUTER PROGRAM                              *
*     I6           WATER TOP B.C.:                                       *
*                       1  READ SEQUENTIAL THETA OR PSI FOR A DAY        *
*                       2  READ SEQUENTIAL FLUX VALUES FOR A DAY         *
*                       3  NO DATA NEEDED.                               *
*                          USE ENERGY BALANCE AS OF HEAT FLOW            *
*     I7           WATER BOTTOM B.C.: 1  READ SEQUENTIAL THETA OR PSI    *
*                                        VALUES FOR A DAY                *
*                                     2  NO FLUX - NO DATA NEEDED        *
*                                     3  UNIT GRADIENT FLUX              *
*                                        -- NO DATA NEEDED               *
*     ISTEP        NUMBER OF TIME STEPS IN A DAY                         *
*     N            NUMBER OF NODES IN THE Z DIRECTION                    *
*     OMEGA        PERIOD OF SINE FUNCTION USED (2*PI/86400)             *
*     ORGAN        FRACTION OF ORGANIC MATTER IN THE SOIL                *
*     PSI          PRESSURE HEAD IN M OF WATER                           *
*     PSIO         PRESSURE HEAD AT THE IMAGINARY POINTS ABOVE SOIL      *
*                  SURFACE                                               *
*     RA           AERODYNAMIC BOUNDARY LAYER RESISTANCE                 *
*     RAMDA        SOIL THERMAL CONDUCTIVITY                             *
*     RGD          DAILY GLOBAL SOLAR RADIATION                          *
*     RN           NET RADIATION HEAT FLUX DENSITY                       *
*     RNCUM        CUMULATIVE NET RADIATION HEAT FLUX                    *
*     SOLID        FRACTION OF SOLID IN THE SOIL                         *
*     RHEAT        VOLUMETRIC LATENT HEAT OF VAPORIZATION                *
*     SIGMA        STEFAN-BOLTZMANN CONSTANT                             *
*     SKY          LONGWAVE SKY IRRADIANCE                               *
*     SNOON        SOLAR NOON                                            *
*     SH           SENSIBLE HEAT FLUX DENSITY                            *
*     SHCUM        CUMULATIVE SENSIBLE HEAT FLUX                         *
*     T            TEMPERATURE                                           *
*     TAIR         AIR TEMPERATURE                                       *
*     TAVE         AVERAGE DAILY AIR TEMPERATURE                         *
*     TAMP         AMPLITUDE OF DAILY TEMPERATURE VARIATION              *
*     TDEW         DEW POINT TEMPERATURE                                 *
*     TDEWA        AVERAGE DAILY DEW POINT TEMPERATURE                   *
*     TDAMP        AMPLITUDE OF DAILY DEW POINT TEMPERATURE VARIATION    *
*     TDRAIN       CUMULATIVE DRAINAGE ACROSS THE BOTTOM BOUNDARY        *
*     TEVAP        CUMULATIVE EVAPORATION ACROSS THE TOP BOUNDARY        *
*     THETA        VOLUMETRIC WATER CONTENT                              *
*     THETAS       SATURATION WATER CONTENT                              *
*     THETAR       RESIDUAL WATER CONTENT                                *
*     TIME1        TIME INTERVAL OF OUTPUT PRINTING AND STORAGE OF DATA  *
*                  FOR LINE CURVE                                        *
*     TMAX         DAILY MAXIMUM SOIL SURFACE TEMPERATURE                *
*     TMIN         DAILY MINIMUM SOIL SURFACE TEMPERATURE                *
```

```
*      TOL         TOLERANCE                                           *
*      TS          TEMPERATURE ON THE SOIL SURFACE NODES               *
*      WSPEED      WIND SPEED                                          *
*      ZO          SOIL SURFACE ROUGHNESS LENGTH                       *
*      INFILE      FILE NAME FOR THE INPUT FILE                        *
*      FILE1       NAME OF ENERGY PARTITIONING OUTPUT FILE             *
*      FILE2       NAME OF TEMPERATURE OUTPUT FILE                     *
*      FILE3       NAME OF WATER CONTENT OUTPUT FILE                   *
*      FILE4       NAME OF PRESSURE HEAD OUTPUT FILE                   *
*                                                                      *
************************************************************************
       IMPLICIT REAL*8 (A-H,O-Z)
       COMMON/WETHER/SIGMA,TAIR,TDEW,ZO,TOL,DELZ,DELT,SKY,HA,RG,RA
       COMMON/SOIL/THETAS,THETAR,SOLID,ORGAN,CONDS,AALPA,EN,B1,B2,B3
       COMMON TIME, OMEGA
       DIMENSION COEF(30,3),RHS(30),T(30),RAMDA(30)
       DIMENSION THETA(2160),CAPA(2160),ALPA(2160)
       DIMENSION COND(2160),PSI(2160),DWDP(2160),THETAN(2160)
       DIMENSION T1(2160),TN(2160),PSI1(2160),PSIN(2160),FLUX1(2160)
       DIMENSION IRAIN(30),TIMEB(5),TIMEE(5),AMOUNT(5)
       DIMENSION TT1(2160),TTN(2160),PSI11(2160),PSINN(2160)
       DIMENSION THETNN(2160),FLUX11(2160),THET11(2160)
       CHARACTER*10 INFIL, FILE1, FILE2, FILE3, FILE4

       WRITE(*,1)
   1   FORMAT(///,25X,'SOIL HEAT AND WATER FLOW  MODEL',/)
       WRITE(*,2)
   2   FORMAT(15X,'ONE -DIMENSION TRANSIENT HEAT AND WATER FLOW PROGRAM')
       WRITE(*,3)
   3   FORMAT(15X,'SOLVE PROBLEM WITH IMPLICIT FINITE DIFFERENCE METHOD')
       WRITE(*,4)
   4   FORMAT(/,35X,'WRITTEN  BY')
       WRITE(*,5)
   5   FORMAT(/,18X,'SANG-OK CHUNG, TAI SONG TAN and ROBERT HORTON')
       WRITE(*,6)
   6   FORMAT(//,34X,'SOIL   PHYSICS')
       WRITE(*,7)
   7   FORMAT(29X,'DEPARTMENT OF   AGRONOMY')
       WRITE(*,8)
   8   FORMAT(29X,'COLEGE  OF  AGRICULTURE')
       WRITE(*,9)
   9   FORMAT(29X,'IOWA   STATE   UNIVERSITY')
       WRITE(*,10)
  10   FORMAT(33X,'AMES IOWA 50011')
       WRITE(*,11)
  11   FORMAT(36X,'U. S. A.',///)

*  Open input and output files

  12   FORMAT(A10)
       WRITE(*,*) 'Enter name of the input file           > '
       READ(*,12) INFIL
       WRITE(*,*) 'Enter name of energy partition file > '
       READ(*,12) FILE1
       WRITE(*,*) 'Enter name of temperature file        > '
       READ(*,12) FILE2

*      OPEN INPUT AND OUTPUT FILES
       OPEN(30,FILE=INFIL,STATUS='OLD')
       OPEN(31,FILE=FILE1,STATUS='NEW')
       OPEN(32,FILE=FILE2,STATUS='NEW')
```

SOIL HEAT FLOW

```
*  Read flow, soil, and boundary conditions indicators

      READ(30,*) I1,I2,I3,I4,I5,I6,I7

      IF (I1 .EQ. 2) THEN
        WRITE(*,*) 'Enter name of water content file    >  '
        READ(*,12) FILE3
        WRITE(*,*) 'Enter name of pressure head file    >  '
        READ(*,12) FILE4
        OPEN(33,FILE=FILE3,STATUS='NEW')
        OPEN(34,FILE=FILE4,STATUS='NEW')
      END IF

      READ(30,*) ZLENG,DELZ,ZO
      READ(30,*) IDAY,DELT
      READ(30,*) TIME1
      READ(30,*) DAYL,SNOON

      WRITE(*,13) I1,I2,I3,I4,I5,I6,I7
      WRITE(*,14) ZLENG
      WRITE(*,15) DELZ
      WRITE(*,16) ZO
      WRITE(*,17) IDAY
      WRITE(*,18) DELT
      WRITE(*,19) TIME1
      WRITE(*,20) DAYL
      WRITE(*,21) SNOON

*     ASSIGN PARAMETER VALUES
*     READ LENGTH OF FLOW REGION AND SPATIAL STEP SIZE
*     READ LENGTH OF SIMULATION AND TIME STEP SIZE
*     READ TIME IN HOURS AT WHICH OUTPUT TO BE STORED
*     READ DAYLENGTH AND SOLAR NOON IN HOURS

 13   FORMAT(1X,'I1 =',I3,5X,'I2 =',I3,5X,'I3 =',I3,5X,'I4 =',I3,
     @       5X,'I5 =',I3,5X,'I6 =',I3,5X,'I7 =',I3,/)
 14   FORMAT(1X,'LENGTH OF FLOW REGION IS ',2X,F7.3,' METER')
 15   FORMAT(1X,'SPATIAL STEP SIZE IS ',2X, F7.3,' METER')
 16   FORMAT(1X,'ZO =',2X,F7.4,' METER')
 17   FORMAT(1X,'LENGTH OF SIMULATION IS ',2X,I4,' DAY')
 18   FORMAT(1X,'TIME STEP SIZE IS ',2X,F7.2, ' SECONDS')
 19   FORMAT(1X,'TIME IS ',2X,F7.2,3X,'HOURS')
 20   FORMAT(1X,'DAY LENGTH IS ',2X,F7.3,3X,'HOURS')
 21   FORMAT(1X,'SOLAR NOON IS ',2X,F7.3,3X,'HOURS')

      TIME1  = TIME1 * 3600.0D0
      DAYL   = DAYL * 3600.0D0
      SNOON  = SNOON * 3600.0D0
      ISTEP  = IDNINT(86400.0D0/DELT)
      N      = IDNINT(DINT((ZLENG+0.001D0)/DELZ)) + 1
      NM1    = N - 1
      OMEGA  = 7.2722D-5
      SIGMA  = 5.67D-8
      TOL    = 0.01D0
      PI     = 4.0D0 * DATAN(1.0D0)
      DETMAX = 0.01D0
      DZ2DT  = DELZ * DELZ / DELT

*  Determine soil parameters
      CALL PARAM(I2)
```

```
************************************************************************
*     STEP 1:  ASSIGN INITIAL VALUES                                    *
************************************************************************
      READ(30,*) (T(J), J=1,N)
      WRITE(*,22) (T(J),J=1,N)

      IF (I1 .EQ. 2) THEN
        IF (I3 .EQ. 1) THEN
          READ(30,*)  (PSI(J), J=1,N)
          WRITE(*,23) (PSI(J), J=1,N)
        ELSE
          READ(30,*)  (THETA(J), J=1,N)
          WRITE(*,24) (THETA(J), J=1,N)
          DO 100 J = 1, N
            PSI(J) = -(((THETA(J)-THETAR) / (THETAS-THETAR)) **
     #                 (EN/(1.0D0-EN))-1.0D0) ** (1.0D0/EN)/(AALPA)
 100      CONTINUE
        ENDIF
      END IF

  22  FORMAT(24F6.2)
  23  FORMAT(8E10.2,/8E10.2)
  24  FORMAT(24F6.2)

      TINFIL = 0.0D0
      TEVAP  = 0.0D0
      TDRAIN = 0.0D0
      RNCUM  = 0.0D0
      SHCUM  = 0.0D0
      ALECUM = 0.0D0
      GSCUM  = 0.0D0

* Determine initial water storage in the system
      CALL STORE(THETA,N,DELZ,VOL)
      VOLUMI = VOL

* Read rainfall indicator for each day
      READ(30,*)  (IRAIN(I), I=1,IDAY)
      WRITE(*,25) (IRAIN(I), I=1,IDAY)
  25  FORMAT('IRAIN IS ',1X, 12I4 )

* Do loop for the length of the simulation in days
      DO 1000 III = 1, IDAY

* Determine Boundary Condition values for a day
      IF (I4 .EQ. 1) THEN
        READ(30,*) NO
        READ(30,*) (TT1(J), J=1,NO)
        WRITE(*,26) NO
  26    FORMAT(1X,'NO =',3X,I4)
        WRITE(*,22) (TT1(J),J=1,NO)
        NOINT = ISTEP/NO
        DO 181 J  = 1, NO
        DO 181 JK = 1, NOINT
          K = (J-1) * NOINT + JK
          T1(K) = TT1(J)
 181    CONTINUE
      END IF
```

SOIL HEAT FLOW

```
* Use SINE FUNCTION to determine soil surface temperature.
* Assume maximum temperature occurs 2 hours after the SOLAR NOON.

       IF (I4 .EQ. 2) THEN
          READ(30,*)  TMAX,TMIN
          WRITE(*,27) TMAX, TMIN
  27      FORMAT(1X,'TMAX =',3X,F6.3,5X,'TMIN =',3X,F6.3)
          DO 182 K = 1, ISTEP
             T1(K) = (TMAX+TMIN)/2.0D0 + (TMAX-TMIN)/2.0D0 *
     #               DSIN(PI/2.0D0 + (DBLE(K)*DELT-SNOON-7200.0D0)*OMEGA)
 182      CONTINUE
       END IF

       IF(I4 .EQ. 3) THEN
          READ(30,*) TAVE,TAMP,TDEWA,TDAMP,WSPEED,RGD
          WRITE(*,505) TAVE,TAMP,TDEWA,TDAMP,WSPEED,RGD
          RA = (DLOG(2.0D0/ZO))**2/ (0.16D0*WSPEED)
       END IF

       IF (I5 .EQ. 1) THEN
          READ(30,*) TNC
          WRITE(*,29) TNC
  29      FORMAT(' TNC = ',F6.2)
       END IF

       IF (I5 .EQ. 2) THEN
          READ(30,*) NO
          READ(30,*) (TTN(J), J=1,NO)
          WRITE(*,26) NO
          WRITE(*,22) (TTN(J), J=1,NO)
          NOINT = ISTEP/NO
          DO 183 J  = 1,NO
          DO 183 JK = 1,NOINT
             K = (J-1) * NOINT + JK
             TN(K) = TTN(J)
 183      CONTINUE
       END IF

       IF (I1 .EQ. 1) GO TO 2000

* If heat flow uses energy balance method, use that for water also

       IF (I4 .EQ. 3) I6 = 3

       IF (I6 .EQ. 1) THEN
          IF (I3 .EQ. 1) THEN
             READ(30,*) NO
             READ(30,*) (PSI11(J), J=1,NO)
             WRITE(*,*) 'NO = ',NO
             WRITE(*,23) (PSI11(J),J=1,NO)
             NOINT = ISTEP/NO
             DO 184 J = 1,NO
             DO 184 JK = 1,NOINT
                K = (J-1) * NOINT + JK
             PSI1(K) = PSI11(J)
 184      CONTINUE
          ELSE
             READ(30,*) NO
             READ(30,*) (THET11(J),J=1,NO)
```

```
              WRITE(*,26) NO
              WRITE(*,24) (THET11(J), J=1,NO)
              NOINT = ISTEP/NO
              DO 200 J = 1,NO
              DO 200 JK = 1,NOINT
                K = (J-1) * NOINT + JK
                THETA(K) = THET11(J)
                PSI1(K)  = -(((THETA(K)-THETAR)/(THETAS-THETAR)) **
     #                   (EN/(1.0D0-EN))-1.0D0)**(1.0D0/EN)/(AALPA)
200         CONTINUE
          END IF
        END IF

        IF (I6 .EQ. 2) THEN
          READ(30,*) NO
          READ(30,*) (FLUX11(J), J=1,NO)
          WRITE(*,26) NO
          WRITE(*,31) (FLUX11(J), J-1,NO)
 31       FORMAT(8D12.4,/8D12.4)
          NOINT = ISTEP/NO
          DO 210 J = 1,NO
          DO 210 JK = 1,NOINT
            K = (J-1)*NOINT + JK
            FLUX1(K) = FLUX11(J)
210       CONTINUE
        END IF

        IF (I7 .EQ. 1) THEN
          IF (I3 .EQ. 1) THEN
            READ(30,*) NO
            READ(30,*) (PSINN(J), J=1,NO)
            WRITE(*,26) NO
            WRITE(*,23) (PSINN(J),J=1,NO)
            NOINT = ISTEP/NO
            DO 250 J = 1,NO
            DO 250 JK = 1,NOINT
              K = (J-1) * NOINT + JK
              PSIN(K) = PSINN(J)
250         CONTINUE
          ELSE
            READ(30,*) NO
            READ(30,*) (THETNN(J),J=1,NO)
            WRITE(*,26) NO
            WRITE(*,24) (THETNN(J),J=1,NO)
            NOINT = ISTEP/NO
            DO 300 J = 1,NO
            DO 300 JK = 1,NOINT
              K = (J-1) * NOINT + JK
              THETAN(K) = THETNN(J)
              PSIN(K) = -(((THETAN(K)-THETAR)/(THETAS-THETAR))**
     #                 (EN/(1.0D0-EN)) -1.0D0)**(1.0D0/EN)/(AALPA)
300         CONTINUE
          END IF
        END IF

2000    CONTINUE
        NUMRN = IRAIN(III)
        IF (NUMRN .NE. 0) THEN
          READ(30,*) (TIMEB(I),TIMEE(I),AMOUNT(I),I=1,NUMRN)
          WRITE(*,507) (TIMEB(I),TIMEE(I),AMOUNT(I),I=1,NUMRN)
        END IF
```

SOIL HEAT FLOW

```
*       DO LOOP FOR THE TIME STEPS IN A DAY
            TIME = 0.0D0

 3000   IF (TIME .LT. 86400.0D0) THEN
            TIME = TIME + DELT
            K = IDNINT(DINT((TIME+0.1D0)/DELT))
            RAIN = 0.0D0

            IF (NUMRN .NE. 0 .AND. TIME .GT. TIMEB(NUMRN) .AND.
        #         TIME .LE. TIMEE(NUMRN)) THEN

*       DETERMINE RAINFALL AMOUNT FOR THIS TIME STEP

              CALL RAINFL(TIME,DELT,NUMRN,TIMEB,TIMEE,AMOUNT,RAIN)
            END IF

* Compute values of coefficients by SUBROUTINE COEFNT
          CALL COEFNT(N,PSI,THETA,DWDP,COND,CAPA,RAMDA,ALPA)

************************************************************************
*       STEP 2:  FOR HEAT FLOW                                          *
************************************************************************

* Set up Coefficient matrix for the top boundary condition

            COEF(1,1) = 0.0D0
            COEF(1,2) = 1.0D0
            COEF(1,3) = 0.0D0
            IF (I4 .NE. 3) THEN
              RHS(1) = T1(K)
            ELSE
              TAIR = TAVE + TAMP*DSIN(OMEGA*TIME+PI)
              IF (DETENT .GT. 0.0D0 .OR. RAIN .GT. 0.0D0) THEN
                CALL INFILT(RAIN,DETENT,DETMAX,PSI(2),CONDS,COND(2),
        #                   DELZ,DELT,FLUX1(K))
                RHS(1) = TAIR
                TINFIL = TINFIL + FLUX1(K)*DELT
              ELSE
                TDEW = TDEWA + TDAMP*DSIN(OMEGA*TIME+PI)
                HA   = 1.323D0*DEXP(17.27D0*TDEW/(237.3D0+TDEW))
        #           /(TAIR+273.16D0)
                SKY  = SIGMA*(TAIR+273.16D0)**4 * (0.605D0 + 1.777D0
        #              * DSQRT(HA))
                TIMESH = (TIME-SNOON+DAYL/2.0D0)/DAYL
                IF (TIMESH .LE. 0.0D0 .OR. TIMESH .GE. 1.0D0) THEN
                  RG = 0.0D0
                ELSE
                  RG = PI/2.0D0 * RGD/DAYL * DSIN(TIMESH*PI)
                END IF
                CALL BARE(THETA(1),T(1),T(2),PSI(1),CAPA(1),CAPA(2),
        #                ALPA(1),ALPA(2),FLUX1(K),TS,RN,SH,ALES,GS)
                RHS(1) = TS

*       COMPUTE CUMULATIVE EVAPORATION

                TEVAP  = TEVAP + FLUX1(K)*DELT
                RNCUM  = RNCUM + RN*DELT
                SHCUM  = SHCUM + SH*DELT
                ALECUM = ALECUM + ALES*DELT
                GSCUM  = GSCUM + GS*DELT
              END IF
            END IF
```

```
*     FOR THE INTERMEDIATE NODES

         DO 400 J = 2,NM1
           COEF(J,1) = -0.5D0 * (ALPA(J-1)+ALPA(J))
           COEF(J,3) = -0.5D0 * (ALPA(J)+ALPA(J+1))
           COEF(J,2) = DZ2DT - COEF(J,1) - COEF(J,3)
           RHS(J)    = DZ2DT*T(J)
 400     CONTINUE

*     FOR THE BOTTOM BOUNDARY POINTS

         IF (I5 .NE. 3) THEN
           COEF(N,1) = 0.0D0
           COEF(N,2) = 1.0D0
           COEF(N,3) = 0.0D0
           IF(I5 .EQ. 1) RHS(N) = TNC
           IF(I5 .EQ. 2) RHS(N) = TN(K)
         ELSE
           HFLUX = -RAMDA(N) * (T(N)-T(N-1))/DELZ
           COEF(N,1) = -2.0D0 * 0.5D0*(ALPA(N-1)+ALPA(N))
           COEF(N,2) = DZ2DT - COEF(N,1)
           COEF(N,3) = 0.0D0
           RHS(N)= DZ2DT*T(N) + 0.5D0*(ALPA(N)+ALPA(N-1))
     #             *HFLUX*DELZ*2.0D0/RAMDA(N)
         END IF

*     SOLVE FOR THE HEAT FLOW BY CALLING TRIDIA

         CALL TRIDIA (N,COEF,RHS)

*     ASSIGN THE SOLUTION TO T(J)

         DO 500 J = 1,N
           T(J) = RHS(J)
 500     CONTINUE

*    SKIP WATER FLOW PART IF THE PROBLEM IS ONLY FOR HEAT FLOW

         IF (I1 .EQ. 1) GO TO 4000
************************************************************************
*     STEP 3:  FOR WATER FLOW                                          *
************************************************************************

*    SEP UP COEFFICIENT MATRIX FOR THE TOP BOUNDARY CONDITION

         IF (I6 .EQ. 1) THEN
           COEF(1,1) = 0.0D0
           COEF(1,2) = 1.0D0
           COEF(1,3) = 0.0D0
           RHS(1)    = PSI1(K)
         ELSE IF (DETENT .GT. 0.0D0) THEN
           COEF(1,1) = 0.0D0
           COEF(1,2) = 1.0D0
           COEF(1,3) = 0.0D0
           RHS(1)    = DETENT
         ELSE
           TERMS = DWDP(1)*DELZ/(2.0D0*DELT)
           CBAR  = 0.5D0*(COND(1)+COND(2))
           PSIO = ((FLUX1(K) - CBAR + CBAR*PSI(2)/DELZ) + PSI(1)*TERMS)
     #            / (TERMS + CBAR/DELZ)
```

SOIL HEAT FLOW

```
              COEF(1,1) = 0.0D0
              COEF(1,2) = 1.0D0
              COEF(1,3) = 0.0D0
              RHS(1)    = PSIO
            END IF

*  For the intermediate rows

            DO 600 J = 2,NM1
              COEF(J,1) = -0.5D0 * (COND(J-1)+COND(J))
              COEF(J,3) = -0.5D0 * (COND(J) + COND(J+1))
              COEF(J,2) = DWDP(J)*DZ2DT - COEF(J,1) - COEF(J,3)
              RHS(J)    = DWDP(J) * DZ2DT * PSI(J) - DELZ/2.0D0 *
     #                    (COND(J+1)-COND(J-1))
 600        CONTINUE

*  FOR THE BOTTOM BOUNDARY NODES

            FLUXN = DSQRT(COND(NM1)*COND(N))*(1.0D0-(PSI(N)-PSI(NM1))/DELZ)
            TDRAIN = TDRAIN + FLUXN*DELT
            IF (I7 .EQ. 1) THEN
              COEF(N,1) = 0.0D0
              COEF(N,2) = 1.0D0
              COEF(N,3) = 0.0D0
              RHS(N)    = PSIN(K)
            ELSE IF (I7 .EQ. 2) THEN
              COEF(N,1) = -1.0D0
              COEF(N,2) =  1.0D0
              COEF(N,3) =  0.0D0
              RHS(N)    = DELZ
            ELSE
              COEF(N,1) = -1.0D0
              COEF(N,2) =  1.0D0
              COEF(N,3) =  0.0D0
              RHS(N)    =  0.0D0
            END IF

*  SOLVE FOR THE WATER FLOW BY CALLING TRIDIA

            CALL TRIDIA(N,COEF,RHS)

*  ASSIGN THE SOLUTION TO PSI(J)

            DO 700 J = 1,N
              PSI(J) = RHS(J)
 700        CONTINUE
4000      CONTINUE

*     PRINT SOLUTION ON SCREEN WHILE CALCULATIONS PROCEED
*     AND CREATE TWO (FOR HEAT FLOW ONLY) OR
*     FOUR (FOR BOTH HEAT AND WATER FLOWS) USER SPECIFIC OUTPUT FILES

          IF (DMOD(TIME,TIME1) .EQ. 0.0D0) THEN
            WRITE(*,99) III, TIME/3600.0D0
 99         FORMAT( ' DAY =',2X,I3,10X, ' TIME =',2X,F5.2)
            WRITE(*,118) RN,RNCUM
            WRITE(*,120) SH,SHCUM
            WRITE(*,122) ALES,ALECUM
            WRITE(*,124) GS,GSCUM
            WRITE(*,111) (T(J), J=1,N)
            WRITE(32,111) (T(J), J=1,N)
            WRITE(31,127) RN,RNCUM, SH,SHCUM, ALES,ALECUM, GS,GSCUM
```

```
              IF (I1 .EQ. 2) THEN
                CALL STORE(THETA,N,DELZ,VOL)
                BALANS = VOL - VOLUMI - TEVAP + TDRAIN - TINFIL
                WRITE(*,114) VOL,BALANS
                WRITE(*,116) FLUXN,TDRAIN
                WRITE(*,126) FLUX1(K),TEVAP,TINFIL
                WRITE(*,111) (THETA(J), J=1,N)
                WRITE(*,113) (PSI(J), J=1,N)
                WRITE(33,111) (THETA(J), J=1,N)
                WRITE(34,113) (PSI(J), J=1,N)
              END IF
            END IF
            GO TO 3000
          END IF
 1000 CONTINUE

  111 FORMAT(13F6.2)
  113 FORMAT(8E10.2,/8E10.2)
  114 FORMAT(' STORAGE =',F8.6, ' BALANCE =',F12.6)
  116 FORMAT(' FLUXN(M/S)=',D12.4, ' CUM. DRAINAGE=',D12.4)
  118 FORMAT(' NET RADIATION (W/M*M)=',F12.0,' CUM. NET RAD. =',F12.0)
  120 FORMAT(' SENSIBLE HEAT FLUX   =',F12.0,' CUM. SEN. HEAT=',F12.0)
  122 FORMAT(' EVAP. HEAT FLUX      =',F12.0,' CUM. EVAP HEAT=',F12.0)
  124 FORMAT(' SOIL HEAT FLUX(W/M*M)=',F12.0,' CUM. SOIL HEAT=',F12.0)
  126 FORMAT(' FLUX1(M/S)=',D12.4, ' TEVAP=',D12.4,'   TINFIL=',D12.4)
  127 FORMAT(8D10.3)
  501 FORMAT(20I2)
  502 FORMAT(24F5.1)
  503 FORMAT(24F6.2)
  504 FORMAT(24F6.3)
  505 FORMAT(5F6.2,E10.4)
  506 FORMAT(24F5.2)
  507 FORMAT(2F7.0,F6.3)
  508 FORMAT(I3,F5.0)
  509 FORMAT(F4.1)
  510 FORMAT(24F5.1)
  511 FORMAT(24F5.2)
  512 FORMAT(24F5.2)

*     CLOSE ANY FILES WHICH ARE OPEN
      CLOSE (30)
      CLOSE (31)
      CLOSE (32)
      CLOSE (33)
      CLOSE (34)
      STOP
      END
```

SOIL HEAT FLOW

```
***********************************************************************
*      THIS SUBROUTINE DETERMINES THE VALUES OF SOIL PARAMETERS FOR A   *
*      GIVEN REPRESENTATIVE SOIL TYPE                                   *
***********************************************************************
       SUBROUTINE PARAM(I2)
       IMPLICIT REAL*8 (A-H,O-Z)
       COMMON/SOIL/THETAS,THETAR,SOLID,ORGAN,CONDS,AALPA,EN,B1,B2,B3

       IF (I2 .EQ. 1) THEN
         AALPA  = 3.28D0
         EN     = 1.54D0
         CONDS  = 2.5D-5
         B1     = 0.228D0
         B2     = -2.406D0
         B3     = 4.909D0
         THETAS = 0.44D0
         THETAR = 0.00D0
         SOLID  = 0.56D0
         ORGAN  = 0.0D0
       ELSE IF (I2 .EQ. 2) THEN
         AALPA  = 1.55D0
         EN     = 1.5D0
         CONDS  = 0.7D-5
         B1     = 0.243D0
         B2     = 0.393D0
         B3     = 1.536D0
         THETAS = 0.48D0
         THETAR = 0.01D0
         SOLID  = 0.52D0
         ORGAN  = 0.0D0
       ELSE IF (I2 .EQ. 3) THEN
         AALPA  = 0.432D0
         EN     = 1.36D0
         CONDS  = 0.2D-5
         B1     = -0.197D0
         B2     = -0.962D0
         B3     = 2.521D0
         THETAS = 0.52D0
         THETAR = 0.04D0
         SOLID  = 0.48D0
         ORGAN  = 0.0D0
       ELSE

* Read user supplied Soil Parameters from input data file

         READ(30,*) B1,B2,B3,THETAS,THETAR,SOLID,ORGAN
         READ(30,*) AALPA,EN,CONDS
       END IF
       WRITE(*,100) B1,B2,B3,THETAS,THETAR,SOLID,ORGAN
       WRITE(*,200) AALPA,EN,CONDS
100    FORMAT(7F8.3)
200    FORMAT(2F8.3,E13.4)
       RETURN
       END
```

```
************************************************************************
* This SUBROUTINE computes volume of water stored in the                *
* one-dimension domain at a given time                                  *
************************************************************************
      SUBROUTINE STORE(THETA,N,DELZ,VOL)
      IMPLICIT REAL*8 (A-H,O-Z)
      DIMENSION THETA(30)

      NM1  = N - 1
      VOL1 = THETA(1)*DELZ/2.0D0
      VOLN = THETA(N)*DELZ/2.0D0
      VOL  = 0.0D0
      DO 10 J = 2, NM1
        VOL  = VOL + THETA(J)*DELZ
 10   CONTINUE
      VOL = VOL + VOL1 + VOLN
      RETURN
      END


************************************************************************
*  SUBROUTINE RAINFL determines the rainfall amount during each         *
*  time step.                                                           *
************************************************************************
      SUBROUTINE RAINFL(TIME,DELT,IRAIN,TIMEB,TIMEE,AMOUNT,RAIN)
      IMPLICIT REAL*8 (A-H,O-Z)
      DIMENSION TIMEB(5),TIMEE(5),AMOUNT(5)

      IF (TIME .LE. TIMEE(1)) THEN
        RAIN = AMOUNT(1)/(TIMEE(1)-TIMEB(1)) * DELT
      ELSE
        DO 10 I = 2, IRAIN
          IF (TIME .GT. TIMEB(I) .AND. TIME .LE. TIMEE(I))THEN
            RAIN = (AMOUNT(I)-AMOUNT(I-1))/(TIMEE(I)-TIMEB(I)) * DELT
            IF (RAIN .GT. 0.0D0) RETURN
          END IF
 10     CONTINUE
      END IF
      RETURN
      END


************************************************************************
*     THIS SUBROUTINE COMPUTES THE VALUES OF THE FLOW EQUATION          *
*     COEFFICIENTS FOR BOTH HEAT AND WATER FLOW                         *
*     INPUTS ARE N, AND PSI, AND THE REST ARE OUTPUTS                   *
************************************************************************
      SUBROUTINE COEFNT(N,PSI,THETA,DWDP,COND,CAPA,RAMDA,ALPA)
      IMPLICIT REAL*8 (A-H,O-Z)
      DIMENSION PSI(30),THETA(30),DWDP(30),COND(30),CAPA(30)
      DIMENSION RAMDA(30),ALPA(30)
      COMMON/SOIL/THETAS,THETAR,SOLID,ORGAN,CONDS,AALPA,EN,B1,B2,B3

*     USE STATEMENT FUNCTION FOR THE SOIL HEAT CAPACITY,
*     HYDRAULIC CONDUCTIVITY AND SPECIFIC WATER CAPACITY.

      CAPACT(SOLID,ORGAN,WATER)
     #          = 1.92D6*SOLID + 2.51D6*ORGAN + 4.18D6*WATER
```

SOIL HEAT FLOW

```
      CONDUC(CONDS,AALPA,EN,PRE)
#            = CONDS*(1.0D0-(-AALPA*PRE)**(EN-1.0D0)*(1.0D0+
#              (-AALPA*PRE)**EN)**((1.0D0-EN)/EN))**2/(1.0D0
#              + (-AALPA*PRE)**EN)**((EN-1.0D0)/EN/2.0D0)

      FUN(WACON,THETAS,THETAR,EN,PRE)
#            = (EN-1.0D0)*(WACON-THETAR)*(1.0D0-((WACON-THETAR)/
#              (THETAS-THETAR))**(EN/(EN-1.0D0)))/(-PRE)

      CONDT(B1,B2,B3,THTA) = B1 + B2*THTA + B3*DSQRT(THTA)
      THET(THETAS,THETAR,AALPA,EN,PRE)
#          = THETAR + (THETAS-THETAR)*(1.0D0+(-AALPA*PRE)**EN)
#              **((1.0D0-EN)/EN)

      DO 100 J=1,N
        IF (PSI(J) .GT. -0.01D0) THEN
          THETA(J) = THETAS
          DWDP(J)  = 0.01D0
          COND(J)  = CONDS
        ELSE
          THETA(J) = THET(THETAS,THETAR,AALPA,EN,PSI(J))
          DWDP(J)  = FUN(THETA(J),THETAS,THETAR,EN,PSI(J))
          COND(J)  = CONDUC(CONDS,AALPA,EN,PSI(J))
        END IF

        CAPA(J) = CAPACT(SOLID,ORGAN,THETA(J))
        RAMDA(J) = CONDT(B1,B2,B3,THETA(J))
        ALPA(J)  = RAMDA(J)/CAPA(J)
  100 CONTINUE
      RETURN
      END


*************************************************************************
*     THIS SUBROUTINE DETERMINES SOIL SURFACE INFILTRATION              *
*     WATER FLUX CONDITION.                                             *
*************************************************************************
      SUBROUTINE INFILT(RAIN,DETENT,DETMAX,PSI2,CONDS,COND2,DELZ,
     #                  DELT,FLUX1)
      IMPLICIT REAL*8(A-H,O-Z)

      PSI1 = 0.0D0
      IF (DETENT .GT. 0.0D0) PSI1 = DETENT
      CAP = ((PSI1-PSI2)/DELZ + 1.0D0) * 0.5D0*(CONDS+COND2)*DELT
      IF (RAIN+DETENT .LE. CAP) THEN
        RINFIL = RAIN + DETENT
        DETENT = 0.0D0
      ELSE
        RINFIL = CAP
        DETENT = DETENT + RAIN - RINFIL
        IF (DETENT .GT. DETMAX) DETENT = DETMAX
      END IF
      FLUX1 = RINFIL/DELT
      RETURN
      END
```

```
***********************************************************************
*      THIS SUBROUTINE DETERMINES SOIL SURFACE TEMPERATURE AND         *
*      EVAPORATION RATE USING ENERGY BALANCE EQUATION BY BISECT ROOT   *
*      FINDING METHOD                                                  *
***********************************************************************
       SUBROUTINE BARE(THETA1,T1,T2,PSI1,CAP1,CAP2,ALP1,ALP2,
      #                FLUX1,TS,RN,SH,ALES,GS)
       IMPLICIT REAL*8 (A-H,O-Z)
       COMMON/WETHER/SIGMA,TAIR,TDEW,ZO,TOL,DELZ,DELT,SKY,HA,RG,RA

       CAPAS = 0.5D0 * (CAP1+CAP2)
       RAMDA = 0.5D0 * (ALP1*CAP1 + ALP2*CAP2)
       EMIS  = 0.9D0 + 0.18D0*THETA1

*      COMPUTE ALBEDO FOR THE GIVEN TOP WATER CONTENT
       IF (THETA1 .GE. 0.25D0) THEN
         ALBE = 0.1D0
       ELSE IF (THETA1 .LE. 0.1D0) THEN
         ALBE = 0.25D0
       ELSE
         ALBE = 0.35D0 - THETA1
       END IF

       CAPAA = 1154.8D0 + 303.16D0/(TAIR+273.16D0)

       A = T1
       B = T1
       DO 100 J = 1,10
         A = A - 0.5D0
         B = B + 0.5D0
         CALL BISEC(A,T1,T2,PSI1,CAPAS,CAPAA,RAMDA,
      #             EMIS,ALBE,RES,EVAP,RN,SH,ALES,GS)
         FA = RES
         CALL BISEC(B,T1,T2,PSI1,CAPAS,CAPAA,RAMDA,
      #             EMIS,ALBE,RES,EVAP,RN,SH,ALES,GS)
         FB = RES
         IF (FA*FB .LT. 0.0D0) GO TO 200
 100   CONTINUE
       WRITE(*,*) ' ITERATION LIMIT EXCEEDED IN BARE'

 200   DO 300 J = 1,10
         AMID = (A+B) / 2.0D0
         IF ((AMID-A) .LE. TOL) GO TO 500
         CALL BISEC(AMID,T1,T2,PSI1,CAPAS,CAPAA,RAMDA,
      #             EMIS,ALBE,RES,EVAP,RN,SH,ALES,GS)
         FMID = RES
         IF ((FA*FMID) .GT. 0.0D0) THEN
           A = AMID
         ELSE
           B = AMID
         END IF
 300   CONTINUE
       WRITE(*,*) ' ITERATION LIMIT EXCEEDED IN SUBROUTINE BISEC'
 500   TS = AMID
       FLUX1 = -EVAP
       RETURN
       END
```

SOIL HEAT FLOW

```
***************************************************************
*     THIS SUBROUTINE SOLVES TRIDIAGONAL MATRIX WITHOUT PIVOTING   *
*     DIAGONALLY DOMINANT MATRICES ARE SAFE                        *
***************************************************************
      SUBROUTINE TRIDIA(N,W,B)
      IMPLICIT REAL*8 (A-H,O-Z)
      DIMENSION W(30,3), B(30)

*  FF IS MULTIPLIER FOR THE FORWARD ELIMINATION STEP
      DO 10 J = 2,N
        FF = W(J,1)/W(J-1,2)
        W(J,2) = W(J,2) - FF*W(J-1,3)
        W(J,1) = FF
   10 CONTINUE

* FORWARD ELIMINATION STEP
* B:  ON INPUT, VECTOR OF RIGHT HAND SIDES
* ON EXIT, SOLUTION VECTOR
      DO 20 J = 2,N
        B(J) = B(J) - W(J,1)*B(J-1)
   20 CONTINUE

*     BACK SUBSTITUTION LOOP
*
      B(N) = B(N)/W(N,2)
  100 DO 30 J = 2,N
        JB = N - J + 1
        B(JB) = (B(JB) - W(JB,3)*B(JB+1)) / W(JB,2)
   30 CONTINUE
      RETURN
      END

***************************************************************
*     THIS SUBROUTINE DETERMINES THE RESIDUAL OF THE ENERGY BALANCE *
*     EQUATION FOR SUBROUTINE BARE                                  *
*     UNITS USED ARE JOULE, METER, AND SECOND                       *
***************************************************************
      SUBROUTINE BISEC(TS,T1,T2,PSI1,CAPAS,CAPAA,RAMDA,EMIS,
     #                 ALBE,RES,EVAP,RN,SH,ALES,GS)
      IMPLICIT REAL*8 (A-H,O-Z)
      COMMON /WETHER/SIGMA,TAIR,TDEW,ZO,TOL,DELZ,DELT,SKY,HA,RG,RA
      COMMON TIME, OMEGA

      RN   = (1.0D0-ALBE)*RG + SKY - EMIS*SIGMA * (TS+273.16D0)**4
      SH   = CAPAA * (TS-TAIR)/RA
      HO   = 1.323D0 * DEXP(17.27D0*TS/(TS+237.3D0)) /(TS+273.16D0)
      IF (PSI1 .GT. 0.0D0) PSI1=0.0D0
      HS   = HO * DEXP(PSI1/(46.97D0*(TS+273.16D0)))
      EVAP = (HS-HA)/RA/1000.0D0
      ALH  = 2.49463D9 - 2.247D6*TS
      ALES = EVAP*ALH
      GS   = RAMDA*(TS-T2)/DELZ +(TS-T1)*CAPAS*DELZ/DELT/2.0D0
      RES  = RN - SH - ALES - GS
      RETURN
      END
```

1. INPUT.DAT

Input data is unformatted; a comma(,) is used to separate data in each line.

```
1, 2, 2, 1, 1, 1, 3
0.6, 0.05, 0.01
2, 100.0
2.0
13.38, 13.27
16.2, 18.2, 19.6, 20.3, 20.4, 20.2, 20.0, 19.8, 19.6, 19.6, 19.7, 19.8, 20.0
0 0 0 0 0 0 0
6
16.2, 12.2, 15.8, 37.6, 35.3, 23.3
20.0
6
18.4, 16.4, 18.4, 37.1, 36.7, 24.0
20.0
6
21.8, 20.2, 19.3, 25.6, 30.2, 23.8
20.0
6
21.1, 19.1, 20.0, 23.1, 25.3, 19.8
20.0
6
15.3, 14.4, 18.4, 22.2, 28.4, 24.4
20.0
```

2. OUT2.DAT

This data file shows temperature at each node in the soil profile. Results in the first two days of every two hours are listed below.

```
16.20 17.64 18.82 19.60 19.97 20.05 19.97 19.85 19.75 19.73 19.77 19.87 20.00
16.20 17.36 18.36 19.11 19.58 19.81 19.88 19.86 19.82 19.81 19.84 19.91 20.00
12.20 14.98 17.10 18.43 19.17 19.56 19.74 19.81 19.83 19.85 19.88 19.93 20.00
12.20 14.37 16.24 17.67 18.63 19.22 19.55 19.72 19.80 19.85 19.90 19.95 20.00
15.80 16.03 16.59 17.41 18.22 18.87 19.30 19.57 19.73 19.83 19.89 19.95 20.00
15.80 16.27 16.81 17.43 18.07 18.64 19.09 19.41 19.63 19.77 19.87 19.94 20.00
37.60 28.30 22.11 19.30 18.55 18.64 18.97 19.28 19.52 19.70 19.83 19.92 20.00
37.60 31.02 25.64 22.04 20.09 19.31 19.16 19.27 19.46 19.64 19.78 19.90 20.00
35.30 31.02 27.07 23.86 21.61 20.30 19.68 19.49 19.51 19.63 19.76 19.88 20.00
35.30 31.49 27.98 25.03 22.78 21.24 20.32 19.86 19.70 19.70 19.77 19.88 20.00
23.30 25.28 25.79 24.89 23.39 21.99 20.95 20.31 19.97 19.84 19.84 19.91 20.00
23.30 24.09 24.40 24.08 23.27 22.29 21.39 20.71 20.27 20.04 19.96 19.96 20.00
18.40 20.94 22.56 23.11 22.88 22.28 21.60 20.99 20.54 20.25 20.09 20.02 20.00
18.40 20.10 21.42 22.14 22.30 22.06 21.62 21.14 20.73 20.42 20.21 20.08 20.00
16.40 18.59 20.28 21.29 21.72 21.74 21.51 21.18 20.84 20.54 20.31 20.14 20.00
16.40 18.09 19.53 20.56 21.15 21.37 21.33 21.13 20.87 20.61 20.38 20.18 20.00
18.40 18.90 19.53 20.19 20.72 21.01 21.09 21.01 20.84 20.62 20.41 20.20 20.00
18.40 18.96 19.52 20.04 20.46 20.75 20.88 20.87 20.77 20.60 20.41 20.20 20.00
37.10 29.21 23.95 21.51 20.77 20.68 20.73 20.74 20.68 20.55 20.38 20.20 20.00
37.10 31.50 26.89 23.76 22.00 21.17 20.84 20.71 20.61 20.50 20.35 20.18 20.00
36.70 32.32 28.44 25.38 23.27 21.97 21.24 20.86 20.64 20.48 20.33 20.17 20.00
36.70 32.90 29.43 26.53 24.32 22.78 21.77 21.15 20.78 20.53 20.34 20.16 20.00
24.00 26.32 27.08 26.30 24.86 23.43 22.31 21.52 21.00 20.64 20.39 20.18 20.00
24.00 25.04 25.56 25.38 24.63 23.64 22.66 21.86 21.25 20.80 20.48 20.22 20.00
```

REFERENCES

Benjamin, J.G., M.R. Ghaffarzadeh, and R.M. Cruse. 1990. Coupled water and heat transport in ridged soils. Soil Sci. Soc. Am. J. 54:963-969.

Bristow, K.L., G.S. Campbell, R.I. Papendick, and L.F. Elliott. 1986. Simulation of heat and moisture transfer through a surface residue-soil system. Agric. Forest. Meterol. 36:193-214.

Campbell, C.S. 1985. Soil physics with BASIC. Elsevier Sci. Publ., Amsterdam.

Chung, S.O. and R. Horton. 1987. Soil heat and water flow with a partial surface mulch. Water Resour. Res. 23:2175-2186.

Davis, J.B. 1977. Simulation of heat and moisture movement during drying of a two-dimensional soil profile. Ph.D. diss. Clemson Univ., Clemson, SC (Diss. Abstr. 77-29688).

De Vries, D.A. 1952. The thermal conductivity of soil. Meded. Landbouwhogesch. Wageningen 52.

De Vries, D.A. 1963. Thermal properties of soils. p. 210-235. *In* W.R. Van Wijk (ed.) Physics of plant environment. North-Holland Publ. Co., Amsterdam.

Forsythe, W.E. 1964. Smithsonian physical tables. Smithsonian Inst. Publ. 4169. Smithsonian Inst., Washington, DC.

Gupta, S.C., J.K. Radke, and W.E. Larson. 1981. Predicting temperature of bare and residue covered soils with and without a corn crop. Soil Sci. Soc. Am. J. 45:405-412.

Hanks, R.J., D.D. Austin, and W.T. Ondrechen. 1971. Soil temperature estimation by a numerical method. Soil Sci. Soc. Am. Proc. 35:665-677.

Hillel, D., and C.H.M. Van Bavel. 1976. Simulation of profile water storage as related to soil hydraulic properties. Soil Sci. Soc. Am. J. 40:807-815.

Horton, R. 1989. Canopy shading effects on soil heat and water flow. Soil Sci. Soc. Am. J. 53:669-679

Horton, R., O. Aguirre-Luna, and P.J. Wierenga. 1984a. Observed and predicted two-dimensional soil temperature distributions under a row crop. Soil Sci. Soc. Am. J. 48:1147-1152.

Horton, R., O. Aguirre-Luna, and P.J. Wierenga. 1984b. Soil temperature in a row crop with incomplete surface cover. Soil Sci. Soc. Am. J. 48:1225-1232.

Horton, R., and P.J. Wierenga. 1984. The effect of column wetting on soil thermal conductivity. Soil Sci. 138:102-108.

Jackson, R.D. 1960. The importance of convection as a heat transfer mechanism in two phase porous materials. Ph.D. diss. Colorado State Univ., Fort Collins.

Jackson, R.D., and D. Kirkham. 1958. Method of measurement of the real thermal diffusivity of moist soil. Soil Sci. Soc. Am. Proc. 22:479-482.

James, M.L., G.M. Smith, and J.C. Wolford. 1977. Applied numerical methods for digital computation with FORTRAN and CSMP. Harper and Row Publ., New York.

Jury, W.A., and B. Bellantuoni. 1976a. Heat and water movement under surface rocks in a field soil: I. Thermal effects. Soil Sci. Soc. Am. J. 40:505-509.

Jury, W.A., and B. Bellantuoni. 1976b. Heat and water movement under surface rocks in a field soil: II. Moisture effects. Soil Sci. Soc. Am. J. 40:509-513.

Kluitenberg, G.J., and R. Horton. 1990. Analytical solution for two-dimensional heat conduction beneath a partial surface mulch. Soil Sci. Soc. Am. J. 54:1197-1206.

Lascano, R.J., C.H.M. Van Bavel, J.L. Hatfield, and D.R. Upchurch. 1987. Energy and water balance of a sparse crop: Simulated and measured soil and crop evaporation. Soil Sci. Soc. Am. J. 51:1113-1121.

Mahrer, Y. 1982. A theoretical study of the effect of soil surface shape upon the soil temperature profile. Soil Sci. 134:381-387.

Mahrer, Y., and J. Katan. 1981. Spatial soil temperature regime under transparent polyethylene mulch: numerical and experimental studies. Soil Sci. 131:82-87.

Milly, P.C.D. 1982. Moisture and heat transport in hysteretic, inhomogeneous porous media: A matric head-based formulation and a numerical model. Water Resour. Res. 18:489-498.

Murray, F.W. 1967. On the computation of saturation vapor pressure. J. Appl. Meteorol. 6:203-205.

Nassar, I.N., and R. Horton. 1989a. Water transport in unsaturated nonisothermal salty soil: I. Experimental results. Soil Sci. Soc. Am. J. 53:1323-1329.

Nassar, I.N., and R. Horton. 1989b. Water transport in unsaturated nonisothermal salty soil: II. Theoretical development. Soil Sci. Soc. Am. J. 53:1330-1337.

O'Neill, K. 1979. Convective heat transport during unsaturated flow. p. 140. *In* Agronomy abstracts. ASA, Madison, WI.

Parton, W.J. 1984. Predicting soil temperatures in a shortgrass steppe. Soil Sci. 138:93–101.

Persaud, N., and A.C. Chang. 1984. Analysis of the stochastic component in observed soil profile temperature. Soil Sci. 138:326–334.

Philip, J.R., and D.A. De Vries. 1957. Moisture movement in porous material under temperature gradients. Trans. Am. Geophys. Un. 38:222–232.

Remson, I., G.M. Hornberger, and F.J. Molz. 1971. Numerical methods in subsurface hydrology. Wiley-Interscience, New York.

Sophocleous, M. 1979. Analysis of water and heat flow in unsaturated-saturated porous media. Water Resour. Res. 15:1195–1206.

Takakura, T., K.A. Jordan, and L.L. Boyd. 1971. Dynamic simulation of plant growth and environment in the greenhouse. Trans ASAE 14:964–971.

Van Bavel, C.H.M., and D.I. Hillel. 1975. A simulation study of soil heat and moisture dynamics as affected by a dry mulch. p. 815–821. *In* Proc. 1975 Summer Computer Simulation Conf., San Francisco, CA. Simulation Councils Inc., La Jolla, CA.

Van Bavel, C.H.M., and D.I. Hillel. 1976. Calculating potential and actual evaporation from a bare soil surface by simulation of concurrent flow of water and heat. Agric. Meteorol. 17:453–476.

Van Genuchten, M.T. 1980. A closed-form equation for predicting the hydraulic conductivity of unsaturated soils. Soil Sci. Soc. Am. J. 44:892–898.

Van Wijk, W.R., and D.A. De Vries. 1963. Periodic temperature variations in a homogenous soil. p. 102–143. *In* W.R. Van Wijk (ed.) Physics of plant environment. North Holland Publ. Co., Amsterdam.

Westcot, D.W., and P.J. Wierenga. 1974. Transfer of heat by conduction and vapor movement in a closed soil system. Soil Sci. Am. Proc. 39:9–14.

Wierenga, P.J., and C.T. de Wit. 1970. Simulation of heat transfer in soils. Soil Sci. Soc. Am. Proc. 34:845–848.

Wierenga, P.J., R.M. Hagan, and D.R. Nielsen. 1970. Soil temperature profiles during infiltration and redistribution of cool and warm irrigation water. Water Resour. Res. 6:230–238.

Wierenga, P.J., D.R. Nielsen, and R.M. Hagan. 1969. Thermal properties of soil based upon field and laboratory measurements. Soil Sci. Soc. Am. Proc. 33:354–360.

18 Runoff and Water Erosion

J. R. WILLIAMS
USDA-ARS
Temple, Texas

Recently, a mathematical model called Erosion-Productivity Impact Calculator (EPIC) was developed to determine the relationship between soil erosion and soil productivity in the USA (Williams et al., 1983). The EPIC model uses physically based components to simulate erosion, plant growth, and related processes, and economic concepts to assess the cost of erosion, determine optimal management strategies, etc. The physical processes involved are simulated simultaneously and realistically using readily available inputs. Since erosion can be a relatively slow process, EPIC is capable of simulating erosion across hundreds of years if necessary.

The EPIC model was used to analyze the relationships among erosion, productivity, and fertilizer needs as part of the Soil and Water Resources Conservation Act (RCA) analysis for 1985. The EPIC model provided erosion-productivity relationships for about 900 benchmark soils and 500 000 crop/tillage/conservation strategies throughout the USA. Thus, the model had to be generally applicable, computationally efficient, and capable of computing the effects of management decisions. The model components include weather simulation, hydrology, erosion-sedimentation, nutrient cycling, plant growth, tillage, soil temperature, economics, and plant environment control.

Only two of the model components, runoff and water erosion, are described here. The runoff model simulates surface runoff volumes and peak discharge rates when provided with daily rainfall amounts. Runoff volume is estimated using a modification of the Soil Conservation Service (SCS) curve number approach (Soil Conservation Service, 1972). The curve number technique was selected because: (i) it is a reliable procedure that has been used for many years in the USA; (ii) it is computationally efficient; (iii) the required inputs are generally available; and (iv) it relates runoff to soil type, land use, and management practices. The use of readily available daily rainfall data is a particularly important attribute to the curve number technique. For many locations, rainfall data with time increments of less than one day are not available. Also, daily rainfall data manipulation and runoff computation are more efficient than similar operations that use shorter time increments. Peak discharge rate is estimated using a modification of the rational formula. A stochastic element is introduced to the rational equation to allow realistic simulation of peak discharge rates when provided with only daily rainfall and monthly rainfall intensity information.

Copyright © 1991 ASA-CSSA-SSSA, 677 S. Segoe Rd., Madison, WI 53711, USA. *Modeling Plant and Soil Systems*—Agronomy Monograph no. 31.

The water erosion model is based on a modification (Onstad & Foster, 1975) of the universal soil loss equation (USLE) (Wischmeier & Smith, 1978). The Onstad-Foster equation's energy factor is composed of both rainfall and runoff variables. Including runoff as an energy source increases prediction accuracy and provides for individual-event sediment-yield estimates instead of annual soil loss.

I. RUNOFF MODEL

A. Runoff Volume

Surface runoff is predicted for daily rainfall using the SCS curve number (Soil Conservation Service, 1972) equation

$$Q = \frac{(R - 0.2s)^2}{R + 0.8s}, \quad R > 0.2s \qquad [1]$$

$$Q = 0, \qquad R \leq 0.2s$$

where Q is the daily runoff, R is the daily rainfall, and s is a retention parameter. The retention parameter, s, varies among watersheds because of variations among soils, land use, management, and slope. It also varies with time because of changes in soil water content. The retention parameter, s, is related to curve number (CN) using the SCS equation (Soil Conservation Service, 1972)

$$s = 254 \left(\frac{100}{CN} - 1\right). \qquad [2]$$

The constant, 254, in Eq. [2] gives s in millimeters. Thus, R and Q are also expressed in millimeters. Moisture condition 2 (CN_2) or the average curve number can be obtained easily for any area using the *SCS Hydrology Handbook* (Soil Conservation Service, 1972). The handbook tables relate CN_2 to soils, land use, and management. An adjustment to express the effect of slope on runoff is developed here by assuming that the handbook CN_2 value is appropriate for a 5% slope. The equation for adjusting the handbook CN_2 value is

$$CN_{2S} = \frac{1}{3}(CN_3 - CN_2)[1 - 2\exp(-13.86\ S) + CN_2] \qquad [3]$$

where CN_{2S} is the handbook CN_2 value adjusted for slope, CN_3 is the moisture condition (wet) curve number 3, and S is the average watershed slope. Values of CN_1, the moisture condition (dry) curve number 1, and CN_3 corresponding to CN_2 are tabulated in the *SCS Hydrology Handbook*

(Soil Conservation Service, 1972). For computing purposes, CN_1 and CN_3 were related to CN_2 with the equations

$$CN_1 = CN_2 - \frac{20(100 - CN_2)}{100 - CN_2 + \exp[2.533 - 0.0636(100 - CN_2)]} \quad [4]$$

$$CN_3 = CN_2 \exp[0.00673(100 - CN_2)] \quad [5]$$

Fluctuations in soil water content cause s to change according to the equation

$$s = s_1 \left\{ 1 - \frac{FFC}{FFC + \exp[w_1 - w_2 (FFC)]} \right\} \quad [6]$$

where FFC, the fraction of field capacity, is computed with the equation

$$FFC = \frac{SW - WP}{FC - WP} \quad [7]$$

where SW is the soil water content in the root zone, WP is the wilting point water content (1500 kPa for many soils), FC is the field capacity water content (33 kPa for many soils), w_1 and w_2 are shape parameters, and s_1 is the value of s associated with CN_1.

Values for w_1 and w_2 are obtained from a simultaneous solution of Eq. [8] assuming that $s = s_1$ at the wilting point and $s = s_3$ at field capacity.

$$w_1 = \ln \left(\frac{1}{1 - \frac{s_3}{s_1}} - 1 \right) + w_2 \quad [8]$$

$$w_2 = \frac{\ln \left(\frac{1}{1 - \frac{s_3}{s_1}} - 1 \right) - \ln \left(\frac{UL}{1 - \frac{2.54}{s_1}} - UL \right)}{UL - 1} \quad [9]$$

where s_3 is the CN_3 retention parameter and UL is defined by the equation

$$UL = \frac{PO - WP}{FC - WP} \quad [10]$$

where PO is the soil porosity. Thus, CN_1 corresponds to the wilting point and CN_3 to field capacity. At saturation, Eq. [8] allows s to approach its lower limit of zero giving $CN \rightarrow 100$.

There is also a provision for estimating runoff from frozen soil. If the temperature in the second soil layer is <0°C, the retention parameter is reduced using the equation

$$s_f = s_1 [1 - \exp(-0.00292\, s)] \qquad [11]$$

where s_f is the frozen ground retention parameter. Equation [8] increases runoff for frozen soils, but allows significant infiltration when soils are dry.

B. Peak Runoff Rate

Peak runoff rate predictions are based on a modification of the rational formula.

$$q_p = (\rho)(r)(A)/360 \qquad [12]$$

where q_p is the peak runoff rate (m^3 s^{-1}), ρ is a runoff coefficient that expresses the watershed infiltration characteristics, r is the rainfall intensity (mm h^{-1}) for the watershed's time of concentration, and A is the drainage area (ha). The runoff coefficient can be calculated for each storm if the amount of rainfall and runoff are known.

$$\rho = \frac{Q}{R} \qquad [13]$$

Since R is input and Q is computed using Eq. [1], ρ can be calculated directly. Rainfall intensity can be expressed with the relationship

$$r = \frac{R_{tc}}{t_c} \qquad [14]$$

where R_{tc} is the amount of rainfall (mm) during the watershed's time of concentration, t_c, in hours. The value of R_{tc} can be estimated by developing a relationship with total R. The National Weather Service (Hershfield, 1961) provides accumulated rainfall amounts for various durations and frequencies. Generally R_{tc} and R_{24} (24-h duration is appropriate for the daily time stem model) are proportional for various frequencies. Thus,

$$R_{tc} = \alpha\, R_{24} \qquad [15]$$

where α is a dimensionless parameter that expresses the proportion of total rainfall that occurs during t_c.

The peak runoff equation is obtained by substituting Eq. [13], [14], and [15] into Eq. [12].

$$q_p = \frac{(\alpha)(Q)(A)}{360(t_c)} \qquad [16]$$

The time of concentration can be estimated by adding the surface and channel flow times.

$$t_c = t_{cc} + t_{cs} \quad [17]$$

where t_{cc} is the time of concentration for channel flow and t_{cs} is the time of concentration for surface flow in hours. The t_{cc} can be computed using the equation

$$t_{cc} = \frac{L_c}{v_c} \quad [18]$$

where L_c is the average channel flow length (km) for the watershed and v_c is the average channel velocity (m s^{-1}). The average channel flow length can be estimated using the equation

$$L_c = [(L)(L_{ca})]^{1/2} \quad [19]$$

where L is the channel length (km) from the most distant point to the watershed outlet and L_{ca} is the distance (km) along the channel to the watershed centroid. Average velocity can be estimated using Manning's equation assuming a trapezoidal channel with 2:1 side slopes and a 10:1 bottom width-depth ratio. Proper substitution of these assumptions yields

$$t_{cc} = \frac{[(L)(L_{ca})]^{1/2} (n)^{0.75}}{0.489 \, (q_c)^{0.25} (\sigma)^{0.375}} \quad [20]$$

where n is Manning's roughness coefficient, q_c is the average flow rate (m^3 s^{-1}), and σ is the average channel slope (m m^{-1}). Assuming $L_{ca} = 0.5L$, and that the average flow rate is about 6.35 mm h^{-1} and is a function of the square root of drainage area, yields the final equation for t_{cc}

$$t_{cc} = \frac{1.1 \, (L)(n)^{0.75}}{(DA)^{0.125} (\sigma)^{0.375}} \quad [21]$$

A similar approach is used to estimate t_{cs} as

$$t_{cs} = \frac{\lambda}{v_s} \quad [22]$$

where λ is the surface slope length (m) and v_s is the surface flow velocity (m s^{-1}). Considering a strip 1 m wide down the sloping surface and applying Manning's equation gives

$$v_s = \frac{(q_s)^{0.4} (S)^{0.3}}{(n)^{0.6}} \quad [23]$$

where S is the land surface slope (m m^{-1}) and q_s is the average surface flow rate. Assuming that the average flow rate is about 6.35 mm h^{-1} and making substitutions into Eq. [22] and [23] to convert from cubic meters per second to millimeters per hour and from seconds to hours, yields the equation for estimating t_{cs}.

$$t_{cs} = \frac{(\lambda n)^{0.6}}{18(S)^{0.3}} \qquad [24]$$

Although some of the assumptions used in developing Eq. [21] and [24] may appear liberal, Eq. [17] generally provides satisfactory results for small homogeneous watersheds. Since Eq. [21] and [24] are based on hydraulic considerations, they are more reliable than purely empirical equations.

To properly evaluate α, variation in rainfall patterns must be considered. Some storms have short durations; most or all rain occurs during t_c, causing α to approach its upper limit of 1.0. Other storms may have uniform intensities that cause α to approach a minimum value. All other patterns have higher α values than the uniform pattern because r_{tc} is greater than r_{24} for all patterns except the uniform. By substituting the products of intensity and time into Eq. [15], an expression for the minimum value of α is obtained.

$$\alpha_{mn} = \frac{t_c}{24} \qquad [25]$$

Thus, α ranges within the limits

$$\frac{t_c}{24} \leq \alpha \leq 1.0.$$

Although the value of α is confined between limits, there is considerable uncertainty in assigning appropriate values given only daily rainfall and simulated runoff amounts. Thus, α is generated from a gamma function, with the base ranging from $t_c/24$ to 1.0.

The peak of the α distribution changes for each month because of seasonal differences in rainfall intensities. The National Weather Service (U.S. Department of Commerce, 1979) provides information on monthly maximum rainfall intensities that can be used to estimate the peak α for each month.

Since the water erosion model estimates the maximum 0.5-h amount of each daily rainfall ($\alpha_{0.5}$), these estimates are used in calculating α. Besides the convenience of avoiding double calculation, it is important to assure that $\alpha_{0.5}$ and α are closely related for each storm. The relationship between $\alpha_{0.5}$ and α can be obtained from Hershfield (1961) by fitting a log function to the 10-yr frequency rainfall distribution.

$$R_t = R_6 \left(\frac{t}{6}\right)^b \qquad [26]$$

where b is a parameter used to fit the Hershfield (1961) relationship at any location, R_t is the rainfall amount (mm) for any time t, and R_6 is the 6-h rainfall amount (mm). The value of α is computed with the equation

$$\alpha = \alpha_{0.5} \left(\frac{R_{tc}}{R_{0.5}}\right) \qquad [27]$$

Details of the procedure for estimating $\alpha_{0.5}$ are provided in the material describing the water erosion model.

II. WATER EROSION MODEL

The EPIC water erosion model simulates erosion caused by rainfall and runoff, and by irrigation (sprinkler and furrow).

A. Rainfall/Runoff

To simulate rainfall/runoff erosion, EPIC contains three equations—the USLE (Wischmeier & Smith, 1978), the modified USLE (MUSLE) (Williams, 1975), and the Onstad-Foster modification of the USLE (Onstad & Foster, 1975). Only one of the equations (user specified) interacts with other EPIC components. The three equations are identical except for their energy components. The USLE equation depends strictly on rainfall as an indicator of erosive energy. The MUSLE equation uses only runoff variables to simulate erosion and sediment yield. Runoff variables increased the prediction accuracy, eliminated the need for a delivery ratio (used with USLE to estimate sediment yield), and provided an equation designed for single-storm application, instead of the USLE equation's annual estimates. The Onstad-Foster equation contains a combination of the USLE and MUSLE energy factors.

Thus, the water erosion model uses an equation of the form

$$Y = \psi \, (K \times CE \times PE \times LS) \qquad [28]$$

$\psi = EI$ for USLE
$\psi = 11.8(Q \times q_p)^{0.56}$ for MUSLE
$\psi = 0.646 \, EI + 0.45 \, Q \times q_p^{0.33}$ for Onstad-Foster

where Y is the sediment yield (Mg ha^{-1}), Q is the runoff volume (mm), q_p is the peak runoff rate (mm h^{-1}), K is the soil erodibility factor, CE is the crop management factor, PE is the erosion control practice factor, and LS is the slope length and steepness factor. The PE value is determined initially by considering the conservation practices to be applied. The value of LS is calculated with the equation (Wischmeier & Smith, 1978)

$$LS = \left(\frac{\lambda}{22.1}\right) \xi \, (65.41 \, S^2 + 4.56 \, S + 0.065) \quad [29]$$

where S is the land surface slope (m m^{-1}), λ is the slope length (m), and ξ is a parameter dependent upon slope. The value of ξ varies with slope and is estimated with the equation

$$\xi = 0.3 \, S/[S + \exp(-1.47 - 61.09 \, S)] + 0.2. \quad [30]$$

The crop management factor is evaluated for all days when runoff occurs using the equation

$$CE = \exp[(\ln 0.8 - \ln CE_{mnj}) \exp(-0.00115 \, CV) + \ln CE_{mnj}] \quad [31]$$

where CE_{mnj} is the minimum value of the crop management factor for crop j.

The soil erodibility factor, K, is evaluated for the top soil layer at the start of each year of simulation using the equation

$$K = \{0.2 + 0.3 \exp[-0.0256 \, SAN \, (1 - SIL/100)]\}$$

$$\left(\frac{SIL}{CLA + SIL}\right)^{0.3} \left[1 - \frac{0.25 \, C}{C + \exp(3.72 - 2.95 \, C)}\right] \quad [32]$$

where SAN, SIL, CLA, and C are the sand, silt, clay, and organic carbon content of the soil in percent. Equation [32] allows K to vary from about 0.1 to 0.5. The first term yields low K values for soils with high coarse-sand content and high values for soils with little sand. The fine sand content is estimated as the product of sand and silt divided by 100. The expression for coarse sand in the first term is simply the difference between sand and the estimated fine sand. The second term reduces K for soils that have high clay to silt ratios. The third term reduces K for soils with high organic carbon content.

The runoff model supplies estimates of Q and q_p. To estimate the daily rainfall energy in the absence of time-distributed rainfall, it is assumed that the rainfall rate is exponentially distributed.

$$r = r_p \exp(-t/k) \quad [33]$$

where r is the rainfall rate at time t (mm h^{-1}), r_p is the peak rainfall rate (mm h^{-1}), and k is the decay constant (h). Equation [33] contains no assumption about the sequence of rainfall rates (time distribution). The USLE energy equation in metric units is

$$E = \Delta R \left(12.1 + 8.9 \log \frac{R}{\Delta t}\right) \quad [34]$$

where ΔR is a rainfall amount (mm) during a time interval Δt (h). The energy equation can be expressed analytically as

$$E = 12.1 \int_0^\infty r \, dt + 8.9 \int_0^\infty r \log r \, dt \qquad [35]$$

Substituting Eq. [33] into Eq. [35] and integrating, yields the equation for estimating daily rainfall energy

$$E = R \, [12.1 + 8.9 \, (\log r_p - 0.434)] \qquad [36]$$

where R is the daily rainfall amount (mm). The rainfall energy factor, EI, is obtained by multiplying Eq. [36] by the maximum 0.5-h rainfall intensity and converting to the proper units.

$$\text{EI} = R \, [12.1 + 8.9(\log r_p - 0.434)](r_{0.5})/1000 \qquad [37]$$

To compute values for r_p, Eq. [33] is integrated to give

$$R = (r_p)(k) \qquad [38]$$

and

$$R_t = R \, [1 - \exp(-t/k)] \qquad [39]$$

The value of $R_{0.5}$ can be estimated using $\alpha_{0.5}$, as mentioned in the runoff model description

$$R_{0.5} = \alpha_{0.5} \, R \qquad [40]$$

To determine the value of r_p, Eq. [38] and [40] are substituted into Eq. [39] to give

$$r_p = -2 \, R \ln (1 - \alpha_{0.5}) \qquad [41]$$

Since rainfall rates vary seasonally, $\alpha_{0.5}$ is evaluated for each month using National Weather Service information (U.S. Department of Commerce, 1979). The frequency of the maximum 0.5-h rainfall amount is estimated using the equation

$$F = \frac{1}{2\tau} \qquad [42]$$

where F is the frequency of occurrence of the largest event of a total of τ events. The total number of events for each month is the product of the number of years of record and the average number of rainfall events for the month. To estimate the mean value of $\alpha_{0.5}$, it is necessary to estimate the mean value of $R_{0.5}$. The value of $R_{0.5}$ can be computed easily if it is assumed that the maximum 0.5-h rainfall amounts are exponentially distributed. From the exponential distribution, the expression for the mean 0.5-h rainfall amount is

$$\overline{R}_{0.5,j} = \frac{R_{0.5F,j}}{-\ln F_j} \qquad [43]$$

where $\overline{R}_{0.5,j}$ is the mean maximum 0.5-h rainfall amount, $R_{0.5F,j}$ is the maximum 0.5-h rainfall amount for frequency F, and subscript j refers to the month. The mean $\alpha_{0.5}$ is computed with the equation

$$\alpha_{0.5j} = \frac{\overline{R}_{0.5,j}}{\overline{R}_j} \qquad [44]$$

where $\overline{R}$ is the mean amount of rainfall for each event calculated as average monthly rainfall/average number of days of rainfall. Daily values of $\alpha_{0.5}$ are generated from a two-parameter gamma distribution. The base of the gamma distribution is established by examining upper and lower limits of $\alpha_{0.5}$. The lower limit determined by a uniform rainfall rate yields $\alpha_{0.5} = 0.5/24$ or 0.0208. The upper limit of $\alpha_{0.5}$ is set by considering a large rainfall event. If the event is large, it is highly unlikely that all the rainfall would occur in 0.5 h ($\alpha = 1.$). The upper limit of $\alpha_{0.5}$ can be estimated by substituting a high value of r_p (250 mm h^{-1} is generally near the upper limit of rainfall intensity) into Eq. [39].

$$\alpha_{0.5u} = 1 - \exp(-125/R) \qquad [45]$$

where $\alpha_{0.5u}$ is the upper limit of $\alpha_{0.5}$. The peak of the $\alpha_{0.5}$ gamma distribution can be computed using the equation

$$\alpha_{0.5P,j} = \frac{\alpha_{0.5,j}(\nu - 1)}{\nu} \qquad [46]$$

where $\alpha_{0.5P,j}$ is the $\alpha_{0.5}$ value at the peak of the gamma distribution and ν is the gamma distribution shape parameter (a value of 10 is generally satisfactory).

B. Irrigation

Erosion caused by applying irrigation water in furrows is estimated with the MUSLE equation (Williams, 1975).

$$Y = 11.8\ (Q \times q_p)^{0.56}\ (K)(CE)(PE)(LS) \qquad [47]$$

where CE, the crop management factor, has a constant value of 0.5. The volume of runoff is estimated using the equation

$$Q = (EIR)(AIR) \qquad [48]$$

where AIR is the volume of irrigation water applied (mm) and EIR is the runoff ratio. The peak runoff rate is estimated for each furrow using Manning's Equation, and assuming that the flow depth is 0.75 of the ridge height and that the furrow is triangular. If irrigation water is applied to land without furrows, the peak runoff rate is assumed to be 0.00189 m^3 s^{-1} m^{-1} of field width.

III. MODEL TESTS

The runoff and water erosion models have been tested extensively. Test results for an early version of the runoff model (Chemicals Runoff and Erosion from Agricultural Management Systems [CREAMS] hydrology) are presented by Knisel (1980) and Williams and Nicks (1982). Because of the extensive CREAMS tests, only supplemental tests were conducted on the modified version used in EPIC (Williams, 1985). However, the EPIC version was installed in the Simulator for Water Resources in Rural Basins (SWRRB) model with little or no modification and was tested thoroughly on several large watersheds throughout the USA (Arnold & Williams, 1985). The SWRRB runoff component was modified to estimate upland runoff in the Simulator for Production and Utilization of Rangeland (SPUR) model and test results were reported.

The water erosion model has also been tested with data from a variety of watersheds. Cooley and Williams (1983) reported test results for the USLE and MUSLE equations using data from several small Hawaiian watersheds. Kramer (1983, unpublished data) tested MUSLE with data from small watersheds located in the midwestern USA. Smith et al. (1984) performed similar tests with data from the southern plains grasslands. Williams (1982) tested MUSLE with data from 102 watersheds located throughout the USA.

Besides these tests that compare runoff and sediment yield estimates with measured values, the EPIC model results were examined thoroughly by experts as part of the RCA review process. The RCA analysis of 1985 required about 20 000 EPIC simulations of 100 yr each. Experts examined the results to ensure that runoff and erosion estimates were realistic for all regions of the USA.

IV. SUMMARY AND CONCLUSIONS

The EPIC runoff and water erosion models were developed to interact with other model components for use in assessing the effect of erosion on soil productivity. Since erosion can be a relatively slow process, the models were designed to operate efficiently. They were also built to operate any place in the USA with readily available data.

Besides their intended purpose (determining the erosion-productivity relationship for the USA), the models should be useful in solving many soil and water resources problems. A few examples are: (i) determining soil-loss tolerance limits; (ii) managing watershed problems involving water and sediment yield estimates; and (iii) providing hydrologic and erosion components to other special purpose models (water quality, crop yield, etc.).

V. APPENDIX

```
      FUNCTION ADSTG (RN1,AI,XN1,KK,NBG)
C     THIS FUNCTION PROVIDES NUMBERS FROM A GAMMA
C     DISTRIBUTION, GIVEN TWO RANDOM NUMBERS.
      COMMON /BK1/ RST(12,3),PRW(2,12),OBMX(12),
     1OBMN(12),OBSL(12),PCF(12),RH(12),SDTMN(12),
     2SDTMX(12),SRD(12),IX(9),IX0(9),IDG(8),XIM(3),
     3AMSR,EXPK,PIT,RHD,SRA,TA,V1,V3,IGN,LW
    1 ADSTG=RN1
      NBG=NBG+1
      XX=RN1*AI
      RN=AUNIF(IDG(KK))
      RN1=RN
      X1=XN1*ALOG10(XX)
      IF(X1.LT.-20.)GO TO 1
      FU=10.**X1*EXP(XN1*(1.-XX))
      IF(FU.LT.RN) GO TO 1
      RETURN
      END

      FUNCTION ADSTN (RN1,RN2)
C     THIS FUNCTION COMPUTES A STANDARD NORMAL
C     DEVIATE GIVEN TWO RANDOM NUMBERS
      ADSTN=SQRT(-2.*ALOG(RN1))*COS(6.283185*RN2)
      RETURN
      END

      FUNCTION ATRI (BLM,QMN,UPLM,KK)
C     THIS FUNCTION GENERATES NUMBERS FROM A
C     TRIANGULAR DISTRIBUTION GIVEN X AXIS POINTS
C     AT START & END AND PEAK Y VALUE.
      COMMON /BK1/ RST(12,3),PRW(2,12),OBMX(12),
     1OBMN(12),OBSL(12),PCF(12),RH(12),SDTMN(12),
     2SDTMX(12),SRD(12),IX(9),IX0(9),IDG(8),XIM(3),
     3AMSR,EXPK,PIT,RHD,SRA,TA,V1,V3,IGN,LW
      U3=QMN-BLM
      RN=AUNIF(IDG(KK))
      Y=2./(UPLM-BLM)
      B2=UPLM-QMN
      B1=RN/Y
      X1=Y*U3/2.
      IF(RN.GT.X1) GO TO 1
      ATRI=SQRT(2.*B1*U3)+BLM
      GO TO 2
    1 ATRI=UPLM-SQRT(B2*B2-2.*B2*(B1-.5*U3))
    2 IF(KK.EQ.8)GO TO 3
      AMN=(UPLM+QMN+BLM)/3.
      ATRI=ATRI*QMN/AMN
      IF(ATRI.GE.1.) ATRI=.99
    3 RETURN
      END

      SUBROUTINE EYSED (*)
C     THIS SUBROUTINE PREDICTS DAILY SOIL LOSS
C     CAUSED BY WATER EROSION AND ESTIMATES THE
C     NUTRIENT ENRICHMENT RATIO.
      COMMON /BK2/ DIR(12,16),HRLT(12),RMX(12),
     1TAMX(12),TAV(12),WVL(12),AL1,ANG,BAS,CF,CO2,
     2ELEV,HR1,PI2,R,RA,RM,SML,SNO,SNOF,TH,TMN,TMX,
     3TX,WAL,WB,WV
```

RUNOFF AND WATER EROSION

```
      COMMON /BK4/CND(30,20),WI(12),AB,AB1,AW,DAT,
     1EI,PR,QP,R1,R2,REP,RN1,RN2,TC
      COMMON /BK5/RR(23),BIR,CAP,CN,CN0,DKHL,DV,
     1FDSF,PDSW,QD,SMX,SQ,VIMX,IAC,NUDK
      COMMON /BK6/ BW(3,11),CVM(12),CX(12),YSD(3),
     1AF,CC,CEJ,CV,DA,DR,EK,FL,FW,PEC,SL,
     2SLT,USL,WK,YM,YW
      COMMON /BK7/ QIN(12),LID(11),FC(10),PO(10),
     1S15(10),SC(10),SEV(10),ST(10),ADEO,ADRF,AWC,
     2QINT,SAT,SW,SWW,WTBL,WTMN,WTMX,IR,LD1,NE,NS
      COMMON /BK9/RTN(10),WMN(10),WN(10),WP(10),
     1 WT(10),ER,THK,Y,YAP,YN,YP
      COMMON /BK11/ BN(4,11),BP(4,11),REG,UN1,UN2,
     1 UNO3,UP1,UP2,UPP,WFX,IRO,JE
      COMMON /BK14/SOL(6,10),SID(11),ALS(10),BD(10),
     1 BDD(10),BDM(10),BDP(10),CAC(10),CBN(10),
     2 CEC(10),CLA(10),PH(10),ROK(10),SAN(10),
     3 SIL(10),SMB(10),Z(10),LORG(10),RZ,TLA,ZF,
     4 ZQT,LC
      COMMON /BK15/ NC(13),IDA,IDS,IFA,IRI,IY,IYT,
     1 KDA,KR1,KR2,KR3,KR4,KR5,KR6,KR7,KW,KW1,KW2,
     2 KW3,LYR,MAT,MO,NBG,NCL,ND,NII,NOP,NRO,NT
      COMMON /BK17/ CNY(11),CPY(11),FON(10),FOP(10),
     1HUM(10),RSD(10),AGP,BTN,BTP,CDG,CMN,HMN,
     2RMNR,STD,STDN,STDP,SUT,TRSD,WDN,WIM,WIP,WMP,JD
      COMMON /BK19/PHU(12,30),DLAP(12,2),FRST(12,2),
     1WAC2(12,2),RIN(20),ALT(12),DLAI(12),DMLA(12),
     2GSI(12),HI(12),HMX(12),RBMD(12),RDMX(12),
     3RLAD(12),SWH(12),SWP(12),TG(12),VPD2(12),
     4VPTH(12),WA(12),WAVP(12),WSYF(12),IDC(12),
     5IHU(12),NCR(12),NHU(12),CAF(11),AJHI,AJWA,
     6CHT,DDM,DM,DM1,HU,RW,STL,SYP,WLV,XDLAI,YLD,
     7YLN,YLP,IG,IHV,IPL,JDHU,JPL,KC
      COMMON /BK22/SMM(58,12),EST(17,10),LY(30,4),
     1SMY(58),STDA(4,12),TIL(20,2),NCP(30),NTL(30),
     2SCRP(14,2),COTL(20),SF(5,4),IHC(20),PARM(14),
     3TRA(12),COSD(11),PST(11),SDW(11),TB(11),
     4WCY(11),SFMO(5),CAW(4),CSTF(4),DMF(4),ETG(4),
     5HUF(4),RDF(4),VIR(4),YLDF(4),YLNF(4),BFT,
     6COIR,COL,CON,COP,EFI,FFC,PSTF,RZSW,SDN,
     7SHM,SHRL,SIM,SIP,SMP,SMR,SN,SP,TSFN,WS,
     8IPD,IRR,NPST
      JJ=JD
      IF(STL.GT.RSD(LD1)+STD) JJ=JE
      CC=EXP((-.2231-CVM(JJ))*EXP(-1.15*CV)+CVM(JJ))
      CC=CC*EXP(-.03*ROK(LD1))
      IF(R.LT.12.7) GO TO 1
      YSD(3)=EI*CC*USL*1.292
      SMM(15,MO)=SMM(15,MO)+CC*EI
      SMM(13,MO)=SMM(13,MO)+EI
    1 IF(QD.EQ.0.) RETURN 1
      D=4.605*R/REP
      F=(R-QD)/D
      REP=REP-F
      YSD(2)=(.646*EI+.45*QD*QP**.3333)*CC*USL
      CX(MO)=CX(MO)+1.
      QX=QP*DAT
      QDF=QD*AF
      YSD(1)=(QX*QDF)**.56*CC*YM
```

```
          DR=(QP/REP)**.56
          CY=YSD(1)*DA/QDF
          DR1=1./DR
          B2=-ALOG10(DR1)/2.699
          B1=1./.25**B2
          ER=B1*(CY+1.E-20)**B2
          IF(ER.LT.1.)GO TO 2
          IF(ER.GT.5.) ER=5.
          GO TO 3
        2 ER=1.
        3 RETURN
          END
          SUBROUTINE HRFEI
C         THIS SUBROUTINE ESTIMATES THE USLE RAINFALL
C         ENERGY FACTOR, GIVEN DAILY RAINFALL.
          COMMON /BK1/ RST(12,3),PRW(2,12),OBMX(12),
         1OBMN(12),OBSL(12),PCF(12),RH(12),SDTMN(12),
         2SDTMX(12),SRD(12),IX(9),IX0(9),IDG(8),XIM(3),
         3AMSR,EXPK,PIT,RHD,SRA,TA,V1,V3,IGN,LW
          COMMON /BK2/ DIR(12,16),HRLT(12),RMX(12),
         1TAMX(12),TAV(12),WVL(12),AL1,ANG,BAS,CF,CO2,
         2ELEV,HR1,PI2,R,RA,RM,SML,SNO,SNOF,TH,TMN,TMX,
         3TX,WAL,WB,WV
          COMMON /BK4/CND(30,20),WI(12),AB,AB1,AW,DAT,
         1EI,PR,QP,R1,R2,REP,RN1,RN2,TC
          COMMON /BK15/ NC(13),IDA,IDS,IFA,IRI,IY,IYT,
         1KDA,KR1,KR2,KR3,KR4,KR5,KR6,KR7,KW,KW1,KW2,
         2KW3,LYR,MAT,MO,NBG,NCL,ND,NII,NOP,NRO,NT
          EI=R-SML
          IF(EI.GT.0.) GO TO 1
          R1=AB
          REP=R1*SML/TC
          GO TO 2
        1 AI=AB1/(WI(MO)-AB)
          AJP=1.-EXP(-125./(EI+5.))
          R1=ADSTG(RN1,AI,10.,4,NBG)
          NCL=NCL+1
          R2=(EI*(AB+R1*(AJP-AB))+SML*AB)/R
          R1=R2*AW
          IF(R1.GE.1.) R1=.999
          PR=2.*R2*EI
          REP=-2.*EI*ALOG(1.-R2)
          EI=EI*(12.1+8.9*(ALOG10(REP)-.4343))*PR/1000.
          IF(EI.LT.0.) EI=0.
        2 RETURN
          END

          SUBROUTINE HVOLQ
C         THIS SUBROUTINE PREDICTS DAILY RUNOFF VOLUME
C         AND PEAK RUNOFF RATE GIVEN DAILY PRECIPITATION
C         AND SNOW MELT.
          COMMON /BK2/ DIR(12,16),HRLT(12),RMX(12),
         1TAMX(12),TAV(12),WVL(12),AL1,ANG,BAS,CF,CO2,
         2ELEV,HR1,PI2,R,RA,RM,SML,SNO,SNOF,TH,TMN,TMX,
         3TX,WAL,WB,WV
          COMMON /BK4/CND(30,20),WI(12),AB,AB1,AW,DAT,
         1EI,PR,QP,R1,R2,REP,RN1,RN2,TC
          COMMON /BK5/RR(23),BIR,CAP,CN,CN0,DKHL,DV,
         1FDSF,PDSW,QD,SMX,SQ,VIMX,IAC,NUDK
          COMMON /BK7/ QIN(12),LID(11),FC(10),PO(10),
```

```
     1S15(10),SC(10),SEV(10),ST(10),ADEO,ADRF,
     2AWC,QINT,SAT,SW,SWW,WTBL,WTMN,WTMX,IR,LD1,NE,NS
      COMMON /BK12/ U(10),COST,GX,RD,S,SX,TYN,TYP,
     1 UB1,UOB,UX
      COMMON /BK14/SOL(6,10),SID(11),ALS(10),BD(10),
     1BDD(10),BDM(10),BDP(10),CAC(10),CBN(10),
     2CEC(10),CLA(10),PH(10),ROK(10),SAN(10),
     3SIL(10),SMB(10),Z(10),LORG(10),RZ,TLA,ZF,
     4 ZQT,LC
      COMMON /BK15/ NC(13),IDA,IDS,IFA,IRI,IY,IYT,
     1KDA,KR1,KR2,KR3,KR4,KR5,KR6,KR7,KW,KW1,KW2,
     2KW3,LYR,MAT,MO,NBG,NCL,ND,NII,NOP,NRO,NT
      COMMON /BK18/ WFT(12),T(10),T0(5),ABD,AVT,BCV,
     1 DD,DST0,ST0
      COMMON /BK19/PHU(12,30),DLAP(12,2),FRST(12,2),
     1WAC2(12,2),RIN(20),ALT(12),DLAI(12),DMLA(12),
     2GSI(12),HI(12),HMX(12),RBMD(12),RDMX(12),
     3RLAD(12),SWH(12),SWP(12),TG(12),VPD2(12),
     4VPTH(12),WA(12),WAVP(12),WSYF(12),IDC(12),
     5IHU(12),NCR(12),NHU(12),CAF(11),AJHI,AJWA,
     6CHT,DDM,DM,DM1,HU,RW,STL,SYP,WLV,XDLAI,YLD,
     7YLN,YLP,IG,IHV,IPL,JDHU,JPL,KC
      COMMON /BK21/HUSC(30,20),LT(30,20),DKH(20),
     1DKI(20),EMX(20),HE(20),ORHI(20),RHT(20),
     2TLD(20),BIG,DHT,DI,RHTT,RINT,IDRL,JP,JT1,
     3JT2,KT
      COMMON /BK22/SMM(58,12),EST(17,10),LY(30,4),
     1SMY(58),STDA(4,12),TIL(20,2),NCP(30),NTL(30),
     2SCRP(14,2),COTL(20),SF(5,4),IHC(20),PARM(14),
     3TRA(12),COSD(11),PST(11),SDW(11),TB(11),
     4WCY(11),SFMO(5),CAW(4),CSTF(4),DMF(4),ETG(4),
     5HUF(4),RDF(4),VIR(4),YLDF(4),YLNF(4),BFT,
     6COIR,COL,CON,COP,EFI,FFC,PSTF,RZSW,SDN,
     7SHM,SHRL,SIM,SIP,SMP,SMR,SN,SP,TSFN,WS,
     8IPD,IRR,NPST
      SUM=0.
      XX=0.
      ADD=0.
      DO 1 JJ=1,NS
         J=LID(JJ)
         IF(Z(J).GT.1.)GO TO 6
         ZZ=(Z(J)-XX)/Z(J)
         SUM=SUM+(ST(J)-S15(J))*ZZ/(FC(J)-S15(J))
         ADD=ADD+ZZ
         XX=Z(J)
    1 CONTINUE
      GO TO 7
    6 ZZ=1.-XX
      SUM=SUM+(ST(J)-S15(J))*ZZ/(FC(J)-S15(J))
      ADD=ADD+ZZ
    7 SUM=100.*SUM/ADD
      SCN=SMX*(1.-SUM/(SUM+EXP(SCRP(4,1)-
     1 SCRP(4,2)*SUM)))
      IF(T(LID(2)).GT.-5.) GO TO 2
      SCN=SCN*(1.-EXP(-SCN/SMX))
    2 CNM=25400./(SCN+254.)
      UPLM=AMIN1(99.5,CNM+5.)
      BLM=AMAX1(1.,CNM-5.)
      CN=ATRI(BLM,CNM,UPLM,8)
```

```
      SCN=25400./CN-254.
      BB=.2*SCN
      PB=R-BB
      IF(PB.LE.0.) GO TO 4
      QD=PB*PB/(R+.8*SCN)
      IF(DHT.EQ.0.) GO TO 4
      CALL HFURD
      IF(NOP.GT.0) WRITE (KW,5) MO,KDA,DHT,RHTT,QD,DV
      IF(QD.GT.DV-AMAX1(0.,ST(LD1)-PO(LD1)))GO TO 3
      QD=0.
      GO TO 4
    3 DHT=0.
      IF(IDRL.EQ.1) GO TO 4
      IF(CHT.GT.1.) GO TO 4
      NUDK=1
    4 QP=R1*QD/TC
      RETURN
C
C
    5 FORMAT (6X,2I4,12X,'DHT = ',F5.0,
     1 ' MM',3X,'RHTT = ',F5.0,' MM',3X,
     2 'Q = ',F5.1,' MM',3X,'DV = ',F5.1,' MM')
      END
```

REFERENCES

Arnold, J.G., and J.R. Williams. 1985. Validation of SWRRB—A simulator for water resources in rural basins. p. 107-114. *In* Proc. Symp., Denver, CO. 30 April-1 May 1985. Irrig. Drain. Div. Am. Soc. Civ. Eng., New York.

Cooley, K.R., and J.R. Williams. 1983. Applicability of the USLE and MUSLE to Hawaiian Agricultural Lands. p. 509-522. *In* S.A. El-Swaify et al. (ed.) Proc. Int. Conf. Soil Erosion and Conserv., Honolulu, HI. 16-22 Jan. 1983. Soil Conserv. Soc. Am., Ankeny, IA.

Hershfield, D.M. 1961. Rainfall frequency atlas of the United States for durations from 30 minutes to 24 hours and return periods from 1 to 100 years. U.S. Dep. Commerce Tech. Paper 40. U.S. Gov. Print. Office, Washington, DC.

Knisel, W.G. 1980. CREAMS, A field scale model for chemicals, runoff, and erosion from agricultural management systems. USDA Conserv. Res. Rep. 27. U.S. Gov. Print. Office, Washington, DC.

Onstad, C.A., and G.R. Foster. 1975. Erosion modeling on a watershed. Trans. ASAE 18(2):288-292.

Smith, S.J., J.R. Williams, R.G. Menzel, and G.A. Coleman. 1984. Prediction of sediment yield from southern plains grasslands with the modified universal soil loss equation. J. Range Manage. 37(4):295-297.

Soil Conservation Service. 1972. National engineering handbook. USDA-SCS, U.S. Gov. Print. Office, Washington, DC.

U.S. Department of Commerce. 1979. Maximum short duration rainfall. National summary, climatic data. U.S. Dep. of Commerce Bull. U.S. Gov. Print. Office, Washington, DC.

Williams, J.R. 1975. Sediment yield prediction with universal equation using runoff energy factor. p. 244-252. *In* Present and prospective technology for predicting sediment yields and sources. USDA-ARS Ser. 40. USDA-ARS, New Orleans, LA.

Williams, J.R. 1982. Testing the modified universal soil loss equation. p. 157-165. *In* Proc. Workshop on Estimating Erosion and Sediment Yield on Rangelands, Tucson, AZ. 7-9 March 1981. Agric. Reviews and Manuals, ARM-W-26. USDA-ARS, Oakland, CA.

Williams, J.R. 1985. The physical components of the EPIC model. p. 272-284. *In* Soil erosion and conservation. Soil Conserv. Soc. Am., Ankeny, IA.

Williams, J.R., P.T. Dyke, and C.A. Jones. 1983. EPIC—A model for assessing the effects of erosion on soil productivity. p. 553–572. *In* W.K. Laurenroth et al. (ed.) Proc. Third Int. Conf. on State-of-the-Art in Ecological Modelling, Fort Collins, CO. 24–28 May 1982. Elsevier Sci. Publ. Co., New York.

Williams, J.R., and A.D. Nicks. 1982. CREAMS hydrology model—option one. p. 69–86. *In* V.P. Singh (ed.) Applied modeling catchment hydrology. Proc. Int. Symp. Rainfall-Runoff Modelling, Mississippi State, MS. 18–21 May 1981. Water Resour. Publ., Littleton, CO.

Wischmeier, W.H., and D.D. Smith. 1978. Predicting rainfall erosion losses, a guide to conservation planning. USDA Agric. Handb. 537. U.S. Gov. Print. Office, Washington, DC.

19 Modified EPIC Wind Erosion Model

EDWARD L. SKIDMORE

USDA-ARS and
Kansas State University
Manhattan, Kansas

J. R. WILLIAMS

USDA-ARS
Temple, Texas

A model proposed by Woodruff and Siddoway (1965), titled a wind erosion equation, has been used extensively with various modifications during the past quarter century. The model was developed as a result of investigations to understand the mechanics of the wind erosion process, to identify major factors influencing wind erosion, and to develop wind erosion control methods. The general functional relationship between the dependent variable, WE (the potential average annual soil loss), and the equivalent variables or major factors is

$$WE = f(I, WK, WC, WL, VE) \qquad [1]$$

where I is soil erodibility index, WK is soil ridge-roughness factor, WC is a climatic factor, WL is the unsheltered median travel distance of wind across a field, and VE is equivalent vegetative cover. These factors will be discussed in more detail later.

The solution of the wind erosion equation yields the expected amount of erosion from a given agricultural field. A second application of the equation involves specifying the amount of erosion that can be tolerated and then solving the equation to determine conditions (i.e., amount of residue, field width, etc.) to limit soil loss to the specified amount. The equation has been used widely for both of these applications.

The USDA Soil Conservation Service has used the equation extensively to plan wind erosion control practices (Hayes, 1966). Hayes (1965) also used the equation to estimate crop tolerance to wind erosion. The equation also is a useful guide to wind erosion control principles (Carreker, 1966; Moldenhauer & Duncan, 1969; Woodruff et al., 1972). Other uses of the equation include: (i) determining spacing for barriers in narrow strip-barrier systems (Hagen et al., 1972); (ii) estimating fugitive dust emissions from agricultural and subdivision lands (PEDCO, 1973; Wilson, 1975); (iii) predicting horizon-

---

Copyright © 1991 ASA-CSSA-SSSA, 677 S. Segoe Rd., Madison, WI 53711, USA. *Modeling Plant and Soil Systems*—Agronomy Monograph no. 31.

tal soil fluxes to compare with vertical aerosol fluxes (Gillette et al., 1972); (iv) estimating effects of wind erosion on productivity (Lyles, 1975; Williams et al., 1984); (v) delineating those croplands in the Great Plains from which various amounts of crop residues may be removed without exposing the soil to excessive wind erosion (Skidmore et al., 1979); and (vi) estimating the erosion hazard in a national inventory (USDA, 1989).

Solving the functional relationships of the wind erosion equation as presented by Woodruff and Siddoway (1965) required the use of tables and figures. The awkwardness of the manual solution prompted a computer solution (Fisher & Skidmore, 1970; Skidmore et al., 1970) and later the development of a slide-rule calculator (Skidmore, 1983). The computer solution not only predicted average annual soil loss, but solved the Woodruff and Siddoway (1965) equation to determine field conditions necessary to reduce potential erosion to a tolerable amount. It also allowed the user to look at many combinations of wind erosion control practices for particular field and climatic conditions.

The model has been adopted for use with personal computers (Halsey et al., 1983) and interactive programs (Erickson et al., 1984). Cole et al. (1983) adapted the Woodruff and Siddoway (1965) model for simulating daily wind erosion soil loss as a submodel in the erosion productivity impact calculator (EPIC) developed by Williams et al. (1984). The latter version, which was simplified by fitting equations to the figures of Woodruff and Siddoway (1965), with additional modifications, is described in this presentation.

The Woodruff and Siddoway (1965) wind erosion model was converted from annual to daily prediction to interface with EPIC. However, two variables, soil erodibility (I) and climatic factor (WC), remained constant for each day of the year. The other variables were subject to daily variation as simulated by EPIC. In this adaption, the user will supply variable inputs for accounting periods, most likely monthly.

I. MODEL DESCRIPTION

A. Soil Erodibility Index

Soil erodibility can be estimated by standard dry sieving and use of Table 19-1. Using sieving results assumes that the percent of soil aggregates >0.84 mm characterize a soil's aggregate status during the erosion period. In current practice, erodibility is often estimated by grouping soils, often according to predominant textural class (Table 19-2).

If the average percentage of dry soil fractions >0.84 mm is 24, the corresponding soil erodibility index from Table 19-1 is 197 Mg/ha.

B. Ridge Roughness Factor

The ridge-roughness factor estimates the fractional reduction of erosion caused by ridges of nonerodible aggregates. It is influenced by ridge spacing and height, and is defined relative to a 1:4 ridge height to ridge spacing ratio.

Table 19-1. Soil erodibility I for soils with various percentages of nonerodible fractions as determined by standard dry sieving (Woodruff & Siddoway, 1965).

Tens	Percentage of dry soil fractions >0.84 mm									
	0	1	2	3	4	5	6	7	8	9
	Mg/ha									
0	--	695	560	493	437	404	381	359	336	314
10	300	294	287	280	271	262	253	244	238	228
20	220	213	206	202	197	193	186	182	177	170
30	166	161	159	155	150	146	141	139	134	130
40	126	121	117	114	112	108	105	101	96	92
50	85	80	75	70	65	61	56	54	52	49
60	47	45	43	40	38	36	36	34	31	29
70	27	25	22	18	16	13	9	7	7	4
80	4	--	--	--	--	--	--	--	--	--

Table 19-2. Descriptions of wind erodibility groups (WEG) (Soil Conservation Service, 1988).

WEG	Predominant soil texture class of surface layer	Dry soil aggregates >0.84 mm	Wind erodibility index (I)
		%	Mg/ha
1	Very fine sand, fine, sand, or coarse sand	1 2 3 5 7	695 560 493 404 359
2	Loamy very fine, sand, loamy fine sand, loamy sand, loamy coarse sand, or sapric soil materials	10	300
3	Very fine sandy loam, fine sandy loam, sandy loam, or coarse sandy loam	25	193
4	Clay, silty clay, noncalcareous clay loam, or silty clay loam with more than 35% clay content	25	193
4L	Calcareous loam and silt loam, or calcareous clay loam, and silty clay loam	25	193
5	Noncalcareous loam and silt loam with less than 20% clay content, or sandy clay loam, sandy clay, and hemic organic soil materials	40	126
6	Noncalcareous loam and silt loam with more than 20% clay content, or noncalcareous clay loam with less than 35% clay content	45	108
7	Silt, noncalcareous silty clay loam with less than 35% clay content and fibric organic soil material	50	85
8	Soils not susceptible to wind	>80	0

$$KR = \frac{4HR^2}{IR} \quad [2]$$

where KR is the ridge roughness (mm), HR is the ridge height (mm), and IR is the ridge interval (mm). The ridge-roughness factor is a function of ridge roughness as expressed by the equations

$$WK = 1.0, \ KR < 2.27 \qquad [3]$$

$$WK = 1.125 - 0.153 \ \ln(KR), \ 2.27 \le KR < 89 \qquad [4]$$

$$WK = 0.336 \ \exp(0.00324 \ KR), \ KR \ge 89 \qquad [5]$$

C. Climatic Factor

Chepil et al. (1962) proposed a climatic factor to determine average annual soil loss for climatic conditions other than those occurring when the relationship between wind tunnel and field erosion was obtained. It is an index of wind erosion as influenced by moisture content in surface soil particles and by average windspeed.

The windspeed term of the climatic factor was based on the rate of soil movement being proportional to average windspeed cubed (Bagnold, 1943; Chepil, 1945; Zingg, 1953). The soil moisture term was developed on the basis that erodibility of soil varies inversely with the square of equivalent water content in the near surface soil, which was assumed to vary with the Thornthwaite index (Chepil, 1956).

The climatic factor was expressed as

$$C = 386 \ \frac{\bar{u}^3}{(PE)^2} \qquad [6]$$

where $\bar{u}$ is the mean annual windspeed corrected to 9.1 m and PE is the Thornthwaite (1931) index. The value of 386 indexes the factor to conditions at Garden City, KS.

As the PE index gets small when precipitation is slight, as in arid regions, the climatic factor in Eq. [6] approaches infinity. In application, an upper limit was established by restricting minimum monthly precipitation to 13 mm (Lyles, 1983). The FAO (1979) modified the Chepil et al. (1962) index so that, as precipitation approaches zero, windspeed dominates the climatic factor. Conversely, as precipitation approaches potential evaporation, the climatic factor approaches zero. The influence of soil water in the FAO version is less than the squared influence of soil water demonstrated by Chepil (1956).

Skidmore (1986) proposed

$$CE = \rho \int_R^\infty (u^2 - R^2)^{3/2} f(u) du \qquad [7]$$

where

$$R = u_t^2 + \gamma'/\rho a^2, \qquad [8]$$

CE is wind-erosion climatic erosivity and is directly proportional to the mass flow rate of an all-erodible material, ρ is air density, u and u_t are windspeed and threshold windspeed, respectively, γ' is cohesive resistance of absorbed water, and a is a combination of constants, $k/\ln(z/z_0)$. When $k = 0.41$, $z = 10$ m, and $z_0 = 0.05$ m, then $a = 0.0774$.

The value of γ' is approximated by

$$\gamma' = 0.5\,\omega^2 \qquad [9]$$

where ω is equivalent soil-water content; fraction of water (by mass or volume) in the soil divided by fraction of water in the same soil at -1500 J/kg (Chepil, 1956; Skidmore, 1986). It was assumed that equivalent surface-water content was approximated by the ratio of precipitation to potential evaporation.

The windspeed probability density function, Eq. [7], may be expressed as the Weibull distribution

$$f(u) = (k/c)(u/c)^{k-1}\exp[-(u/c)^k] \qquad [10]$$

where c and k are scale and shape parameters, respectively. Parameter c has units of velocity and k is dimensionless (Justus et al., 1976; Apt, 1976). Weibull parameters have been determined from windspeed distribution summaries for many locations in the Great Plains (Hagen et al., 1980).

Equation [7] expresses wind power (W m^{-2}). When multiplied by the time duration in the accounting period represented by $f(u)$, it yields erosive wind energy. This is the energy of the wind in excess of that necessary to overcome threshold shear stresses represented by R. It indicates the time-average values for meteorological elements tendency to cause wind erosion. It accounts for the meteorological influence of both wind and precipitation. It also overcomes a threshold windspeed for surface particles. Thus, erosive wind energy becomes a useful parameter to evaluate climatic factor for the wind erosion equation. As the climatic factor in the wind erosion equation, the result of Eq. [7] must be converted to an annual basis and referenced to the standard (Skidmore, 1986).

Therefore, using the summation notation for evaluating Eq. [7]

$$WC = \frac{100}{GC}\sum_{\substack{i=1 \\ u_{i+0.5}>R}}^{n}(u_{i+0.5}^2 - R)^{3/2}[F(u_{i+1}) - F(u_i)] \qquad [11]$$

where $F(u_i)$ is the cumulative distribution function

$$F(u_i) = 1 - \exp[-(u_i/c)^k]. \qquad [12]$$

Choose an n large enough so that $F(u_{n+1}) \approx 1.0$.

The notation $u_{i+0.5}$ refers to a windspeed midway between u_{i+1} and u_i. GC indexes WC to reference conditions and has the value as calculated by Eq. [11] and using long-term climatic averages for Garden City, KS.

D. Field Length

Originally, field length was considered as the length of a field in the prevailing wind erosion direction (Woodruff & Siddoway, 1965). However, sometimes almost as much wind occurs from one direction as from another, so there is essentially no prevailing wind erosion direction. In these cases, the preponderance of wind erosion forces in the prevailing wind erosion direction was used to assess equivalent field length (Skidmore & Woodruff, 1968; Skidmore, 1965). Later, from a more detailed analysis, tables were prepared that provide wind-erosion direction factors, numbers that when multiplied by field width, yield median travel distance. The factors are functions of preponderance of wind erosion forces in the prevailing direction and the deviation of prevailing wind erosion direction from perpendicular to the direction of field length (Skidmore, 1987).

In some of the modeling efforts, the procedure for determining L for use in the wind erosion equation was simplified by ignoring wind direction distributions. Cole et al. (1983) suggested

$$WL - FW \sec \theta \; WL \leq (FL^2 + FW^2)^{1/2} \quad [13]$$

$$WL = WL \csc \theta \text{ in all other cases} \quad [14]$$

where FW and FL are the small and large dimensions, respectively, of a rectangular field, and θ is the angle between side w and the prevailing wind erosion direction. As θ varies through $\pi/2$ radians, WL will range from FW to FL with a maximum equal to the main diagonal of the field. The procedure Williams et al. (1984) used in EPIC was

$$WL = \frac{(FL)(FW)}{FL \left|\cos\left(\frac{\pi}{2} + \theta - \phi\right)\right| + FW \left|\sin\left(\frac{\pi}{2} + \theta - \phi\right)\right|} \quad [15]$$

where FL is the field length (m), FW is the field width (m), θ is the wind direction clockwise from north in radians, and ϕ is the clockwise angle between field length and north in radians.

II. VEGETATIVE FACTOR

The value of crop residue for controlling wind erosion was recognized early (Chepil, 1944). Siddoway et al. (1965) quantified the specific properties of vegetative covers influencing soil erodibility. They also developed regression equations relating soil loss by wind to selected amounts, kinds, and orientation of vegatative covers, wind velocity, and soil cloddiness. Their studies led to the relationship developed by Woodruff and Siddoway (1965) showing the influence of an equivalent vegetative cover of small grain and sorghum [*Sorghum bicolor* (L.) Moench] stubble for various orientations (flat, standing) and heights, and then relating soil loss to equivalent vegetative cover.

Efforts to evaluate the protective role of additional crops have continued. Lyles and Allison (1980, 1981), in wind-tunnel tests, determined equivalent

EPIC WIND EROSION MODEL

wind-erosion protection from selected range grasses and crop residue. They found high simple-correlation coefficients from an equation of the form

$$(SG)_e = aX^b \qquad [16]$$

where $(SG)_e$ is flat small-grain equivalent (kg/ha), X is the quantity of residue or grass to be converted, and a and b are constants. Prediction equation coefficients are given in Tables 19-3 and 19-4.

It is not practical in testing all combinations of crops and residues to determine their protection value as flat small-grain equivalents. Therefore, a method is needed to estimate the protection values of crops and residues not tested.

Until recently, all small-grain equivalence data have been limited to dead crop residue or dormant grass. Armbrust and Lyles (1985) reported flat small-grain equivalents for five growing crops (corn [*Zea mays* L.], cotton [*Gossypium hirsutum* L.], grain sorghum, peanut [*Arachis hypogaea* L.], and soybean [*Glycine max* (L.) Merr]).

$$(SG)_e = a_1 Rw^{b_1} \qquad [17]$$

where Rw is the aboveground dry weight of the crop to be converted (kg/ha), and a_1 and b_1 are constant coefficients for each crop. They found that if only rough estimates of $(SG)_e$ are needed, an average coefficient could be used. An average equation determined from pooling all crop data with rows running perpendicular to wind direction yielded 8.9 and 0.9 for a_1 and b_1, respectively.

Suppose we wish to know the equivalent, flat, small-grain residue for a field with grain sorghum growing in 400 kg/ha of flat-random winter wheat (*Triticum aestivum* L.) residue when the dry weight of the growing grain sorghum is 83 kg/ha and the grain sorghum is growing in rows perpendicular to the expected wind. In this case, the $(SG)_e$ for the growing sorghum Eq. [11] would be

$$(SG)_e = 8.9(83)^{0.9} = 475 \qquad [18]$$

and from Table 19-3 for the wheat residue

$$(SG)_e = 7.3(400)^{0.8} = 880. \qquad [19]$$

However, because of nonlinear relationships, the flat small-grain equivalents are not strictly additive. When more than one crop contributes to the residue, it is better to use a single equation of the form

$$(SG)_e = a_1^{p_1} a_2^{p_2} (Rwt)^{b_1 p_1 + b_2 p_2} \qquad [20]$$

where p_1 and p_2 are fractions of total residue, Rwt; a_1, a_2, b_1, and b_2 are constant coefficients for respective crops as in Eq. [11].

Table 19–3. Coefficients in the prediction equation, $(SG)_e = aR_w^b$, for conversion of crop residues to equivalent quantity of flat, small-grain residue (kg/ha) (Lyles & Allison, 1981).

Crop residue	Surface orientation	Height	Length	Row spacing	Row orientation to flow	Prediction equation coefficients		
						a	b	r^2
		cm						
Winter wheat	Standing	25.4	--	25.4	Normal	4.306	0.970	0.997
Rape†	Standing	25.4	--	25.4	Normal	0.103	1.400	0.990
Cotton	Standing	34.3	--	76.2	Normal	0.188	1.145	0.998
Sunflower	Standing	43.2	--	76.2	Normal	0.021	1.342	0.994
Winter wheat	Flat random	--	25.4	--	--	7.279	0.782	0.993
Soybean	Flat random	--	25.4	--	--	0.167	1.173	0.993
Rape	Flat random	--	25.4	--	--	0.064	1.294	0.997
Cotton	Flat random	--	25.4	--	--	0.077	1.168	0.998
Sunflower	Flat random	--	43.2	--	--	0.011	1.368	0.993
Forage sorghum	Standing	15.9	--	76.2	Normal	0.353	1.124	0.995
Silage corn	Standing	15.9	--	76.2	Normal	0.229	1.135	0.998
Soybean }	1/10 standing 9/10 flat random	6.4 --	-- 25.4	76.2 --	Normal --	0.016	1.553	0.991

† Rape (*Brassica rapa*), sunflower (*Helianthus annuus* L.).

Table 19-4. Coefficients in the prediction equation, $(SG)_e = aX^b$, for conversion of range grasses to equivalent quantity of flat, small-grain residue (kg/ha) (Lyles & Allison, 1980).

Grass species[†]	Grazing management	Grass height	Prediction equation coefficients		
			a	b	r^2
		cm			
Blue grama	Ungrazed	33.0	0.60	1.39	0.98
Buffalograss	Ungrazed	10.2	1.40	1.44	0.97
Big bluestem	Properly grazed	15.2	0.22	1.34	0.99
Blue grama	Properly grazed	5.1	1.60	1.08	0.99
Buffalograss	Properly grazed	5.1	3.08	1.18	0.99
Little bluestem	Properly grazed	10.2	0.19	1.37	0.99
Switchgrass	Properly grazed	15.2	0.47	1.40	0.99
Western wheatgrass	Properly grazed	10.2	1.54	1.17	0.99
Big bluestem	Overgrazed	2.5	4.12	0.92	0.99
Blue grama	Overgrazed	2.5	3.06	1.14	0.99
Buffalograss	Overgrazed	1.5	2.45	1.40	0.99
Little bluestem	Overgrazed	2.9	0.52	1.26	0.99
Switchgrass	Overgrazed	2.5	1.80	1.12	0.99
Western wheatgrass	Overgrazed	2.5	3.93	1.07	0.99

[†] Blue grama (*Bouteloua gracilis*), buffalograss (*Buchloë dactyloides* [Nutt.] Engelm), big bluestem (*Andropogon gerardi* [Vitman]), little bluestem (*Andropogon scoparius* [Mich.], switchgrass (*Panicum virgatum*), and western wheatgrass (*Agropyron smithii* [Rybd.]).

Either the equivalent, flat, small grain or the vegetative factor is needed for the various procedures to estimate wind erosion. The relationship between equivalent, flat, small grain and vegetative cover was demonstrated graphically by Woodruff and Siddoway (1965). Williams et al. (1984) fit an equation to the graphical relationship to give

$$VE = 0.2533 \, (SG)_e^{1.363}. \qquad [21]$$

A. Model Execution

Several approaches to finding the solution are possible using graphs, figures, tables, slide rule, computer, etc. For this example, we will use the procedure presented by Williams et al. (1984). This procedure is done stepwise, but has been simplified computationally by fitting equations to the figures of Woodruff and Siddoway (1965). The first step (E1) is to determine soil erodibility, I. Steps E2 and E3 are determined by multiplying the value from the previous step by the ridge-roughness and climatic factors, respectively. Accounting for ridge roughness

$$E2 = (I)(WK) \qquad [22]$$

and climatic factor

$$E3 = (I)(WK)(WC). \qquad [23]$$

The inclusion of field length is

$$E4 = (WF^{0.348} + E3^{0.348} - E2^{0.348})^{2.87} \qquad [24]$$

where

$$WF = E2[1.0 - 0.122(WL/WL_o)^{-0.383} \exp(-3.33\ WL/WL_o)] \quad [25]$$

and

$$WL = 1.56 \times 10^6 (E2)^{-1.26} \exp(-0.00156\ E2) \quad [26]$$

where WF is a field length factor and accounts for the influence of field length on reducing erosion estimate, and WL_o is maximum field length for the reducing wind-erosion estimate.

The role of equivalent vegetative cover is expressed by

$$E5 = \psi_1 E4^{\psi_2}. \quad [27]$$

Parameters ψ_1 and ψ_2 are functions of vegetative cover factor described by the equations:

$$\psi_1 = \exp(-0.759 VE - 4.74 \times 10^{-2}\ VE^2 + 2.95 \times 10^{-4}\ VE^3) \quad [28]$$

$$\psi_2 = 1.0 + 8.93 \times 10^{-2}\ VE$$
$$+ 8.51 \times 10^{-3}\ VE^2 - 1.5 \times 10^{-5}\ VE^3 \quad [29]$$

where VE is (Mg ha^{-1}) determined by Eq. [21].

The estimate from Eq. [27] represents the annual rate of expected erosion during the period represented by the climatic factor WC. To determine the expected soil loss during an accounting period, it is necessary to multiply the estimate from Eq. [27] by the fraction of the year occurring during the accounting period.

III. SUMMARY

The EPIC version of the wind erosion model is presented with two major modifications. In EPIC, soil erodibility and climatic factor are held constant for each day of the year. Soil erodibility is calculated at the start of each year using the soil textural triangle. In this modified version, the soil erodibility as influenced by tillage, weather, crop, etc. may be input by the user for subyearly accounting periods. Similarly, a climatic factor may input for each accounting period. The soil loss during each subyearly accounting period is calculated for the conditions (values of equivalent variables or major factors) representative of the accounting period and summed to obtain a yearly soil loss.

The other major modification is the method for calculating the climatic factors. The former method, which gave an inordinately large value when precipitation was low, was replaced with a more physically based procedure. Both methods yield the same values for the Garden City, KS reference.

A. Verification

The fundamental relationship between soil loss and independent variables is that proposed by Woodruff and Siddoway (1965). Continued research has furnished equivalent vegetative cover information for additional crops and crop residues (Lyles & Allison, 1980, 1981; Armbrust & Lyles, 1985). A major limitation is the uncertainty in the conversion of relative field erodibility to annual soil loss and that this conversion is developed for average annual soil loss rather than individual wind storms.

B. Research Needed

Research is needed and in progress (Hagen, 1988) to develop submodels to furnish the values of input variables in time and space. These variables include: surface soil wetness; distributions of leaf and stem silhouette; biomass of residue in standing, flat, and buried categories; soil surface configuration; aggregate size distribution; and stability. A wind simulator is needed that provides friction velocity and wind direction at subhourly intervals on specified surfaces, separated various distances from the weather station that furnished the data base for the simulator.

More research is needed to formulate and verify soil-loss flux equations that are sensitive to changing values of input variables during individual wind storms. Additional development is needed to determine soil-, climate-, and crop-specific conservation tillage and residue management systems that are most cost effective for sustaining productivity and protecting the environment.

REFERENCES

Apt, K.E. 1976. Applicability of the Weibull distribution to atmospheric activity data. Atmos. Environ. 10:777–782.

Armbrust, D.V., and L. Lyles. 1985. Equivalent wind erosion protection from selected growing crops. Agron. J. 77:703–707.

Bagnold, R.A. 1943. The physics of blown sand and desert dunes. William Morrow and Co., New York.

Carreker, J.R. 1966. Wind erosion in the Southeast. J. Soil Water Conserv. 11:86–88.

Chepil, W.S. 1944. Utilization of crop residues for wind erosion control. Sci. Agric. 24:307–319.

Chepil, W.S. 1945. Dynamics of wind erosion: III. The transport capacity of the wind. Soil Sci. 60:475–480.

Chepil, W.S. 1956. Influence of moisture on erodibility of soil by wind. Soil Sci. Soc. Am. Proc. 20:288–292.

Chepil, W.S., F.H. Siddoway, and D.V. Armbrust. 1962. Climatic factor for estimating wind erodibility of farm fields. J. Soil Water Conserv. 17:162–165.

Cole, G.W., L. Lyles, and L.J. Hagen. 1983. A simulation model of daily wind erosion soil loss. Trans. ASAE 26:1758–1765.

Erickson, D.A., P.C. Deutsch, D.L. Anderson, and T.A. Sweeney. 1984. Wind driven interactive wind erosion estimator. p. 247. *In* Agronomy abstracts. ASA, Madison, WI.

Fisher, P.S., and E.L. Skidmore. 1970. WEROS: A FORTRAN IV program to solve the wind erosion equation. USDA-ARS 41-174. U.S. Gov. Print. Office, Washington, DC.

Food and Agriculture Organization. 1960. Soil erosion by wind and measures for its control on agricultural lands. FAO of the United Nations Agric. Develop. Paper 71, FAO, Rome, Italy.

Gillett, D.A., I.H. Blifford, Jr., and C.R. Fenster. 1972. Measurements of aerosol size distribution and vertical fluxes of aerosols on land subject to wind erosion. J. Appl. Meteorol. 11:977–987.

Hagen, L.J. 1988. Wind erosion prediction system: An overview. ASAE Paper 88-2554. Am. Soc. Agric. Eng., St. Joseph, MI.

Hagen, L.J., L. Lyles, and E.L. Skidmore. 1980. Method of determination and summary of Weibull parameters at selected Great Plains weather stations. *In* Application of wind energy to Great Plains irrigation pumping. Adv. in Agric. Tech. Sci. and Educ. Admin., USDA, AAT-NC-4.

Hagen, L.J., E.L. Skidmore, and J.D. Dickerson. 1972. Designing narrow strip barrier systems to control wind erosion. J. Soil Water Conserv. 27:269–270.

Halsey, C.F., W.F. Detmer, L.A. Cable, and E.C. Ampe. 1983. SOILEROS—A friendly erosion estimation program for the personal computer. p. 20. *In* Agronomy abstracts. ASA, Madison, WI.

Hayes, W.A. 1965. Wind erosion equation useful in designing northeastern crop protection. J. Soil Water Conserv. 20:153–155.

Hayes, W.A. 1966. Guide for wind erosion control in the northeastern states. USDA-SCS.

Justus, C.G., W.R. Hargraves, and A. Mikhail. 1976. Reference windspeed distributions and height profiles for wind turbine design and performance evaluation applications. Energy Res. Develop. Admin., Div. Solar Energy Tech. Rep., Georgia Inst. of Technol., Atlanta, GA.

Lyles, L. 1975. Possible effects of wind erosion on productivity. Soil Water Conserv. 30:279–283.

Lyles, L. 1983. Erosive wind energy distributions and climatic factors for the West. J. Soil Water Conserv. 38:106–109.

Lyles, L., and B.E. Allison. 1980. Range grasses and their small grain equivalents for wind erosion control. J. Range Manage. 33:143–146.

Lyles, L., and B.E. Allison. 1981. Equivalent wind-erosion protection from selected crop residues. Trans. ASAE 24(2):405–408.

Moldenhauer, W.C., and E.R. Duncan. 1969. Principles and methods of wind-erosion control in Iowa. Special Rep. 62, Iowa State Univ., Ames, IA.

PEDCO-Environmental Specialists, Inc. 1973. Investigations of fugitive dust-sources, emissions, and control. U.S. EPA Rep., Natl. Tech. Info. Serv., U.S. Dep. of Commerce, Springfield, VA.

Siddoway, F.H., W.S. Chepil, and D.V. Armbrust. 1965. Effect of kind, amount, and placement of residue on wind erosion control. Trans. ASAE 8:327–331.

Skidmore, E.L. 1965. Assessing wind erosion forces: Directions and relative magnitudes. Soil Sci. Soc. Am. Proc. 29:587–590.

Skidmore, E.L. 1983. Wind erosion calculator: Revision of residue table. J. Soil Water Conserv. 38:110–112.

Skidmore, E.L. 1986. Wind-erosion climatic erosivity. Clim. Change 9:195–208.

Skidmore, E.L. 1987. Wind-erosion direction factors as influenced by field shape and wind preponderance. Soil Sci. Soc. Am. J. 51:198–202.

Skidmore, E.L., P.S. Fisher, and N.P. Woodruff. 1970. Wind erosion equation: Computer solution and application. Soil Sci. Soc. Am. Proc. 34:931–935.

Skidmore, E.L., M. Kumar, and W.E. Larson. 1979. Crop residue management for wind erosion control in the Great Plains. J. Soil Water Conserv. 34:90–96.

Skidmore, E.L., and N.P. Woodruff. 1968. Wind erosion forces in the United States and their use in predicting soil loss. USDA-ARS Agric. Handbook 346. U.S. Gov. Print. Office, Washington, DC.

Soil Conservation Service. 1988. Wind erosion. National agronomy manual 190. USDA-SCS, U.S. Gov. Print. Office, Washington, DC.

Thornthwaite, C.W. 1931. Climates of North America according to a new classification. Geogr. Rev. 25:633–655.

U.S. Department of Agriculture. 1989. The second RCA appraisal, soil, water, and related resources on nonfederal land in the United States. USDA 242-141/03004. U.S. Gov. Print. Office, Washington, DC.

Williams, J.R., C.A. Jones, and P.T. Dyke. 1984. A modeling approach to determining the relationship between erosion and soil productivity. Trans. ASAE 27:129–144.

Wilson, L. 1975. Application of the wind erosion equation in air pollution surveys. J. Soil Water Conserv. 30:215-219.

Woodruff, N.P., L. Lyles, F.H. Siddoway, and D.W. Fryrear. 1972. How to control wind erosion. USDA-ARS, Agric. Info. Bull. 354.

Woodruff, N.P., and F.H. Siddoway. 1965. A wind erosion equation. Soil Sci. Soc. Am. Proc. 29:602-608.

Zingg, A.W. 1953. Wind-tunnel studies of the movement of sedimentary materials. p. 111-135. *In* Proc. Fifth Hydraulic Conf., Iowa Inst. of Hydraulic Res. Bull. 34, State Univ. of Iowa.

20 Promises and Problems of Extending Models to the Grass Roots Level

V. PHILIP RASMUSSEN
Utah State University
Logan, Utah

Modeling or systems analysis was first described in detail by Newton in the *Principia* almost 300 yr ago (1687). The recent advances in electronic computers have helped inaugurate systems management, a method of management that uses models to optimize systems. Systems management techniques are now widely used in industry to optimize resources for a given task (Weinberg, 1975). With the advent of powerful personal computers, the capability to perform true systems management at the farm and ranch level has become a reality. A new era in agricultural management is starting in which the outcome of management options can be simulated before these options are actually implemented. The use of personal-computer systems management at the farm or ranch level is not without problems or pitfalls, however. Systems management techniques require at least a rudimentary knowledge of the models used and their critical assumptions.

Today's models have been touted as potential solutions to many of our agricultural woes (Joseph, 1987; USDA, 1982). Many Cooperative Extension Service and research professionals also acknowledge their limitations (Tweeten, 1987). There is a danger that a user may not understand the limitations of the models being used because they deal with extremely complex processes that have been simplified. A user, unaware of these limitations, can create a myriad of problems, due to overconfidence in the model. Both the promises and problems of systems management should be considered as this technology is moved to the farm and ranch level.

Numerous researchers have described the possible uses of systems modeling in actual management situations. Norman (1981), Sakamoto and LeDuc (1981), and Sampson (1984) discuss many of these uses, among which are the following:

1. Managing scarce resources by evaluating various options through nondestructive (computer simulation) testing.
2. Evaluating management options that are too costly or too dangerous to actually perform, such as the influence of a nuclear winter on worldwide wheat production.
3. Evaluating management options that have prohibitively large time-scales for actual testing, such as the influence of 100 years of irrigation with saline water at a given site.

Copyright © 1991 ASA-CSSA-SSSA, 677 S. Segoe Rd., Madison, WI 53711, USA. *Modeling Plant and Soil Systems*—Agronomy Monograph no. 31.

Many other uses for systems modeling can be listed. However, the emphasis here will be on the possible uses of systems management by educators, Cooperative Extension Service specialists or agents, or trained farm or ranch managers to evaluate the profitability and crop yield potential of their decisions before implementation.

These techniques may be difficult for those who are unfamiliar with microcomputers or systems analysis. Many Extension personnel, educators, and other users are capable of using the models described in this text. The key problem is to deliver the model in a form that can be used effectively. Numerous models for various crops, irrigation schemes, and soil phenomena abound, but few are presented in a form that is usable by any but highly trained scientists or technicians. In addition, the required data sets are often arranged in confusing arrays or lists. The task, then, becomes one of technology transfer. Transferring new technology to a relatively untrained audience is often a slow and difficult process.

I. TECHNOLOGY TRANSFER

The productivity of 20th-century agriculture has traditionally been driven by technological innovations. Hybridization of germplasm (hybrid vigor), automated irrigation systems, and the use of mineral fertilizer, selective pesticides, and large-scale mechanization have all contributed to the high productivity (yield/man-hour) ratios found in North America and elsewhere. This green revolution in the USA has often been attributed to the unique partnership among federal, state, and county governments, and agribusiness in agricultural research, teaching, extension, and production (Rasmussen, 1989). The role of Extension in this partnership has often been defined as one of technology transfer. There have been dramatic changes in the traditional roles of Extension and land-grant institutions. The effect of these changes on the implementation of systems management at the grass roots level are the focus of this chapter.

In a landmark publication, Bliss et al. (1952) state that the partnership of federal, state, and county extension services was

> ...a new type of higher education that is now recognized as one of this country's greatest contributions to democracy. It provided a way to put scientific knowledge to work for *all* the people.

Bliss (1952) traced the common history of the Cooperative Extension Service and land-grant college agricultural experiment stations to the 1862 Land Grant College Act, the 1887 Hatch Act, and the 1914 Smith-Lever Cooperative Extension Act. The Smith-Lever Act formalized the widely held concept that all people should have access to the latest scientific information on agriculture, rather than just the chosen few who could attend institutions of higher education. Since then, Extension has been viewed as the agency that would extend research findings from public universities and the USDA to the public. Although the methods of technology transfer have changed over the years, Extension personnel are still assigned primary responsibility for the task.

This is not to say that the fourth partner in the green revolution, private industry, has not or will not continue to contribute significantly to the improved technology of agriculture. In many areas where patent laws protect significant developmental investments, great contributions have been and will be made. However, the American ethic of free public education and public access to information mandate that research, teaching, and extension be available equally to all (Davenport, 1952). In fact, the partnership between these three elements of a land-grant university and industry is getting stronger. Many spin-off business ventures originated in land-grant university teaching, research, and extension. Sine (1986) explains industry's view of the imperative to keep all three elements "equally yoked together" in combination with industry and production agriculture. Technology transfer of agronomic models to the end user must include all elements of this shared partnership. This chapter will present several approaches in which Extension, research entities, and agribusiness have worked together to extend systems management to the end user.

Models can and must be disseminated to users other than those who are in the research establishments that created the models. However, it is not easy for Extension and others to meet their mandate of technology transfer to the agricultural community, particularly since there are more competing uses for federal, state, and county government funds (Sine, 1986). Hence, many agencies responsible for the transfer of technology have investigated the feasibility of using high-technology tools to ease the time constraints of limited staff. Movies, audio-slide presentations, automated audio cassette/phone information systems, video tape, telephone networking, and computer technology have all been investigated to increase the efficiency of the front-line technology-transfer staff. It has long been recognized that computers and associated agricultural software can increase productivity. James H. Smith (1979), past-president of Control Data Corp., speculated on the possible uses of large computers to assist small, productive farms in his book *Back to the Farm—With Technology*. However, his view was couched in the era of large, mainframe computers. The current view of technology transfer by computer-assisted management is now affected by the ever increasing availability and power of personal microcomputers (Rasmussen et al., 1985). Most full-time farmers own or plan to purchase personal computers within the next 2 yr (Burhoe, 1988). Therefore, many agencies on both a state and federal level have placed a major priority on delivering extension information to users via computer software and systems (ECOP, 1986).

II. BRIDGING THE MODEL/USER GAP—CLOSING THE LOOP

The USDA and the land-grant system are committed, in principle, to actively disseminating agronomic models as they are developed. This is easy to support in principle, but very difficult to achieve in practice (Rasmussen et al., 1985). Most agronomic models are mechanistic simulations of specific processes developed by highly specialized researchers. In addition, most require input data in a specified form, which often require difficult conversion or extensive empirical interpretation. Some models even require users to subjectively evaluate boundary conditions to prevent computation problems. Many models require, at the very least, initial measurements of climate, crop, and soils data that are often unavailable (or not easily available) in a rural area. Any systems model also has implicit assumptions made by the creator to simplify its operation or data requirements. These assumptions are often ignored by other users. It is not surprising that many researchers are reluctant to release their models to the general public. Even if users are adequately trained in providing the proper input data, there is the danger that they might become so confident in the model's results, that they may ignore the variability of the system and certain weaknesses in the model. This is often termed the black box syndrome (Weinberg, 1975).

Several key problems must be addressed before agronomic systems models can attain widespread application:

1. The user needs to accomplish the difficult, and sometimes impossible, task of identifying sources of desired models.
2. The user must ensure that the model, if found, is technically correct for the chosen situation.
3. Documentation must be available so the user can determine:
    a. Which input data are required. The permissible limits of timing, spatial variability, and data measurement precision should be adequately documented.
    b. The assumptions made concerning boundary conditions and the level of output error that is created by a given error in each variable.
    c. The accuracy, precision, and limitations of the output data.
4. The specific hardware requirements of the model and whether it can be adapted to other systems should be determined.
5. The availability of input data in the region where this model is used should be considered. If certain data are estimated, the model or documentation should adequately indicate this fact.
6. The model should ease the interface between a novice user and a complex systems model.

A. Identifying Sources of Models

The first, and often most difficult, problem that a would-be user must face is to locate sources and descriptions of possible models. This is a long-standing problem and has been addressed in several ways. Folks et al. (1972) and, later, Ffolliott (1987) surveyed university personnel to identify those who used computers and simulation models in agricultural research, teaching, and extension activities. However, their studies did not directly address the problem of identifying software models. Ffolliott's (1987) directory identifies people to contact regarding possible new, unpublished models. It lists the contact person's interests as well as those who are working on artificial intelligence (AI) applications in agriculture.

Mainframe computer systems were initially used to house agronomic models and offered a controlled form of releasing these models to the end user. Pioneering systems were the University of Nebraska-Lincoln AGNET system (Griffith & Wright, 1982), Purdue's FACTS system (Fredericks, 1982), and others. However, the use of the microcomputers at the grass-roots level has skyrocketed. Several reports predict that most medium- and large-size farms will soon have microcomputer capability (Burhoe, 1988; Dobbins & Suter, 1981; Rasmussen et al., 1985; Smith, 1984). The use of large systems to house models for downloading, large data bases for model use, and large directories of available models cannot be discounted, but the microcomputer revolution has literally brought many systems-management capabilities out of the mainframe computer centers and into the home, rural office, and classroom. Hence, an effort was, and still is, needed to constantly survey available models and software for small system use.

Strain (1982) made an extensive attempt to address the proliferation of microcomputer systems-management when he, with assistance from USDA Extension in Washington, DC., published two directories of available agricultural software (Strain & Fieser, 1982; Strain & Simmons, 1984). However, Strain (1982) cautions that these directories are limited and do not evaluate the quality of the models. Extension later contracted with the Virginia Polytechnic Institute to place this directory on line for phone modem access (see the VPI—CMN listing in Appendix 1). Although these directories are soon outdated, they do list hundreds of programs and make it possible to contact the authors of models to identify more recent versions. Appendix 1 of this chapter lists federal, state, and other agency systems that can be accessed to search for possible applications/models.

The private sector is another source of information regarding applications/models. Several private companies have provided invaluable service to end-users by compiling directories of both private and public applications software, such as that currently provided by Doanes Information Services, Inc. (Burhoe, 1988). In addition, some of these groups provide on-line computer-accessed bulletin boards that allow users to exchange accounts of both good and bad experiences with applications software. Appendix 2 provides a list of private groups that provide data or models directly to the end-user. Rasmussen et al. (1985) also provide a compilation of commercial, institutional, and several other sources of software and evaluations in the appendices of their text.

B. Obtaining Evaluations of Models

The second key problem of implementing systems management, ascertaining the technical propriety of applications/models, can be very difficult. Researchers and scientists can obtain critical review of their own or other models by searching the appropriate refereed literature. However, this is very difficult for many end-users. Some (Coates, 1985) have attempted to offer directories and evaluations of available software. Donald A. Holt of the University of Illinois (Gage, 1987) has suggested a novel approach by involving private industry in software packaging and distribution after the applications/models have been developed, tested, and evaluated by universities and USDA researchers. This might, indeed, yield a solution to this problem; however, such a system may be difficult to implement.

The Kellogg Foundation proposed a potential solution to both problems discussed above. The Foundation helped initiate feasibility studies and inaugurate four regional centers in the USA for agricultural computing in cooperation with the extension/land-grant system (Schmidt, 1982). These centers were all established by the respective Extension regional directors as noted in Appendix 3. Many of these regional centers have closed or substantially reduced their service as Kellogg funding was phased out. Indeed, many of them still offer a source of information on the applicability and availability of systems-management models.

The now defunct Western Computer Consortium (WCC) published the *Journal of Computer Applications.* This journal attempted to address the void in refereed systems-management software. The journal is currently being restructured by the successor of the WCC, the Western Extension Computer Application Committee (WECAC). This will provide an additional source of information, if it continues to be published. The National Agricultural Library also published several reports, regarding the availability of systems-management information (USDA, 1984).

C. Other Problems

Some of the other problems outlined above (problem no. 3, 4, and 5) must be addressed by the creator of the model in question. Most models are developed in a quick and dirty fashion, which generates a stream of data points with little regard for the untrained user. Most model developers are now becoming more sensitive to the user and adequate documentation is required by the sponsoring agency or institution.

D. Optimizing the User/Model Interface

The last problem (problem no. 6), easily interfacing agronomic systems models with the end-user, can be mediated by several means. One of the most promising methods of bridging the gap between the user and complex physical models is to incorporate AI modules into the models.

The field of AI is so new that its tasks are often not clearly defined. However, it can include a variety of tasks (Morrison, 1985) including (i) theorem proving, (ii) game playing, (iii) machine learning, (iv) knowledge systems (expert systems), (v) pattern recognition, (vi) natural language processing, (vii) robotics, and (viii) machine cognition. The most probable uses of AI for agronomic modeling are expert systems, the intelligent query, sorting, and processing of knowledge (data) bases, and natural language processing, plain English command processing (Morrison, 1985). The potential and applications of AI systems have often been overstated by the researchers involved (Winograd, 1987). Since this often occurs with new technology, it should not deter our examination of the true possibilities that AI holds for improving our management techniques.

Expert systems usually consist of large data bases of information and rules that are cognitively manipulated by inference schemes in a manner that emulates human reasoning (forward chaining, backward chaining, and pattern recognition). These can be used to make decisions, based on a defined data set and a series of logical rule evaluations. They have been used in several agronomic systems (Balas, 1987; Joseph, 1987). Their most likely use in systems (modeling) management is to define the correct data set for a given model. For example, consider any inexperienced person who is using an irrigation-scheduling evapotranspiration model. A novice user may not comprehend the sophistication of the meteorological data set required or may not have available many of the individual soil parameters (volumetric water holding capacity, permanent wilting point volumetric-water content, field capacity volumetric-water content, etc.). The novice may have a limited set of data (temperatures and wind values, but little or no solar radiation data) and some data may be expressed in the wrong units (gravimetric water content of permanent wilting point, field capacity, etc.). An expert system could guide the user as he constructs a data set and would automatically correct incorrect units of measure. The expert system could also warn the user on the possible accuracy and prediction problems that may occur due to estimation of missing data (such as solar radiation estimated by a maximum/minimum temperature deviation formula).

Expert systems are, admittedly, only as good as the rules and knowledge bases from which they are constructed. They can take the place of an expensive and often inaccessible tutor in many real-world applications of systems models. Their use is limited at present, but is growing rapidly as systems modelers realize their potential to extend the common numerical generators of the lab into the real world. A recent conference in the application of expert systems to agriculture was dominated by agronomic modelers who were anxious to extend systems management to the grass-roots level (Gage, 1987).

III. NATURAL LANGUAGE SYSTEMS

An additional category of AI that may hold great promise for the extension of models to the nonresearch user is that of natural language processing. Natural language processing (NLP) attempts to let the user converse with the computer in his or her own language or slang. The key element in an NLP system is the transformation of potentially ambiguous phrases into an unambiguous phrase that the computer can use internally (Obermeier, 1987). Thus, a person might respond to a computer with "What in the world is on my disk?," rather than "DIR A:" or another specific command syntax.

The transposition of the ambiguous phrase into an unambiguous one is usually known as parsing, which is actually derived from the Latin phrase pars orationis, meaning part of speech. Most natural language processors combine the symbols of a phrase into a group that can be successively replaced by others, until an understandable (to the computer) structure results (Obermeier, 1987). A newer form of NLP takes a somewhat different approach in using bit-image pattern-recognition systems (Tulino, 1982). These systems use bit mapping (Frey, 1986) to examine a phrase in several pseudorandom clusters and map it into a histogram-like structure. The phrase is then compared, in pattern, against computer-recognized phrases and then statistically evaluated. Hence, a user that types "YSE" would get the same result as "YES," because their bit-level histogram patterns resembled each other. Of course, this feature can work against the user, at times, if undesired commands resemble desired commands in bit pattern.

Some programmers have combined the Bit-Image Pattern Recognition System (BIPS) feature with associative memory techniques (Tulino, 1982) to produce a system that learns as you use it. Thus, if a person responds "What is my profit?," the computer provides several possible options. Once an option is matched with a natural query, they are permanently associated in computer memory. Thereafter, when "What is my profit?" is typed in, the computer immediately responds with the proper answer that was previously associated with that query. This has tremendous implications in extending models to users of other cultures and languages. Association of the initially programmed commands with similar commands of another language would allow a model to be used by those who speak another language. The computer memory and speed requirements escalate dramatically as more and more common commands are associated with uncommon ones. In addition, users of this feature are reluctant to lend their computer to other users who burden the machine with their memory-associated slang and vernacular.

Additions, such as those just noted, to various systems-management models will allow their use by the broadest possible audience. In addition, the associative memory option will allow adaptation to foreign users much more quickly (Fig. 20-1). Several land-grant institutions have implemented AI in one or more educational software systems. It is agreed that this trend is certain to continue and be improved upon (Rasmussen et al., 1985).

The extension of systems-management models to the grass-roots level is not a readily accomplished endeavor. Mechanisms for identifying available models and evaluating them have been established by federal, state, commercial, and other entities. However, some of them are in serious jeopardy, due to financial and other constraints. It will take the combined resources of the traditional land-grant system and the USDA, combined with the financial assistance of commercial ventures, to allow the full and optimum use of systems management at all levels of our diverse agricultural system. Expert systems offer the promise of easing the user/model interface, but cannot take the place of wise and prudent construction of the models themselves. It is assured that systems management with agronomic models will increase at the grass-roots level, but not without considerable expense and effort by the agricultural community as a whole.

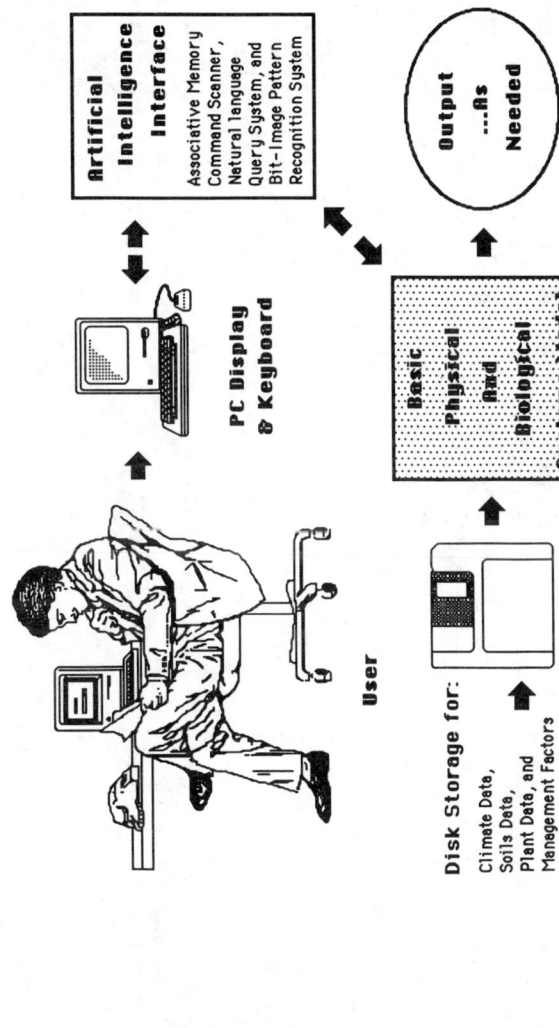

Fig. 20–1. Artificial intelligence to bridge the model/user gap.

IV. APPENDIX 1

Some on-line (modem or videotext access) information systems available from federal and state government agencies that provide models or useful model data. Information from Burhoe, 1988; Coates, 1985; Strain and Simmons, 1984; and Ffolliott, 1987.

System name	Agency	Users	Type of services
AGNET Contact: Pam Peters AGNET, Univ. of Nebraska Lincoln, NE 68583-0713 (402) 472-1892	University of Nebraska South Dakota State Univ. North Dakota State Univ. Washington State Univ.	Nationwide agribusiness producers, and education	Management models for crops and livestock; information retrieval, market prices and analysis, weather, USDA reports; electronic mail.
AGRICOLA Contact: Gary McCone National Agricultural Library Information Systems Division 10301 Baltimore Blvd. Beltsville, MD 20705 (301) 344-3813	USDA/NAL	Farmers, agricultural and life science researchers nationwide	Information retrieval (primarily agricultural research information), Extension publications.
ANSER Contact: Dr. Robert Fehr Ag Data Center-Univ. of Kentucky Lexington, KY 40546-0092 (606) 257-7207	Univ. of Kentucky	Farmers and 25 Extension offices in Kentucky	Problem solving for farm management, information retrieval.
ATI-Net Contact: Jeff Ennen California Agricultural Technology Inst. California State Univ. Fresno, CA 93740-0079 (209) 294-2361	CATI	California farmers and nationwide	Information on California agriculture, public domain software, bulletin boards, electronic mail, teleconferencing, free communications software.

EXTENDING MODELS TO THE GRASS ROOTS LEVEL 481

DRAGNET
    Contact: Michael McGrath
    Delaware Dep. of Agriculture
    2320 South DuPont Hwy
    Dover, DE 19901
    (302) 736-4811

Delaware Dep. of Agriculture in cooperation with Extension

Delaware agricultural professionals

Bulletin board, public domain farm management and agronomics software for downloading.

ESTEL
    Contact: ESTEL, MD Coop. Ext. Serv.
    Room 1326, Synons Hall
    Univ. of Maryland
    College Park, MD 27042
    (301) 454-4515

Maryland Cooperative Extension Service

Maryland farmers, Extension offices, and mariners

National Weather Service reports for Maryland and the East Coast.

EXNET
    Contact: Deborah Coates
    108 Computation Center
    Iowa State Univ., Ames, IA 50011
    (515) 294-8658

Iowa State Univ. Extension Service

Iowa Extension services and farmers

Integrated pest management, Extension and home economics information, weather reports.

FACTS
    Contact: Bill Simmons
    Ag Computer Network
    105 Smith Hall, Purdue Univ.
    W. Lafayette, IN 47907
    (317) 494-3492

Purdue Univ.

Indiana farmers and Extension offices

Agricultural data bases for Indiana, decision aid programs, problem solving programs, information retrieval.

Grassroots
    Contact: Anthony Batiaanssen
    550 Barry St.
    Winnipeg, Manitoba R3H 0R9
    1-800-665-0302

Grassroots Information Services

Nationwide, mainly agricultural and financial

Market prices, stock market quotes, advice, and analysis, various reports and newsletters.

(continued on next page)

IV. APPENDIX 1 CONTINUED

System name	Agency	Users	Type of services
National Weather Service Contact: Joanna Dionne, W/NMC53 Climate Analysis Center World Weather Bldg., Room 811 Washington, DC 20233 (301) 763-6614	National Weather Service	Nationwide agricultural and energy applications	Temperature and precipitation data, various weather reports, textural forecase.
NPIRS Contact: User Services, NPIRS Entomology Hall, Purdue Univ. West Lafayette, IN 47907 (317) 494-6614	Purdue Univ.	Nationwide, plus 100 international users	Federal and state pesticide registrations, related chemical information, all hazardous chemicals, MSDS (Material Safety Data Sheets), EPA's PDMS (Pesticide Documentation Management System).
SCAMP Contact: Jim Tete IPM Program NYS Agricultural Experiment Station Geneva, NY 14456 (315) 787-2208	Cornell Univ.	New York Extension offices, New York State agencies, cooperative institutions, and farmers nationwide.	Information retrieval, pest forecasting, weather forecasting, pest/crop/weather reporting, electronic mail. Macintosh and IBM bulletin boards.

UMC BBS Contact: John Travlos 325 Mumford Hall Univ. of Missouri Columbia, MO 65211 (314) 882-4827	Univ. of Missouri-Columbia	Utility programs, spreadsheets, daily and weekly information from Univ. Agric. Dep., information from Soil Conservation SErv., USDA, Integrated Pest Management daily Missouri agri-weather forecast, more.
	Missouri and mid-western agricultural professionals	
VPI-CMN Contact: Thomas McAnge, Jr. Plaza I—Building D Blacksburg, VA 24061 (703) 961-6910	Virginia Polytech. Inst.	On-line, USDA-ES supported library of models, etc.
	Professionals and educators	
USDA ON-LINE Contact: Russell Forte Room 536-A, Office of Information USDA Washington, DC 20250	USDA	USDA news releases, crop and livestock markets, analysis reports, etc.
	Distributors of information to farmers and ranchers	

V. APPENDIX 2

Some on-line (modem or videotext access) information systems available from commercial sources that provide models or useful model data. Information from Burhoe, 1988.

System name	Agency	Users	Type of services
ACRES Contact: Kim Wells American Agric. Communications Systems 225 Touhy Park Ridge, IL 60068 (312) 399-5870	American Farm Bureau Federation	Members of 36 participating state farm bureaus	Information retrieval, market data analysis and AgriVisor trading advice, weather, legislative developments, all USDA market service reports.
AgLine† Contact: Connie Nelson Doane Information Services 11701 Borman Drive, Suite 100 St. Louis, MO 63146	Doane Information Services	Agricultural Computing newsletter subscribers	Electronic mail, agricultural software reviews, public domain agricultural software, cash, futures, and options prices, market advice and analysis, major USDA reports, Washington reports.
Agribusiness U.S.A. Contact: Meg Ullestad Pioneer Hybrid Data Base Manage. 5608 Merle Hay Rd. Johnston, IA 50131 (515) 253-5860	Pioneer Hi-Bred Intl., Inc.	Agribusiness	Indexes and abstracts from more than 300 businesses and government publications, complete from Jan. 1985 to present. Full text of USDA statistical reports since Jan. 1986. Updated every week.

EXTENDING MODELS TO THE GRASS ROOTS LEVEL

AgriData Network Contact: Beth Koehler 330 E. Kilbourn Milwakee, WI 53202 (800) 558-9044; WI (800) 242-6001	AgriData Resources, Inc.	Nationwide	Business decision information; education service, Ag Ed Network, market advisories, cash & futures prices, government and commodities news.
AQS II Contact: Steve Reyhle Commodity News Services, Inc. P.O. Box 6053 Leawood, KS 66206 (800) 255-6490 or (913) 642-7373	Commodity News Services	Nationwide	Grain and Livestock futures and cash prices, weather, and financial information.
CAMP Contact: Dr. Carey Moxley CAMP, Inc. 1 Progress Blvd., Box 19 Alachua, FL 32615 (904) 462-1956	CAMP, Inc.	Brokers, shippers, and buyers of agricultural produce, nationwide	Electronic marketing service, plus complete market news for any item at any terminal.
FAST Contact: Lauren Curry 1 Progress Blvd., Box 13 Alachua, FL 32615 (904) 462-3492	FAST (Florida Agricultural Services and Technologies)	Southeastern farmers and horticulturists	National Weather Service data, updated every 30 min. Includes forecasts. IR tracking of hurricanes, severe thunderstorms.
Instant Update Contact: Instant Update 219 Parkade Cedar Falls, IA 50613 (800) 553-2910; IA (800) 772-2106	Professional Farmers of America, Inc.	Nationwide	Information retrieval, commodity prices, market advice and analysis, options, weather, news wire.

V. APPENDIX 2

System name	Agency	Users	Type of services
The Source† Contact: Source Telecomputing Communications and Marketing P.O. Box 1305 McLean, VA 22102 (800) 336-3366	Source Telecomputing Corp.	Nationwide and international	Information retrieval, commodity prices, USDA reports, electronic mail, news, business and investing, travel, education, shopping, games.
TELECOT P.O. Box 2827 Lubbock, TX 79408 (806) 763-8011; (800) 868-4532 in states adjoining Texas; (800) 692-4459 in Texas	Plains Cotton Coop Assoc.	Cotton gins and buyers	Electronic trading of cotton.

† Indicates those commercial sources that provide software evaluations.

VI. APPENDIX 3

Regional land-grant computer institutes (Partially sponsored by the Kellogg Foundation.)†

The North Central Computer Institute
Univ. of Wisconsin
667 WARF Office Building
610 Walnut Street
Madison, WI 53705

Serving:

Univ. of Wisconsin (Sponsoring)	Univ. of Missouri
Univ. of Illinois	Univ. of Nebraska
Iowa State Univ.	North Dakota State Univ.
Kansas State Univ.	Ohio State Univ.
Michigan State Univ.	Purdue Univ. (Indiana)
Univ. of Minnesota	South Dakota State Univ.

The Northeast Computer Institute
Pennsylvania State Univ.
1315 South Allen Street
State College, PA 16801

Serving:

Pennsylvania State Univ. (Sponsoring)	Univ. of Maine at Orono
Univ. of Connecticut	Univ. of Massachusetts
Univ. of Delaware and Del. State College	Univ. of Rhode Island
Univ. of the Dist. of Columbia	Rutgers (State Univ. of New Jersey)
Univ. of New Hampshire	West Virginia Univ.

The Southern Regional Computer Coordinator
Univ. of Georgia
Hoke Smith Building
Athens, GA 30602

Serving:

Univ. of Georgia (Sponsoring)	Prairie View A&M Univ.
Alabama A&M Univ.	Southern Univ.
Alcorn State Univ.	South Carolina State College
Auburn Univ.	Tennessee State Univ.
Clemson Univ.	Texas A&M Univ.
Florida A&M Univ.	Tuskegee Univ.
Ft. Valley State College	Univ. of Arkansas
Kentucky State Univ.	Univ. of Florida
Langston Univ.	Univ. of Maryland
Louisiana State Univ.	Univ. of Puerto Rico
Mississippi State Univ.	Univ. of Tennessee
North Carolina A&T Univ.	Virginia State Univ.
North Carolina State Univ.	Virginia Polytechnic Institute and
Oklahoma State Univ.	State Univ.

(continued on next page)

VI. APPENDIX 3 CONTINUED

The Western Computer Consortium
Univ. of Arizona
1132 East Mabel
Tucson, AZ 85719

Serving:

Univ. of Arizona (Sponsoring)	Univ. of Nevada
Univ. of Alaska	New Mexico State Univ.
Univ. of California	Oregon State Univ.
Colorado State Univ.	Utah State Univ.
Univ. of Hawaii	Washington State Univ.
Univ. of Idaho	Univ. of Wyoming

† As of the date of publication:
1. The North Central Computer Institute is operating at a reduced service level.
2. The Northeast Computer Institute had been disbanded by its governing institutions.
3. The Western Computer Consortium has been disbanded by its governing institutions. However, it has been restructured as an unfunded regional extension committee; and, can be contacted through:
   V. Philip Rasmussen
   WECAC Extension Computer Committee Chair
   USU UMC#4840
   Logan, UT 84322-4840
4. The Southern Regional Computer Coordinator (Margaret P. Ezell) is still functioning.

REFERENCES

Balas, S.L. 1987. The use of expert systems in soil acidity management. p. 1-3. In Agrotechnology transfer 5. Int. Benchmark Sites Network for Agrotechnology Transfer (IBSNAT), Honolulu, HI.

Bliss, R.K. 1952. The new education. p. 1-7. In The spirit and philosophy of extension work. USDA and Epsilon Sigma Phi, Washington, DC.

Burhoe, S.A. (ed.). 1988. Agricultural computing: 1988 Agricultural software directory. Doanes Information Services, St. Louis, MO.

Coates, D.M. 1985. NECI directory of evaluated software. Northeast Computer Inst., State College, PA.

Davenport, E. 1952. Spirit of the land-grant college. p. 8-29. In The spirit and philosophy of extension work. USDA-Epsilon Sigma Phi, Washington, DC.

Dobbins, C.L., and R.C. Suter. 1981. A microcomputer for the farm family. Purdue Univ. Ext. Serv. Circular 527.

Extension Committee on Organization and Policy (ICOP). 1986. Cooperative extension system program priorities for agriculture and natural resources. USDA-CES ECOP Bull., U.S. Gov. Print. Office, Washington, DC.

Ffolliott, L.M. 1987. Computer contacts in land grant universities: 1987 survey results. Univ. of Arizona, Tucson, AZ.

Folks, H., J.K. Pasto, W.R. Thomas, and I.J. Weiss. 1972. Instructional use of computers in agriculture and natural resources. Resident Instruction Committee on Policy committee report. Univ. of Nebraska-Lincoln, Lincoln, NE.

Fredericks, E.E. 1982. FACTS, six years later. Ext. Rev. 53(2):34-35.

Frey, P.W. 1986. A bit-mapped classifier. Byte 11(11):161-172.

Gage, S.J. 1987. Summary of the university-industry conference on agricultural expert systems in agriculture, Bloomington, MN. 19-20 Oct. 1987. Midwest Technology Develop. Inst., St. Paul, MN.

Griffith, D.W., and M.A. Wright. 1982. AGNET-American network for agriculture. Ext. Rev. Summer, p. 10-12.

Joseph, C. 1987. AI is keeping the small farmer competitive. Appl. Artif. Intell. Rep. 4(4):19-20.

Morrison, J.B. 1985. The forthcoming role of artificial intelligence in the agricultural science and education system. Purdue Univ., West Lafayette, IN.

Newton, Sir Isaac. 1687. The Principia: The mathematical principles of natural philosophy. Translated into English by Andrew Motte. Printed for B. Motte, London, England, 1972.

Norman, J.M. 1981. Data requirements for generalized physiological models. *In* A. Weiss (ed.) Computer techniques and meteorological data applied to problems of agriculture and forestry: A workshop. Am. Meteorol. Soc., Boston.

Obermeier, K.K. 1987. Natural language processing. Byte December, p. 225-232.

Rasmussen, W.D. 1989. Taking the university to the people: 75 years of cooperative extension. Iowa State Univ. Press, Ames, IA.

Rasmussen, W.O., C.T.K. Ching, L.A. Linden, P.A. Meyer, V.P. Rasmussen, R.S. Rauschkolb, and C.B. Travieso. 1985. Computer applications in agriculture. Westview Press, Boulder, CO.

Sakamoto, C., and S. LeDuc. 1981. Sense and nonsense. *In* A. Weiss (ed.) Computer techniques and meteorological data applied to problems of agriculture and forestry: A workshop. Am. Meteorol. Soc., Boston.

Sampson, J.R. 1984. Biological information processing: Current theory and computer simulation. John Wiley and Sons, Inc., New York.

Schmidt, J. 1982. Regional computer efforts. p. 17-19. *In* The Computer—Management power for modern agriculture (The educational role of the cooperative extension service). Purdue Univ. Coop. Ext. Serv., West Lafayette, IN.

Sine, C. 1986. Save the troika. Farm Chem. 149(4):70.

Smith, D. 1984. Computers on the farm. USDA Farmers Bull. 2277. U.S. Gov. Print. Office, Washington, DC.

Strain, J.R. 1982. Agricultural needs for computers: Available software. p. 77-83. Proc. Computers in Agric. Exp. Programs Conf., Tampa, FL. 1-2 Dec. 1982. Inst. of Food and Agric. Sci., Univ. of Florida, Gainesville, FL.

Strain, J.R., and S. Fieser. 1982. The cooperative extension service inventory of agricultural computer programs. Florida Coop. Ext. Serv. Bull. 531.

Strain, J.R., and S. Simmons. 1984. The cooperative extension service updated inventory of computer programs. Florida Coopo. Ext. Serv. Bull. 531-A.

Tulino, M. 1982. Writing your own programs using SAVVY—the personal language. SAVVY Marketing International. Excalibur Technologies Corp., Albuquerque, NM.

Tweeten, L. 1987. No great impact on rural areas expected from computers and telecommunications. Rural Development Perspectives (June):7-10.

U.S. Department of Agriculture. 1982. The computer: Management power for modern agriculture. USDA Ext. Committee on Organization and Policy (ECOP), Washington, DC.

U.S. Department of Agriculture (USDA). 1984. NAL quick bibliography series: Mini and micro computers in agriculture—1979-1983 (261 Citations). March 1984. Natl. Agric. Library, Washington, DC.

Weinberg, G.M. 1975. An introduction to general systems thinking. Wiley-Interscience Publ., New York.

Winograd, T. 1987. Natural language: The continuing challenge. AI Expert (May):7-8.

21 Irrigation Scheduling

ROBERT W. HILL
Utah State University
Logan, Utah

The purpose of irrigation scheduling is to inform the farmer when to irrigate and how much water to apply to obtain a desired objective. The objective usually involves maximizing crop yield or profit, but it may also include taking advantage of irrigation opportunities, minimizing deep percolation or leaching, and soil and water salinity management. Improving crop quality, and maximizing the utility of rainfall and preseason-stored soil moisture are other possible results. In areas where pumping is prevalent or drought is experienced, energy and water conservation are recognized benefits of irrigation scheduling. The irrigator should have control of the irrigation interval, and water application flow rate and duration for complete flexibility in scheduling. The irrigation system, including both delivery (conveyance) and field application methods, is also an important factor that may place constraints on the irrigation schedule. Even those canal systems operated by demand, upon closer inspection, usually have constraints imposed by capacity, gate opening times, and other operating rules. For example, in areas where canal water is delivered to the farm on a set rotation pattern, the irrigator does not have control of the interval nor the flow rate, but may be able to reduce the duration. On the other hand, a variable-speed center pivot supplied with water pumped from a reservoir or from groundwater allows considerably more flexibility in interval and duration with a set flow rate. Similarly, solid-set sprinkle or trickle irrigation provides greater scheduling flexibility (and higher capital cost) than hand-move or side-roll sprinklers.

Through proper irrigation scheduling, it should be possible to apply only the water that matches the crop evapotranspiration (ET) or to meet some other desired criteria. Possible savings to pump irrigators are particularly important if power costs are high.

For irrigation scheduling to be most useful at a specific location, the following should be done:

1. Evaluate the irrigation system. Determine application depth, efficiency, and operating capabilities and constraints.
2. Select an appropriate irrigation scheduling method.
3. Monitor performance at intervals during the growing season.
4. Perform a post-season evaluation and determine changes for next year.

Copyright © 1991 ASA-CSSA-SSSA, 677 S. Segoe Rd., Madison, WI 53711, USA. *Modeling Plant and Soil Systems*—Agronomy Monograph no. 31.

While each step is important and should not be ignored, the method of irrigation scheduling is emphasized herein.

Basic scheduling methods that have been used include:

1. Calculating a soil water budget using soils, crop, weather, and irrigation management information. This can be done simply by using hand calculations (checkbook method) or by the use of computer models such as shown in Ch. 11.
2. Monitoring soil water content with instruments or sampling techniques such as feel, gravimetric, gypsum blocks, tensiometers, and neutron probes.
3. Observing and measuring plant indicators, such as when the crop shows visible evidence of stress by color change or leaf wilt, or by use of canopy temperature measurements.
4. A combination of any of the above methods to reduce field visits.

While irrigators have used these and other means to schedule irrigations, for instance when it is their turn to water or when the neighbor irrigates, the purpose of this chapter is to describe the use of a simple weather-based computer model as an irrigation scheduling method. With this, as well as other methods, a considerable amount of assistance is required to train users to successfully apply it to their situation. It also requires a continuing time commitment to maintain current information.

I. RECENT EXPERIENCE IN IRRIGATION SCHEDULING ACTIVITIES

Cooperative extension and other agencies in many of the western United States (and some of the other states) provide publicly available, crop water-use related information through various media. Additionally, there are many private and custom irrigation scheduling programs in use. While it is beyond the scope of this chapter to list all these activities, an example of a public information approach will be given.

Attempts at providing irrigation scheduling information to farmers in Utah were first made through a simple water-balance model. The county extension agent produced (from a program on his microcomputer) a weekly crop water-use table, which was placed in the local newspaper. Farmers were provided with a worksheet for using data from the newspaper and their crop and soil to calculate a weekly soil water budget. The date and amount of the next irrigation was also estimated. Many farmers indicated considerable interest in this program when discussed at farmers' meetings. However, follow-up visits revealed that, while the general crop water-use data was noticed, no one was calculating the specific soil water budgets on the worksheets provided. Subsequent addition of these calculations to the microcomputer programs has led to more use by farmers for specific fields, if the county agent continues to be involved week to week.

Three computer programs are presently in use: SET UP, PCET, and SOIL WATER BUDGET. The SET UP program contains a list of crops that can be included in the table developed by the crop water-use program (PCET). The SET UP program must be correctly run to produce the region set up file at least once prior to running the other programs. The PCET program calculates crop water use values for selected crops for a given region, as specified by the SET UP program. The PCET program uses ETr calculated from weather data, usually for a 7-d update period. At least one region set up file must exist before the program will run successfully. The output table was designed to be included in a county weekly newspaper as Utah State Univ. Extension Avisory information. The SOIL WATER BUDGET program determines when to irrigate an individual field based on output from the PCET program and site-specific soil, rain, and irrigation data, input by the irrigator. A user's manual and program diskette (IBM-PC compatible or Apple II family) are available.

These models require the use of weather related information to estimate crop water use. Consequently, microprocessor-controlled electronic weather stations (Datapod DP219, Omnidata Int., Logan, UT) were located in some irrigated areas of the state. These stations measure air temperature and solar radiation, and calculate a reference-crop potential evapotranspiration (ETr) which is displayed as an accumulated value. Accumulated ETr for each update period, usually 7 d is obtained from the Datapod display for estimating crop water use. In some locations, more complete weather data (air temperature, solar radiation, humidity, precipitation, and wind speed) are available from datalogger stations (e.g., CR21, CR21X, or CR10 microloggers, Campbell Scientific, Inc., Logan, UT). This allows the use of a modified Penman equation to calculate ETr. Class A pan evaporation, with the appropriate pan coefficient, could also be used for ETr. The equations that use ETr values to calculate crop water requirements and soil water budgets for various crops and areas are described hereafter.

Currently, a combined approach using the neutron probe for field checks, at 3- to 4-wk intervals, and the model to estimate soil water content in between the intervals is being tested in a few areas. This could reduce costs and permit additional irrigators to be involved. The information delivery mechanism between field checks is a major problem that has not yet been resolved. The accuracy of the model appears to be sufficient if irrigation and rain can be satisfactorily measured. This is particularly true if the soil water content is reset to the most recent field check.

II. SOIL WATER BUDGET MODELS

The purpose of the irrigation scheduling model is to estimate current and future soil water contents. When the soil water content drops to a specified level, the program indicates that irrigation is needed. Estimated soil water (SW) content is determined by:

$$SW_2 = SW_1 + IRR + RAIN - ET - DP \qquad [1]$$

where SW_1 and SW_2 are the beginning and ending total available soil water contents, respectively; IRR and RAIN are the respective amounts of infiltrated irrigation and rain; ET is the calculated crop water use, or evapotranspiration; and DP is the deep percolation or drainage out of the root zone. The time interval (update period) between Points 1 and 2 is generally 1 wk, although it could be as short as 1 or 2 d. The model for a specific field is performed by the SOIL WATER BUDGET program.

Evapotranspiration is determined from:

$$ET_{cp} = k_{cm} \times ETr \qquad [2a]$$

where ET_{cp} is the crop evapotranspiration with adequate soil water, k_{cm} is the mean crop coefficient described by Wright (1981), and ETr is evapotranspiration for an appropriate reference crop (i.e., alfalfa with 40 cm or more top growth and nonlimiting soil water).

Estimated actual ET was obtained from:

$$ET = (ET_{cp}/FA) \times SW/(FC - WP), \text{ if } SW/(FC - WP) < FA \qquad [2b]$$

or

$$ET = ET_{cp}, \text{ if } SW/(FC - WP) \geq FA \qquad [2c]$$

where FA is equal to the fraction of available soil water in the root zone, where stress is assumed to begin; SW is the current available soil-water content (assumed to be zero if equal to WP); FC is field capacity; and WP is wilting point in the root zone. The value of FA is assumed to be 0.50 for crops such as alfalfa (*Medicago sativa* L.), wheat (*Triticum aestivum* L.), and corn (*Zea mays* L.) and 0.65 for potato (*Solanum tuberosum*) and vegetables in the intermountain USA. This reflects the rule of thumb that irrigation is needed when the soil water content (SW) drops to a depletion rate of 50% of available water (field capacity minus wilting point, FC − WP) in the root zone for such crops as alfalfa, corn, and spring grains. Potato and other vegetables may produce better when the soil water is maintained in the upper 35% of available water [(FC − SW) ≤ 0.35 (FC − WP)]. This assumes that crop growth is limited if ET < ET_{cp}. Values of FA are summarized for many crops by Doorenbos and Pruitt (1977, Table 39, where $p = 1 - FA$).

The modified Penman equation as calibrated at Kimberly, ID (Jensen, 1974) with a wind term limit of 161 km/d (100 miles/d) was used as the ETr standard. A temperature radiation equation (Jensen & Haise, 1963) has been used instead of the modified Penman equation where sufficient data of humidity and wind are not available.

$$ETr = C_t(T - T_x)R_s \times CF \qquad [3]$$

where C_t and T_x were treated as site-specific calibration coefficients; T is the average daily temperature; R_s is total daily solar radiation; and CF is a units conversion factor. The value of CF is about 0.734 when T is measured in degrees Celsius, R_s is measured in megajoules per square meter, and ETr is measured in millimeters (CF is 0.000673 when T is measured in degrees Farenheit, R_s is measured in calories per square centimeter, and ETr is measured in inches). Equation [3] was calibrated against the modified Penman equation (Kimberly, ID parameters) at a few sites in northern Utah and southern Idaho, as described by Grabow (1984). The calibration was accomplished by identifying the values of C_t and T_x that minimized the sum of squared differences between Eq. [3] and the Penman equation for consecutive 5-d ETr totals. This resulted in a T_x value of -17.78 ($T_x = 0.0$ if T is measured in °F) with C_t equal to 0.009 for valleys with extensive irrigated or wetland area, and 0.011 in predominately arid lands.

The k_{cm} values, or mean crop coefficients, inherently include the wet-surface evaporation adjustment for normal irrigation and rain patterns (Wright, 1981) at Kimberly, ID, a semi-arid intermountain western USA research site. Lacking locally calibrated crop coefficients, the mean Kimberly, ID coefficients are acceptable for most purposes; however, caution must be exercised. If the anticipated field irrigation frequency is different from the schedule at the research site (in this case Kimberly, ID), then adjustments to account for a different evaporation from a wet soil surface may be appropriate. This would be particularly true for high frequency irrigation by center pivots.

The k_{cm} crop curves are represented in the model by polynomial equations of the form:

$$k_{cm} = a_0 + a_1 r + a_2 r^2 + a_3 r^3 \quad [4]$$

or

$$k_{cm} = b_0 + b_1 d + b_2 d^2 + b_3 d^3 \quad [5]$$

where k_{cm} is the estimated daily value of the mean crop coefficient, ET/ETr, and is generally constrained by maximum and minimum limits; $a_0 \ldots a_3$, and $b_0 \ldots b_3$ are polynomial coefficients determined by regression analysis; r is the fraction of time from planting to effective cover [i.e., days from planting to the present divided by days from planting to effective cover] for estimating k_{cm} before effective cover; and d is days after cover for estimating k_{cm} after effective cover.

At the beginning of the season, the planting date may be known, but the effective cover and harvest dates will be estimated from previous years experience. These estimated dates should be revised, by re-running SET UP, as the season progresses. Field-observable conditions relating crop growth state to SET UP program effective cover and to last harvest dates are given in Table 21-1.

Information identifying planting or beginning growth, effective cover, and harvest dates are obtained by PCET from a diskette file established for

Table 21-1. Observable crop growth states for the SET UP program.†

Crop‡	Planting or beginning growth	Effective cover	Last harvest
Spring grain	Planting date	At heading (when head escapes from sheath)	Crop is ready to harvest
Corn	Planting date	Tassel emergence above leaves	Killing frost
Alfalfa	When field looks green from pickup	Crop is 40 cm high (about 20-25 d prior to first cutting)	After last cutting (31 Oct. is default)
Potato	Planting	Interleafing of rows or no distinctive rows visible	Vine kill, killing frost, early dying, or ripening
Winter grain	Out of dormancy, effective emergence, or green up	At time of heading (when head escapes from sheath)	When crop is ready to harvest
Bean	Planting	Interleafing of rows	Killing frost, plant is dead
Pasture	When growth is evident in spring	Grass is about 20 cm tall	Killing frost
Apple, cherry, and peach	Full bloom	Terminal bud set, no further leaf or stem growth	Harvest of fruit
Garden§	Planting	Interleafing of rows	Last harvest of garden

† The assistance of J.L. Wright, USDA-ARS, Kimberly, ID, and A.H. Hatch, Utah State Univ. Ext. Horticultural Specialist, Provo, UT is acknowledged.
‡ Dry pinto bean (*Phaseolus vulgarus* L.), apple (*Malus pumila* Mill.), cherry (*Prunus avium* L.), and peach (*Prunus persica* L.).
§ A typical garden is a composite of several crops. In northern Utah, planting occurs about 1 May, after actual planting of pea (*Pisum sativum* L.) and carrot (*Daucus carota* L.), but prior to sweet corn and tomato (*Lycopersicon esculentum* Mill.). Effective cover is assumed when the midseason sweet corn tassels near the end of July.

each region (or farm) by the SET UP program. The calculation of ET (Eq. [2]), using k_{cm} determined from Eq. [4] and [5] for various crops) is performed by the program PCET. Coefficients for the polynomial equations, Eq. [4] and [5], for various crops are presented in Table 21-2. These were derived using judgement and regression analysis to obtain a reasonable fit with published tabular values. Crop coefficient curves for spring grain and field corn, as examples, are shown in Fig. 21-1, along with tabular values for comparison.

III. OPERATION OF IRRIGATION SCHEDULING PROGRAMS

As mentioned earlier, the SET UP program must be correctly run at least once at the beginning of the season for each region, i.e., county, subcounty area, or farm. The user specifies a region name, such as VERNAL, from which SET UP creates a region file name, VERNAL.RDAT. Polynomial equations have been developed for several areas of the state represent-

Table 21-2. Coefficients for polynomial equations describing the mean crop coefficients (k_{cm}).†

Crop§	Before effective cover, r (percent/100)‡				Limits on k_{cm}	
	a_0	a_1	a_2	a_3	Maximum	Minimum¶
Spring grain	0.1580	−0.3084	2.966	−1.756	1.00	0.20
Bean	0.2078	−0.4680	2.562	−1.340	0.95	0.20
Pea	0.1983	−0.1972	1.667	−0.7323	0.93	0.20
Potato	0.1970	−0.6813	3.148	−1.884	0.78	0.20
Sugar beet#	0.1965	0.9984×10^{-2}	−0.1929	1.000	1.00	0.20
Corn	0.2010	−0.7520	2.476	−0.9615	0.95	0.20
Alfalfa (before first cutting)	0.3113	1.248	−0.5599	0	1.00	0.30
Winter wheat	−0.1540	2.252	0.3415	−1.438	1.00	0.30
Pasture	0.35	0.30	0	0	0.65	0.35
Orchard, w/cover†† apple-cherry	0.1855	1.503	0.3093	−0.2894	1.10	0.30
peach	0.1862	1.527	−0.7531	−0.04011	1.00	0.30
Garden‡‡	0.2160	−0.02259	0.8178	−0.1877	0.85	0.25

Crop	After effective cover (d)§§				Limits on k_{cm}¶¶	
	b_0	b_1	b_2	b_3	Minimum	Days after
Spring grains	1.004	0.02925	-0.1100×10^{-2}	0.7079×10^{-5}	0.1	70
Bean	0.9806	0.01745	-0.1295×10^{-2}	0.1232×10^{-4}	0.05	60
Pea	1.105	−0.02512	0.1799×10^{-3}	0.3741×10^{-6}	0.1	70
Potato	0.7879	0.8353×10^{-4}	-0.3561×10^{-4}	-0.2817×10^{-6}	0.2	100
Sugar beet	1.187	-0.7856×10^{-2}	0.7628×10^{-4}	-0.4643×10^{-6}	0.0	100
Field corn	0.9740	-0.2157×10^{-2}	0.8658×10^{-4}	-0.1741×10^{-5}	0.25	90
Sweet corn	0.9514	0.4186×10^{-2}	-0.2097×10^{-3}	0.5963×10^{-6}	0.1	90
Alfalfa##	0.245	0.0378	0	0	--	--

(continued on next page)

Table 21-2. Continued.

Crop	After effective cover (d)§§				Limits on k_{cm}¶¶	
	b_0	b_1	b_2	b_3	Minimum	Days after
Winter wheat	1.052	0.0530	-0.1972×10^{-2}	0.1456×10^{-4}	0.1	70
Pasture	0.65	0	0	0	0.65	250
Orchard, w/cover						
apple-cherry	1.090	-0.3638×10^{-3}	0.3000×10^{-4}	-0.6004×10^{-6}	0.8	90
peach	1.001	-0.1422×10^{-2}	0.8196×10^{-4}	-0.1210×10^{-5}	0.8	90
Garden	0.8288	0.008379	-0.8141×10^{-3}	0.8243×10^{-5}	0.2	55

† These equations were developed at Utah State Univ. Agric. and Irrig. Eng. Dep., based on tabular values presented by Wright (1981), except for orchard and garden, which were adapted to an alfalfa reference from FAO 24 (Doorenbos & Pruitt, 1977, rev.).
‡ Before effective cover: $k_{cm} = a_0 + a_1 r + a_2 r^2 + a_3 r^3$, r in percent/100.
§ Some crop equations are not included in present microcomputer version.
¶ Wright suggests that the minimum k_{cm} value of 0.2 is "appropriate for relatively dry-surface soil conditions from planting until significant crop development. For moderately wet-surface soil, as with preemergence irrigation(s) or some precipitation, use 0.35 and for very wet conditions, 0.50.
# Sugar beet (*Beta vulgaris* L.).
†† Begin growth 1 April; estimated effective cover 15 July.
‡‡ Garden represents a composite of several crops, see Table 21-1.
§§ After effective cover: $k_{cm} = b_0 + b_1 d + b_2 d^2 + b_3 d^3$, (d).
¶¶ Same maximum as before cover. Minimum value is imposed following the indicated days after effective cover.
Time base is days since previous cutting, same maximum and minimum k_{cm} values as before effective cover.

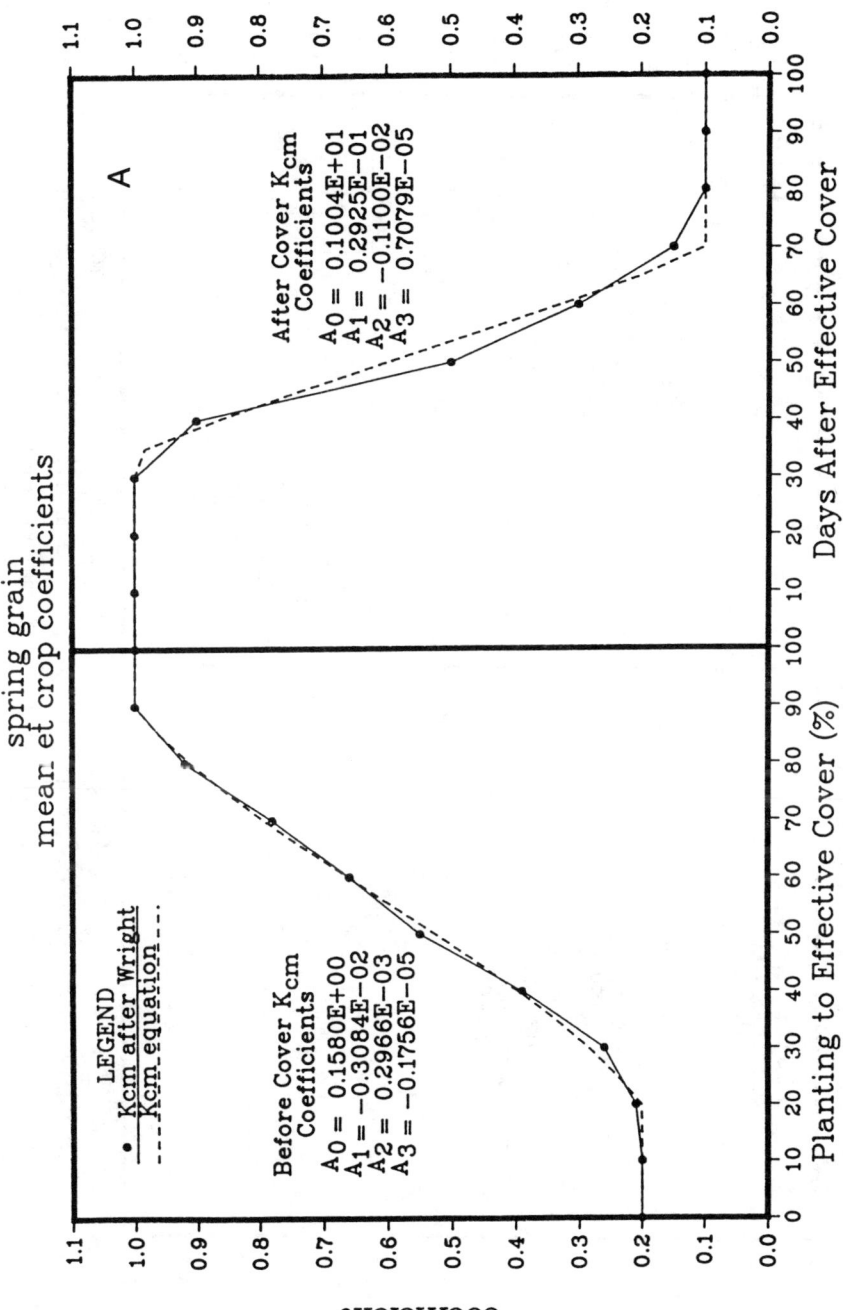

Fig. 21-1a. Crop coefficient curves for spring grain from polynomial equations compared with tabular values of Wright (1981).

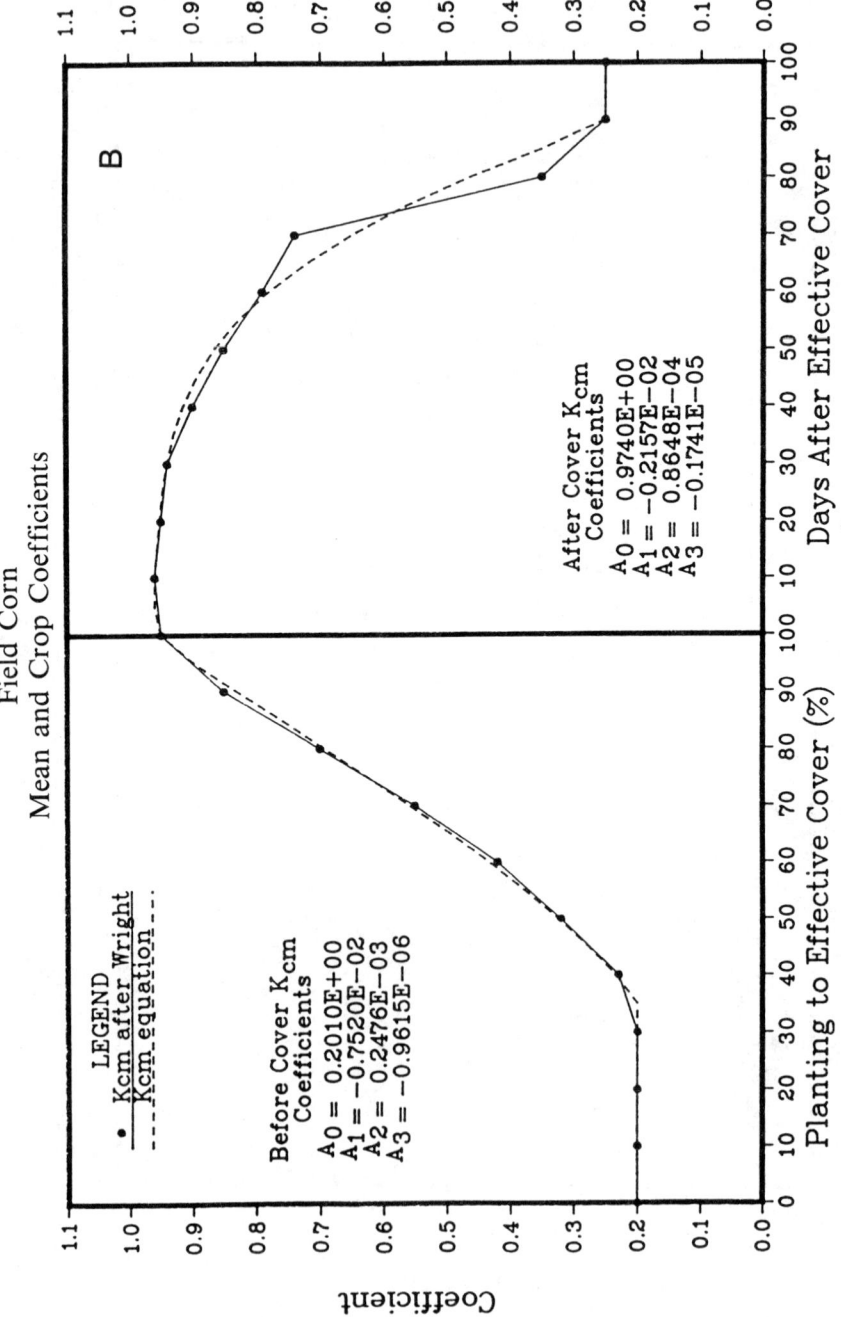

Fig. 21-1b. Crop coefficient curves for field corn from polynomial equations compared with tabular values of Wright (1981).

ing average ETr vs. time. These are used with a forecast adjustment to predict future ETr values in PCET. The user selects the appropriate one for the area from a list presented by SET UP. A table of available crops with default growth progress dates is displayed on the screen. The user identifies which crops should be included in the crop water-use output table from PCET. Planting, effective cover (100% of ground shaded at noon or as per Table 21-1) and harvest dates are input for each crop in the table. A sample SET UP hardcopy output for the VERNAL.RDAT file is shown in Fig. 21-2. At anytime during the season when observed crop-growth progress differs from that estimated at the beginning of the season, SET UP should be rerun to incorporate these changes in the region file.

Crop water-use tables are obtained from PCET for weekly, or shorter time interval, updates of the estimated historical ET and the ET predicted for the following week and the following two weeks for crops selected in SET UP. The program requires the name of an existing region file (i.e.,

```
              P C E T    S C H E D U L I N G    S E T    U P
REGION: VERNAL UINTAH CO.    DATE: 7/7/87

                        BEG-GROWTH   FULL        LAST
                        OR PLANT     COVER       HARV
   C R O P              FIRST   DUR  DATE        DATE
   #    NAME       STS  DATE    WKS

   0   REFERENCE   IN   APR 10   1   MAY 30      JUL 31
   1   SPR BARLEY  IN   APR 15   2   JUN 15      AUG 20
   2   SPR WHEAT   OUT  MAR 27   3   JUN 15      JUL 31

   3   ALFALFA     IN   APR 15   2   MAY 15      OCT 15
 CUT JUN 10  AUG   1    SEP 15       OCT 31    OCT 31

   4   CORN        IN   MAY  5   2   JUL 20      SEP 30
   5   WNT BARLEY  OUT  MAR 16   2   JUN 14      AUG  8

   6   WNT WHEAT   OUT  MAR 16   2   JUN 14      AUG  8
   7   PASTURE     OUT  MAR 23   3   MAY  2      OCT 27
   8   GARDEN      OUT  MAY  5   2   JUN 29      SEP 27

   9   SWEET CORN  OUT  MAY  5   2   JUN 29      SEP 27
  10   BEANS       OUT  MAY 30   2   JUL 24      SEP  7
  11   SUGAR BEETS OUT  MAY  5   2   JUN 29      OCT 27

  12   PEAS        OUT  MAR 16   2   MAY 10      JUN 29
  13   APPLES      OUT  APR 15   2   MAY 30      OCT 27
  14   PEACHES     OUT  APR 15   2   MAY 30      SEP 27

  15   CHERRIES    OUT  APR 15   2   MAY 30      AUG 28
  16   POTATOES    OUT  MAY 10   2   JUL 24      OCT 17
  17   OATS        OUT  MAR 27   2   JUN  5      AUG  8

THE ABOVE ARE CURRENT VALUES

THE FORECAST AREA IS VERNAL
```

Fig. 21-2. Hardcopy output from the SET UP program for region file VERNAL.RDAT.

VERNAL), the beginning date of the update period, and number of days. It then prompts for the Datapod display (accumulated reference ET) at the beginning and end of the update. Alternatively, the sum of ETr for the update period could be substituted. This is echoed to the screen and printer. The user must then enter a forecast percent, which is used to adjust the forecast equation ETr values up or down to match expected weather conditions for the coming week. The program determines the crop coefficient value from Eq. [4] or [5] for each crop and planting date combination, for each day of the update period, and for 14 d following. The resulting output for an update ending 7 July at Vernal, UT is shown in Fig. 21-3. This output was designed to be camera ready for use in the local weekly newspapers. The county agent obtains the ETr data and runs PCET on the morning of the

```
        AGRICULTURAL   CROP   WATER   REQUIREMENTS
                   ADVISORY  INFORMATION
             AT  VERNAL      UPDATE FOR WEEK ENDING JUL  7
              UTAH STATE UNIVERSITY EXTENSION, UINTAH COUNTY
                   AND USU IRRIGATION ENGINEERING


                     C R O P   W A T E R   U S E, MILLIMETERS;
                                             P R E D I C T I O N
                      WEEK       7  DAY     --------------------
           CROP      PLANTED     UPDATE     FOR THIS   FOR NEXT
                    OR BEGIN     ENDING       WEEK     14 DAYS
                     GROWTH      JUL  7      JUL 14     JUL 21
                                 (ETLS)      (ETHS)

        REFERENCE                  51          53         98

        SPR BARLEY  APR  15        51          53         97
                    APR  22        51          53         98

        ALFALFA     APR  15        51          53         98
                    APR  22        48          53         98

        CORN        MAY   5        35          42         84
                    MAY  12        32          40         80

        LAWN  USE:  MM             41          42
              LITERS/10 SQM       410         420

        NOTE: BEFORE PLANTING OR BEGIN GROWTH, ONLY EVAPORATION
              FROM SOIL SURFACE IS ESTIMATED.
        ======================INFORMATION FOR HOME GARDENERS===================
                                                MM OF
                                FOR LAWNS,      USABLE
                                DAYS BETWEEN    WATER TO
            SOIL TYPE           WATERINGS       30 CM. DEEP
            ---------           ------------    ---------
            LIGHT (SANDY LOAM)       3              19
            MEDIUM (LOAM)            4              25
            HEAVY (CLAY)             5              28
```
Fig. 21-3. The PCET program output table for the Vernal, UT area.

IRRIGATION SCHEDULING

newspaper press deadline. Rural newspaper schedules are such that the information may be 2 or 3 d old when the farmer receives it. However, farmers with their own computers can get the ETr values directly from the county agent or other sources and eliminate the delay by running the programs themselves.

Irrigation scheduling is accomplished by using the SOIL WATER BUDGET program with input that describes past irrigation and rain for a specific field, and ET values from Fig. 21-3 for the crop. The initial execution establishes a new file, e.g., VERNAL.SWB, and asks for the producers' name, field identification, crop, soil type, irrigation efficiency, and whether or not a neutron probe is being used for a field check. If a neutron probe is used, then the root-zone measured soil-water content at field capacity and wilting point can be entered. Otherwise, the available moisture is calculated from typical root-zone depth and water-holding values matching the crop and soil type. This information, plus carry-over and summary data from each weekly update run, is saved on file.

The daily soil-water budget calculations are performed by the SOIL WATER BUDGET program as part of each weekly update run. This is done for a 14 d period; 7 d previous (last week, the update) and 7 d future (this week). The soil water content in the root zone at the end (SW_2) of the previous update (i.e., the end of Day 7) is carried over in the file to become the beginning soil water content (SW_1) for the next update. The user may change SW_1 if needed as part of each update run. Other user-entered data include: (i) crop ET for the last week (ETLS of Fig. 21-3); (ii) predicted crop ET for the coming week (ETHS of Fig. 21-3); (iii) number, day of occurrence, and amount of each irrigation or rainfall, if any, during the past week. The program divides ETLS and ETHS by seven as an estimate of average daily ET_{cp} for the prevous and future weeks. The program computations are performed for each of the 14 d using Eq. [1] and [2]. Soil water content at the end of the upcoming week (i.e., Day 8-14) is estimated, assuming that no irrigation or rain occurs. Figure 21-4 shows the screen graph of soil water

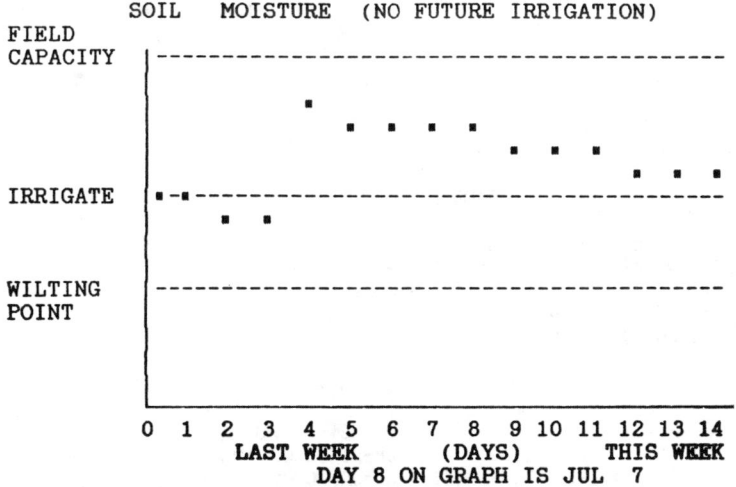

Fig. 21-4. Sample print-screen of soil water graph at completion of the SOIL WATER BUDGET program update calculations for the Vernal, UT area.

content in which an irrigation occurred on Day 4 of the last week. However, since the prior soil water content was below the irrigate line, some stress days occurred and the root zone was not refilled to field capacity. Evapotranspiration rates were such that another irrigation would be suggested on Day 13 or 14, which is 5 or 6 d from the present (the update is run as if on Day 8). If desired, the user can obtain a printer output summary of the update calculations as shown in Fig. 21-5. This summary includes accumulated seasonal totals of irrigation, rain, deep percolation, ET_{cp}, ET, and crop stress days. At the conclusion of the update, the program revises file information for carry-over seasonal totals and sets SW_2 on Day 7 to equal SW_1 for the next update.

```
               S O I L    W A T E R    B U D G E T    7 / 7 / 8 7

    FOR VERNAL IN FIELD : ALFALFA 1
    ALFALFA WITH A ROOTING DEPTH OF   152 CM.


               D A I L Y   C R O P   W A T E R   U S E   S U M M A R Y
               IRRIGATION              EVAPOTRANSPIRATION  SOIL      DEEP
    DAY        TOTAL     NET    RAIN   POTENTIAL  ACTUAL   MOISTURE  PERCOLATION
    -------------------------------------------------------------------------
                        L A S T        W E E K
    JUN 30                                7.3      7.0       344
    JUL  1                                7.3      6.5       338
    JUL  2                                7.3      6.0       332
    JUL  3      178      125              7.3      7.3       449
    JUL  4                                7.3      7.3       442
    JUL  5                                7.3      7.3       434
    JUL  6                                7.3      7.3       427
                        T H I S        W E E K
    JUL  7                                7.6      7.6       419
    JUL  8                                7.6      7.6       412
    JUL  9                                7.6      7.6       404
    JUL 10                                7.6      7.6       397
    JUL 11                                7.6      7.6       389
    JUL 12                                7.6      7.6       382
    JUL 13                                7.6      7.6       374
    -------------------------------------------------------------------------

    TOTAL       178      125              104      102


                        IRRIGATION SCHEDULING SUMMARY

    IRRIGATE WHEN WATER DROPS TO    356 MM
    GROSS IRRIGATION SHOULD BE LESS THAN 180 MM TO AVOID DEEP PERCOLATION
    LAST WEEK THE CROP WAS STRESSED 3 DAYS
    IRRIGATION SHOULD HAVE OCCURED 7 DAYS AGO
    IRRIGATION NOT REQUIRED THIS WEEK

                        S E A S O N A L    S U M M A R Y
                              AS OF JUL. 7, 1987
    TOTAL IRRIGATION.........>   325    EFFECTIVE IRRIGATION......>  228
    RAIN.....................>   114    DEEP PERCOLATION..........>   12
    POTENTIAL CROP WATER USE..>  427    ACTUAL CROP WATER USE.....>  412
    CROP STRESS DAYS.........>     9
```

Fig. 21-5. Optional summary output of the SOIL WATER BUDGET program update calculations.

IV. COMPARISON WITH FIELD DATA

Calculated soil water content from these programs, agrees closely with measured values from farm fields when accurate irrigation and rainfall data are available. This is illustrated in Fig. 21-6 and 21-7 for neutron-probe readings in commercial alfalfa fields. Figure 21-6 was developed from daily weather data, with ETr calculated in the Datapod by Eq. [3], as calibrated for use in the particular region. Irrigation data was missing for some of the weeks. Thus, estimates were made, which accounts for the early and late-season divergence between soil water contents by the neutron-probe and calculated values from the program. The fit was improved after the county agent made a correction on 11 June. However, the model values departed significantly from observed soil water content after late July when some catch-up irrigations were applied, but not communicated. This is contrasted with the much better fit between the model and neutron-probe soil water contents shown in Fig. 21-7. Rainfall and irrigation amounts were measured at the access tube site and ETr was obtained from a Penman equation. The model initial soil water content was set equal to the late-April neutron-probe reading and was not reset for the remainder of the season.

V. OTHER SIMILAR MODELS

Numerous models exist that use soil water budget and ET calculation procedures similar to those presented here. These models include public, com-

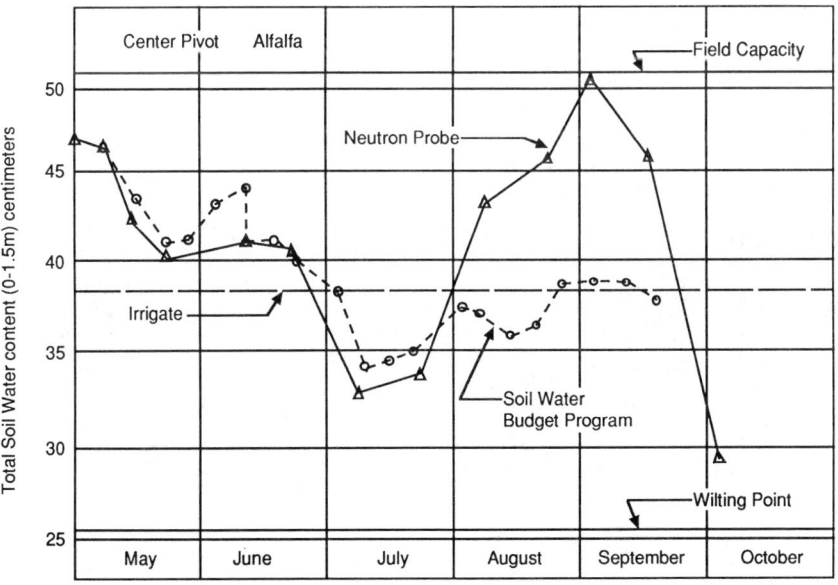

Fig. 21-6. Calculated and measured soil water budget components for a commercial alfalfa field in Uintah County, Utah, 1984, irrigated by center pivot.

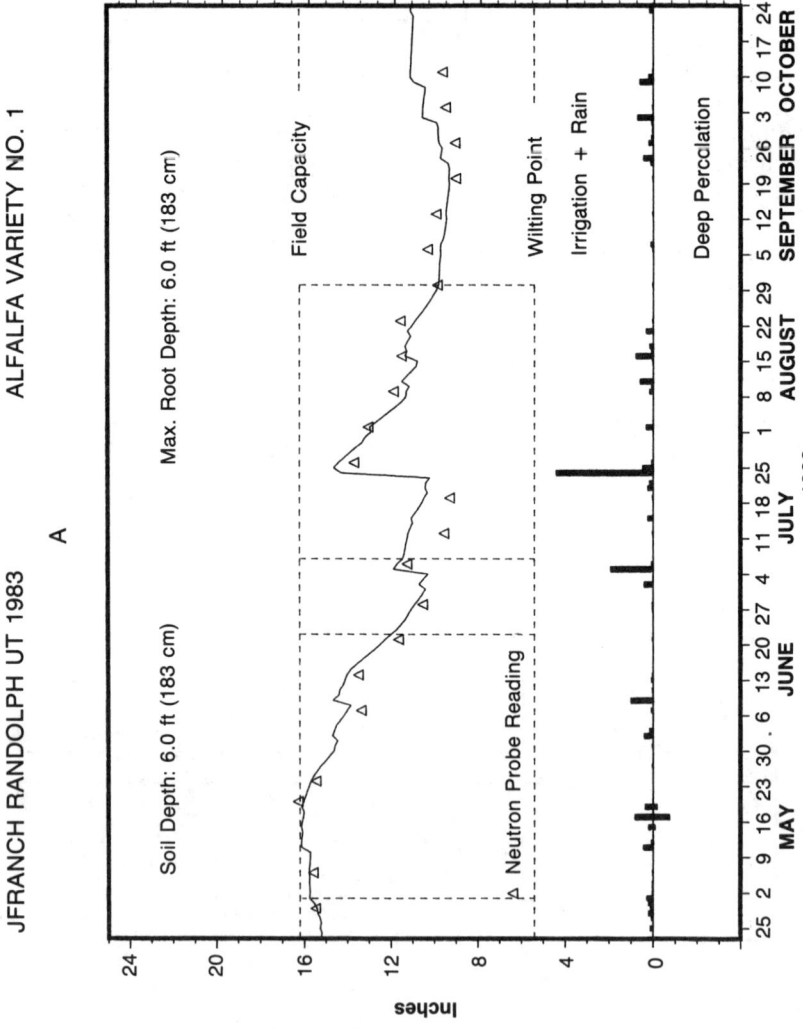

Fig. 21-7. Calculated and measured soil water budget components for a commercial alfalfa field in Rich County, Utah, 1983, irrigated by sideroll sprinkler.

mercial, and research developed irrigation scheduling programs along with crop yield models. The USDA and Univ. of Nebraska (AGNET) irrigation scheduling programs are two widely used programs. A satisfactory cataloging of the available programs is beyond the scope of this paper. Some models relating crop response to irrigation, and soils, climate, and salinity are presented by Hanks and Hill (1980).

A soil water model similar to Eq. [1] was described by Hanks (1974). He demonstrated its usefulness in predicting the influence of soil type and irrigation frequency on the amount of yield. The CRPSM model reported by Hill et al. (1984) incorporates crop phenology and yield prediction subroutines combined with a soil water budget model, using empirical equations for calculating crop ET for alfalfa, corn, dry bean (*Phaseolus vulgaris* L.), potato, soybean [*Glycine max* (L.) Merr.] (Hill et al. (1979) and wheat. Calculation procedures have been adapted from these models for use in microcomputers as described herein. Two games, IRRIGMAT (Hanks, 1981, unpublished data) and IRRIGAME (Hill, 1987, unpublished data), also based on these models have been developed at Utah State University. The games, which have been valuable education tools, incorporate economic aspects of crop production and irrigation, with maximum profit as the desired objective.

The irrigation scheduling programs (on Apple II computers) as described by Stegman and Coe (1984) are particularly relevant to this topic. They use a similar approach in calculating ETr and ET, but relate the soil water budget to a soil-moisture deficit concept. Crop coefficients were calibrated to field- and research-measured crop water use rather than calibrating the ETr equation. Thus, while there are slight differences in philosophy between Stegman and Coe (1984) and the approach described herein, the objectives are the same.

VI. CRITICAL ASSUMPTIONS

1. Weekly estimates of reference crop ET (alfalfa, 40 cm or more in height, pest free, and not short of water) are available from other sources; i.e., adjusted pan evaporation or from equations based on weather data.
2. Crop coefficients representing the irrigation method, crops, and region of use will be substituted for those contained herein, which are applicable for southern Idaho and other intermountain western United States.
3. Deep percolation occurs within one day whenever the soil water content exceeds field capacity.
4. Surface runoff is accounted for outside of the soil water budget model (Eq. [1]).
5. Upward water movement is not included and there is negligible lateral water movement.
6. The soil is assumed to be uniform throughout the field.
7. Crop growth is limited if $ET < ET_{cp}$, which occurs when the soil water content drops below the desired level.

VII. SUMMARY

Calculated soil water content using the PCET and SOIL WATER BUDGET computer programs matches neutron-probe readings well. However, this is realized only if accurate measurements of irrigation and rain are made and reported for use in the programs. Crop growth progress, as observed in the field during the summer, is also needed to modify what was assumed in the spring as a starting point for the SET UP and PCET programs. Thus, continued communication between the county agent and the farmer is essential throughout the irrigation season.

A couple of farmers who had Apple II microcomputers obtained the programs, user manual, and some individual instruction. However, they were also cooperating in the neutron-probe activity and had little incentive to use the computer for estimating irrigation dates. Irrigators seem more inclined to use and understand soil-water content data determined each week with the neutron probe and plotted on a graph at the farm. The only counties where the computer information has continued to be available have been those in which the county agent or a staff member have dedicated 0.5 d or more per week to this activity. In some of those areas, cost savings in pumping bills have ranged from $25 to $100 per hectare per year, compared with previous years, for irrigators following a scheduling program. Statewide potential, annual, energy-cost savings estimates range from $3 to $5 million as a result of improved irrigation management.

According to county agents and farmers, one of the most valuable benefits of this program has been learning about fundamental soil-water-plant relationships. Irrigators have learned about their own soils and crops, and how their irrigation management fits in. This activity has also promoted cooperation among extension, the state department of agriculture, the Soil Conservation Service, and Soil Conservation Districts.

VIII. ACKNOWLEDGMENTS

Contributions and suggestions of many of the cooperative extension personnel in Utah, particularly Lynn Esplin, Brent Gledhill, Richard Griffin, Harold Lindsay, Robert Morris, and Steve Cox, and the cooperation of farmers for field sites is gratefully acknowledged. Special appreciation is extended to Niel Allen for his IBM-PC compatible adaptions of the Apple II programs and to Karen Hammer, Lori Johnson, and Cheryl Johnson for the manuscript typing.

The assistance of Kelvin Anderson, Richard Humphreys, Garry Grabow, Brian Boman, and Steve Barfuss in developing crop coefficient polynomials, program modifications, data analysis, and in the field work is appreciated. Also appreciated is the cooperation of local Soil Conservation Service personnel for sharing their neutron-probe data.

REFERENCES

Doorenbos, J., and W.O. Pruitt. 1977. Crop water requirements (revised). FAO Irrig. and Drain. Paper 24. United Nations, Rome, Italy.

Grabow, G. 1984. Transferability of selected consumptive use equations in the intermountain U.S. M.S. thesis. Utah State Univ., Logan.

Hanks, R.J. 1974. Model for predicting plant yield as influenced by water use. Agron. J. 66:660-664.

Hanks, R.J., and R.W. Hill. 1980. Modeling crop response to irrigation in relation to soils, climate and salinity. Pergamon Press, New York.

Hill, R.W., R.J. Hanks, and J.L. Wright. 1984. Crop yield models adapted to irrigation scheduling programs. Utah Agric. Exp. Stn. Res. Rep. 99.

Hill, R.W., D.R. Johnson, and K.H. Ryan. 1979. A model for predicting soybean yields from climatic data. Agron. J. 71:251-256.

Jensen, M.E. (ed.). 1974. Consumptive use of water and irrigation water requirements. Am. Soc. Civ. Eng., New York.

Jensen, M.E., and H.R. Haise. 1963. Estimating evapotranspiration from solar radiation. J. Irrig. Drain. Div. Am. Soc. Civ. Eng. 89:15-41.

Stegman, E.C., and D.A. Coe. 1984. Water balance irrigation scheduling based on Jensen-Haise equation: Software for Apple II, II+ and IIe computers. North Dakota State Univ. Agric. Exp. Stn. Res. Rep. 100.

Wright, J.L. 1981. Crop coefficients for estimates of daily crop evapotranspiration. Proc. 1981 Irrig. Scheduling Conf., Chicago, IL. 14-15 Dec. 1981. ASAE, St. Joseph, MI.

22 Biophysical Simulation of Wheat and Soybean to Assess the Impact of Timeliness on Double-Cropping Economics

LUCAS D. PARSCH, MARK J. COCHRAN,
KALVEN L. TRICE, AND H. DON SCOTT

University of Arkansas
Fayetteville, Arkansas

In recent years, the double cropping of wheat (*Triticum aestivum* L.) and soybean [*Glycine max* (L.) Merr.] has been a popular enterprise for producers in the southern USA. Between 1982 and 1986, the proportion of total soybean acreage that was double cropped was 24% in Arkansas, 39% in Georgia, 33% in North Carolina, and 36% in South Carolina (Arkansas Agricultural Statistics Service, 1987). Smaller, but nevertheless significant, proportions of soybean acreage were also double cropped in Louisiana and Mississippi (9 and 14%, respectively) during the same period. A double-cropped system ideally begins with winter wheat seeded in mid to late October and harvested by early June. Soybean is then planted to the same acreage and harvested in autumn resulting in two crops per acre per year. Typically, the benefits of double cropping, in comparison with full-season soybean, consist of the additional net returns from the wheat crop and the improved cash flows, which accrue to the producer in June at wheat harvest.

In spite of these advantages, there is a potential increase in risk to the producer with a double cropped system. Any delays in the wheat harvest will subsequently delay soybean planting and increase the probability of reduced soybean yields due to the diminished photoperiod with the later plantings. The Arkansas Cooperative Extension Service (CES) recommends early soybean planting to avoid yield losses of 2%/d for soybean planted between 15 June and 1 July, and even greater losses of 2.5%/d after 1 July (Nester, 1983). These recommendations are based on research that found that early maturing soybean varieties (Group V) attain highest yields if planted in early June, compared with midseason and late midseason varieties (Groups VI and VII) whose yields are highest with mid- to late-June plantings (Caviness & Thomas, 1979). Since virtually all wheat in Arkansas is double cropped, the average date of wheat harvest provides one indication of the approximate date of double-crop soybean planting in the state. On average, only 3% of winter wheat acreage was harvested by 1 June compared with 40%

Copyright © 1991 ASA-CSSA-SSSA, 677 S. Segoe Rd., Madison, WI 53711, USA. *Modeling Plant and Soil Systems*—Agronomy Monograph no. 31.

on 15 June and 89% on 29 June for the 5-yr period from 1982 to 1986 (Arkansas Agricultural Statistics Service, 1987). These figures suggest that the wheat harvest in Arkansas occurs over a range of dates, which could effectively contribute to reduced double-crop soybean yields.

One risk-reducing decision rule that the double-crop producer has the option to use is to harvest wheat earlier. This can be done by initiating the wheat harvest when the grain moisture content of the standing mature crop is relatively high, in lieu of waiting for field drydown to near-storage-safe moisture levels. By reducing the field drydown period of wheat, the entire time window of double-cropping field operations can be moved forward in the season, but not without a cost. Harvesting wheat earlier at a higher moisture content necessitates artificial drying costs for wheat. In the case of direct sales to an elevator, a graduated discount is applied, which increases with the moisture level of the wheat and which compensates for the implicit cost of commercial drying.

This chapter assesses the risk-return tradeoffs that are incurred by the double cropping producer when the harvest of winter wheat is initiated at an earlier date and a higher moisture content in order to avert reduced soybean yields due to delayed planting. The results of the study are based on a computer simulation experiment in which two phenological crop growth models, CERES-Wheat and SOYGRO, are interfaced to reflect the biophysical interdependencies of soybean following wheat. The key management decision variable evaluated by the study involves the timeliness of wheat harvest, which is defined in the simulation model as the threshold grain moisture content (MCHVST) at which wheat harvest is initiated. Eight alternative management strategies reflecting the spectrum between an early high moisture (20%) and a late low moisture (13%) wheat harvest are each simulated over 20 "states of nature" using Arkansas weather data and environmental conditions. Performance measures of each of the eight systems, including simulated yields, and harvest and planting dates over the 20-yr simulation, are presented and compared. Empirical cumulative distributions of double-cropping net returns are developed and subsequently ranked for risk-return tradeoffs using stochastic dominance ordering. The study provides an example of how crop growth simulation models can be adapted for use to assist in farm management decision making.

I. SIMULATION MODELS

A. Model Description

The CERES-Wheat (Ritchie & Otter, 1985) and SOYGRO (Wilkerson et al., 1983) models are physiologically based crop growth models which simulate daily dry-matter growth, vegetative and reproductive development, leaf area index, and final yield as a function of both management and environmental variables specified by the user. User inputs to each of the models that describe crop environmental conditions include daily climatological data (maximum and minimum temperature, precipitation, and solar radiation) and parameters reflecting characteristics of the soil profile to be simulated. The SOYGRO 5.0 (Wilkerson et al., 1985) version of the soybean model used in the present study, differs from earlier versions of SOYGRO in that it includes the subroutine WATBAL, which incorporates the soil water balance model developed by Ritchie (1972). Hence, simulated crop growth in both CERES-Wheat and SOYGRO is sensitive to moisture in the root profile based on the same underlying soil moisture logic. Management inputs to each of the models include planting date, seed density and spacing, genotype, and irrigation management. Neither version of the models used in this study addresses issues of crop stress due to disease and insect infestation.

B. Model Testing and Refinement

Both CERES-Wheat and SOYGRO were extensively tested under Arkansas conditions and statistically validated for final yield prediction using field data collected in Arkansas over a multiple year period (Trice, 1986; Prickett, 1985). Initially, tests were conducted with CERES-Wheat to assess sensitivity of model output to planting date, seed density, and genetic coefficients associated with alternative genotypes. Model calibration was based on the modification of two constants in the GROSUB subroutine. These constants affect the values of the daily amount of biomass produced (PCARB) and the daily maximum possible area expansion of the wheat stem (CLG). All model testing was done by introducing site-specific user-specified input parameters for a Crowley silt-loam soil (fine, montmorillonitic, thermic Typic Albaqualf). Details of these tests are reported in Trice (1986).

Model testing of SOYGRO included sensitivity of the model output to planting date, genotype, and nonirrigated versus irrigated management for a wide variety of irrigation strategies. Model calibration and changes were subsequently incorporated into the phenology and water balance components of the model.

Using historical data from 4 yr of field measurements taken at five locations in Arkansas, the passage through the simulated plant growth stages was modified in SOYGRO. Both the vegetative stage of growth (VSTAGE) and the maximum number of nodes at harvest (VSMAX) in subroutine PHEN1 were modified to become functions of variety, planting date, and daily maximum and minimum temperatures. The remaining stages are estimated from VSTAGE and VSMAX. A complete discussion of these modifications can be found in Prickett (1985).

The soil-water balance subroutine, WATBAL, was also modified to allow the simulation of tensiometer strategies to schedule irrigation applications. The procedure used was based on an algorithm described in Brown et al. (1985). Using regression analysis, the relationship between water retained in the soil at a set pressure (pascals) was estimated for each of four depths of the soil profile. The four regression equations were estimated from experimental field data on Crowley silt loam soil. User-specified management inputs in the modified SOYGRO identify the depth at which the tensiometer reading is taken and the set pressure at which irrigation commences. Documentation for this procedure is found in Prickett (1985).

The top 37 cm of Crowley soil has a silt loam texture that is porous and allows water to flow through easily. Below this layer is clay. For all simulations, the soil profile in WATBAL was divided into seven layers with Layers 1 and 2 representing the silt loam portion (0-37 cm) and Layers 3 through 7 representing clay (37-180 cm). Due to a problem with excessive runoff, calibration efforts were directed at managing those layers representing the transition from the silt loam to the clay portion of the soil profile. Refinements were necessary on those parameters that measure the soil water constant (SWCON) for drainage and the lower limit (LL) on the soil water content for the third soil layer. Infiltration and drainage were increased, whereas runoff was decreased, in this process.

C. Model Interface for Double Cropping

The CERES-Wheat and SOYGRO models were initially designed to be used independently of each other. In order to accommodate the simulation of a double-cropped system for the present study, the two models were interfaced to enable tandem simulation of wheat followed by soybean. The computational flow of this interface is diagrammed in Fig. 22-1. Of paramount importance to the design of this interface was that the resulting combined model reflect the interdependencies of a double cropped system. The two algorithms that accomplish the most important aspects of this interdependence are (i) the addition of a grain drydown component, which simulates field drydown of the wheat crop between physiological maturity and harvest; and (ii) the linkage of the soil moisture components of the wheat and soybean models. Each is discussed in turn.

D. Drydown Component

The grain drydown subroutine (DRYDWN) was developed to simulate the moisture content of wheat between physiological maturity and a user-specified harvest maturity. Physiological maturity in CERES-Wheat is reached when all simulated growth processes terminate and there is no further dry matter accumulation of grain or vegetative plant components (ISTAGE = 6). Harvest maturity designates that the state variable for the grain moisture content of the mature wheat on Day t (MC_t) has attained a user-selected threshold level (MCHVST) that permits initiation of harvest. Hence, the grain drydown algorithm updates the daily status of MC beginning at physiological maturity and triggers the harvest of the wheat crop whenever MC ≤ MCHVST. For all simulations in this study, the initial moisture content at physiological maturity was assigned a value of 38%. Implicitly, lower values for MCHVST will result in a lengthier drydown period because greater amounts of moisture must be removed before harvest can begin.

The grain drydown algorithm is based on the model developed by Chen and McClendon (1984):

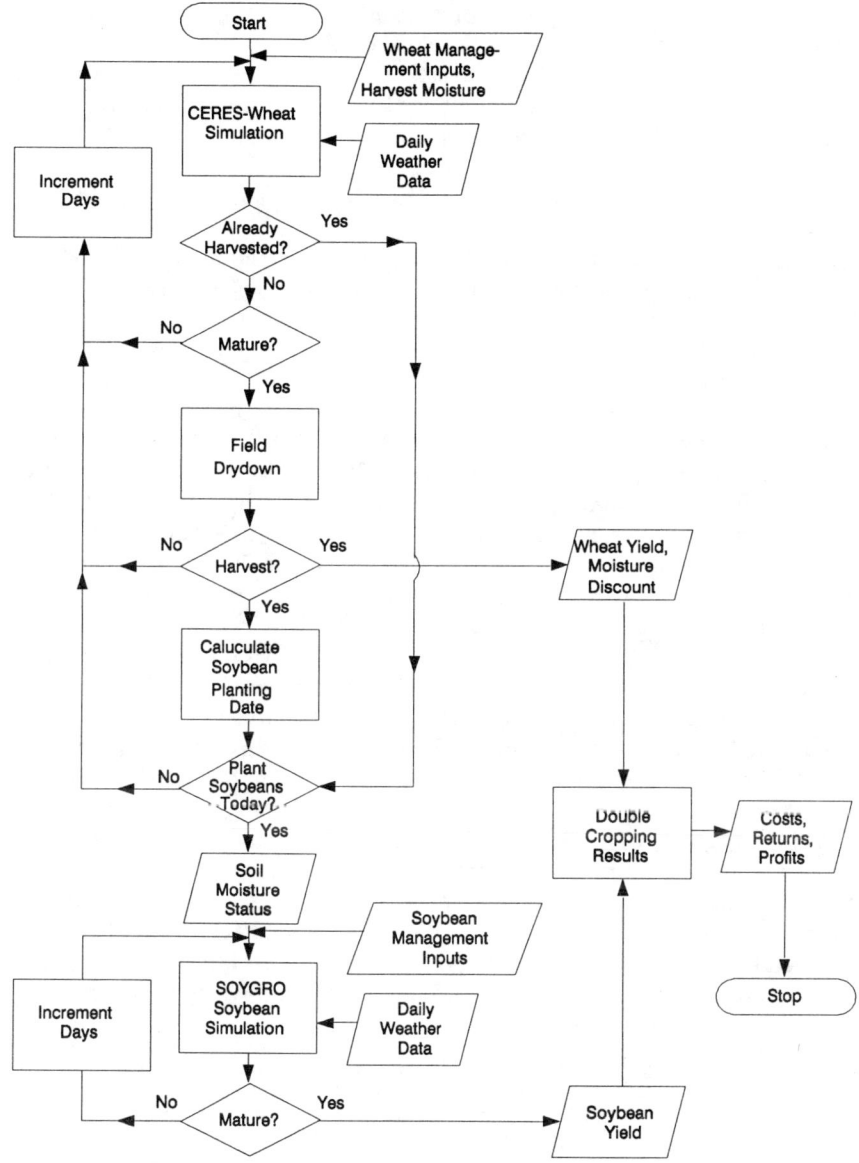

Fig. 22-1. Daily computational flow of the CERES-Wheat/SOYGRO models' interface during one double-cropping season.

$$MC_t = MC_{t-1} + \Delta MC \qquad [1]$$

where MC, on any simulation Day t, is a function of the absolute level of moisture on the previous day and the daily change in moisture content, ΔMC. If no rainfall occurs, ΔMC is a negative constant whose value is determined solely by the level of MC on the previous day ($t - 1$). For $MC_{t-1} \geq 20\%$, $\Delta MC = -4\%$; for $20\% > MC_{t-1} \geq 14\%$, $\Delta MC = -2\%$; and for $MC_{t-1} < 14\%$, $\Delta MC = -1\%$. By contrast, if rainfall does occur, ΔMC incorporates a rewetting function that allows for an increase in moisture content as a function of the previous day's MC and precipitation (PRECIP):

$$\Delta MC = 5.08 + 5.77 \; PRECIP_{t-1} - 0.33 \; MC_{t-1} \qquad [2]$$

The drydown model developed by Chen and McClendon (1984) was originally estimated for moisture reduction in mature soybean, based on Mississippi data. Nevertheless, it provides results consistent with those observed for wheat drydown in Arkansas (Adams, 1984). The Arkansas CES has observed that wheat harvested at 20 to 22% moisture will occur 2 to 10 d earlier than wheat harvested at 13% moisture, depending on whether conditions are dry and favorable or cloudy and wet (Adams, 1984). Over the 20-yr period simulated in this study, wheat harvested at 20% moisture occurred, on average, 6.25 d earlier than the 13% moisture harvest. In 18 out of the 20 yr simulated, this difference ranged from as little as 2 d to 9 d earlier in moderately wet weather. In the remaining two years, the 20% harvest preceded the 13% harvest by 16 and 33 d, respectively, due to unusually excessive moisture during the drydown period.

A summary of the performance of the drydown model over the 20-yr simulation is provided in Table 22-1. As hypothesized, the simulated mean number of drydown days decreases monotonically with an increase in moisture level at harvest. Likewise, Table 22-1 demonstrates that the length of drydown is less variable at higher moisture levels in that the standard deviation decreases from just over 9 d to under 4 d for 13 and 20% wheat harvests, respectively. This decrease in variability can be more easily understood in terms of the minimum and maximum number of drydown days simulated for the various strategies (Table 22-1). With favorable drying conditions, the minimum number of drydown days does not differ greatly between a low and high moisture harvest. In other words, there is little field time gained by harvesting wheat at high moisture during favorably dry weather. Under

Table 22-1. Summary statistics for the number of days required to field-dry mature wheat to a specified moisture level at harvest (MCHVST).

Statistics†	MCHVST, %						
	13	14	15	16	17	18	19	20
	drydown period, d							
$\bar{x}$	14.45	13.25	11.60	10.65	10.30	9.30	9.10	8.20
s	9.10	9.23	6.53	6.45	6.21	6.18	6.02	3.99
CV	0.63	0.70	0.56	0.60	0.60	0.66	0.66	0.49
Min	9	8	8	7	7	6	6	5
Max	42	41	33	32	31	31	30	18

† $\bar{x}$ represents sample mean; s represents standard deviation; CV represents coefficient of variation; and Max and Min are the maximum and minimum values simulated over the 20-yr period.

wet conditions, however, the low 13% moisture strategy requires more than twice the number of days (42) to reach harvest maturity compared with the 20% strategy, which requires a maximum of 18 d. Based on Table 22-1, one can speculate that, for all strategies, the frequency distribution of drydown days is skewed to the right, but that the low moisture strategies are especially prone to wheat harvest delays, which will sufficiently delay soybean plantings and may potentially reduce yields.

E. Linkage of Soil Moisture Components

Figure 22-1 shows that the CERES-Wheat logic simulates daily development of the wheat crop until physiological maturity is attained. Once the wheat is mature, the drydown subroutine is called daily to determine the status of MC and to test whether the threshold moisture level MCHVST has been reached so that the crop can be harvested. If this test is negative (i.e., if the wheat is too wet to be harvested) the day loop continues to call the appropriate subroutines in CERES-Wheat, including WATBAL, which updates the status of the soil moisture (SW) in the soil profile. During this preharvest drydown period, the subroutine GROSUB is bypassed and the maturity stage remains constant at ISTAGE=6 (physiological maturity to harvest), with no further biomass being produced.

On the day that the wheat is harvested (WHVST), the simulated values for yield and MC are tabulated for further processing (Fig. 22-1). In addition, the soybean planting date (SBPLANT) is set to occur DELAY days after wheat harvest, where DELAY is a user-specified value. This value would implicitly reflect the impact of tillage system (traditional vs. minimum tillage or no-till), suitable days for fieldwork, and machinery capacity on the time interval between wheat harvest and soybean planting. For all runs in this study, DELAY was set to 10.

After the wheat is harvested, but prior to the established soybean planting date, the daily simulation loop continues to call the appropriate subroutines in CERES-Wheat to update the moisture status (SW) in the seven layers of the soil profile. On the Day t that soybean is to be planted, simulation program control is passed from CERES-Wheat to SOYGRO, and ending soil-moisture status coefficients for each of the seven layers in the entire soil profile of wheat are passed into the soybean subcomponent:

$$SW(L)_{t,SOYGRO} = SW(L)_{t,CERES} \quad \text{for } L = 1, 2, \ldots 7 \qquad [3]$$

where $SW(L)$ is the soil water content of layer L (volume fraction) and t is the Day of Year of soybean planting (SBPLANT). These SW coefficients passed from CERES-Wheat to SOYGRO become the initialized values of the soybean soil profile for that cropping season. The importance of this linkage is that significant amounts of moisture can be extracted from the soil profile by the wheat crop. This could inadvertently affect soybean emergence and/or germination in a double-cropped system if adequate amounts of moisture are not available to the soybean at planting.

Table 22-2. Summary statistics for soil-water status coefficients (SW) passed from CERES-Wheat to SOYGRO over a 20-yr simulation for wheat harvested at 16% moisture content.

Statistics[†]	Date	Soil layer						
		1	2	3	4	5	6	7
		Soil depth, cm						
		0–15	15–37	37–57	57–87	87–117	117–147	147–180
		soil moisture, volume fraction						
x̄	10 June	0.218	0.245	0.302	0.298	0.313	0.319	0.307
s	8.2	0.050	0.048	0.036	0.023	0.016	0.014	0.004
CV	0.051	0.231	0.195	0.119	0.080	0.051	0.045	0.014
Min	29 May	0.120	0.148	0.224	0.241	0.276	0.286	0.299
Max	26 June	0.278	0.296	0.351	0.328	0.338	0.342	0.321

† x̄ represents sample mean; s represents standard deviation; CV represents coefficient of variation; and Max and Min are the maximum and minimum values simulated over the 20-yr period.

Table 22-2 indicates the level and variability of soil moisture coefficients (SW) passed from CERES-Wheat to SOYGRO over a 20-yr simulation for an intermediate (MCHVST = 16%) strategy. Because these coefficients are passed on the date of soybean planting, they represent the status of the soil moisture (volume fraction) on only one day in any simulation year. For Table 22-2, this 20-yr mean date was 10 June, but ranged between 29 May and 26 June. Note that in Table 22-2, both the variability (as measured by the standard deviation and coefficient of variation) and range (as measured by the difference between minimum and maximum values) of soil-water coefficients decreases monotonically from upper to lower layers of the soil profile. Implicitly, the availability of moisture to the newly planted soybean in Layer 1 is also highly variable from year to year as a function of the SW coefficients passed from CERES-Wheat to SOYGRO.

F. Potential Interactions

Potential interaction between the soil moisture interface and the grain drydown may result in simulated scenarios bounded by the following two extremes: (i) excessive precipitation after physiological maturity of wheat may result in a lengthy drydown, which delays wheat harvest and soybean planting, but which reduces the risk of poor soybean emergence due to moisture deficit; and (ii) a droughty period after physiological maturity of wheat may shorten the wheat drydown period, permitting an earlier wheat harvest and soybean planting, but at greater risk of insufficient moisture for the newly planted soybean.

II. EXPERIMENTAL DESIGN

The simulation study was designed to evaluate the impact on double cropping risk returns of reducing the field drydown period. This is done by initiating wheat harvest at each of eight alternative grain moisture threshold levels (MCHVST) ranging from storage safe (13%) to high moisture (20%) in 1% increments. Each of the eight harvest initiation strategies was simulated over 20 alternative states of nature using CERES-Wheat and SOYGRO in tandem runs. Daily historical weather data for Stuttgart, AR for the years 1966 to 1985 inclusive (20 yr) were used to simulate these alternative weather scenarios.

Each scenario began with wheat (Coker 68-15) planted on 21 October and harvested when grain moisture reached the specified threshold level (MCHVST) that defined that strategy. The Arkansas CES recommends wheat planting between 10 October and 10 November, which is late enough to avoid pests, yet sufficiently early to obtain hardy tillering (Adams, 1984).

After an arbitrary 10-d delay following wheat harvest, to allow for conventional tillage practices, soybean (Forrest) was planted. Forrest is a Maturity Group V soybean, which comprised 26% of all soybean acreage in Arkansas over the 1982 to 1986 period (Arkansas Agricultural Statistics Service, 1987). Research in Arkansas has shown that Group V soybean suffers a yield decrease if planted after 1 June compared with Groups VI and VII, which experience yield declines only after mid-June or 1 July plantings (Caviness & Thomas, 1979). However, this same research showed that Forrest produced yields as high as the later maturing varieties even when planted in late June or early July.

For all scenarios, soybean was irrigated throughout the growing season as necessary based on a simulated tensiometer reading. These soybean irrigations were scheduled in 2.54 cm applications whenever soil moisture fell below -50 kPa measured at a depth of 30 cm. This irrigation strategy is consistent with Arkansas CES recommendations (Nester, 1983).

A. Economic Model

Simulation model output for wheat and soybean yield was used to calculate net returns ($/ha) for each of the 160 strategy years (20 yr for each of eight strategies) simulated. Net returns (NR) were defined as gross returns (GR) from the sale of each of the two crops minus selected costs (TC) of crop production associated with each crop:

$$NR_w = GR_w - TC_w \qquad [4]$$

$$NR_s = GR_s - TC_s \qquad [5]$$

$$NR_{dc} = NR_w + NR_s \qquad [6]$$

where w, s, and dc refer to wheat, soybean and double cropping, respectively. The net returns value measures the dollar contribution to overhead labor, management, land, and overhead capital.

Gross returns (GR) in this analysis are the product of simulated yield and price received. For wheat, gross returns were adjusted by imposing a discount on production for moisture content in excess of 13.5%:

$$GR_w = PRICE_w \times [YIELD_w \times (1 - DISCOUNT_{mchvst})] \qquad [7]$$

Certain elevators in Arkansas follow a schedule by which they discount the volume of a grain purchase by a factor (DISCOUNT) that increases with the moisture content of the grain to be purchased. The discount schedule used in this study was obtained from a major elevator in Arkansas. This nonlinear discount is zero at a moisture content of 13.5% or less and reaches 0.19 for 20% moisture in wheat. Implicitly, this discount defrays the cost of artificial drying.

Prices of wheat ($0.113/kg) and soybean ($0.218/kg) used in the analysis are 5-yr average Arkansas commodity prices indexed to 1986 levels (Arkansas Agricultural Statistics Service, 1987). Production costs of wheat ($233.96/ha) and double-crop soybean ($247.49/ha) are based on Arkansas enterprise budgets (Stuart et al., 1986a, b). Total costs (TC) for both crops in Eq. [4] and [5] include a charge for hauling ($0.006/kg), which varies with the annual simulated yield. Additional costs for soybean include an annual overhead charge ($134.71/ha) for centerpivot irrigation equipment and a variable charge ($2.60/ha-cm) for actual irrigation water applied.

III. RESULTS

A. Wheat and Soybean Yields

Table 22-3 provides summary statistics for the key variables generated under each of the eight strategies over the 20 simulated states of nature. Consistent with the hypotheses stated above, the results show that initiating wheat harvest at a higher moisture content (Row 1) decreases wheat yields (Row 2) while simultaneously increasing soybean yields (Row 5) due to an earlier planting date (Row 6). For example, a wheat harvest at 20% moisture results in soybean being planted an average of 6 d earlier (8 June) than with a wheat harvest at 13% moisture (14 June). Consequently, the high moisture wheat harvest strategy (20%) results in an average 122 kg increase in soybean yields compared with the low moisture strategy (2767 vs. 2645 kg/ha, respectively), but wheat yields are discounted by 672 kg (2870 vs. 3542 kg/ha at 20 and 13%, respectively).

Because the discount on wheat increases with moisture level in Table 22-3, effective wheat yields are seen to monotonically decrease as the wheat is harvested progressively earlier in the season (Row 4). Soybean yields do not exhibit a correspondingly monotonic yield vs. planting date tradeoff in Table 22-3, however. As wheat is harvested at higher moisture levels, average soybean planting date moves up from 14 to 10 June resulting in a 20-yr mean yield of 2811 kg/ha for a MCHVST of 17%. Higher moisture wheat harvests result in even earlier soybean plantings, but soybean yields show no improvement. This is seen more clearly in Fig. 22-2. Simulated soybean yields begin their decine after a 10 June planting date under the 17% wheat harvest strategy. By delaying planting until 14 June (MCHVST = 13%), mean

Table 22-3. Summary statistics for simulated wheat and soybeans at alternative moisture levels (MCHVST) at wheat harvest initiation.

1. MCHVST, %	Statistics†	13	14	15	16	17	18	19	20
Wheat									
2. Yield, kg/ha	x̄	3542	3509	3435	3361	3294	3153	3012	2870
3. Maturity date	x̄	20 May	20 May	20 May	20 May	20 May	20 May	20 May	20 May
4. Harvest date	x̄	4 June	3 June	1 June	31 May	31 May	30 May	30 May	29 May
Soybean									
5. Yield, kg/ha	x̄	2645	2668	2742	2800	2811	2788	2793	2767
	s	527	518	438	406	402	417	414	450
	CV	0.20	0.19	0.16	0.15	0.15	0.15	0.15	0.16
6. Plant date	x̄	14 June	13 June	11 June	10 June	10 June	9 June	9 June	8 June
7. Maturity date	x̄	29 Oct.	28 Oct.	26 Oct.	25 Oct.	24 Oct.	23 Oct.	23 Oct.	22 Oct
Net returns									
8. Wheat, $/ha	x̄	146.10	142.30	134.69	127.11	119.50	104.28	89.09	73.90
9. Soybean, $/ha	x̄	119.60	123.75	138.76	149.73	152.72	147.80	149.02	143.04
10. Double crop, $/ha	x̄	265.70	266.07	273.48	276.84	272.22	252.09	238.11	216.94
	s	102.55	105.42	92.77	86.13	86.08	85.26	84.72	95.71
	CV	0.39	0.40	0.34	0.31	0.32	0.34	0.36	0.44
	Min	24.85	51.72	43.27	60.79	52.34	31.10	14.18	−2.72
	Max	416.37	413.90	405.82	408.14	400.07	348.69	333.75	323.27

† x̄ represents sample mean; s represents standard deviation; CV represents coefficient of variation; and Max and Min are the maximum and minimum values simulated over the 20-yr period.

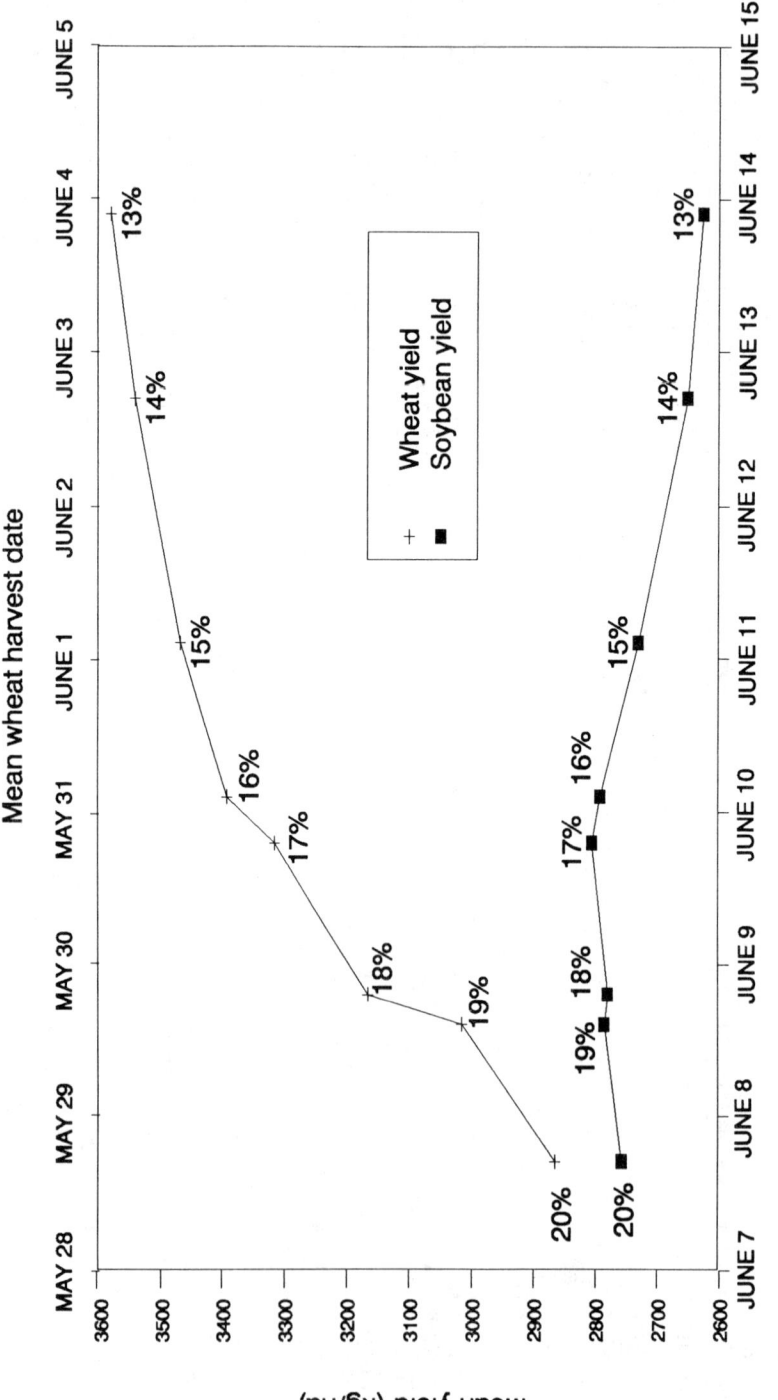

Fig. 22-2. Simulated mean yield, wheat harvest date, and soybean planting date. The numbers represent the percent moisture content at harvest. Observations appear at uneven intervals because all calendar dates of field operations are the simulated means for each strategy.

yields decrease to 2645 kg. This 167-kg yield decrease over a 4-d delay represents a daily yield loss of 1.5%, which is consistent with Arkansas CES expectations for Group V irrigated soybean (Nester, 1983).

B. Wheat Harvest and Soybean Planting Dates

The soybean yields depicted in Fig. 22-2 are 20-yr means. Fig. 22-3 provides the empirical cumulative frequency of soybean yields for a representative low (13%), intermediate (16%), and high moisture (19%) wheat harvest strategy. Graphs of all cumulative frequencies for this study use the fractile method suggested by Schlaifer (1959). The wider spread between the cumulative frequencies at low yields indicates that there is a higher cumulative probability of incurring low yields with the 13% strategy in comparison with the 16% or 19% strategies. This occurs in years in which soybean plantings are sufficiently delayed causing a yield decline. By contrast, at high yield levels the three cumulative frequencies converge, indicating that under favorable drydown conditions very little if any yield advantage is gained with higher moisture strategies. Since the difference in the length of the drydown period between a low and high moisture strategy is minimal in dry warm weather (Table 22-1), soybean plantings will be early and only a few days apart for both the low and high moisture strategies.

These relationships are seen more clearly in Fig. 22-4 and 22-5, which show the cumulative frequencies of wheat harvest dates and soybean planting dates, respectively. For all simulated scenarios, the average date of physiological maturity of wheat is 20 May (Table 22-3, Row 3) with maturity occurring as early as 12 May during some years and as late as 2 June (Fig. 22-4) during others. Under optimal conditions, the drydown requires a minimum of 9, 7, and 6 d (Table 22-1) for the 13, 16, and 19% strategies resulting in wheat harvested on 18, 19, and 21 May, respectively (Fig. 22-4). Given the arbitrary 10-d interval (DELAY) used in this simulation, the corresponding soybean planting dates are 28, 29, and 31 May for these three strategies, respectively (Fig. 22-5).

Under wet conditions, the length of drydown increases, resulting in wheat maturing as late as 14, 16, and 22 June for the same three strategies (13, 16, and 19%, respectively, Fig. 22-4), with corresponding soybean planting dates of 24 and 26 June, and 2 July (Fig. 22-5). Figure 22-2 showed that simulated mean soybean yields decreased whenever plantings were delayed after 10 June. The cumulative frequencies in Fig. 22-5 show that the lower yields with the 13% strategy result, in part, because there is only a 0.35 probability that soybean will be planted by 10 June, compared with approximately a 0.50 probability that planting will occur before 10 June for the 16 and 19% strategies. In other words, there is a higher probability (0.65) of soybean plantings being delayed beyond 10 June with a low moisture strategy than there is with an intermediate or high moisture strategy (0.50).

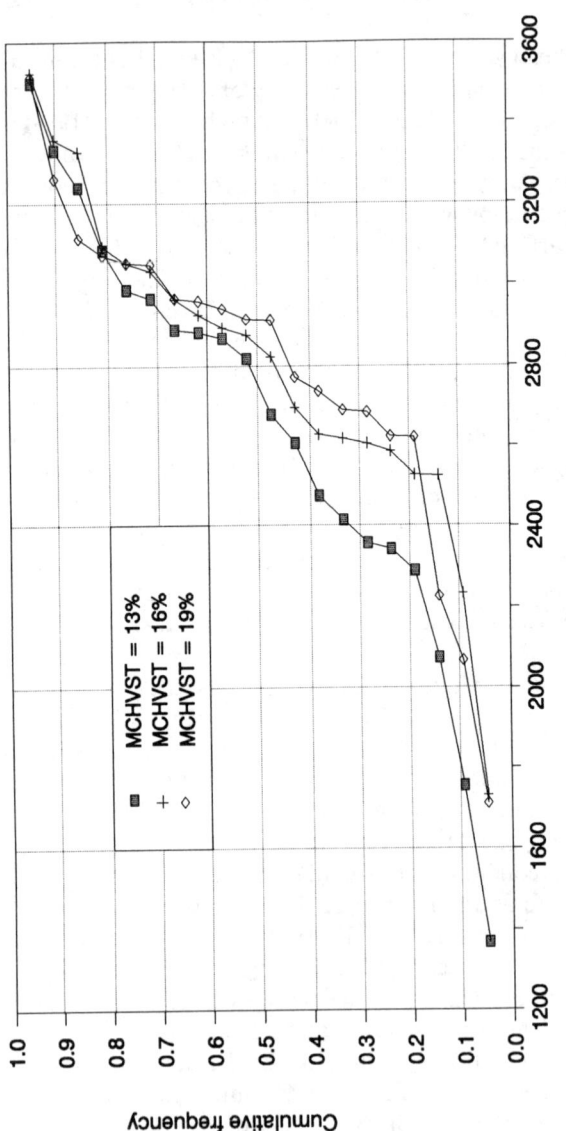

Fig. 22–3. Cumulative frequency of soybean yields for three wheat harvest strategies using various moisture contents at harvest (MCHVST).

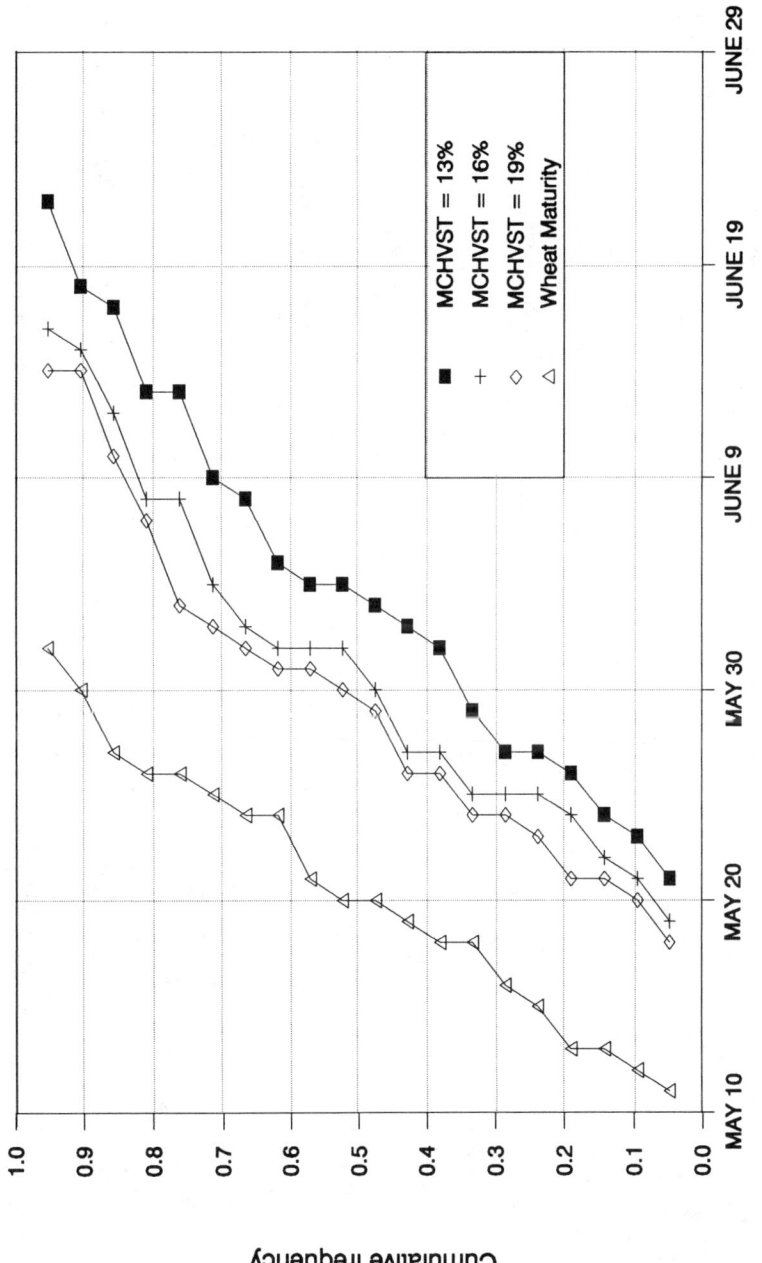

Fig. 22–4. Cumulative frequency of date of wheat maturity and wheat harvesting for three strategies using various moisture contents at harvest (MCHVST).

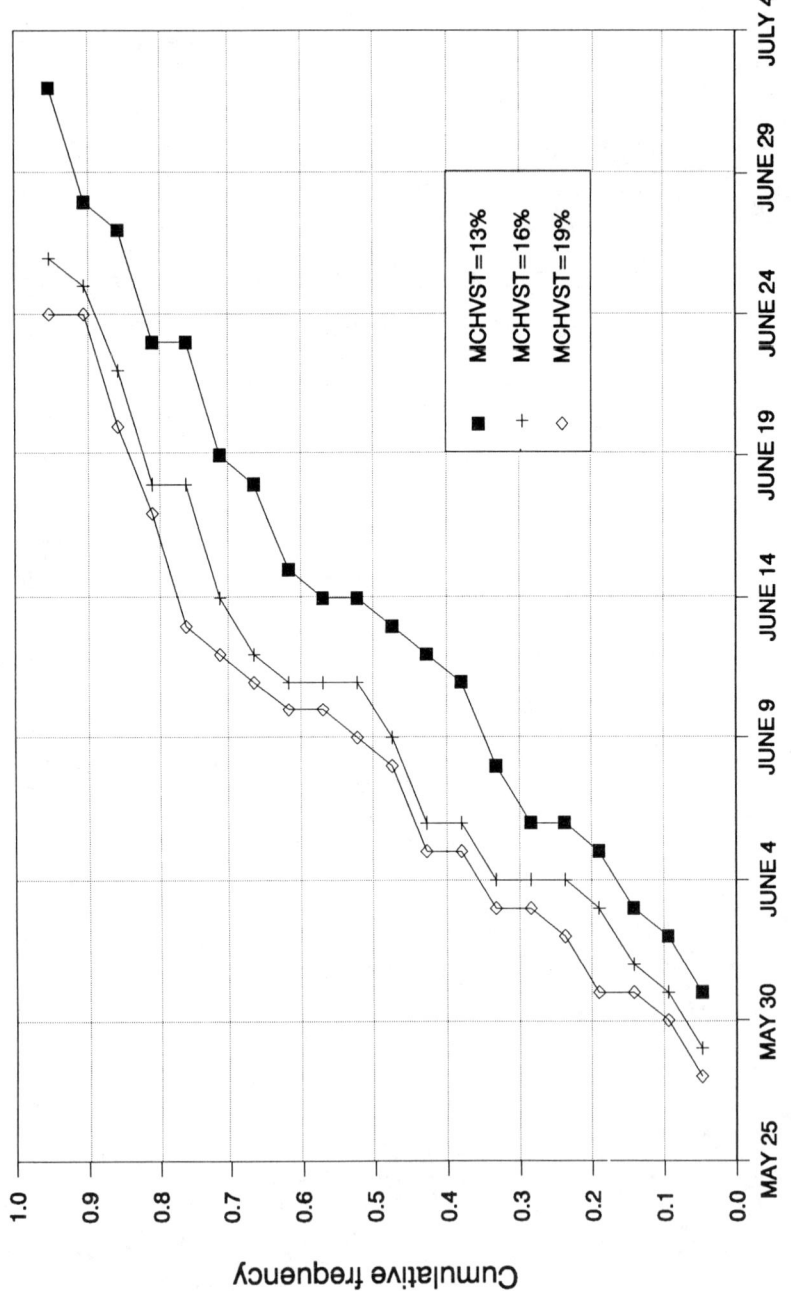

Fig. 22–5. Cumulative frequency of soybean planting date for three wheat harvest strategies using various moisture contents at harvest (MCHVST).

C. Net Returns

Lines 8 and 9 in Table 22-3 show that mean net returns over the 20-yr simulation roughly reflect (i) the monotonically decreasing wheat yields and (ii) the simultaneously increasing soybean yields, which initially result when wheat is harvested at progressively higher moisture levels. When summed, the double-cropping net returns (Table 22-3, Row 10) are highest when wheat is harvested at an intermediate moisture level. The 16% strategy renders the producer a $11.14/ha average annual increase over the 13% low moisture strategy and a $59.90/ha increase over the high-moisture early harvest strategy. These results are depicted graphically in Fig. 22-6. Larger declines in double-cropping net returns with the 18, 19, and 20% strategies correspond to lower average yields for both wheat and soybean with these high moisture strategies (Fig. 22-2).

D. Weather Risk

One measure of the sensitivity of a management strategy to weather risk is indicated by the standard deviation of its performance variable when subjected to a broad range of weather scenarios. Table 22-3 demonstrates that soybean yield risk is greatest at low moisture wheat harvests and decreases with higher moisture harvests up to 17%. Both the standard deviation and coefficient of variation of soybean yield are greatest at a MCHVST of 13% and reach a minimum at a MCHVST of 17%. This tendency is mirrored in double-crop net returns (Table 22-3, Row 10) whose standard deviations and coefficients of variation are likewise lowest at intermediate harvest moisture levels.

An additional consideration when evaluating the risk inherent in any strategy is the potenital for very low or very high net returns. Minimum values for net returns in Table 22-3 (Row 10) are highest for the 16% strategy and decrease toward either end of the MCHVST spectrum. This indicates that with either a low or high moisture strategy there is increased potential for poor economic performance in any given year.

This potential for poor economic performance with high or low moisture harvests is depicted graphically in Fig. 22-7, which illustrates the year-to-year variability of simulated net returns for the 13, 16, and 19% strategies. In the simulation year of 1974, soybean is planted prior to 4 June under the 16 and 19% strategies, and results in moderaterly high yields of 2820 and 2958 kg, respectively. Net returns for both strategies are near $296.00/ha. Due to high rainfall during the drydown period, however, soybean plantings under the 13% strategy are delayed until 2 July and result in a 1368 kg yield with net returns of only $24.82/ha. In contrast, during the simulation year of 1980, soybean for all three strategies is planted by 13 June, resulting in nearly identical low yields of 1710, 1731, and 1755 kg (for the 19, 16, and 13% strategies, respectively), due to excessively hot and dry weather throughout the growing season. In spite of nearly identical yields, however, the large discount on the high-moisture wheat results in a low net return of $14.18/ha under the 19% strategy compared with $60.79/ha and $87.17/ha for the 16 and 13% strategies, respectively.

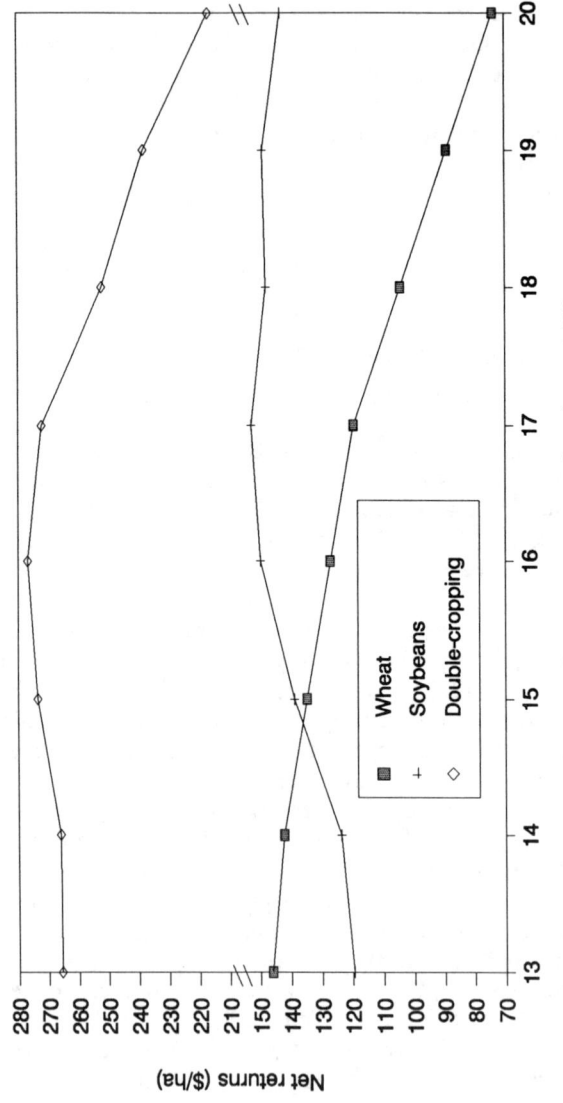

Fig. 22-6. Simulated mean net returns for wheat, soybean, and double cropping by wheat harvest strategy.

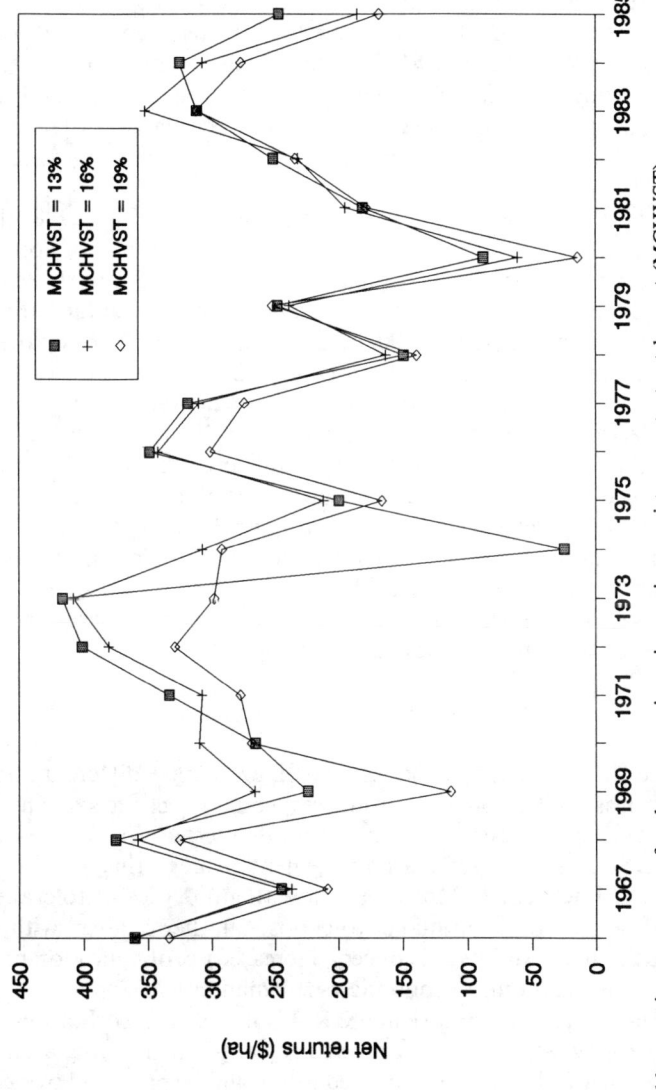

Fig. 22-7. Double-cropping net returns for three strategies using various moisture contents at harvest (MCHVST).

The net returns information in Fig. 22-7 is replotted with cumulative frequency distributions in Fig. 22-8. It is noteworthy that over the broad range of net returns below $235.00/ha, the cumulative frequency of the intermediate strategy lies to the right of the other distributions. This implies that there is a reduced probability of attaining low net returns with the 16% strategy in comparison with the other two strategies. For example, in one year out of ten (i.e., at a cumulative frequency of 0.10) a producer can expect net returns below $62.00/ha with the 13% strategy, compared with net returns below $106.00/ha and $163.00/ha for the 19 and 16% strategies, respectively. Conversely stated, the intermediate strategy will result in net returns above $163.00/ha in nine years out of ten, compared with net returns above $62.00/ha at the same 0.90 probability level for the low-moisture 13% strategy.

At the upper end of the cumulative distributions in Fig. 22-8, the ordering of the three strategies is partially reversed. The probability of attaining high economic performance is greatest for a low-moisture strategy and lowest for a high-moisture strategy. For example, the probability of attaining net returns in excess of $370.00/ha is approximately 0.16 for the 13% strategy compared with 0.12 and 0.0 for the 16 and 19% strategies, respectively. Table 22-3 (Row 10) shows that maximum net returns over the 20-yr simulation decrease monotonically with progressively higher wheat moisture harvests. A maximum net return for a low-moisture strategy will exceed the maximum for a high-moisture strategy during years in which there is little difference in the drydown period across strategies. This results in nearly identical soybean planting dates and yields. Under these circumstances, the high-moisture strategies are penalized in that there is little or no soybean-yield advantage due to early planting. Nevertheless, wheat yields are highly discounted due to their greater moisture content.

E. Risk-Efficient Strategies

Because alternative crop producers are likely to have different attitudes toward risk, it is unlikely that all producers would select the same management strategy when presented with a set of alternatives. For example, producers may typically prefer a management strategy that results in the highest expected net return. However, a low (high) degree of tolerance for risk may influence a decision maker to choose another strategy with lower average profits if it exhibits a reduced (increased) probability of poor or minimal (top or maximum) economic performance.

A risk-efficiency criterion (King & Robison, 1984) incorporates information concerning both risk and expected returns in identifying a management strategy that will maximize expected utility and be preferred by decision makers who have a known risk attitude. Stochastic-dominance ordering (Meyer, 1977a, b) is one risk-efficiency criterion that identifies a preferred set of strategies for decision makers with differing levels of tolerance for risk. With stochastic dominance, a pair-wise comparison of the cumulative frequency distributions of net returns results in the efficient strategy or efficient set

BIOPHYSICAL SIMULATION OF WHEAT & SOYBEAN

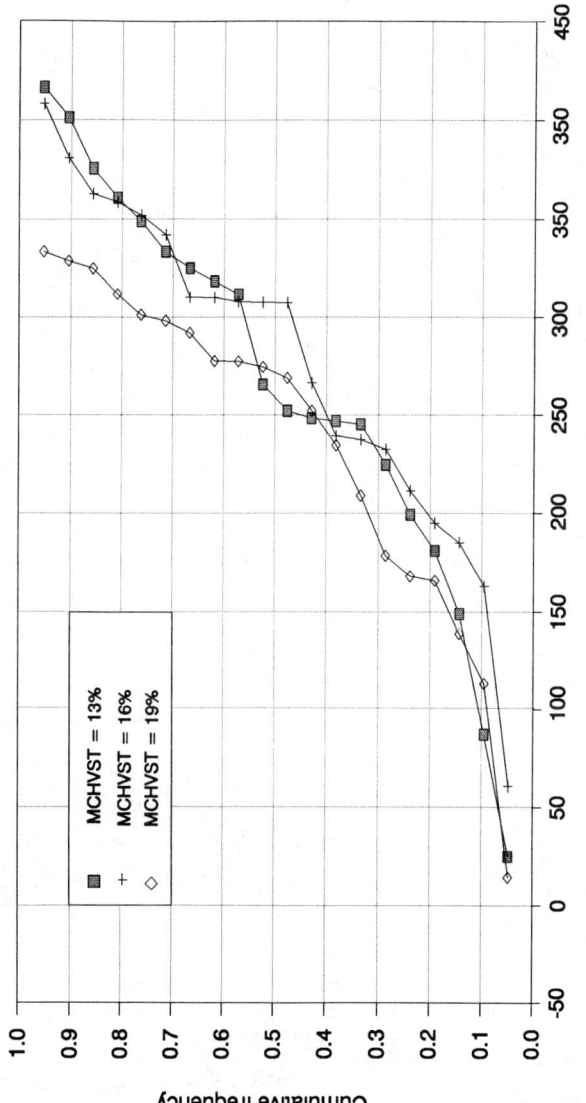

Fig. 22-8. Cumulative frequency of double-cropping net returns for three wheat harvest strategies using various moisture contents at harvest (MCHVST).

of strategies. In this process, the risk attitude of the decision maker is designated by a risk interval, which is bounded by an upper and lower Pratt-Arrow coefficient of absolute risk aversion. Recent examples of the use of stochastic dominance in conjunction with biophysical simulation models include McGuckin (1983), Schoney and McGuckin (1983), Boggess et al. (1985), Anaman and Boggess (1986), Harris and Mapp (1986), and Parsch and Loewer (1987).

Rankings of the eight MCHVST strategies using stochastic-dominance ordering are presented in Table 22-4. The risk aversion intervals presented in Table 22-4, including approximations for first (FSD) and second degree stochastic dominance (SSD), were arbitrarily defined for this study and are approximations of empirical estimates of Pratt-Arrow coefficients from the literature (Love & Robison, 1984; Wilson & Eidman, 1983).

The 16% intermediate-moisture strategy will be the preferred choice of producers who are characterized solely as being risk-averse without specifying their degree of risk-aversion (SSD). Table 22-3 demonstrated that this 16% strategy resulted in the highest mean net returns, the highest minimum net returns, and the lowest relative measure of variability (coefficient of variation) of all strategies analyzed in this study. Consequently, the trade-offs between down-side risk and returns are minimal with this strategy. It is likewise preferred by decision makers more precisely characterized as being either slightly or strongly risk averse (Table 22-4). By contrast, decision makers with higher tolerance for risk (i.e., those who are risk-neutral or risk preferring) will maximize expected utility with one of the low-moisture strategies, which affords the possibility of higher maximum net returns even though mean net returns are somewhat lower.

Table 22-4. Ranking of eight wheat harvest initiation strategies for alternative risk-efficiency criteria.†

Efficiency criterion‡	Risk interval§	Efficient set, %¶
FSD	−0.0010, 0.0010	13, 14, 15, 16, 17
SSD	0.0000, 0.0010	16
Risk preferring	−0.0008, −0.0001	13, 14
Risk neutral	−0.0001, 0.0001	13, 14, 15, 16
Slightly risk averse	0.0001, 0.0004	16
Strongly risk averse	0.0004, 0.0010	16

† All rankings were calculated for a 283 ha farming operation, which is the average area of wheat/soybean farms in Arkansas County, Arkansas, based on a scaling procedure described by Raskin and Cochran (1986).
‡ Risk attitudes, including first degree (FSD) and second degree (SSD) stochastic dominance are arbitrarily defined based on empirical studies (Love & Robison, 1984; Wilson & Eidman, 1983).
§ Lower and upper bound of Pratt-Arrow coefficient of absolute risk aversion.
¶ Efficient set of strategies defined by wheat moisture content at harvest initiation.

IV. CONCLUSIONS

This study interfaces two physiologically based crop growth models in order to assess the impact of delayed wheat harvest on the performance of a double-cropping wheat and soybean system in a risk-returns setting. The preliminary results showed that: (i) higher expected net returns are attained when wheat harvest is initiated at an intermediate (16%) moisture level than with either a low or high moisture harvest; and (ii) an intermediate moisture harvest (16%) results in minimal risk of poor economic performance in comparison with high or low moisture strategies. Both the low-moisture and high-moisture strategies were shown to result in potentially inferior performance because: (i) an increased drydown period with a low-moisture wheat harvest delayed soybean plantings, causing reduced soybean yields in comparison with an intermediate strategy; and (ii) earlier soybean planting dates under a high-moisture harvest resulted in little or no yield gains over low or intermediate MCHVST strategies, yet wheat yields were greatly discounted.

Although the study ignores market issues (i.e., price risk) and certain other production aspects (i.e., impact of suitable days for fieldwork, machinery capacity, and number of field operations on the timeliness of field events), which are important in a double cropping perspective, it demonstrates the potential for using biophysical crop growth models to assess: (i) cropping-system production risk using multiple-year simulations; and (ii) within-year interactions between two crops in a multiple-crop system.

REFERENCES

Adams, D. (ed.). 1984. Wheat production and marketing in Arkansas. Univ. of Arkansas Coop. Ext. Serv. Bull.

Anaman, K.A., and W.G. Boggess. 1986. A stochastic dominance analysis of alternative marketing strategies for mixed crop farms in northern Florida. South. J. Agric. Econ. 18(2):257-266.

Arkansas Agricultural Statistics Service. 1987. 1986 Arkansas agricultural statistics. Univ. Arkansas Agric. Exp. Stn. Rep. Ser. 300.

Boggess, W.G., D.T. Cardelli, and C.S. Barfield. 1985. A bioeconomic simulation approach to multi-species insect management. South. J. Agric. Econ. 17(2):43-56.

Brown, L.G., J.W. Jones, J.D. Hesbeth, J.D. Hartsog, F.D. Whisler, and F.A. Harris. 1985. COTCROP: Computer simulation of cotton growth and yield. User's manual. Mississippi Agric. and For. Exp. Stn. Info. Bull. 69.

Caviness, C.E., and J.D. Thomas. 1979. Influence of planting date on three soybean varieties. Arkansas Farm Res. 28(2):8.

Chen, L.H., and R.W. McClendon. 1984. Selection of planting schedule for sobyeans via simulation. Trans. ASAE 27(1):29-32.

Harris, T.R., and H.P. Mapp. 1986. A stochastic dominance comparison of water-conserving irrigation strategies. Am. J. Agric. Econ. 68(2):298-305.

King, R.P., and L.J. Robison. 1984. Risk efficiency models. p. 68-81. *In* P.J. Barry (ed.) Risk management in agriculture. Iowa State Univ. Press, Ames, IA.

Love, R.O., and L.J. Robison. 1984. An empirical analysis of the intertemporal stability of risk preference. South. J. Agric. Econ. 16(1):159-165.

McGuckin, T. 1983. Alfalfa management strategies for a Wisconsin dairy farm—an application of stochastic dominance. North Cent. J. Agric. Econ. 5(1):43-49.

Meyer, J. 1977a. Second degree stochastic dominance with respect to a function. Int. Econ. Rev. 18(2):477-487.

Meyer, J. 1977b. Choice among distributions. J. Econ. Theory 14:326-336.

Nester, R.P. (ed.). 1983. Soybean production handbook. Univ. of Arkansas Coop. Ext. Serv. MP197.

Parsch, L.D., and O.J. Loewer. 1987. Economics of simulated beef-forage rotational grazing under weather uncertainty. Agric. Systems 25(4):279-295.

Prickett, M. 1985. Irrigation scheduling for Arkansas soybean production. M.S. thesis. Univ. of Arkansas, Fayetteville.

Raskin, R., and M.J. Cochran. 1986. Interpretations and transformations of scale for the Pratt-Arrow absolute risk aversion coefficient: Implications for generalized stochastic dominance. West. J. Agric. Econ. 11(2):204-210.

Ritchie, J.T. 1972. Model for predicting evaporation from a row crop with incomplete cover. Water Resour. Res. 8:1204-1213.

Ritchie, J.T., and S. Otter. 1985. Description and performance of CERES—Wheat: A user oriented wheat yield model. p. 159-175. *In* W.O. Willis (ed.) ARS wheat yield project. UDSA-ARS-38.

Schlaifer, R. 1959. Probability and statistics for business decisions. McGraw-Hill, New York.

Schoney, R.A., and J.T. McGuckin. 1983. Economics of the wet fractionation system in alfalfa harvesting. Am. J. Agric. Econ. 65(1):38-44.

Stuart, C., C.R. Garner, C. Farler, and R.E. Coats, Jr. 1986a. Maintaining financial viability in the farm business by estimating production costs (wheat). Univ. of Arkansas Cooop. Ext. Serv. Econ. Anal. Ser. EA1-WT8601.

Stuart, C., B. Herrington, J.D. Clark, C.R. Garner, C. Farler, and R.E. Coats, Jr. 1986b. Maintaining financial viability in the farm business by estimating production costs (soybean). Univ. of Arkansas Coop. Ext. Serv. Econ. Anal. Ser. EA8-SOY8606.

Trice, K.L. 1986. The economics of double cropping wheat and soybeans: A simulation analysis using WHEATSOY. M.S. thesis. Univ. of Arkansas, Fayetteville.

Wilkerson, G.G., J.W. Jones, K.J. Boote, K.T. Ingram, and J.W. Mishoe. 1983. Modeling soybean growth for management. Trans. ASAE 26(1):63-73.

Wilkerson, G.G., J.W. Jones, K.J. Boote, and J.W. Mishoe. 1985. SOYGRO V5.0: Soybean crop growth and yield model. Technical documentation. Univ. of Florida, Gainesville, FL.

Wilson, P.N., and V. Eidman. 1983. An empirical test of the interval approach for estimating risk preferences. West. J. Agric. Econ. 8(2):170-182.

SUBJECT INDEX

Absorbed radiation, 127
ACRES system, 484
Activity coefficient, 367-368, 370
Aeration, soil, 96-97
AgLine system, 484
AGNET system, 475, 480
Agribusiness, 473
Agribusiness U.S.A. system, 484
AGRICOLA system, 480
Agricultural experiment station, 472
AgriData Network, 485
Agropyron smithii (Rybd.). *See* Western wheatgrass
Air entry value, soil, 147
Alfalfa
   evaporation and transpiration, 249-252, 254
   fraction available soil water in root zone (FA value), 494, 496-497
   soil water budget, 505-506
Aluminum toxicity, root growth and, 92-93
Ammonia
   soil solution, 347-348
   volatilization, 287-288, 346
Andropogon gerardi (Vitman). *See* Big bluestem
Andropogon scoparius (Mich.). *See* Little bluestem
ANSER system, 480
Anthesis, temperature response, 13-15
Apple
   crop coefficient, 496-498
   light-use efficiency, 126
AQS II system, 485
Arachis hypogaea L. *See* Peanut
Artificial intelligence, 475-479
Associative memory, 478
ATI-Net system, 480
Avena sativa L. *See* Oat
Axial resistance, 276

Barley
   light-use efficiency, 126
   phyllochron, 21
   temperature response, 20-21
Base temperature, 6, 11-17, 19-20, 33, 58-60, 74
BASIC computer code, 2
Basin irrigation, 184
Bean, crop coefficient, 496-497
Beta vulgaris L. *See* Sugar beet
Big bluestem, small-grain equivalent, 465
Biomass accumulation, 5

Bio-photo-thermal model, 55
Bit-Image Pattern Recognition System, 478
Black box syndrome, 474
Black layer, corn. *See* Maturity, corn
Blue grama, small-grain equivalent, 465
Boussinesq equation, 212-213
Bouteloua gracilis. See Blue grama
Brassica oleracea. See Kale
Buchloë dactyloides (Nutt.) Engelm. *See* Buffalograss
Buffalograss, small-grain equivalent, 465
Bulk density, soil, 343
   root growth and, 94-96, 104-107

Calcium, soil, 374-375
   deficiency, root growth and, 93
Calcium carbonate. *See* Lime
CAMP system, 485
Canopy
   architecture, 128
      photosynthesis and, 129-130
Canopy model, light-use efficiency, 132-133. *See also* Light-use efficiency
Carbon budget, 125-141
Carbon dioxide
   soil, 376-377
   uptake by plants, 127-128, 133
Cation exchange, 365-377
   model, 368-370
      critical assumptions, 371
      validation, 371-376
   rate
      canopy, 128
      leaf, 128
Cereal, temperature response, 18-19
CERES-Maize model, 55, 288-289, 304, 306
CERES-N model, 287-308
   computer program listing, 308-319
   denitrification, 296-297
   evaluation, 304-307
   grain N concentration, 302-304
   inputs, 304
   limitations and uses, 304-308
   N fertilization, 290-291
   N immobilization, 291-295
   N mineralization, 291-295
   N uptake by plants, 298-301
   NO_3 and urea movement, 289-290
   nitrification, 295-296
   plant N concentration, 297-299
   soil N transformations, 290
   urea hydrolysis, 290-291

SUBJECT INDEX

CERES-Wheat model, 42–43, 288–289, 292, 304–305, 307
  computer program listing, 44–53
  interfaced with SOYGRO, 511–533
  testing, 513–514
  validation, 43
Channel flow, 443
CHEM subroutine, 367–371, 373, 376, 380–384
Cherry, crop coefficient, 496–498
Chloride, soil, 372, 374–375
Climate
  evaporation and transpiration and, 217–218, 247, 250
  wind erosion and, 457, 460–461, 465
Climatic erosivity, 460
Clover, light-use efficiency, 126
Clumping factor, 133
Coarse soil fragments, root growth and, 94, 101, 109–110
Coefficient of variation, 150–151
Cold hardening, wheat, 40–42
Computer program listing
  CERES-N model, 308–319
  CERES-Wheat model, 44–53
  EPIC model, P submodel, 335–337
  evaporation model, 256–271
  heat flow model, 401–404, 420–436
  infiltration model, 192–203
  N transformation and transport model, 351–363
  root growth model, 111–119
  runoff model, 450–454
  solute transport model, 164–179, 377–394
  transpiration model, 256–271
  water erosion model, 450–454
  water flow model, 164–179, 192–203
  water uptake by roots model, 278–280
Conduction, 397–398
Controlled drainage, 217
Convection, 397–398
Convergence, 213, 216, 404
Cooperative Extension Service, 471–479
Corn
  development, 55–69
  emergence, 56, 60–62
  evaporation and transpiration, 249, 251
  fraction available soil water in root zone (FA value), 494, 496–497, 500
  germination, 56
  grain filling, 56, 61–63
  juvenile phase, 56
  leaf appearance, 55, 58–60
  leaf initiation, 55, 59–61
  leaf number, 55, 60–61
  leaf reflectance and transmittance, 133
  light-use efficiency, 126, 134–138, 140–141
  maturity, 25–26, 56, 61–63, 67–68
  N fertilization, 347–350
  P deficiency, 331–332
  P fertilization, 330–333
  P uptake, 324
  photoperiod response, 23–24, 55–69
  photosynthesis, 128, 130–132
  phyllochron, 20
  root growth, 93, 97, 100, 104–110
  silking, 55–56, 60–68
  small-grain equivalent, 463–464
  tasseling, 55–56, 60–62, 134–135
  temperature response, 8–19, 24, 26, 55–69
  water uptake, 281
  yield, 228–238
Corn phenology model
  assumptions, 58
  inputs and operation details, 56–57
  research needs, 68
  validation, 63–68
CORNF model, 55
Cotton
  light-use efficiency, 126, 137
  root growth, 96–97, 99
  small-grain equivalent, 463–464
Critical concentration, N, 297–298
Critical night length, 74
Critical soil water-filled porosity, 97
Crop coefficient, 251–252, 495–500
Crop development. *See* Development
Crop management factor, 445–446, 448
Crop residue, 291–295, 327, 462–465
Crop susceptibility factor, 229
Crop water-use table, 492–493, 501
Crown temperature, wheat, 35, 38, 42
CRPSM model, 507
CULEAF model, 128
Cupid model, 125–141

Daily cycle, temperature, 22–23
Darcy's law, 157
Dark respiration, 132, 134
Day-degree, 6
Decide model, 325
Deep seepage, 207–208, 212. *See also* Drainage
Deficiency demand, N, 298
Degree-day, 6, 55
Dehardening, wheat, 42
Denitrification, 287–288, 292, 296–297, 343–347, 350
Depression storage, 211, 232
Development
  corn, 55–69
  maturity type and, 25
  photoperiod response, 23–24
  soybean, 71–89
  temperature response, 5–27, 33
  timing of stages, 13–22
  wheat, 31–53
Dicot, temperature response, 20
Discharge rate, peak, 439
Dispersion-convection equation, 151
Dispersivity, 152–153, 157

SUBJECT INDEX

Double-cropping, wheat and soybean, 511–533
  drydown, 514–517
  model
    description, 512–513
    economic, 519–520
    experimental design, 518–520
    interface, 514–515
    soybean planting date, 512, 523–526
    testing and refinement, 513–514
    wheat harvest date, 512, 523–526
    yields, 520–523
  net returns, 527–528
  risk-efficient strategies, 530–532
  soil moisture components, 517–518
  weather risk, 527–530
DRAGNET system, 481
Drain spacing, 211, 232–233, 238
Drainable porosity, 212, 219
Drainage, 205–240, 253, 289, 442
  controlled, 217
  research needs, 239
  subsurface, 206, 232, 239
    DRAINMOD model, 211–216
  surface, 206, 232–233
    DRAINMOD model, 211
Drainage coefficient, 216
DRAINMOD model, 206
  background, philosophy, and equations, 207–209
  evapotranspiration, 217–219
  field evaluation, 224–225
  infiltration, 209–211
  objectives, 225–228
  precipitation, 209
  printout for sandy loam soil, 234–237
  problem definition, 206–207
  rooting depth, 224
  soil water distribution, 219–224
  subirrigation, 216–217
  subsurface drainage, 211–216
  surface drainage, 211
  yield, 228–238
Drain-tube radius, 213
Drought stress, 239, 302
Dry day, 226–227
Dry matter, 127
Dry zone, 218, 223–224
Drydown component, wheat and soybean double-cropping, 514–517
Dupuit-Forchheimer assumption, 212–216, 221

Ear growth, wheat, 33, 39–40
Economics, double-cropping of wheat and soybean, 511–533
Educational software, 478
Effective rooting depth, 224
Electrical analogs, water uptake by roots, 273
Electrical conductivity, soil solution, 367, 372–373, 376
ELM model, 325

Emergence
  corn, 56, 60–62
  photoperiod response, 73
  soybean, 73, 78–80
  temperature response, 15–17, 60–62, 73, 80
  wheat, 33–34
Energy Crop Growth Model, 55
Energy partitioning, soil surface, 397, 399–400
EPIC model, 289
  P submodel
    computer program listing, 335–337
    inputs, 327–328
    performance, 328–334
    structure, 326–327
  runoff, 439–449
  water erosion, 439–449
  wind erosion, 457–467
Equivalent conductivity, 214–215
Erodibility factor, soil, 445–446
Erodibility index, soil, 457–459
Erosion
  water. See Water erosion
  wind. See Wind erosion
Erosion control practice factor, 445
Erosion-Productivity Impact Calculator. See EPIC model
Erosive wind energy, 461
ESP. See Exchangeable sodium percentage
ESTEL system, 481
Evaporation, 222, 245–271
  climate and, 247, 250
  constant rate, 236–237
  model, 246–248
    computer program listing, 256–271
    computer requirements, 255
    critical assumptions, 255
    irrigation in, 252–255
    temperature response, 246–247
    variable rate, 236–237
Evapotranspiration, 207–208, 212, 217–219, 221–224, 227, 245
  in irrigation scheduling, 491–508
  potential, 217–218
  salinity and, 252–253
Excess water, sum of, 226–227
Exchange coefficient, 388–390
Exchangeable sodium percentage (ESP), 368, 376
  model, 368–370
EXCOEF model, 369
EXNET system, 481
Expansion growth, 31
Expectation, 153–154
Expert system, 476–477
Extension. See Cooperative Extension Service

FACTS system, 475, 481
FAST system, 485

SUBJECT INDEX

Fertilizer
  N, 287-288, 290-291, 347-350
  P, 323, 325-327, 330-333
  strategies, 288
Field capacity, 441, 494
Field length, 461-462, 465-466
Field-effective value, 239
Finite difference equations, 154, 401
Flowering
  latitude response, 86-88
  photoperiod response, 71, 73-74
  soybean, 71-74, 79, 81, 83-88
  temperature response, 25-26, 72-74
  wheat, 34, 40
Fluoride, 376
FORTRAN computer code, 2
Frozen soil, 442

Game playing, 476
Garden, crop coefficients, 496-498
Genetic coefficient
  photoperiod in wheat, 37-38
  root growth, 99-101
  soybean, 82
  vernalization of wheat, 34, 36
Germination
  corn, 56
  temperature response, 15-17
  wheat, 33-34
Glycine max (L.) Merr. *See* Soybean
GOSSYM model, 43
Gossypium hirsutum L. *See* Cotton
GPHEN model, 75
Grain, N uptake, 305-306
Grain filling, 32
  corn, 56, 61-63
  temperature response, 11, 16-17, 302-304
  wheat, 33, 40
"Grass roots" modeling, 471-479
Grassroots system, 481
Green-Ampt equation, 210
Gross returns, 519-520
Growing degree-day, 6, 58
Growth
  duration, 5, 31
  N demand, 299
  rate, 31
  temperature response, 5
Growth respiration, 125, 128, 133-135
Gypsum reactions, 365
  model, 367-368, 370

Hadley soybean model, 73-74
Hardening index, 42
Hardware, 474
Heat capacity, soil, 398, 405-406
Heat flow, soil, 246-247, 397-414
  energy partitioning at soil surface, 399-400
  model, 398-399
    computer program description, 401-404

computer program listing, 420-436
inputs, 402
measured vs. predicted temperatures, 407-414
user manual, 415-419
validation, 404-407
Heat index, 218
Heat sum, 6
Heat unit, 6
Helianthus annuus L. *See* Sunflower
Hooghoudt analysis, 213
Hooghoudt equation, 208
Hordeum L. *See* Barley
Hydraulic conductivity, soil, 146-147, 155, 181-182, 206, 209-212, 223, 248, 274, 342, 376, 403-404
  equivalent lateral, 214-215
  estimation, 183-185
Hydraulic head, soil, 248
Hydrodynamic dispersion coefficient, 152
Hysteresis, 191

Immobilization, N, 287, 291-295, 343-344, 346
Infiltration, 181-192, 206-208, 442
  boundary conditions, 184-185
  equations, 181
  model, 145-161, 182-185
    computer program listing, 164-179, 192-203
    critical assumptions, 161-162, 190-191
    DRAINMOD, 209-211
    input and output data, 190
    results, 150
    verification, 186-189
  research needs, 162-163, 197
  theoretical approach, 146-149
  time steps, 185
Infiltration time, 146
Information systems, 480-486
Instant Uptake system, 485
Intercepted radiation, 127
Ipomoea batatas Lam. *See* Sweet potato
IRRIGAME game, 507
Irrigation, 156, 190. *See also* Subirrigation
  basin, 184
  evaporation and transpiration model, 252-255
  root growth and, 107
  sprinkler, 184, 191, 347
  surface, 206, 217
  volume, 226
  wastewater volume, 227-228
  water erosion model, 448-449
Irrigation scheduling program, 491-508
  comparison with field data, 505-506
  critical assumptions, 507
  operation, 496-504
  public information approach, 492-493
  soil water budget model, 493-496
IRRIGMAT game, 507

SUBJECT INDEX

Journal of Computer Applications, 476
Juvenile phase, 25
  corn, 56
  photoperiod response, 80
  soybean, 78-80

Kale, light-use efficiency, 126

Land-grant system, 473, 476, 478
  computer institutes, 487-488
Latent heat flux, 399, 406, 408
Latitude response
  flowering, 86-88
  maturity, 87-88
  soybean, 86-88
Layer-weighting factor, 99, 101
Leaching, 152
  N, 287-288, 350
  steady gravitational, 157
Leaf appearance
  corn, 55, 58-60
  photoperiod response, 68
  soybean, 78-79
  temperature response, 18-23, 60
  wheat, 34, 38-39
Leaf extension, temperature response, 10-12
Leaf growth, wheat, 33
Leaf initiation
  corn, 55, 59-61
  temperature response, 18-22, 25-26, 59-61
Leaf model, light-use efficiency, 128-132
Leaf number
  corn, 55, 60-61
  maturity type and, 25-26
  photoperiod response, 23-24
  temperature response, 25-26, 60-61
  vernalization and, 24-25
Leaf reflectance, 132-133
Leaf size, temperature response, 10-12
Leaf transmittance, 132-133
Leaf water potential, 280, 282
Light-use efficiency
  canopy model, 132-133
  corn, 134-138, 140-141
  cotton, 137
  Cupid model, 125-141
    comparison with measurements, 135-136
    interpretations from, 137-141
    definitions, 126-127
    leaf characteristics and, 125-141
    leaf model, 128-132
    measurements, 136-137
    model, 127-135
    respiration and, 133-135
    soybean, 136, 138-140
    specific crops, 126
    wheat, 136-137
Lime reactions, 365
  model, 367-368, 370
Little bluestem, small-grain equivalent, 465

Lolium perenne L. *See* Rye
Longwave sky irradiance, 399
Luxury consumption, N, 298
Lysimeter data, 252, 371

Machine cognition, 476
Machine learning, 476
Macropore, 239
Maintenance respiration, 125, 128, 133-135
Maize. *See* Corn
Major et al. soybean model, 72-73
Malus pumila Mill. *See* Apple
Manning's equation, 443, 448
Mass balance, 151-152
Mass growth, 31-32
Matric potential, soil, 181-182, 276, 365-366
Maturity
  corn, 25-26, 56, 61-63, 67-68
  development and, 25
  latitude response, 87-88
  leaf number and, 25-26
  photoperiod response, 71, 73
  rice, 25
  sorghum, 25
  soybean, 71, 73, 79, 83-88
  temperature response, 13-15, 73
Medicago sativa L. *See* Alfalfa
Microorganisms, soil, 291-295
Millville silt loam, 186-189, 191
Mineralization
  N, 291-295, 343-344, 346-347
  P, 325, 327, 332-333
Model
  identifying source, 474-475
  obtaining evaluations, 475-476
Model/user interface, 471-479
Moisture content, soil. *See* Water content, soil
Moisture retention curve, soil, 154-155
Monte Carlo method, 149
Morphological development, 31-32
MUSLE equation, 445, 448-449

National Weather Service, 482
Natural language processing, 476-478
Net radiation, 246-247, 399, 406, 408
Net returns, 519, 527-528
Newton-Raphson iteration, 280
Nicollet clay loam, 407-414
Night length, 23, 74
Nitrate
  movement in soil, 289-290
  soil solution, 347, 349
  uptake by plants, 346
Nitrification, 292, 295-296, 343-344, 346
Nitrification capacity index, 295
Nitrification inhibitor, 292
Nitrogen
  dynamics in soil-plant system, 287-308. *See also* CERES-N model
  fertilizer, 287-288, 290-291, 347-350

SUBJECT INDEX

Nitrogen (cont.)
  grain, 302–306
  immobilization, 287, 291–295, 343–344, 346
  leaching, 287–288, 350
  mineralization, 291–295, 343–344, 346–347
  plant N concentration, 297–299
  transformation and transport model
    computer program listing, 351–363
    critical assumptions, 345–347
    description, 341–347
    input, output, and verifiable parameters, 344–345
    validation, 347–350
  transport in soil, 341–351
  uptake by plants, 288, 298–301, 305–306, 346, 350
North Carolina Coastal Plains, 224–225
NPIRS system, 482
Nutrient, 239

Oat, evaporation and transpiration, 252
Onstad-Foster equation, 445
Organic matter, soil, 291–295, 327, 333–334, 346
Oryza sativa L. *See* Rice
Osmotic potential, soil, 365–366
Overdrainage, 232, 238

Palouse silt loam, 280–281
Panicum virgatum. *See* Switchgrass
Panoche loam, 220
PAPRAN model, 288, 292
PAR. *See* Photosynthetically active radiation
Pasture, 496–498
Pattern recognition, 476
Pea
  crop coefficient, 497
  root growth, 97
Peach, crop coefficient, 496–498
Peak development temperature, 11–15, 19
Peanut, small-grain equivalent, 463
Pearl millet, temperature response, 18–19
Penman equation, 494–495
Pennisetum glaucum L. *See* Pearl millet
Personal computers, 471–479
Phaseolus vulgarus L. *See* Bean
Phasic development
  corn, 55–69
  wheat, 31–53
Phosphorus
  deficiency, 331–332
  dynamics in soil-plant systems, 323–335
  fertilizer, 323, 325–327, 330–333
  mineralization, 325, 327, 332–333
  model, 323–326. *See also* EPIC model, P submodel
  soil test, 330–332
  transformations in soil, 324–325
  transport in soils, 324–325

uptake by plants, 324–325, 327
Photoperiod response
  corn, 23–24, 55–69
  development, 23–24
  emergence, 73
  flowering, 71, 73–74
  juvenile phase, 80
  leaf appearance, 68
  leaf number, 23–24
  maturity, 71, 73
  pod fill, 73
  pod set, 71
  soybean, 71–89
  tassel initiation, 60–62
  wheat, 36–40
  winter cereal, 24
Photosynthesis, 125–141
  canopy architecture and, 129–130
  corn, 128, 130–132
  sorghum, 128–129, 132
  soybean, 128–130, 132
  temperature response, 130–131
  wheat, 129–130
Photosynthetically active radiation (PAR), 126
  absorbed, 127
  intercepted, 127
  measurement, 136
Photothermal coefficient, SOYGRO model, 80–83
Photothermal time, 6
  SOYGRO model, 75–77
Phyllochron
  barley, 21
  corn, 20
  wheat, 20–21, 38–40
Piston water flow, 153, 155–156
Pisum sativa L. *See* Pea
Plant cover, 248–250, 457, 462–466
Planting, delayed, yield response, 230–231
Pod fill
  photoperiod response, 73
  soybean, 73
  temperature response, 73
Pod set
  photoperiod response, 71
  soybean, 71–72, 79
  temperature response, 72
Pollution, 152, 239, 288
Ponding, 146, 184–188, 215–216
Porosity, soil, 441
Portsmouth sandy loam, 232–238
Potato
  fraction available soil water in root zone (FA value), 494, 496–497
  light-use efficiency, 126
Potential evapotranspiration, 217–218
PRECIP subroutine, 367–368, 384–385
Precipitation, DRAINMOD model, 209
Precipitation-dissolution reaction, 365–377
  model, 367–368

SUBJECT INDEX

Precipitation-dissolution reaction model (cont.)
  critical assumptions, 370
  validation, 371-376
Probability density function, 147-149, 162
  uniform, 157
Prunus avium L. *See* Cherry
Prunus persica L. *See* Peach
Pyrus L. *See* Apple

Radial flow, 213
Radiation, 397
Rainfall
  intensity, 442, 444
  rate, 446
  water erosion model, 445-448
Rainfall energy factor, 447
Range grass, small-grain equivalent, 465
Rape, small-grain equivalent, 464
Redistribution, 181-192
  boundary conditions, 184-185
  equations, 181
  model, 145-161, 182-185
    computer program listing, 164-179, 192-203
    critical assumptions, 161-162, 190-191
    input and output data, 190
    results, 150
    verification, 186-189
  research needs, 162-163, 197
  theoretical approach, 146-149
  time steps, 185
Regionalized variable, 161
Relative vernalization effectiveness factor, 34-35
Reproductive stage, 72
Research needs, 473
  corn development, 68
  drainage, 239
  infiltration, 162-163, 197
  water flow, 162-163, 197
  water uptake by roots, 283
  wind erosion, 467
Respiration
  dark, 132, 134
  growth, 125, 128, 133-135
  light-use efficiency and, 133-135
  maintenance, 125, 128, 133-135
Retention parameter, 440-441
  frozen soil, 442
RHIZOS model, 43
Rice
  light-use efficiency, 126
  maturity type, 25
  root growth, 96, 100
Richard's equation, 146, 208-210, 212, 221
Ridge-roughness factor, 457-460, 465
Robotics, 476
Root density function, 248, 253
Root endodermis resistance, 273-274

Root extraction, 245, 248-250, 253, 365
Root growth
  aeration and, 96-97
  Al toxicity and, 92-93
  Ca deficiency and, 93
  coarse soil fragments and, 94, 101, 109-110
  corn, 93, 97, 100, 104-110
  cotton, 96-97, 99
  depth of rooting front, 104-105
  evaporation and transpiration model, 249
  irrigation and, 107
  length density and length/weight ratio, 99-102, 105-110
  model, 91-119
    computer program listing, 111-119
    root length density and length/weight ratio, 105-110
    rooting depth, 104-105
  pea, 97
  rice, 96, 100
  rye, 97
  soil bulk density and, 104-107
  soil factors affecting
    dynamic factors, 94-98
    static factors, 92-94
  soil strength and, 94-95, 101
  sorghum, 93, 100
  soybean, 96, 100
  sunflower, 96, 100
  temperature response, 97-98
  tillage response, 109
  wheat, 93, 96, 100
Root resistance, 276, 366
Root senescence, 102-103
Rooting depth, 98-99, 104-105, 275
  DRAINMOD model, 224
Runge-Benci model, corn, 55
Runoff, 187-191, 206-207, 239, 439-449
  model
    computer program listing, 450-454
    peak rate, 442-445
    runoff volume, 440-442
    testing, 449
  in water erosion model, 445-448
Runoff coefficient, 442
Rye, root growth, 97

Salinity
  evapotranspiration and, 252-253
  solute transport and, 365-394
Salt-affected soil, solute transport, 365-394
SAR. *See* Sodium adsorption ratio
Sarpy loam, 186-187
SCAMP system, 482
SCS curve number, 439-440
Seed germination. *See* Germination
Seedling
  emergence. *See* Emergence
  expansion, temperature response, 14
Selectivity coefficient, 365, 369-371

SUBJECT INDEX

Senescence, root, 102–103
Sensible heat exchange, 246–247
Sensible heat flux, 399, 406, 408
Sensitivity analysis, 332–333
Silking
  corn, 55–56, 60–68
  temperature response, 14–15, 60–62
SIMAIZ model, 55
SIMTAG model, 43
Simulator for production and Utilization of Rangeland (SPUR) model, 449
Simulator for Water Resources in Rural Basins (SWRRB) model, 449
Slope, runoff and, 440
Slope length, 443
Slope length and steepness factor, 445–446
Small-grain equivalent, 462–465
Snow depth, 38, 40
  estimation, 41
Sodium, exchangeable, 376
Sodium adsorption ratio (SAR), 368, 370–374
Soil
  properties, 145–179. See also specific properties
  spatial variability, 145–179
Soil factors, root growth and
  dynamic factors, 94–98
  static factors, 92–94
Soil heat flux, 399–400
Soil resistance, 273–277
Soil solution, 152
Soil strength, root growth and, 94–95, 101
Soil surface
  changes, 191
  energy partitioning, 397, 399–400
Soil test, P, 330–332
Soil texture, root growth and, 94–96
Soil-root interfacial resistance, 273–274
Soil-water potential, weighted mean, 277, 280
Solanum tuberosum L. *See* Potato
Solar radiation (SR), 126
  absorbed, 127
  intercepted, 127
Solute transport
  model, 151–161, 365–367
    computer program listing, 164–179, 377–394
    critical assumptions, 161–162, 370
    results, 155–156, 159–161
    validation, 371–376
  N, 341–351
  P, 324–325
  research needs, 162–163
  salt-affected soils, 365–394
  in steady gravitational flow, 156–159
Sorghum
  light-use efficiency, 126
  maturity type, 25
  photosynthesis, 128–129, 132
  root growth, 93, 100

  small-grain equivalent, 462–464
  temperature response, 18–19
Sorghum bicolor (L.) Moench. *See* Sorghum
Sorghum vulgare L. *See* Sorghum
The Source system, 486
Soybean. *See also* Double-cropping, wheat and soybean
  development, 71–89
  emergence, 73, 78–80
  flowering, 71–74, 79, 81, 83–88
  juvenile phase, 78–80
  latitude response, 86–88
  leaf appearance, 78–79
  leaf reflectance and transmittance, 133
  light-use efficiency, 126, 136, 138–140
  maturity, 71, 73, 79, 83–88
  P uptake, 324
  photoperiod response, 71–89
  photosynthesis, 128–130, 132
  pod fill, 73
  pod set, 71–72, 79
  root growth, 96, 100
  small-grain equivalent, 463–464
  temperature response, 15–17, 20, 71–89
  vegetative development, 72
SOYGRO model, 72
  estimation of coefficients, 79–83
  interfaced with CERES-Wheat, 511–533
  phenology model, 74–79
  photothermal time, 75–77
  testing, 513–514
  thermal time, 77–78
  validation, 83–88
SOYPHEN model, 74
Spatial variability, soil, 145–179
Splinter model, corn, 55
Spring grain, crop coefficient, 496–497, 499
Sprinkler irrigation, 184, 191, 347
SPUR model, 449
SR. *See* Solar radiation
Stability of model, 404
Statistical moment, 145
  concentration, 154–155
  flow variables, 149–150
Stochastic modeling, 145
Stomatal closure, 250, 280
Stomatal conductance, 131–132, 277
Stress day factor, 229
Stress-day index, 228–230
Subirrigation, 205–207, 213, 232–238
Subsurface drainage, 206, 232, 239
  DRAINMOD model, 211–216
Sugar beet
  crop coefficient, 497
  light-use efficiency, 126
Sum of excess water, 226–227
Sunflower
  light-use efficiency, 126
  root growth, 96, 100
  small-grain equivalent, 464

SUBJECT INDEX

Surface drainage, 206, 232–233
  DRAINMOD model, 211
Surface flow, 443–444
Surface irrigation, 206, 217
Surface storage, 211
Sweet corn, crop coefficient, 497
Sweet potato, light-use efficiency, 126
Switchgrass, small-grain equivalent, 465
SWRRB model, 449
Systems management, 471–472

TAMW model, 43
Tassel initiation
  corn, 55–56, 60–62, 134–135
  photoperiod response, 60–62
  temperature response, 60
Technology transfer, 471–479
TELECOT system, 486
Temperature
  daily cycle, 22–23
  soil, 397
  soil surface, 400, 406, 414
Temperature response
  anthesis, 13–15
  barley, 20–21
  cereal, 18–19
  corn, 8–19, 24, 26, 55–69
  determination, 9–11
  development, 5–27, 33
  dicot, 20
  emergence, 15–17, 60–62, 73, 80
  evaporation, 246–247
  flowering, 25–26, 72–74
  germination, 15–17
  grain filling, 11, 16–17, 302–304
  grain growth, 11
  growth, 5
  idealized, 11–13
  leaf appearance, 18–23, 60
  leaf extension, 10–12
  leaf initiation, 18–22, 25–26, 59–61
  leaf number, 25–26, 60–61
  leaf size, 10–12
  maturity, 13–15, 73
  nitrification, 296
  pearl millet, 18–19
  photosynthesis, 130–131
  pod fill, 73
  pod set, 72
  root growth, 97–98
  seedling expansion, 14
  silking, 14–15, 60–62
  sorghum, 18–19
  soybean, 15–17, 20, 71–89
  tassel initiation, 60
  terminology, 6–7
  vernalization, 34–36
  wheat, 16–17, 20–22, 24–25, 34–36, 39–42
Temperature sensor, accuracy, 7–9

Terminal spikelet initiation, wheat, 32–34, 36, 39
Thermal conductivity, soil, 398, 405–414
Thermal diffusivity, soil, 405–406
Thermal threshold, SOYGRO model, 79–80
Thermal time, 6–7, 13, 33, 56
  calculations, 7–9
  SOYGRO model, 77–78
Thermal unit, 6
Thornthwaite method, 218, 460
Threshold killing temperature, 42
Tillage response, root growth, 109
Time of concentration, 443
Trafficable conditions, 230
Transpiration, 245–271, 277, 280, 282
  climate and, 247, 250
  model, 248–252
    computer program listing, 256–271
    computer requirements, 255
    critical assumptions, 255
    irrigation in, 252–255
    verification, 252–255
Triticum aestivum L. *See* Wheat

UMC BBS system, 483
Universal soil loss equation (USLE), 440, 445–446, 449
  modified, 445
Unsaturated flow, 152
  equation, 182
Upward flow, 218–219, 252–253, 289
Urea
  hydrolysis, 290–291, 346
  movement in soil, 289–290
USDA, 472, 475, 478
USDA ON-LINE system, 483
User/model interface, 471–479
USLE. *See* Universal soil loss equation

Variance, 153–154
Vegetative cover. *See* Plant cover
Vegetative development, soybean, 72
Vegetative factor, wind erosion model, 462–465
Vegetative growth, 32
Vernalization
  leaf number and, 24–25
  temperature response, 34–36
  wheat, 34–36, 39–40
  winter cereal, 24–25
Vernalization day, 34–35
VPI-CMN system, 483

Wagram loamy sand, 222
Wastewater
  disposal, 206
  volume irrigation, 227–228
Water balance, 208
Water budget, soil, model, 493–496
Water capacity, soil, 182, 403–404
  estimation, 183–185

Water characteristics, soil, 403-404
Water content, soil, 146-147, 150-151, 181-189, 209-210, 218, 248, 254, 282, 342, 405, 441, 460-461, 493-496, 505, 517-518
  deficient, 225, 229-230
  excessive, 225, 228-229
  root growth and, 94-96
Water content front, 149
Water distribution, soil, 219-224
Water erosion, model, 439-449
  computer program listing, 450-454
  irrigation, 448-449
  rainfall/runoff, 445-448
  testing, 449
Water flow, 181-192, 366-367, 414. *See also* Drainage
  boundary conditions, 184-185
  channel, 443
  model, 145-161, 182-185, 341-343, 347
    computer program listing, 164-179, 192-203
    critical assumptions, 161-162, 190-191
    input and output data, 190, 344-345
    results, 150
    verification, 186-189
  piston, 153, 155-156
  radial, 213
  research needs, 162-163, 197
  statistical moment of flow variables, 149-150
  steady gravitational, 156-159
  surface, 443-444
  theoretical approach, 146-149
  time steps, 185
  unsaturated, 152
    equation, 182
  upward, 218-219, 252-253, 289
Water management. *See* Drainage; DRAINMOD model
Water potential
  leaf, 277, 280, 282
  plant crown, 276
  root, 248, 250, 253, 366
  soil, 274, 276, 280, 282
Water table
  control, 205-240. *See also* DRAINMOD model
  depth, 208-212, 216-224, 227
  drawdown, 219-221
  shape, 213
Water uptake
  corn, 281
  by roots
    computer program listing, 278-280
    critical assumptions in model, 282-283
    electrical analogs, 273
    model input and output, 280-282
    model validation, 280-282
    numerical algorithms, 278-280
    research needs, 283
    from soil with spatially varying root density, 275-277
    from uniformly rooted soil layer, 273-275
  wheat, 281-282
Water-filled pore, soil, root growth and, 97
Water-use table, 492-493, 501
Weather risk, wheat and soybean double-cropping, 527-530
Weather station, 493
Weibull distribution, 461
Weir setting, 217, 238
Western Computer Consortium, 476
Western Extension Computer Application Committee, 476
Western wheatgrass, small-grain equivalent, 465
Wet zone, 223-224
Wetting front, 149, 181, 186, 188-189, 210
Wheat. *See also* Double-cropping, wheat and soybean
  cold hardening, 40-42
  development, 31-53
  ear growth, 33, 39-40
  emergence, 33-34
  evaporation and transpiration, 251
  fraction available soil water in root zone (FA value), 494
  flowering, 34, 40
  germination, 33-34
  grain filling, 33, 40
  leaf appearance, 34, 38-39
  leaf growth, 33
  light-use efficiency, 126, 136-137
  P deficiency, 331-332
  P fertilizer, 330-331
  photoperiod response, 36-40
  photosynthesis, 129-130
  phyllochron, 20-21, 38-40
  root growth, 93, 96, 100
  small-grain equivalent, 463-464
  temperature response, 16-17, 20-22, 24-25, 34-36, 39-42
  terminal spikelet initiation, 32-34, 36, 39
  vernalization, 34-36, 39-40
  water uptake, 281-282
  winter kill, 40-42
Wilting point, 441, 494
Wind erosion
  control, 457-458
  equation, 457
  model, 457-467
    climatic factor, 460-461
    execution, 465-466
    field length, 461-462
    ridge-roughness factor, 458-460
    soil erodibility index, 458-459
    vegetative factor, 462-465
    verification, 467
  research needs, 467

SUBJECT INDEX

Winter cereal. *See also* Wheat
  photoperiod response, 24
  vernalization, 24-25
Winter grain, crop coefficient, 496-498
Winter kill
  estimation of plant survival, 41-42
  wheat, 40-42
Working day, 226

XCHANG subroutine, 369, 371, 385-388
XCOEF program, 369-370, 391
Xylem resistance, 273-274

Yield response, 245
  corn, 228-238
  to deficient soil water, 229-230
  to delayed planting, 230-231
  DRAINMOD model, 207, 228-231
  to excessive soil water, 228-229
  wheat and soybean double-cropping, 520-523

Zea mays L. *See* Corn